EXPLANATION IN THE SCIENCES

BOSTON STUDIES IN THE PHILOSOPHY OF SCIENCE

VOLUME 128

… τὴυ θατέρου φύσιν δύσμικτον
οὐσαυ εἰς ταὐτὸν ξυναρμόττων βία.
Plato, *Timaeus*

ÉMILE MEYERSON

(dessin de A. Bilis)

ÉMILE MEYERSON

EXPLANATION IN THE SCIENCES

Translated from the French by
Mary-Alice and David A. Sipfle

SPRINGER SCIENCE+BUSINESS MEDIA, B.V.

Library of Congress Cataloging-in-Publication Data

Meyerson, Emile, 1859-1933.
[De l'explication dans les sciences. English]
Explanation in the sciences / Emile Meyerson ; translated from the French by Mary-Alice and David A. Sipfle ; with an introduction by .
p. cm. -- (Boston studies in the philosophy of science ; v. 128)
Translation of: De l'explication dans les sciences.
Includes bibliographical references.
ISBN 978-0-7923-1129-4 **ISBN 978-94-011-3414-9 (eBook)**
DOI 10.1007/978-94-011-3414-9
1. Science--Methodology. 2. Science--Philosophy. 3. Explanation. 4. Logic. I. Title. II. Series.
Q174.B67 vol. 128
[Q175]
501 s--dc20
[501] 90-28236

ISBN 978-0-7923-1129-4

printed on acid-free paper

TABLE OF CONTENTS

BOOK FOUR

SCIENTIFIC AND PHILOSOPHIC REASON

APPENDICES

FOREWORD

Émile Meyerson's writings on the philosophy of science are a rich source of ideas and information concerning many philosophical and historical aspects of the development of modern science. Meyerson's works are not widely read or cited today by philosophers or even philosophers of science, in part because they have long been out of print and are often not available even in research libraries. There are additional *chevaux de frise* for all but the hardiest scholars: Meyerson's books are written in French (and do not all exist in English versions) and deal with the subject matter of science – ideas or concepts, laws or principles, theories – and epistemological questions rather than today's more fashionable topics of the social matrix and external influences on science with the concomitant neglect of the intellectual content of science.

Born in Lublin, Poland, in 1859, Meyerson received most of his education in Germany, where he studied from the age of 12 to 23, preparing himself for a career in chemistry.[1] He moved to Paris in 1882, where he began a career as an industrial chemist. Changing his profession, he then worked for a time as the foreign news editor of the HAVAS News Agency in Paris. In 1898 he joined the agency established by Edmond Rothschild that had as its purpose the settling of Jews in Palestine and became the Director of the Jewish Colonization Association for Europe and Asia Minor.

These activities represent Meyerson's formal career. Informally, he was a voracious reader of philosophy and the literature of science and its history; and he soon became a major figure among French scholars interested in questions of philosophy, especially the philosophy of science, and the history of science. Although he held no academic appointments, he did become a member of an important circle of French intellectuals concerned with major issues of philosophy, the nature and history of science, philosophy of science, and social problems. His weekly "intellectual salon" in Paris became a pivotal point in discussions of all sorts of major questions of philosophy, notably those dealing with science. This intellectual community included the philosophers Léon Brunschvicg and Lucien Lévy-Bruhl, along with the scientists Paul

Langevin and Louis de Broglie. Others who became part of his circle were the philosopher André George, the historian of religions Salomon Reinach, and the philosopher Henri Gouhier. Younger scholars attending these intellectual sessions came to include Hélène Metzger-Bruhl (the niece of Lucien Lévy-Bruhl) and Alexandre Koyré, both of whom were to make their mark as historians of science.[2] Koyré's first major contribution to the history of science, his celebrated *Études Galiléennes* was dedicated "to the memory of Émile Meyerson."[3] In the opening paragraph, Koyré referred to "the *philosophical interest and fruitfulness*" of "the historical study of science," a facet of such studies that he said could be taken for granted "after the magisterial work of those such as Duhem and Émile Meyerson, Cassirer and Brunschvicg."

An autodidact in philosophy, Meyerson was initiated into philosophy by reading the works of Charles-Bernard Renouvier, a philosopher also not much read nowadays. Renouvier was one of the two most original philosophical thinkers in France in the 19th century, the other being Auguste Comte. We may note that both of these philosophers were graduates of the École Polytechnique. The rigorous training they received at the Polytechnique gave their writings on science a high degree of authenticity. This factor may explain the particular fascination that Renouvier had for Meyerson.

As a trained chemist, Meyerson was naturally interested in theoretical and philosophical reflections on chemistry and the theory and properties of matter. In particular, he became deeply impressed by studies on the history of early chemistry. He was also influenced by two philosophical works, Kristian Kroman's *Naturerkenntnis* (1883) and the writings of the Danish philosopher Harold Höffding, who became a close friend and correspondent and was responsible for Meyerson's election to the Royal Danish Academy of Sciences and Letters in 1926.[4]

Meyerson's first philosophical book, perhaps the most influential of all his writings, was *Identity and Reality*.[5] This work, originating in Meyerson's reflections on the early development of chemistry, is basically a study in ontology. His program, as it developed from his work, has been described by H.W. Paul as "the unfulfilled one of Comte: to discover *a posteriori* the *a priori* principles guiding thought in the search for the nature of reality." George Boas has explained Meyerson's goal as follows: "to discover inductively the *a priori* principles of human thinking. By '*a priori*' É. Meyerson...means...those principles by which the human mind has operated to date and which are not discovered by it in experience itself."[6]

Another significant book by Meyerson is his study of Einsteinian relativity and its philosophical significance. The original French edition appeared in 1925 under the title *La déduction relativiste*. An English translation, by D. and M.-A. Sipfle, was published in 1985, together with an important review by Albert Einstein.[7] For many historians of science, Meyerson's most interesting and important work is his *Du cheminement de la pensée* (1931), a rich source of insight and information concerning almost every possible aspect of the development of science and its philosophy. Many of the notes or supplements outline significant topics in the history of scientific thought that have not as of now been explored much beyond the outlined suggestions made by Meyerson six decades ago. It is earnestly to be hoped that some intrepid translators will make this valuable resource available to readers in English.

The present work commends itself to us for a number of features. First of all, it is – like Meyerson's other writings – a tremendous repository of useful information and significant insights. Unlike most other philosophical works drawing on science and its history, *Explanation in the Sciences* draws heavily on the history of chemistry for its examples. One feature that every reader will remark is the importance given to Hegel, whose ideas are discussed at greater length and in more variety than those of other philosophers. In the English and American philosophical pantheon, Hegel does not occupy the same high place that he is accorded by German and French thinkers. In fact, one of the remarkable aspects of the present work is the merciless way in which Meyerson exposes what he finds to be the basic scientific ignorance and lack of understanding of scientific principles that invade every aspect of Hegel's writings relating to scientific thought.

Like Meyerson's other books, *Explanation in the Sciences* is not just a philosophical work that draws heavily on a knowledge of science itself, since it also displays a deep and sure acquaintance with science's history. As such, it belongs to the great tradition of philosophers of science who really knew their scientific subject and who were serious students of the history of science, among them Ernst Mach (whose philosophical position Meyerson abhorred and attacked), Pierre Duhem, William Whewell, Federigo Enriques, and Léon Brunschvicg (who was especially learned in mathematics and its history).

Meyerson's contributions to the philosophy and the history of science are still esteemed in France. In 1960, the session on 26 November of the Société Française de Philosophie was devoted to a commemoration of the centenary of the birth of "two French epistemologists" Meyerson and Gaston Milhaud (who, like Meyerson, was a deep and creative scholar in the area of the history of science as well as philosophy).[8]

In estimating the lasting influence of Meyerson, we must include his importance in the development of the thought and research of Alexandre Koyré who became Meyerson's leading protégé, and of Hélène Metzger-Bruhl, whose pioneering and penetrating studies of the development of chemistry in relation to the rise of modern science remain unparalleled interpretations of the founding of modern science. H.W. Paul has observed that "through the critical mediation and example of Koyré, it is likely that Meyerson has exerted more influence than is generally recognized." He reminds us that Koyré admitted that "this influence is not to be found in fidelity to the subtle dogma of the basic identity of human thought," as Meyerson believed; most of us "follow Koyré in recognizing" that human thought exhibits "different structures in different historical periods." We may thus applaud Paul's sentiment to the effect that we should follow Meyerson's "great precept" and as historians "respect our predecessors who made errors and ... seek reasonable explanations of their mistakes as carefully as the explanations of their successes."

Although the problems raised by Meyerson may not always seem at the forefront of current philosophical debates, his influence still penetrates many aspects of the history of thought. Most recently, Philip Mirowski has shown how Meyerson's ideas illuminate the most fundamental questions of modern "neo-classical" economics in relation to the borrowing of concepts and methods from physics and mathematics.[9] Not only does Mirowski draw heavily on Meyerson's ideas as a guide to his interpretation but he even finds that Meyerson himself thought that the basic issues that he raised might have implications for economics, giving examples.[10] Here we see how fundamental contributions to thought continue to have a life of their own and to influence the development of many disciplines even beyond the range that founders like Meyerson himself might have envisaged.

I. BERNARD COHEN

NOTES

1. For details concerning Meyerson's life, I have drawn heavily on the article on Meyerson by H. W. Paul in the *Dictionary of Scientific Biography*, vol. 15, pp. 422–425.
2. Although Hélène Metzger became known primarily for her seminal studies of eighteenth-century Newtonianism, theory of matter, and chemical theory, her earliest work was in the philosophy of science. See her *Les concepts scientifiques*, avec une préface de André Lalande (Paris: Alcan, 1925).
3. Published in French in Paris in 1939 (by Hermann), this work has been translated by John Metham under the title "*Galileo Studies*" (Hassocks, Sussex: The Harvester Press, 1978). Sad to relate, in this English version the translator has misspelled Meyerson's name as "Myerson."
4. See *Correspondence entre Harald Höffding et Émile Meyerson* (Copenhagen: Einar & Munksgaard, 1939).
5. *Identité et réalité* (Paris: Vrin, 1908; 2d ed., revised and enlarged, 1912; 3d ed., 1926) has been translated into English by Kate Loewenberg (London/New York: Macmillan, 1930). According to H. Paul (D.S.B., vol. 15, p. 425), Meyerson "considered the German translation of 1930, which has a long introduction by the mathematician Leon Lichtenstein, who spread Meyerson's ideas in Germany, better than the English."
6. George Boas: *A Critical Analysis of the Philosophy of Émile Meyerson* (Baltimore: Johns Hopkins Press, 1930); see also Thomas Kelly: *Explanation and Reality in the Philosophy of Émile Meyerson* (Princeton: Princeton University Press, 1937). Also Silvestro Marcucci: "Filosofia, scienza e storia della scienza in Emilio Meyerson," *Physis* **3** (1961): 5–19 and *Émile Meyerson – Epistemologia e filosofia* (Turin, 1962).
7. *The Relativistic Deduction*, with an introduction by Milic Čapek, *Boston Studies in the Philosophy of Science*, vol. 83 (1985).
8. See *Bulletin de la Société Française de Philosophie* **55** (1961): 51–116.
9. Philip Mirowski: *More Heat than Light – Economics as Social Physics: Physics as Nature's Economics* (Cambridge/New York: Cambridge University Press, 1989), esp. pp. 5–8, 314–316. Although Mirowksi calls *Identity and Reality* "a now-dated book," he also declares that it was the most important work (the one whose "influence is most felt") in his analysis. In particular, Mirowski made use of Meyerson's discussions of conservation laws in *Identity and Reality* and his discussions of the tension in physics between "a search for identity and invariance" and "the acknowledgement of diversity and change." In Meyerson's view, according to Mirowski, "conservation laws were just a special case of the more sweeping postulate of the identity of things in time, a postulate he insisted was central to all human thought."
10. "The expression [of equivalence] is borrowed from the language of economics. When I affirm that such a thing is worth such a price, that means that I can buy it or sell it at that price..." Quoted by Mirowski (p. 7) from *Identity and Reality* (1962 ed.), p. 283.

TRANSLATORS' ACKNOWLEDGMENTS

This translation was supported in part by grants from the Faculty Development Endowment and the Faculty Research and Assistance Fund of Carleton College. We are very grateful to the College for its support.

We wish to thank Jackson Bryce, Roy Elveton, Jerry Mohrig, Richard Noer, Ross Shoger, and our many other friends and colleagues at Carleton College who gave freely of their time and wisdom and offered valuable advice when we shared translation problems with them. We also thank Sandra Allen, Carol Jenkins, Karen Menghini, Lisa Orlowski, Cindy Seger, and scores of anonymous librarians and library workers at the *Bibliothèque Nationale* and other libraries of Paris, and at libraries throughout the United States, for untold hours of work on our behalf.

M.-A. and D. A. SIPFLE

BIBLIOGRAPHIC ABBREVIATIONS

The following frequently cited works are referred to in the text or the notes by abbreviated titles, as indicated below. The corresponding English translations are identified in the text and notes by the translator's name only. Other works, after an initial full citation in each chapter, will be identified by the author's name and/or short title only.

JOURNALS:
Bull. Soc. fr. phil.: *Bulletin de la Société française de philosophie.*
Rev. gén. sci.: *Revue générale des sciences.*
Rev. de méta.: *Revue de métaphysique et de morale.*
Rev. phil.: *Revue philosophique de la France et de l'étranger.*
Scientia: *Scientia*, including early volumes published under the title *Rivista di scienza.*

COLLECTIONS:
Brussels Conf.: *La Théorie du rayonnement et les quanta*, Reports and discussions of the conference at Brussels, 30 Oct. - 3 Nov. 1911, under the auspices of E. Solvay, published by Paul Langevin and Louis De Broglie (Paris, 1912).
Idées modernes: Edmond Bauer, et al., *Les Idées modernes sur la constitution de la matière* (Paris: Gauthier-Villars, 1913).
Enc. méth.: *Encyclopédie méthodique*: *Chimie, pharmacie et métallurgie* (Paris: H. Agasse, Year 4 [of the First French Republic; 1796]).
Enc. méth., 1786: *Encyclopédie méthodique*: *Chymie, pharmacie et métallurgie* (Paris: Panchoucke, 1786).

BERNARD, Claude:
Méd. expér.: *Introduction à l'étude de la médecine expérimentale*, ed. Sertillanges (Paris: F. Levé, 1900) [*An Introduction to the Study of Experimental Medicine*, trans. Henry Copley Greene (U.S.A.: Henry Schuman, Inc., 1927].

BERTHELOT, René:
'Sur la nécessité': 'Sur la nécessité, la finalité et la liberté chez Hegel,' *Bull. Soc. fr. phil.* 7 (1907) 115–118, and the discussion that follows, pp. 119–184).

COMTE, Auguste:
Cours: *Cours de Philosophie positive*, 4th ed. (Paris: J.-B. Baillière et Fils, 1887) [*The Positive Philosophy of Auguste Comte*, freely translated and condensed, from the first ed., by Harriet Martineau (New York: Calvin Blanchard, 1858)]. We use Martineau's translation where it does not differ substantially from the original quoted by Meyerson. In cases where there is a substantial difference we provide a

translation from the original; "cf. Martineau" identifies the corresponding passage in Martineau's free or abridged translation. Where there is no reference to Martineau at all, the quoted passage is not included in her translation.

CROCE, Benedetto:
Ce qui est vivant: *Ce qui est vivant et ce qui est mort de la philosophie de Hegel*, trans. Henri Buriot (Paris: Giard et E. Brière, 1910) [*What is Living and What is Dead in the Philosophy of Hegel*, trans. Douglas Ainslie (London: Macmillan, 1915)].

CUVIER, Georges:
Histoire: *Histoire des progrès des sciences naturelles depuis 1789 jusqu'à nos jours*, in *Oeuvres complètes de Buffon, Complément* (Paris: Baudouin frères et N. Delangle, 1826).

DESCARTES, René:
Oeuvres: *Oeuvres*, ed. Adam and Tannery (Paris: Léopold Cerf, 1904).
Principes: *Les Principes de la philosophie*, *Oeuvres*, Vol. 9 [*Principles of Philosophy*, trans. Valentine Rodger Miller and Reese P. Miller (Dordrecht: Reidel, 1983)]. Meyerson used the French version, while Miller & Miller worked from the Latin. We follow the Millers when possible, but give preference to the French when it diverges.

FISCHER, Kuno:
Geschichte: *Geschichte der neueren Philosophie* (Heidelberg: Carl Winter, 1899 [Vol. 7] and 1901 [Vol. 8]).

HAYM, Rudolf:
Hegel: *Hegel und seine Zeit* (Berlin: Rudolph Gaertner, 1857).

HEGEL, Georg Wilhelm Friedrich:
Meyerson usually cites the 18-volume edition of Hegel's *Werke* prepared by a group of his friends after his death (Berlin: Duncker und Humblot, 1832–1840) and, for the correspondence, Vol. 19 (Leipzig: Duncker und Humblot, 1887). The 1832–1840 Berlin edition of the *Werke* was republished in a facsimile edition, the 20-volume *Jubiläumsausgabe*, edited by Hermann Glockner (Stuttgart: Fr. Frommann, 1927–1930). The works have been rearranged in the Glockner edition, but the original pagination is provided at the top inner margin of each page. We refer to specific works by the abbreviated titles provided below, followed by the volume and page number of the 1832–1840 Berlin edition of the *Werke* as reproduced in the *Jubiläumsausgabe*. We do so even in the one case in which Meyerson uses a different source. Although (see note 6, p. 141) Meyerson in fact worked from *Wissenschaft der objectiven Logik* (Stuttgart: Cotta, 1832), this edition is so rare that it would serve little purpose for us to reproduce its pagination.
N. B.: To conform to Meyerson's usage (see Appendix 7) we have substituted "concrete reason" for "reason" and "abstract reason" for "understanding" where appropriate in the English translations listed below. Other such changes are identified in the notes.

Briefe: *Briefe von und an Hegel*, *Werke*, Vol. 19 (Leipzig: Duncker und Humblot, 1887), Parts I and II (19_1 and 19_2) [*Hegel*: *The Letters*, trans. Clark Butler and Christiane Seiler (Bloomington: Indiana University Press, 1984)].

De Orbitis: *De Orbitis Planetarum*, *Werke* (Berlin: Duncker und Humblot, 1832–1840), Vol. 16 ['G.W.F. Hegel: Philosophical Dissertation on the Orbits of the Planets (1801),' trans. Pierre Adler, *Graduate Faculty Philosophy Journal* [New York: New School for Social Research] **12** (1987–88) 269–309].

Enc., *Logik*: *Encyclopädie der philosophischen Wissenschaften im Grundrisse*, Part I: *Die Wissenschaft der Logik*, *Werke*, Vol. 6 [*The Logic of Hegel*, trans. William Wallace (Oxford: Oxford University Press, 1904)].

Hegel: The Letters: See *Briefe*, above.

Naturphilosophie: *Die Naturphilosophie*, *Werke*, Vol. 7, Pt. 1 [7_1] [*Hegel's Philosophy of Nature*, trans. A. V. Miller (Oxford: Clarendon Press, 1970)]. For the Foreword, see Michelet below.

Phänomenologie: *Phänomenologie des Geistes*, *Werke*, Vol. 2 [*The Phenomenology of Mind*, trans. J. B. Baillie, 2nd ed. (London: George Allen & Unwin, 1949)].

Phil. der Geschichte: *Philosophie der Geschichte*, *Werke*, Vol. 9 [*Lectures on the Philosophy of History*, trans. J. Sibree (London: George Bell and Sons, 1902)].

Phil. des Geistes: *Philosophie des Geistes*, *Werke*, Vol. 7, Pt. 2 [7_2] [*Hegel's Philosophy of Mind*, trans. A. V. Miller (Oxford: Clarendon Press, 1971)].

Phil. des Rechts: *Grundlinien der Philosophie des Rechts*, *Werke*, Vol. 8, [*Hegel's Philosophy of Right*, trans. T. M. Knox (Oxford: Clarendon Press, 1942)].

Wiss. der Logik: *Wissenschaft der Logik*, *Werke*, Vols. 3–5 [*Hegel's Science of Logic*, trans. A. V. Miller (London: George Allen & Unwin, 1969)].

LEIBNIZ:

Opera: *Opera philosophica quae exstant Latina Gallica Germanica Omnia*, ed. Johannes Eduardus Erdmann (Berlin: G. Eichler, 1840).

Opuscules: *Opuscules et fragments inédits de Leibniz*, ed. Louis Couturat (Paris: Félix Alcan, 1903).

Translations used are from

[Parkinson: *Philosophical Writings*, trans. Mary Morris and G.H.R. Parkinson (London: Dent / Totowa, N.J.: Rowman and Littlefield, 1975)].

[Huggard: *Theodicy*, trans. E. M. Huggard (London: Routledge & Kegan Paul, 1951)].

[Alexander: *The Leibniz-Clarke Correspondence*, trans. H. G. Alexander (New York: Philosophical Library, 1956)].

LOEB, Jacques:

La Dynamique: *La Dynamique des phénomènes de la vie*, trans. from the 1906 German edition by A. Daudin and G. Schaeffer, with additions by the author (Paris: Félix Alcan, 1908) [*Dynamics*: *The Dynamics of Living Matter* (New York: Columbia University Press, 1906)]. Neither the English nor the German text is a translation of the other, but the two texts deal with the same material and are quite similar. Where Meyerson's quotations have no exact counterparts in the English text we direct the reader to the corresponding passages with "cf." Where the English text is not mentioned there is no English counterpart.

LUCRETIUS:
De rerum nat.: *De rerum natura*. We have used the Ronald Latham translation, *On the Nature of the Universe* (Baltimore: Penguin Books, 1951).

McTAGGART, John McTaggart Ellis:
Studies: *Studies in the Hegelian Dialectic* (Cambridge: Cambridge University Press, 1896).

MEYERSON, Émile:
IR: *Identité et réalité* (Paris: F. Alcan, 1926) [*Identity and Reality*, trans. Kate Loewenberg (New York: The Macmillan Company, 1930)].

MICHELET, Karl:
Foreword to *Naturphilosophie*: Foreword to Hegel's *Die Naturphilosophie*, in Hegel, *Werke*, Vol. 7, Pt. 1 [7_1] [*Hegel's Philosophy of Nature*, trans. M.J. Petry (London: George Allen and Unwin, 1970), 1:179–190].

PASCAL, Blaise:
Pensées: *Pensées et opuscules*, ed. Brunschvicg (Paris: Hachette, 1917) [*Pensées*, trans. A.J. Krailsheimer (Baltimore: Penguin, 1966)].

PLITT, Gustaf Leopold:
Aus Schelling's Leben: *Aus Schelling's Leben in Briefen* (Leipzig: G. Hirzel, 1869–70).

ROSENKRANZ, Karl:
Hegel als deut. Nat.: *Hegel als deutscher Nationalphilosoph* (Leipzig: Duncker & Humblot, 1870) [*Hegel as the National Philosopher of Germany*, trans. George S. Hall (St. Louis: Gray, Baker, 1874)]. Where there is no reference to Hall the German cited is not included in his translation.
Hegel's Leben: *Hegel's Leben* (Berlin: Duncker & Humblot, 1844).
Schelling: *Schelling: Vorlesungen, gehalten im Sommer 1842 an der Universität zu Königsberg* (Danzig: Fr. Sam. Gerhard, 1843).

SCHELLING, Friedrich Wilhelm Joseph von:
Meyerson cites *Friedrich Wilhelm Joseph von Schellings Sämmtliche Werke* (Stuttgart and Augsburg: I.G. Cotta, 1856–1861. This was published in two series, which we will designate by I and II.
Aus den Jahrbüchern: *Aus den Jahrbüchern der Medicin als Wissenschaft*, I, 7:131–259.
Darlegung: *Darlegung der wahren Verhältnisses des Naturphilosophie zu der verbesserten Fichteschen Lehre*, I, 7:1–126.
Einleitung zu dem Entwurf: *Einleitung zu dem Entwurf eines Systems der Naturphilosophie*, I, 3:269–326.
Erster Entwurf: *Erster Entwurf eines Systems der Naturphilosophie*, I, 3:1–268.
Ideen: *Ideen zu einer Philosophie der Natur*, I, 2:1–343 [*Ideas for a Philosophy of Nature*, trans. Errol E. Harris and Peter Heath (Cambridge: Cambridge University Press, 1988)].

Phil. der Offenbar.: *Philosophie der Offenbarung*, II, 3:1–530. The first Book of this work (II, 3:1–174) is the *Einleitung in die Philosophie der Offenbarung*.
Transc. Idealismus: *System des transcendentalen Idealismus*, I, 3:327–634 [*System of Transcendental Idealism*, trans. Peter Heath (Charlottesville, Virginia, 1978)].
Weltseele: *Von der Weltseele*, I, 2:345–583.
Zur Geschichte: *Zur Geschichte der neueren Philosophie*, I, 10:1–200.

SPINOZA, Benedict:
Ethics: *The Ethics*, *The Chief Works of Spinoza*, trans. R.H.M. Elwes (Bohn Library ed.; reprint New York: Dover, 1951).

TAINE, Hippolyte:
Les Philosophes classiques: *Les Philosophes classiques du XIXe siècle en France*, 11th ed. (Paris: Hachette, 1912).

TRENDELENBURG, Adolf:
Log. Untersuch.: *Logische Untersuchungen* (Berlin: Gustav Bethge, 1840).

WALLACE, William:
Prolegomena: *Prolegomena to the Study of Hegel's Philosophy and especially of his Logic*, 2nd ed. (Oxford: Clarendon Press, 1894).

ZELLER, Eduard:
Phil. der Griechen 2_1: *Die Philosophie der Griechen*, 3rd ed. (Leipzig: Fues, 1875), Vol. 2, Pt. 1 [*Plato and the Older Academy*, trans. Sarah Frances Alleyne and Alfred Goodwin (London: Longmans, Green, 1888)].
Phil. der Griechen 2_2: *Die Philosophie der Griechen*, 3rd ed. (Leipzig: Fues, 1875), Vol. 2, Pt. 2 [*Aristotle and the Earlier Peripatetics*, trans. B. F. C. Costelloe and J. H. Muirhead (London: Longmans, Green, 1897), Vols. 1 and 2].

COMMENTS ON DOCUMENTATION

Émile Meyerson's *De l'Explication dans les sciences* was first published in 1921, by Payot. Our translation follows the second (and final) edition (Paris: Payot, 1927).

Although Meyerson was a prodigious reader – in many languages – and obviously cared about his sources, there are a surprising number of errors in his quotations and his documentation. One has the impression that he worked from imperfect handwritten notes at best or from memory (impressive though that may be) at worst and that he did not recheck his sources. Since reproducing careless errors seems to us to serve no scholarly purpose, we have corrected quotations freely and without comment unless there is a significant difference between Meyerson's version and the original, in which case we have so noted. We have freely added bibliographic information not provided by Meyerson himself and have silently corrected obvious errors where possible.

In a few cases we have been unable to locate Meyerson's quotations on the basis of his documentation. In these cases we have simply reproduced his citations, noting obvious errors in brackets. Although we have often corrected or completed bibliographic material without introducing brackets, all unbracketed explanatory notes are Meyerson's. Where Meyerson himself has supplied bracketed material, we have so indicated. Bracketed material not otherwise designated has been added by the translators.

Where Meyerson quotes English sources in French, we have of course gone directly to the English source wherever possible, introducing ellipses or brackets as necessary. Where he quotes non-English works in French we have used standard English translations from the original language wherever possible, introducing ellipses or brackets as necessary. If his French version differs significantly from the standard English translation, we have substituted his language, explaining the discrepancy in the notes or, in some frequently recurring cases, in the note on bibliographic abbreviations above.

We have reproduced only Meyerson's italics, omitting both italics he has ignored in the sources he cites and italics introduced in the translations we use.

M.-A. and D. A. SIPFLE

PREFACE

We believe we can make our work more accessible by summarizing our general plan in advance. In Book One we seek to establish that the current conception of science, which is positivistic, neglects two observations that seem to us to be fundamental, namely, on the one hand, the incontestable fact that science is essentially ontological, that it cannot dispense with a reality posited outside the self (Ch. 1) and, on the other hand, its equally clear tendency to go beyond the search for law to the search for explanation (Ch. 2). While suggesting that the scientific ontology itself can, in the last analysis, be considered as resulting from the need for explanation, we nevertheless begin Book Two by considering the two tendencies insofar as they conflict with one another, showing how scientific explanation actually ends up dissolving the external world into undifferentiated space. Indeed, what explanation seeks first and foremost is the *deduction* of the phenomenon from its antecedents, of which it must be the logical consequence (Ch. 3). That process obviously rests on a postulate, which is none other than the belief in the rationality of nature (Ch. 4), a rationality that can be realized only by applying to scientific reality the schema or process of identification (Ch. 5). This application is limited by the existence of the irrational, both in the physical sciences (Ch. 6) and – at least insofar as we are justified in formulating hypotheses in this area – in the biological sciences (Ch. 7). Within the vast area left for its action, scientific explanation is essentially spatial, and we study its forms (Ch. 8) and future possibilities (Ch. 9), going on to show how science manages, by the "state of potentiality," to create a semblance of explanation where identity is clearly lacking (Ch. 10).

In order better to grasp the nature and necessity of the process of destruction of reality by explanatory reason – so paradoxical at first sight – in Book Three we study the attempt at global explanation of nature due to Hegel (Ch. 11). The objections Schelling formulated against his rival then disclose the obstacle inevitably encountered by any purely idealistic interpretation of scientific reality (Ch. 12). A glance at the relations between Hegel's doctrine and that of Auguste Comte leads us to recognize that the two positions may have more points of contact than generally seems to be acknowledged (Ch. 13), while a comparison between the edifice built by Hegel and the analogous constructions of Descartes and Kant allows us to discern the feature these theories have in common, which is none other than the continuity of their deduction (Ch. 14).

Book Four, finally, is devoted to an even closer study of the functioning of scientific reason through a comparison with philosophic reason. After investigating science's true attitude toward philosophic systems (Ch. 15), we return (Ch. 16), in the light of what we have established, to the primordial question of the agreement between reason and reality already touched upon in Chapter 4. Then, having shown how the seemingly paradoxical character of our theory explains its belated appearance and discontinuous historical development (Ch. 17), we conclude by making an effort to establish that human reason, though antinomic by nature, is nevertheless one and the same in all domains and in all eras (Ch. 18).

In skimming through this summary table of contents, a reader at all familiar with our previous work (*Identité et réalité*, 2nd ed., Paris: Payot, 1912 [*Identity and Reality*, trans. Kate Loewenberg, New York: Macmillan, 1930]) will easily perceive that the two books have many points in common. Indeed, our area of research has remained unchanged: our concern is still the theory of knowledge. Nor has our method varied: we again seek, insofar as possible, to identify the essential principles of thought by considering the processes followed by scientific reason. We try, however, to reach our conclusions (which are somewhat broadened) by a different route than the one we followed in our first work. At that time we had treated the schema of identification as merely heuristic, taking pains to go on to demonstrate, by an analysis of scientific theories, that our understanding of the role of this principle in fact offered the key explaining both the present state of science and its historical evolution. This way of presenting the thesis, which attempts to reach a conclusion about the logic of the sciences entirely *a posteriori*, seems to have led more than one reader astray, and highly competent critics, while in general receiving our efforts favorably, appeared to disapprove of our method. For this reason we asked ourself if it was not possible to arrive at the same result in a more strictly logical way, trying to analyze the mechanism of scientific thought more directly. Since in this way the process of identification comes to be connected with the general tendency to *deduce* nature, to understand it as rational, as necessary, perhaps its role in scientific reasoning will be found to be better justified.

In any case, this way of envisaging the problem has the advantage of posing it in a more general form than we had done previously. Thus we have been able to turn our attention in the present work to modes of explanation that do not strictly fit into the framework of science, at least as it is understood today. It is true that in our earlier work we had already examined nonmechanical theories, especially qualitative theories, but these were still conceptions whose rigorously scientific character was unquestionable. On the other hand, we had completely omitted any attempt at *logical* (or pseudological) explanation of nature, such as that which provides the basis for Aristotle's theory or for the attempts of the German *philosophers of nature*; or at least we had considered the Peripatetic doctrine only insofar as, by an obvious detour, it had given rise to a true qualitative physics. Our procedure in the present work allows us to fill this gap, if only in part.

One could no doubt claim that since these processes of explanation are very remote from those we today recognize as valid, an examination of them could teach us nothing useful about contemporary scientific reasoning. But that would be to disregard the essential principle of the oneness of reason, a principle that will receive new confirmation through our final observations – or so we believe. Thus we dare to hope that the reader will be willing to recognize with us that modes of reasoning which at first sight seem completely strange, altogether inconsistent with those to which we are accustomed, nevertheless often have more than one thing in common with them and can thus serve to reveal motives that might be likely to remain hidden to us. This is the case, in particular, for the theories set forth by the man whose philosophic reputation – to say the least – is one of the most resounding of all: Hegel.

The reader will see in the ensuing pages what we mean to gain from an examination of Hegel's work. He will see how this powerful mind was able, straight away so to speak, to fathom, at least partially, the real guiding principles of scientific thought, and how then, carried away, as it were, by the very strength of his intellect and by his boundless confidence in his own powers, he used these accurate and profound views as the foundation for a monstrous monument. But even the errors in his thought sometimes contain valuable lessons. For as a result of the seriousness and the tireless tenacity with which Hegel pursues his ideas to their ultimate conclusion, as a result of the sincerity with which he expounds them and his profound disdain for any consideration drawn from common sense –

one of his English disciples has aptly said that Hegel seeks "to enlighten by provoking us"[1] – his deductions often lay bare the true and hidden motives of our thoughts.

A philosophy is an attempt to reconcile us with ourselves or, if one prefers, given that our reason is what it is, to reconcile the "realities" that assail us from various sides. Thus it has value above all in terms of the whole, the system, and one cannot profitably criticize it or attack it except by considering the system in its totality, at least by its main features. Now that is not at all what we try to do insofar as Hegel's system is concerned; on the contrary, we study only a strictly limited part of it from a particular point of view. In other words, we in no way pretend to have *refuted* him. In the pages that follow, the reader will no doubt find more than one passage that might seem to suggest a pretension of this sort; but that is a simple lack of perspective, so to speak, which we have not been able to rectify. We mean to set it right here, once and for all, by begging the reader to add the necessary reservations wherever they seem indicated. Repeating them each time would have been tedious and would only have further complicated a subject already sufficiently difficult to elucidate.

The same remarks are at least as applicable for other writers we mention in the course of our work. The reader familiar with these great names will sometimes think our portraits of these men bear little resemblance to those he remembers from his studies, that our treatment in a manner of speaking deforms these figures by too one-sided a vision, by the bias of an artificial perspective that distorts the proportions, exaggerating one particular trait, generally considered secondary, at the expense of what the best critics deem the most essential content of the doctrine. But these are not after all meant to be mirror images. To use a metaphor borrowed from the world of art, we do not intend to shape figures in the round; what we want, rather, is to make a rough sketch that captures an attitude toward this problem of scientific explanation, which is the only thing that interests us here. We thus dare lay claim to some indulgence on the part of the reader and beg him not be too quick to condemn us if, at first glance, the gesture seems exaggerated, overstated – as in the outline of a Quaternary animal or in a Japanese drawing. Even the titles of our chapters are sometimes more elliptical than a title has the right to be. When he reads 'Hegel, Descartes and Kant,' the reader will kindly remember that we shall by no means study the relations between these three thinkers in general; rather we shall limit ourself to comparing their epistemologies and, in particular, the way in which they considered the explanation of the physical phenomenon, especially insofar as this conception seems likely to cast light on the attitude of contemporary science.

Our book, as the title indicates, is based on a theory of science. Consequently, when we venture beyond this, onto the terrain of pure metaphysics, it is always the conceptions of science that provide our starting points and our supporting evidence, and, insofar as possible, we consider everything from this perspective. Thus, when we refute this or that doctrine, when we declare it inadmissible, what we really mean is that it could not be reconciled with the way in which science considers these matters. Now there do exist other "realities" than those of the material world and of science – the English neo-Hegelians in particular never tire of insisting on this point, and with good reason. And since, on the other hand, our reason never fully resigns itself to not understanding, the efforts of a monistic metaphysics, one that seeks to conceptualize the world from a single point of view, are and always will be ageless. We do not at all aspire to put an end to them – even if they should plunge into the paths of romantic idealism, indeed even into those of Hegelianism. We only wish to make as clear as possible to future creators of systems the obstacles they will have to overcome, and our highest ambition will have been fulfilled if our works are recognized as being a part of the *prolegomena to any future metaphysics*.

* * *

In view of the close connection between the present work and our previous book, we are frequently obliged to refer the reader to the latter. In other cases we have felt the need to summarize briefly what we had presented there and finally, in a few cases, especially when we found it necessary to add new developments to these earlier treatments, we were able to find no better alternative than to reproduce passages almost word for word. We apologize for this, as well as for the unavoidable disparity in the procedures we have used here.

Nor have we been able to avoid a multiplicity of citations and references, any more than we could in our earlier book for that matter; they constitute a necessary evil in a study where one means to seek the inner mechanism of thought by examining the thought of others and its historical evolution. This is especially true because, since our point of view differs from that generally taken by historians of science and of philosophy, and since our attention is frequently drawn to questions that did not interest them much at all, we are, in cases of this sort, forced to go beyond textbooks and résumés to the original works themselves. Here erudition is not an extraneous matter nor a vain ornament; it is an integral part of the very substance of the search.

We also realize that the history of the sciences, as we are obliged to present it, will appear chaotic and disconcerting. Indeed, it is not possible to recount it in any continuous way; nevertheless, we have tried to do so in Appendix 2, with regard to a precise moment in the evolution of chemistry, and there the reader will see what led us to make this exception. But everywhere else historical detail is cited only as an illustration of the working of this or that deep-seated reasoning process. Now these processes, which we have done our utmost to isolate, constantly combine with one another in real thought processes, so that a given phase of history is capable of providing examples of quite distinct mental tendencies. Therefore the reader will frequently have the impression of hearing the same things discussed in very different terms. Obviously that is also an unavoidable drawback of the method we have chosen. If it is even more noticeable here than in our previous work, it is because we mean to go somewhat more deeply into the procedures of scientific thought and to analyze them more thoroughly, so that we are forced by that very fact to show that these procedures are more tightly intertwined.

The present book incorporates (in Chs. 1 and 15) a large part of the work published in the *Revue de métaphysique et de morale* **23** (January 1916) under the title 'La Science et les systèmes philosophiques.'

In preparing the present work for publication, we have received evidence of esteem and friendship which we could not value more highly. André Lalande and Désiré Roustan both consented to read over our manuscript in its entirety and have provided extremely important critical commentary from which we have done our best to profit. Léon Robin kindly helped us with the interpretation of Plato's texts. May we express our gratitude to them here.

July 1920 ÉMILE MEYERSON

NOTE

1. [James Hutchison Stirling, *The Secret of Hegel* (Edinburgh: Oliver & Boyd, 1898), p. xlix.]

BOOK ONE

THE TWO FUNDAMENTAL OBSERVATIONS

CHAPTER 1

SCIENCE DEMANDS THE CONCEPT OF THING

Although language is full of figures of speech that have become somewhat obscure, and words have in some cases come to mean something quite different from what they were originally intended to denote, it is not without value, when one wants to establish the content of a term, to begin by looking into its etymology. The etymology of the word *explication*[1] is perfectly clear. The Latin word *plica*, which became *pli* [fold] in French, has the same meaning as its derivative, and to explicate is thus more or less the equivalent of *to unfold* [*déplier*], with the nuance (sufficiently emphasized by use of the prefix[2] *ex* as opposed to *de*) that it is less a question of making the material flat and smooth than of bringing out and revealing what was hidden in its folds. Bossuet uses the term in this literal sense: "We shall be forced to acknowledge that there is within the seed a secret principle of order and arrangement, since one sees the branches, leaves, flowers and fruits explicate themselves and develop from it with such regularity ...." It is in an already derivative yet still closely related sense that Boileau says of a tragedy that it "takes hold, moves forward and explicates itself."[3] Moreover, this usage is consistent with the tradition of medieval philosophy, as we see in Nicholas of Cusa, who defines a line as "the explication of a point."[4]

Some foreign philosophers, in borrowing the term from the Latin, have at times also retained this primitive meaning. Hegel, for example, when he speaks of "the World-Spirit ... whose nature is always one and the same, but which unfolds this its one nature in the phenomena of the World's existence," uses the term *explicirt* (*Phil. der Geschichte*, 9:14 [Sibree 11]).[5] Similarly, James Ward, commenting on the German philosopher, refers to the "process of explication" of the Absolute One,[6] while William Wallace speaks of the "explication ... of a microscopic organism"[7] with regard to theories of preformation.

However, as we all know, this sense of the term, while it comes close to the original etymological meaning, is not its customary sense; specifically, it is not the one used in the language of science. For us, to explicate means "to make clear that which is obscure"; this is Littré's definition, and the two excellent *Dictionaries of Philosophy*, with which the French language is – perhaps uniquely – blessed, agree with him on this point. Goblot's definition is almost identical,[8] and André Lalande writes that "to explicate, in all its senses, is to make understood."[9]

What is the connection between this usage and the etymological meaning? Goblot quite rightly points to what might be called an intermediate sense of "making manifest what was shrouded and hidden, making *explicit* what was *implicit*." Thus, when we have closely examined and probed a phenomenon, when we have revealed its most subtle details, those most hidden in its "folds" [*plis*], we shall have explicated, that is to say that our comprehension of the phenomenon in question will be found to be perfect and our intelligence will thenceforth declare itself entirely satisfied.

It should be noted that the etymological sense of the words used to express the concept of explication in other European languages does not contradict this interpretation. The Italian *spiegare* is etymologically identical to the French verb, and the English *to explain*[10] is dependent upon the same image. The German *erklären* – to *make clear*, *enlighten*, *cast light upon* – proceeds from a different physical image, but in the end amounts to a quite analogous conception, since the increase of light is obviously designed to bring out details that may have escaped more superficial examination. The Russian and Polish terms *ob"iasnit'* and *objasnić* have recourse to the same figure of increased light, whereas two other verbs used by these Slavic languages (*rastolkovat'* and *wytłomaczyć*) express the idea of a translation – a meaning, moreover, shared by the verb *expliquer* in French, as, for example, when we speak of "explicating a text."

Can we be satisfied with the meaning we have just determined? Yes, if we grant that the sole task of science is to *describe* phenomena. This is essentially the position of Auguste Comte, taken up by Kirchhoff and especially by Mach, who made it the basis for a complete theory of science. Indeed, it is obvious that everything in science must depend on this point: the form and content of science will necessarily be modified according to what its fundamental task is considered to be.

The position to which we have just referred bears, of course, the name *positivism*, and this term, at least insofar as it is applied to scientific theory, has an altogether precise meaning. As a matter of fact, Auguste Comte showed great merit in being able to create – amidst philosophical concepts that are by nature rather fluid and blurred – a clear doctrine with fixed angles, something rigid which can truly be grasped and which can doubtless be broken, but which is much less likely to be deformed, or at least in which any deformation, any excessive stress, is easily discernible. Positivism signifies (if we limit ourselves to the epistemological content of the term, since the social aspect of the doctrine does not interest us here) "abstention from all metaphysics." Thus it is claimed that science need not penetrate into the true being of things, that this being can be disregarded. Science, positivism goes on to say, seeks and needs to know only relations; it is a set of relations, and its only essential part is the rules, the laws that formulate the relations in question. Even if we are called upon to formulate suppositions, hypotheses, they must have as their unique object an as yet unknown empirical rule: "every scientific hypothesis, to be a matter of judgment, must relate exclusively to the laws of phenomena, and never to their mode of production."[11]

We hope to be able to show, as a conclusion to the study we are about to undertake, that the positivistic affirmation contains a measure of truth. Taken literally, however, it does not stand up to examination.

First, we must beware of a concept that is precisely one of the above-mentioned excessive stresses in the doctrine. Man does metaphysics as he breathes, involuntarily and, above all, usually without realizing it. In the present work will be found more than one proof to support this claim. However, it would be difficult to find a better way to illustrate this mental tendency than by observing that the very formula intended to exclude all metaphysics quite often serves as a basis for the construction of a sort of metaphysics *sui generis*. Indeed, if we look a little more closely at the way in which we most often speak of these laws, we see that they are established as veritable entities, existing in themselves, independently of the mind that conceived them or applies them. They are *laws of nature*. The relations they serve to express are the true relations of things among themselves, relations we can thus know while abstaining from any attempt to know the things in themselves.

This statement is not inherently contradictory. For instance, to use a mathematical example, expressions bearing an imaginary factor can provide entirely real relationships among themselves, and the transcendence of π, as Cournot has pointed out,[12] disappears in the relation of the surface of a sphere to that of its great circle. Therefore, it would not be impossible *a priori* that by relating entities that are unknowable because they are indissolubly linked to a subjective element, we might end up with a totally objective datum, the subjective having been eliminated by the operation. And in fact this is undoubtedly the line of reasoning followed, more or less consciously, by those who believe in the objectivity, in the innate reality, of the laws of nature.

There are many who hold such a view, which is understandable, since, at first glance, the opinion seems quite in accordance with the spirit of modern science. "The mathematical laws of the motion of the stars," says Ampère, "had ... been regulating this movement since the beginning of the world and long before Kepler had discovered them."[13] Similarly, an extremely able contemporary physicist states that the laws were in effect before men formulated them and will still be in force when there are no more men,[14] while a contemporary philosopher of history declares that "Archimedes' principle existed in the essence of things before solids had floated on liquids."[15] But as soon as one grasps the true implications of the theory, it begins to appear somewhat shocking. To be sure, nature does seem to us to be ordered. Each observation more firmly convinces us of this order, and each of our acts, each of our gestures, insofar as they aim at a goal, bear eloquent witness to our confidence in its existence. They also attest to our conviction that this general order of nature is so constituted that it is possible for us to fathom it (cf. Ch. 4, p. 79 below). However, it is clear that these general propositions exhaust all we can know about the order of nature. As soon as we set forth a particular formula, it will contain elements that clearly belong, not to nature, but to ourselves, elements it will be totally impossible to eliminate.

Let us think, for example, of the mathematical form assumed by the rules in the most advanced chapters of science. It seems to us to be the most perfect form of the law, the one toward which law must tend, and in fact seems to tend. It is certainly quite plausible to suppose that nature itself, at its deepest level, is essentially mathematical, and in one of our later chapters we shall deal with this metaphysics of *panmathematicism*. But, quite obviously, concrete mathematical processes, those we really use, are the product of a historical development in which chance must have played a part. Thus, in formulating the law of refraction, we use the sine function, which seems very simple to us because we are familiar with it and even have tables for rapidly determining its value. Yet if we were to use series to express the law, it would seem rather complex, and we would give an altogether different expression to the relation in question. When an astronomer, by means of one laborious operation after another, calculates by approximation the "perturbations" the heavenly bodies cause in one another as they move, he has no doubt that nature resolves this problem instantaneously and that, as Fresnel remarked, nature is not bothered by analytical difficulties. *Dum Deus calculat fit mundus.*[16] That may be. But, if so, He certainly does not calculate with the aid of logarithmic tables, and His mathematics can in no wise resemble our own, the formulas of which necessarily bear the imprint of our minds.

Would it be possible to preserve the objective value of laws by abandoning their mathematical form? Attempts have indeed been made to establish that the laws designated as *qualitative* are invested with particular dignity. It seems likely that any attempt to introduce this distinction would encounter strong resistance on the part of the physicist for whom (with good reason, to be sure) the true expression of the law can only be mathematical. But in the present case we can directly establish that the sacrifice would be fruitless. Let us consider a proposition defining the properties of sulfur. What did the chemist intend to designate by this term? He certainly did not have in mind any particular piece of the familiar yellow material. Sometimes what he says refers to the average among the pieces likely to be encountered commercially, and at times even (when he says "pure sulfur") to an almost ideal material, which can be approximated only after multiple operations; the properties of a random sample of sulfur can differ considerably from those of the material in question. We know what formidable labors Stas had to undertake in order to obtain silver that was nearly chemically pure. We know, moreover, that he had chosen this material as the starting point for his determinations because it promised to be particularly easy to work with, and we also know that the silver he obtained was not really pure, so that his data later had to be corrected. One could doubtless point out that since silver and sulfur are well-defined elements, the pure substance must necessarily exist in the sample I have in my possession, call by the same name, yet know to be impure. But the existence of a physical element is only a hypothesis reached by means of multiple deductions, and *pure* silver or sulfur are only entities created by theories. They are *genera* in the sense given the term by the Schools, and the fact that we attribute a substantial existence to this genus in no way modifies the situation. Thus a glance at the history of the sciences allows us to ascertain that for many centuries all of humanity firmly believed in the existence of substances, even substances considered to be elements, which have since vanished, so to speak. This is the case for the four elements of Empedocles and Aristotle. Actually, for a medieval scientist water was a collection of liquids which for us have nothing in common with our combination of oxygen and hydrogen except its state of aggregation, and the same is obviously true for earth and air. Indeed, one could show that these three elements were only representatives, symbols of what we today call *state of aggregation* and that therefore, from this point of view, the concept is not outdated. But as a matter of fact it has been able to persist only by transforming itself, by renouncing any pretensions to existence as a substance: for us, solids, liquids or gases are *genera*; we do not believe that there is a unique and primordial solid, liquid or gas as the basis for each of them (which was the meaning of the Peripatetic theory of the elements). But the case is even clearer with respect to the fourth element, fire. For here the genus itself has dissolved; the phenomena science classified under this category belong, for us, to quite diverse branches of physics and chemistry. Sometimes they are phenomena of combustion and at other times phenomena of radiation. The *substance* of the element fire at times seems to suggest vaguely the modern physicist's ether; then again, we might be tempted to compare this concept with what we call energy. Or else it is something altogether different, at times even (in alchemy, for example, when fire enters into the composition of substances) something modern science would have great difficulty defining.

To be sure, the existence of our *substances* seems to us, at the present time, to be on much firmer ground. Is it, however, safe from future upheavals? We must remember that chemists have encountered some real surprises in such cases. For example, the material with which the chemistry of dyestuffs had long preferred to work – benzene – suddenly turned out to contain significant quantities of a substance (thiophene) that, in various properties, bears an astonishing resemblance to it, though having a very different composition, since it even contains an element foreign to benzene, namely, sulfur. Similarly, sugar refineries had for generations been producing (or at least isolating) prodigious quantities of what they claimed to be almost pure saccharose, and had been *analyzing* it with all their might – it is no exaggeration to state that for many years thousands of these analyses had been carried out daily in all these factories taken together – without noticing that it contained significant quantities of a different sugar, *raffinose*.

However, the most striking example is undoubtedly that of water. Without speaking of Thales, who saw it as the prime substance from which all others were derived, water was of course considered for centuries to be an element. The discovery of its composition was the turning point in Lavoisier's fight against the advocates of phlogiston, but in spite of experimental proofs to which the genius of this incomparable man lent a well-nigh irresistible demonstrative force, people had so much difficulty renouncing the idea that water had to be an element that in 1798, fifteen years after these experiments, the noted chemist Baumé was still able to speak of the "alleged decomposition and recomposition of water" (see below, Appendix 2, p. 555). Nevertheless, water, though it is a compound, remained, of course, a *unique* and perfectly well-defined substance – perhaps even the best-defined of all – because of the innumerable experiments to which it was continually being subjected. The theory of Laurent and Gerhardt considered the formula of water one of the fundamental *types* of the composition of substances in general, and no one, to be sure, ever thought of doubting at the time, or indeed years later, that the water molecule was really H_2O. Now this is no longer the view of contemporary physical chemists.[17] For them, the liquid we know so well is a solution, and, to begin with, the solvent, which is called *hydrol*, is not H_2O but a polymer $(H_2O)_2$ or $(H_2O)_3$ – the point is not completely settled. Dissolved in it are ice crystals, more strongly polymerized than hydrol: the suppositions range from $(H_2O)_3$ to $(H_2O)_{12}$. Thus the genus *water*, which seemed so secure, disappears by being broken down into its components.

The same is true for the substances we call elements. In general, what was defined as nitrogen was, until Lord Rayleigh's discovery, nothing more than a mixture containing variable but not insignificant quantities of argon (not to mention the other "rare gases"). Indeed, many chemists at the present time seem rather inclined to believe that at least a certain number of our "elements" could in fact be only mixtures of very analogous and yet distinct substances. This might be the source of the difficulty in determining certain constants, notably the one that chemistry considers the most important of all, that is, atomic weight, and Soddy, for example, believes that what we call lead (which is, according to him, the final stage of the transmutation of radioactive bodies) consists of a group of substances with atomic weights varying from 206 to 208.5.[18]

In the case of physics, the situation is clearer yet. We shall never encounter in nature the "ideal gas" of theory, or crystals as they are depicted in crystallographic models. All that is nothing but generalization, abstraction, object of thought, *idea* in the Platonic sense of the term. In order to see to what extent this circumstance is inseparable from the very concept of a rule, let us choose an example from outside the domain of the physical sciences. Let us imagine a historian of the Austerlitz campaign. As long as he limits himself to telling what happened, he can try to capture the personality of Napoleon in all its complexity, attempt to endow the figure with all the character traits that make it unique, all its *Napoleoneity*, as the Schools would have it. But if, immediately thereafter, he behaves like a military theoretician, if he wishes to reason about this campaign, to draw information from it – that is, to establish rules – the Napoleoneity of the emperor will necessarily become blurred, the emperor will become simply *the commander* and his actions will be examined in the same way we would examine decisions that any army commander could be called upon to make under analogous conditions.

Furthermore, it is clear that, in this sense, everything that happens is a historic event. No phenomenon ever truly repeats itself. There are always, there *must be* – we are assured of this by *the law of indiscernibles* – circumstances that distinguish its different appearances; each of its appearances must have its *quid proprium*, which the law deliberately puts aside.

The distinction between real phenomenon and scientific fact seemed so essential to the medieval Peripatetics that one of the philosophical masters of the epoch, the Jewish scholar Levi ben Gerson (Gersonides), whose works, translated into Latin by the order of Pope Clement VI, enjoyed great authority in the West, came to use it as the basis for a solution to the troublesome problem of free will. He reconciles free will with divine foreknowledge by declaring that God does not know the particular as such, but only insofar as it is subject to universal law. He knows the universal order and the particular that is subsumed under this order and foresees the events determined by this order. But man's free will can act in a sense contrary to that of the universal order, in which case the predicted events can fail to occur. This is what is meant by the scriptural expression that "God repents." In other words, God is ignorant of the particular because the particular is contingent. This is, to be sure, a singularly daring doctrine from the religious point of view: the esteemed authority on the spiritual life of the epoch to whom we owe these facts calls it a "theological monstrosity."[19] It brings home to us how essential the distinction must have seemed to the leading minds of the period.

These medieval analogies enlighten us on the true nature of the doctrine of the ontological existence of laws. To believe that the abstractions, the *ideas* dealt with by science actually exist outside us, in things, exist prior to the things of which they are the essence, is to profess a *realism* in the medieval sense of the term. Now, it is evident, on the contrary, that on this question the true convictions of the man schooled in modern science are very clearly nominalistic or conceptualistic. He believes, with St. Thomas, that what really exists is not the general but the particular, infinitely diverse by virtue of the law of indiscernibles: *existentia est singularium*. But since, as the same philosopher rightly observes, science is concerned only with the general – *scientia est de universalibus*[20] – the true *haecceity* of things, to use the medieval term, totally escapes it, with the result that its laws cannot directly apply to the real phenomenon. At this point there is a veritable gap between science and reality, between our understanding and our sensation, since, according to Roustan's judicious formula, "all that is *perceived* by our senses is broken up into particular sensations," whereas "all that is *conceived* by our understanding takes the form of general idea."[21] If we sometimes have the contrary illusion, we owe it solely to the coarseness of our senses and the imperfect means of investigation used, which do not permit us to perceive all that differentiates the particular phenomena from one another. In fact, law, with regard to the directly observed phenomenon, can never be anything except an approximation. Law is an ideal construct expressing not what is happening, but what would happen if certain conditions (more or less unrealizable in full) were to be met: it can only be what logicians would classify as a hypothetical judgment. To be sure, if nature were not ordered, if it did not present us with similar objects likely to furnish general concepts, we could not formulate laws, and we shall see below (p. 82), that, extended a bit further, this proposition actually forces us to suppose that the nominalism of modern science is less complete than it would seem at first glance. But what we must remember here is the indubitable fact that the laws we formulate can be only an image of the real ordering of nature; they correspond to it only to the extent that a projection can correspond to a body with *n* dimensions; they express it only as much as a written word expresses the thing, for, in both cases, it is necessary to proceed through the intermediary of our understanding. The particular law did not exist, in the most literal sense of the term, before it was formulated, and it will cease to exist the day it is incorporated into a more general law. And one must bear in mind that very often the law will disappear, not because it is henceforth a particular case of a more general law, but because it is actually abolished and is recognized as being only a first rough approximation, contradicted by more precise determinations. Since Newton's law, we know that Kepler's laws can be only more or less exact, and the kinetic theory teaches that no gas can rigorously follow Mariotte's law.

The same is true, appearances to the contrary, for certain very simple propositions that, as part of "rational" mechanics, appear to us to be, in some sense, revelatory of the reason that governs things and thus to be inherent in things themselves. No one has ever seen a real body follow uniquely and strictly the rule of inertial motion. For terrestrial objects, friction, among other things, intervenes, and for celestial bodies the motion is complicated by the action of gravity (not to mention the at least very probable, though still elusive, action of the ethereal medium). Moreover, we do not know whether inertia is really a property of things; it could well be that this is only an illusion, that in fact it is itself due to an action of the "medium" and that its apparent rigor is only "statistical," that it results only from the largeness of the numbers coming into play. And as to Archimedes' principle (brought up by Xenopol), we must note that, as a matter of fact, not only will its manifestation always be "disturbed" by the adhesion, viscosity, etc., of liquids, but that furthermore the particular form in which we must conceive of it can in no way be objective. Indeed, this principle implies a whole series of concepts such as volume, weight, etc., which arise, one might say, spontaneously and uniformly in all normal human intelligence; but they are most certainly concepts of our reason. Thus if one maintains that there is, in the principle, a relation existing prior to things, it will be necessary to formulate this relation in a very indeterminate manner. One must say that the essence of things is characterized by a particular trait that, in an intelligence contemplating it, could have inspired the idea of a formula analogous to Archimedes' principle. But it must be understood that the intelligence thus involved is not mere human intelligence, but a sort of *intellectus angelicus*, an intelligence that, while being of the same order as man's, would be infinitely superior to it, since it would have been possible for it to conceive the existence of the relation before the very existence of liquids and solids.

Therefore we cannot foster the illusion that the laws we discover are truly "laws of nature." They are only laws of nature in its relations to our sensation and our intelligence. And it certainly remains true that we can know only relations. But we must state this proposition more precisely, restrict it, by specifying that the only relations we can really know are those in which we ourselves form one of the terms. If things do exist in the external world, it is clear that relations among them must exist; but these relations – like the things themselves – we can know only in relation to ourselves.

However, as we have pointed out, the theory we have just criticized is, in short, merely a deviation from the true positivistic position. True positivism, essentially hostile to all metaphysics, would therefore lend no transcendental character to laws, conceiving of them simply as rules governing our relations with the external world, the ensemble of our sensations. But then a new and serious difficulty arises, namely a profound divergence between this schema and the image that science, even reduced to a mere collection of laws, actually presents.

Here we have an aspect of the doctrine that, it would appear, rather tends to elude its initiates. Certainly Auguste Comte himself never seems to have envisaged the consequences of his system, though in this case they are so blatant; he has, on the contrary, spoken quite appropriately of the "rough but judicious indications of popular good sense, which will ever be the true starting-point of all wise scientific speculation."[22] But on this point his supporters have sometimes seen more clearly. Thus John Stuart Mill declared that the "ultimate Laws" which science would one day attain and to which, in the meantime, it came closer each day, refer to the qualitatively distinguishable sensations we experience and are therefore at least as numerous as these sensations.[23]

There is no doubt that here Mill is entirely correct against Comte; all he does is state a consequence ineluctably entailed by the foundations of the Comtian doctrine. Indeed, on one hand science is defined as a collection of rules designed to facilitate prediction and based on experience, on the phenomena that are known to us, and on the other hand we are expressly forbidden to seek out what lies behind these phenomena. Obviously then, all that remains is to connect the phenomena themselves directly to one another, that is to say – since here the term *phenomenon* can only be synonymous with the term *sensation*, by reason of the repudiation of all search for causes – that what one must really seek is relations between sensations taken as pure, deprived of all ontology. In other words, it cannot be a question of the *things* we perceive when we open our eyes in the morning, for such perception, which seems passive, is in fact a result of the activity of our minds, and the world of common sense things is most certainly a metaphysical speculation concerning the causes of our sensations, that is, an ontology. To avoid this, it would be necessary to begin with perceptions (given all at once, since their elaboration remains unconscious), and to penetrate all the way to the elements that constitute them, to those "immediate data of consciousness" Bergson had so much difficulty extricating.[24] It is among these elements that one would then have to establish relationships. This would thus be a kind of psychophysics, but in some sense infinitely more extreme than the science we know under this name, which obviously presupposes physics with its whole conception of reality. Can such a science be established? Lucretius, although he had quite a clear conception of science, explicitly denied it, claiming that without the prior establishment of this world of things we could do no science at all.[25] Malebranche, in developing the same point of view, strove to show we could never claim to measure sensations, insofar as they are subjective phenomena, directly against each other, and that all comparison between them presupposes a reduction to causes existing outside us and thus subject to the conditions of time and space.[26] That position seems very hard to contest.

First of all, there can be no doubt about the mental attitude of the physicist who studies *nature*: he is certainly not the least bit convinced that he is seeking only relations between sensations; on the contrary, he is convinced that he is probing a mystery independent of his own sensation. When Claude Bernard defines an experiment as a "control" of hypotheses by means of "reasoning and facts" and declares that it is "the only process that we have for teaching ourselves about the nature of things outside us" (*Méd. expér.* 12 [Greene 5])," he is undoubtedly expressing the instinctive faith of all scientists. One can of course claim that the scientist is mistaken about his own beliefs (this is the positivistic stance), and that in any case such an attitude has no influence at all on the course of his work. But the claim is manifestly false.

Furthermore, the entire evolution of modern physics clearly shows how far the program outlined by John Stuart Mill diverges from reality. Indeed, this science obviously proceeds in a manner directly opposed to that leading to the establishment of direct relations among sensations. One of the most eminent theorists of contemporary physics, Max Planck, considers science to be characterized precisely by the fact that it is moving further and further away from what he terms "anthropomorphic" considerations, that is, from those in which the person of the observer intervenes,[27] or to restate this in philosophical terms, from that which has reference to the *self*. Höffding is only stating the obvious when he declares that "physicists have until recently been metaphysicians on this point."[28] They will most likely be metaphysicians as long as there is a physics, because for them this attitude appears to be necessary.

In any case, all we seem to be able to say on the subject is that we know nothing about it, since no one has ever tried to construct such a science. Physics today is plainly nowhere near that point. To be convinced of this, one need only follow a professor's exposition. It will be futile to look at the beginning for the slightest trace of the subtle analyses of which we have spoken. On the contrary, each sentence, each affirmation, if we examine it, will bear witness to an unshakable faith in the existence of things, in their independence from sensation. Anything that concerns the intervention of the *subject* will be treated parenthetically, so to speak, as an "error of judgment," or else will be relegated to one of the final chapters, which at the present time hardly seems to be part of science, that is, to physiology.

It would be at least superfluous to dwell on this situation, which is so obvious and which each reader can verify for himself with the help of any textbook of "experimental physics" whatsoever.

The demonstration seems less pointless in the case of certain well-defined areas of *theoretical* or *mathematical* physics. What physics designates by the term "theory" is, of course, a general conception from which the phenomena with which one is concerned follow or are deduced. In Chapter 3 we shall treat this process of deduction, which, at least in contemporary science, is carried out mathematically. All we need to remember here is that these theories are of two kinds. One kind (often termed hypotheses) calls into play representations designed to teach us (as do the kinetic theories in particular) that phenomena are the consequence of a particular spatial arrangement or a particular movement of more or less material particles, or even (as in the physics of the eighteenth century) of semi-material or immaterial fluids. The other kind, on the contrary, begins with abstract propositions, with *principles*. The occurrence of the first kind of theories – representational theories – in science has always been a sort of embarrassment for the positivistic doctrine; we shall see shortly how it has attempted to extricate itself from this difficulty. On the other hand, positivism has found abstract theories, those based on principles, to be the very ideal toward which science should aim. Now, there is (or at least not too long ago was) a branch of physics that actually appears to fit this model: thermodynamics. Indeed, this science, strongly permeated with mathematical theory, can be entirely reduced to two very general principles, the principle of the conservation of energy and Carnot's principle, neither of which seems to use any material or semi-material image. Furthermore, it cannot be denied that thermodynamics plays a considerable role in physics in general, one that – particularly during the epoch of which we are speaking – seemed destined to grow even larger. It was therefore not unreasonable at that time to hope that science as a whole would eventually come closer and closer to this model, so consonant, at least on the surface, with the positivistic program.

Let us hasten to add that these hopes were nipped in the bud, so to speak. Thermodynamics was found to be incapable of accounting for certain phenomena, such as the blue of the sky, which kinetic theory succeeded in explaining perfectly. Moreover, it was impossible to reconcile it with the observations of Brownian motion which anyone who uses a microscope had known for nearly a century, but to which scientists turned their attention only after Gouy connected it with molecular motion. In the words of Smoluchowski, one of the scientists whose works greatly contributed to clearing up this area, "Brownian motion is indeed a phenomenon that clearly demonstrates the correctness of kinetic arguments and at the same time the inexactness of the notions of thermodynamics,"[29] namely, those whose generalization and transposition into ontological concepts constitute the metaphysics of energy. Lord Rayleigh's theory of the blue of the sky is a proof in the same sense.[30]

It is quite characteristic that the triumph of kinetics over thermodynamics, despite the fact that it was accompanied by considerable progress from the standpoint of the general course of science, should have seemed altogether regrettable to the loyal supporters of positivism. Their sentiments were expressed very well as early as 1898 by the scientist Lucien Poincaré:

> Is the history of physics, like the history of mankind, only an eternal new beginning, and must we periodically return to the concepts philosophers have imagined since antiquity? The progress of thermodynamics had given rise to other hopes; all by itself it seemed to be able to guide us in the physical domain, relying only on arguments and principles formed by the natural generalization of a few experimental laws. Must we then always have recourse to representations, to mechanical interpretations, doubtless corresponding so imperfectly to nature?[31]

We know how unavailing his regrets were: no sooner had Lucien Poincaré's statements appeared (he is to be admired for having, to some extent, anticipated events, since the transformation was just beginning at that time) than additional facts came to light to underscore the triumph of atomism, which became complete.[32]

Given the special role thus attributed to thermodynamics, one is led to examine more closely the foundations of this science, in order to find out whether it conforms as much to the positivistic program as seems to have been claimed, or at least suggested. Now, a superficial glance at a textbook is enough to convince us how untenable such claims are. Let us take the book of Henri Poincaré, whose authority is unquestionable, and examine the way in which the two basic principles are explained. For the principle of the conservation of energy, we find that from the outset[33] we are dealing with "material points." Let us suppose for an instant that the existence of these material points depends on our sensation and then try to redo the demonstration as it is presented by the textbook: it will have lost all sense. And that is just as obvious, a few pages later, with regard to the second principle. The fundamental notion of Carnot's principle is, of course, the notion of temperature. Now, here is how Henri Poincaré defines it: "By definition, two bodies are at equal temperatures or their temperatures are in equilibrium when upon contact they show no change in volume." Thus, in order to establish the concept of temperature, the concept of *body*, and of body having a definite volume, is indispensable. Is it really necessary to belabor the point? Who could actually imagine a temperature as anything other than the attribute of a material body? To invoke Lotze's famous image, it would be almost like a toothache no one has.

Thus thermodynamics is no less ontological in nature than any other part of physics, and the conviction to the contrary is an illusion. We shall see presently (p. 27) how this illusion could have arisen.

Furthermore, the physicist is so incapable of detaching himself from the concept of thing that when common sense things are not sufficient, he creates others entirely in their image. Such is the case, for example, for telescopic or microscopic objects. Obviously, the scientist – and probably, by repercussion, the majority of ordinary men – today has just as much faith in their existence as he does in the existence of directly perceived objects. An eminent histologist, Nageotte, has quite recently observed that a particular class of errors in judgment arises from the fact that observers forget that microscopic images are not on the same scale as the objects that surround them and that they have in fact only perceived phantoms.[34] As for the sun, there is not the least doubt that, of the two images we have of it according to Descartes,[35] namely that of a small speck of light and that of a body immensely larger than our earth, the second is completely substituted for the first by the contemporary astronomer; for him the sun is an incandescent mass, almost analogous to those flowing from a Bessemer converter. Moreover, this elaboration is not limited to objects we cannot *see* without the aid of instruments; it extends to those whose existence is only inferred. Consider an electrician who is studying a current. If we hide the galvanometer behind a screen and ask him if the current is still flowing, he will probably think we are asking whether a switch has accidentally been turned off. Let us persist and ask him whether he believes that the current has stopped merely because he cannot see the indicator of the galvanometer. If the man has no philosophical background, if he has remained sheltered from "metaphysical doubt," and if we have actually succeeded in making him understand the point of our question (which will not be easy, since he is so unaccustomed to connecting these two kinds of considerations) – well, if he is sincere, he will laugh in our faces! Doubt in this case will seem as unjustified to him as if we were to ask him whether he doubts the existence of his wife or his shop simply because he does not perceive either of them at a given moment. His belief in the two categories of *objects* is apparently analogous and flows from the same source. Electricians have always believed so firmly in electrical current, they have *seen* it to such an extent, that they have finally "materialized" it, almost in the same way a spiritualistic medium claims to materialize his thought. Anyone doubting the reality of electrical current as an object need only refer to certain recent theories; here the current consists of a veritable flow of electrons. Furthermore, it is impossible to doubt that the electrons are considered real, since they are what makes up matter and are consequently supposed to constitute the source of all reality.

Thus, not only is the starting point of science ontological, since it is the world of common sense *objects*, but when science abandons these conceptions or transforms them, what it adopts is just as ontological as what it abandons.[36] Cournot had already clearly recognized this peculiar nature of scientific theories. "Whatever one may say," he declares, "in the modern schools of science, where nothing is more to be feared than the appearance of doing metaphysics, mitigated atomism, as well as pure atomism, implies the hope of somehow understanding the essence of things and their innermost nature."[37]

It is even easy to establish that the hypothetical entities of science are actually more *things* than the things of common sense. Indeed, what constitutes the thing is the fact of being independent of sensation: the thing remains what it is whether I am looking at it or not. Now, the hypothetical entity is clearly more independent, further from sensation, than the thing of common sense, since it was never part of our direct sensation, and at least for many of these entities, such as chemical atoms or electrons, such sensation even appears to us to be more or less impossible. Furthermore, what distinguishes the thing from the sensation is that it is less fleeting, more *permanent*; but here again the theoretical being surpasses the thing of common sense, for it is considered to be immutable: energy, material mass, the atom, the electron are absolutely constant, eternal, whereas all that we perceive directly is, without exception, subject to the influence of time.

That is exactly the evolution we mentioned above, which, according to Planck, causes science to move further and further from "anthropomorphic" considerations.

Obviously, a science that would conform to the positivistic ideal, that would be truly phenomenalistic, attempting to relate sensations directly to each other, could never allow itself to engage in the task of creating new *things*. If, however, we care to speculate about how far the sacrifice would have to be pushed, it is conceivable that it could reach as far as geometry. Is it in fact quite certain that, deep within itself, even geometry does not harbor some substantialistic conceptions? Let us weigh the opinion of a great mathematician, one certainly little disposed to exaggerations of an ontological bent: "Geometry would not exist if there were in nature no solid bodies" that move without being modified.[38] Now such bodies could not arise from our direct sensation, which, in the case of visual sensation, shows the moving bodies continually changing their aspect and size. If, on the other hand, we refer to tactile sensation alone, their attributes can appear simultaneously only in exceptional cases, for bodies of very small size and very simple shape; usually for a person blind from birth these attributes appear only in succession and reappear only after long intervals. The geometric solid spoken of by Poincaré, which is certainly indispensable to the establishment of geometry, can thus be only a representation. Furthermore, one need only consider other processes by which our understanding reacts against our sensations and transforms reality (or forms it, if you will) to be convinced that here we are not dealing with a unique reaction but, on the contrary, with a process analogous to a whole series of others, with a habitual and constant action of the intellect. Not only will the body remain unmodified in space insofar as its geometric shape is concerned, it will retain *all* its physical and chemical properties while it is moving; a piece of sulfur cannot change its thermal or electrical conductivity, nor its melting point, as a result of being transported from one place to another, and if it could be transported to a planet of another system and if, under specific conditions of temperature, etc., it came into contact with oxygen, I maintain that it would form a body called sulfurous acid whose properties I can determine in advance. Furthermore, when I burn a piece of sulfur, I shall maintain that something in it – something very essential, its *matter* – will persist, not only in weight but also (although one sometimes hesitates to formulate it so explicitly) qualitatively, since sulfur is an *element*, of which sulfurous acid is a *compound*, according to the conviction expressed by the formula SO_2. To see how analogous this conception is to that of the geometer, we have only to imagine we know the *shape* of the atoms – perhaps not really such an extravagant supposition in light of recent discoveries. Moreover, to make it more tangible, so to speak, let us for the moment assume that it is not a question of sulfur but of carbon (simply because from the standpoint of chemistry this is the best-known element of all). Well, supposing that the carbon atom actually has the shape of a tetrahedron (the famous tetrahedron of Le Bel and Van't Hoff), it is certain that it will retain this shape when I have burned the coal.

It is also important to note in this regard that science is *compelled* to undertake this creation of new *things*, not only in the sense that new sensations acquired through more and more refined methods of investigation are transformed, spontaneously one might say, into things – every day the astronomer discovers new specks of light in the sky and no one doubts that they are new stars – but also because the things with which science began cannot be sustained and must therefore be replaced by others. To become convinced of this, one need only glance at any branch of the physical sciences. What could a physicist who is studying a steel bar from the standpoint of its elasticity, its expansion under heat and its electrical capacity possibly do with the common sense concept of this bar, that is, with the concept of a rigid and coherent piece of matter? He is obliged to suppose that it has lacunae, pores, and finally to resolve it into a complex of discrete particles, molecules. And how could the chemist possibly confine himself to the concept of sulfur as a continuous yellow mass? He must deal with the atoms of this sulfur, atoms whose properties will be quite different from those of the piece presented to him by direct perception.

Moreover, it is clear that if this were not so, if science really stuck to the ontology of common sense and if one discovered this ontology again at the end of scientific elaborations, the positivistic claim would in fact be inexplicable. Now, there is no need to belabor the fact that an important doctrine professed by so many and such good minds is always justified in some respect. This (at least partial) justification results here from the fact that science, while adopting common sense as its starting point, itself subsequently destroys this ontology. But – and this is the important point – it always destroys it in favor of a new ontology.

That is the process to which science devotes itself in most of its branches. It does not do so in thermodynamics, since, as we have seen, thermodynamics does not advance any particular hypothesis, kinetic or otherwise. But that simply means that in this case the common sense ontology remains in effect; when thermodynamics speaks to us of a *body*, it is a body as we know it in our everyday lives, and likewise the *matter* of material points is more or less (for it is assumed to have no dimensions) matter like that we habitually handle. But precisely because we are accustomed to this ontology, it tends no longer to seem like an ontology, a metaphysical hypothesis, but like an established fact, whereas the unaccustomed ontology of hypotheses reveals itself as such at first glance. And that is surely the sole source of the illusion noted above as to the *positivistic* character of thermodynamics.

What we have just recognized helps us better understand why, as we have observed, the positivistic program of science is so easily transformed into a veritable metaphysics of laws. Earlier we attributed this transformation to the influence of the general metaphysical tendency characteristic of human reason, and that is certainly the underlying cause of the phenomenon. However, it must be admitted that here this tendency manifests itself under very special conditions. If, indeed, as positivism would have it, the concept of reality upon which science is based were clearly opposed to the metaphysical conception, it would be astonishing that a notion whose origin is as incontestably scientific as that of law could have been distorted in that fashion. To be more explicit, if law as science knows it were to be understood (as seems to be claimed) as a *relation without relata*, if it really cut itself off from all existence outside consciousness, it would be difficult to understand how one might have sought to attribute a metaphysical existence to that law itself. But such is not the case. The whole of science rests on the bed-rock – not exposed of course (since attempts have been made to deny the existence of such a foundation), but nevertheless solid and deep – of belief in a being independent of consciousness. And it is on this very belief and not, despite appearances, on positivistic theory that the concept of the metaphysical existence of laws is actually founded: the existence of the world of objects seems so secure that we come to suppose that even the relationships between objects, as determined by the human intelligence that contemplates them, must nevertheless exist independently of that intelligence. Thus, if the metaphysics of laws does not conform to the true spirit of science, still, as metaphysics, all it is doing is cultivating a seed unquestionably contained in science. We can also turn the proposition around: the very existence of this metaphysics of laws will then provide us with a new confirmation of the fact that science, even purely legalistic science,[39] is in reality saturated with metaphysics.

Thus true science, the only one we know, is not in any way, nor in any of its parts, consistent with the positivistic program. What a strict application of positivism would produce, we do not know, since, here again, we do not even know whether such an application is possible, that is to say, whether one can really use this program to construct anything resembling a science, and we rather tend to suspect that the enterprise would be totally chimerical. What seems certain, however, is that if it were to be established, this really positivistic science could in no way resemble our own.

NOTES

1. [In passages such as this where the etymology of the French "*expliquer/explication*" plays a role, we have chosen to use their English cognates. At all other times we translate "*expliquer/explication*" by their standard English equivalents, "explain/explanation," as we have in the title of the book itself (see p. 10 and note 10 below). Throughout the text, whether the reader encounters "explain/explanation" or "explicate/explication," Meyerson's French reads "*expliquer/explication.*"]
2. [Reading "*préfixe*" for "*suffixe.*"]
3. We borrow these two quotations from Littré's *Dictionnaire de la langue française* (Paris: Hachette, 1863), 2:1571, under the entry '*Expliquer.*'
4. Rudolf Eucken, *Geschichte der philosophischen Terminologie im Umriss* (Leipzig: Veit, 1879), pp. 82, 187.
5. See "Bibliographic Abbreviations," p. xxvii above.
6. "It is an immanent and self-determining process of explication of the Absolute One" (James Ward, *The Realm of Ends*, Cambridge: Univ. Press, 1911, p. 101).
7. "Growth is thus not accretion, but explication and enlargement of a microscopic organism subsisting in the germ" (Wallace, *Prolegomena* 152).
8. Edmond Goblot, *Le Vocabulaire philosophique* (Paris: Armand Colin, 1901), p. 227.
9. André Lalande, 'Vocabulaire philosophique,' *Bulletin de la société française de philosophie*, July 1905, p. 244 [reprinted in André Lalande, *Vocabulaire technique et critique de la philosophie* (Paris: Presses Universitaires de France, 1980, p. 325].

10. [This suggests that in general Meyerson himself would have chosen to translate *"expliquer/explication"* as "explain/explanation," rather than using the English cognates.]
11. Auguste Comte, *Cours* 2:312 [Martineau 204]. Cf. below the quotations in note 15, p. 39.
12. Antoine Cournot, *Essai sur les fondements de nos connaissances et sur les caractères de la critique philosophique* (Paris: Hachette, 1851), 2:21, § 215 [*An Essay on the Foundations of Our Knowledge*, trans. Merritt H. Moore (New York: The Liberal Arts Press, 1956), p. 320].
13. Edmond Goblot, *Essai sur la classification des sciences* (Paris: Félix Alcan, 1898), p. 17.
14. Max Planck, *Die Einheit des physikalischen Weltbildes* (Leipzig: S. Hirzel, 1909), p. 32.
15. Alexandru Xenopol, 'L'Idée de loi scientifique et l'histoire,' *Scientia* **12** (1912) 40.
16. [While God calculates the world is being made.]
17. Cf. Jacques Duclaux, 'La constitution de l'eau,' *Journal de chimie physique* [Geneva and Paris] **10** (1912) 71–109.
18. Frederick Soddy, 'The Periodic Law from the Standpoint of Radioactivity,' *Scientia* **13** (1913) 369.
19. Isaac Husik, *A History of Medieval Jewish Philosophy* (New York: Macmillan, 1916; reprint Philadelphia, The Jewish Pub. Soc. of America, 1944), pp. xli, 346, 388, 395–396.
20. Cf. Zeferino González, *Histoire de la Philosophie*, trans. de Pascal (Paris: Lethielleux, 1890), 2:254. Moreover, St. Thomas has only clarified a fundamental concept of Aristotle; cf. Zeller, *Phil. der Griechen* 22:306, 309, 348 [Costelloe 1:331, 335, 377], and Pierre Duhem, *Le Système du monde* (Paris: A. Hermann, 1913), 1:132, 146.
21. Désiré-Auguste Roustan, *Leçons de philosophie*, Vol. 1: *Psychologie*, 3rd ed. (Paris: Ch. Delagrave, 1911), p. 349.
22. Comte, *Cours* 3:205 [Martineau 306]. Nevertheless, it is possible that Comte vaguely sensed these implications of his premises, which may be the source of his assertion that the different branches of physics (corresponding more or less to our qualitative sensations) are entirely irreducible. Cf. below, Ch. 16, p. 456.
23. John Stuart Mill, *A System of Logic*, 3rd ed. (London: J. W. Parker, 1851), 2:4.
24. [Bergson's first book, translated into English as *Time and Free Will*, was entitled *Essai sur les données immédiates de la conscience* (*Essay on the Immediate Data of Consciousness*).]
25. *De Rerum Nat.* I, 423–426. *Corpus enim per se communis dedicat esse / sensus; cui nisi prima fides fundata valebit, / haut erit occultis de rebus quo referentes / confirmare animi quicquam ratione queamus.* [The existence of bodies is vouched for by the agreement of the senses. If a belief resting directly on this foundation is not valid, there will be no standard to which we can refer any doubt on obscure questions for rational confirmation.]
26. Nicolas Malebranche, *De la recherche de la vérité* (Paris: Christophe David, 1721), Éclaircissement 11, 4:277 ff. [*Elucidations of the Search after Truth*,

trans. Thomas M. Lennon, bound with *The Search after Truth* (Columbus: Ohio State Univ. Press, 1980), pp. 636 ff.]. Among contemporary philosophers, F. H. Bradley in particular has pointed out in his meticulous study how difficult it is to arrive at a coherent concept of the physical world, and particularly of its laws, starting from pure phenomenalism, without the "transcendent" (*Appearance and Reality*, London: Swan Sonnenschein, 1893, Ch. 11: 'Phenomenalism,' pp. 123 ff.). Hegel understood quite well that science is attached to the notion of the thing as it is delivered to us by immediate perception and accomplishes its task by the very same method as common sense (*Enc.*, *Logik*, *Einleitung*, § 1 [Wallace 3]).

27. Max Planck, *Acht Vorlesungen über Theoretische Physik* (Leipzig: S. Hirzel, 1910), p. 3 [*Eight Lectures on Theoretical Physics* (New York: Columbia Univ. Press, 1915), p. 3].
28. Harald Höffding, *La Pensée humaine*, trans. Jacques de Coussanges (Paris: Félix Alcan, 1911), p. 279.
29. Marian Smoluchowski, 'Théorie cinétique de l'opalescence des gaz à l'état critique et de certains phénomènes corrélatifs,' *Bulletin international de l'Académie des sciences de Cracovie* (1907), p. 1059.
30. Jean Perrin, 'Les Preuves de la réalité moléculaire,' *Brussels Conf.* 224–225. Cf. Paul Langevin, 'Les Grains d'électricité et la dynamique électromagnétique,' *Idées modernes* 97–98.
31. Lucien Poincaré, 'Revue annuelle de physique,' *Revue générale des sciences* **9** (1898) 429.
32. For the history of this transformation, see Ch. 6, pp. 163 ff.
33. Henri Poincaré, *Thermodynamique* (Paris: George Carré, 1892), pp. 10–11.
34. Jean Nageotte, *Notice sur les travaux scientifiques* ... (Paris, 1911), p. 9 [citation unverified].
35. Descartes, *Méditations*, *Oeuvres*, 9:31 [*The Philosophical Works of Descartes*, ed. Elizabeth S. Haldane and G.R.T. Ross (Cambridge: University Press, 1935; reprint, New York: Dover, 1955), 1:161].
36. On this subject, cf. Urbain's observations below, Ch. 15, p. 431.
37. Antoine Cournot, *Traité de l'enchaînement des idées fondamentales dans les sciences et dans l'histoire* (Paris: Hachette, 1861), 1:264.
38. Henri Poincaré, 'L'espace et la géométrie,' *Revue de métaphysique et de morale* **3** (1895) 638. Cf. also his 'Les géométries non-euclidiennes,' *Revue générale des sciences* **2** (1891) 772. Paul Painlevé is even more emphatic: the geometric axioms concerning invariable figures "state in a purified form the properties of the form of material solids" ('Mécanique,' in Henri Bouasse et al., *De la méthode dans les sciences*, 1st series, 2nd ed. (Paris: Félix Alcan, 1910), p. 77 [1909 ed., p. 367].
39. [*la science, même purement légale*. Meyerson introduces the term *légalité* (which, when possible, we translate, with Loewenberg, as "lawfulness") at the beginning of *Identity and Reality*: "Although the use of this term is not customary in the sense we are giving to it, yet we believe it is clear: it signifies the supremacy of law." In a purely legalistic science law is to be understood in the manner of Berkeley or Hume: "Here we find a complete assimilation of the two concepts of 'cause' and 'law,' the second entirely absorbing the first" (IR 2

[Loewenberg 18]). A purely legalistic science would limit itself to strictly empirical laws, eschewing causal explanation.]

CHAPTER 2

SCIENCE SEEKS EXPLANATION

What is more, it is easy to see that at bottom the positivistic theory rests on a palpable psychological error: it is not true that when we do science our sole aim is action. This conception, of course, originates in particular with Francis Bacon, who made it the basis of his philosophy and tirelessly stressed it in innumerable passages. "The true and lawful goal of the sciences," he said in one of his aphorisms "is none other than this: that human life be endowed with new discoveries and powers."[1] Hobbes took up the theory, affirming that "the end of knowledge is power,"[2] and Comte characterizes science in quite analogous terms: "all science has prevision for its end" and "From Science comes Prevision: from Prevision comes Action."[3]

Now, this conception has been substituted for another earlier one, which, as we see especially in Bacon, it consciously opposed and sought to supplant. Indeed, Plato had already pointed out that geometry, appearances to the contrary, pursues no practical goal and "the real object of the entire study is pure knowledge," the starting point of all science being the wonder man feels with regard to nature,[4] and Aristotle, reiterating the concept of the "wonder" and admiration because of which "men both now begin and at first began to philosophize," declares that "all men by nature desire to know" and that there is a knowledge which does "not aim at giving pleasure or at the necessities of life," a knowledge of which mathematics in particular is a part.[5]

In the Middle Ages, needless to say, given the prestige then enjoyed by Aristotelian thought, this view reigns supreme. But even in the Renaissance, Montaigne expresses it unequivocally: "There is no desire more natural than the desire for knowledge. We try all the ways that can lead us to it. ... truth is so great a thing that we must not disdain any medium that will lead us to it."[6]

For Pascal as well, it seems to go without saying that "the curious and the scholars ... have wit as their object" and that in things of the mind "curiosity properly" reigns (*Pensées* 541 [erroneous citation]). Spinoza in his turn declares that "Whatsoever we endeavour in obedience to reason is nothing further than to understand," and that "neither does the mind, in so far as it makes use of reason, judge anything to be useful to it, save such things as are conducive to understanding" (*Ethics*, Pt. 4, Prop. 26).

Contrary to what current positivism would tend to suggest, this conception has not been abandoned by modern epistemology. Jacobi, who, along with Abel, was reproached by Fournier for studying highly abstract mathematics instead of turning to the motion of heat, responded that "the unique goal of science is the honor of the human mind,"[7] and Lowell, the famous American astronomer, gives the following characterization: "The whole object of science is to explain and so make more comprehensible the universe about us."[8] Auguste Comte, while not absolutely denying the spirit of scientific curiosity, declares this penchant to be "one of the least imperative of our nature." This is the inevitable consequence of his system, and it is at this point that the psychological error leaps out at us, because the thirst for knowledge is something each of us feels in himself. So strong is it that Littré, in the "Disciple's Preface" to the *Cours*, does not feel he can adhere to his master's teaching on this point. He does not contest "that the object of an ideal science is to satisfy a need of the human mind borne by an imperious necessity to say the last word about things, or at least to seek it" and that this is "an observed fact proved by the study of each epoch, each people, each individual; one cannot refuse to perceive it; it is a fact like any other; the necessity of its existence obviates the need to discuss its legitimacy." He only pleads that whoever indulges in such research realize that he is trying "in vain to resolve insoluble problems" (*Cours* 1:xxxv-xxxvi). Be that as it may, this is not the issue; furthermore, Comte, as we have seen, presented the foundations of positivism quite differently.

If one wishes to know the opinion of authoritative scientists on this point, one hardly knows where to start; we shall merely cite a few passages by authors more or less contemporary with Comte or following him. Cuvier, in the preface to his great work, the *Lectures on Comparative Anatomy*, declares that facts call for facts. "However rich may be the acquisitions that are made, more will still be desired."[9] Claude Bernard speaks of the "constant stimulation by the spur of the unknown" and of "constantly recurring thirst" (*Méd. expér.* 353 [Greene 222]). Obviously neither Cuvier nor Claude Bernard is thinking of the *utility*, even in a very general sense, of the research he is going to undertake, and what drives them both is pure scientific curiosity. Similarly, Henri Poincaré states that not only are we not easily resigned "to be forever ignorant of the foundation of things," but that to his mind this sentiment is more powerful than the one that moves us to act: "In my eyes ..., it is knowledge that is the goal and action that is the means."[10] In the scientist the desire to know becomes a veritable passion. At the moment when Pasteur, still a student at the École Normale, arrived at the final stage of his research on tartaric acids and was about to proceed to the decisive verification, he was so overcome by emotion that he could no longer look in the polarimeter.[11] Now need we point out that his work, of prime theoretical importance, aimed at no practical result?

It is not true that our intelligence declares itself to be satisfied with the mere description of a phenomenon, however detailed it may be. Even if science is able to submit a problem, in all its details, to empirical laws, it still seeks to go further; it has always done so and continues to do so today. If anyone claims otherwise, the entire course of science, past and present, becomes an enigma, or rather a sort of gigantic and monstrous absurdity. In an earlier work we have cited in this regard the case of Newtonian gravitation (IR 46 ff. [Loewenberg 49 ff.]). The law governing these phenomena is of unsurpassable clarity and simplicity. How can it be that astronomers and physicists, from the very moment it was formulated, sought to go further; how can they have considered gravitation to be an enigma? To this example could be added others. For instance, at the other end, so to speak, of the domain of science (since it is a question of a region where, unlike what has taken place in astronomy, mathematics has as yet hardly penetrated), namely in the biological sciences, we see that the search for what goes beyond the pure and simple rule, beyond empirical law, is just as active. Some would like to establish explanations modeled entirely on the type currently used in the physical sciences; others declare this task unrealizable and appeal to finality; but all of them seem to agree in admitting that empirical law is not at all sufficient to explain the phenomena. In this regard, the continual debates between the two parties are in fact of particular interest; one can actually see how close the two kinds of explanation are to each other, to the point that one seems to grasp immediately what escapes the other. The demonstrations of the finalist all come down to the same type: a given phenomenon cannot be reduced to the exclusive action of antecedent causes; therefore it is necessary to appeal to the concept of purpose. On the other hand, the anti-finalist himself, each time he does not expressly appeal to mechanistic and other such considerations when speaking of an organ, instinctively reasons as if this organ had been created with an eye to its functions. The physiologist, as has been pointed out by Claude Bernard, although he himself is so little inclined toward finalism, is driven to use finalistic concepts in spite of himself, as it were.[12]

Thus biologists seem to be irresistibly attracted by one system or the other. Is it necessary to point out how paradoxical this situation appears if one adopts the positivistic position? Since law is supposed to satisfy our minds, one cannot account for such an intrusion of the concept of *purpose*, clearly so foreign to the spirit of modern science. Even if, because of a sort of traditional training (obviously vicious) or because of the desire (more legitimate according to the Comtian credo) to unify nature, the biologist were at first prompted to seek physical explanations, he ought, as soon as he has established that this is impossible in a particular case, to return automatically, so to speak, to the purely empirical rule. But in fact he most often acts quite differently. It probably would not be impossible to cite cases in which someone, confronted with the obvious absurdity of a teleological concept, has returned to an empirical law. But whenever a finalistic theory actually achieves any kind of consistency, it generally yields only in the face of physical explanation, and, inversely, when physical explanation is lacking, the idea of a purpose seems to impose itself irresistibly on the mind of the biologist. That is evidently the source of the constant accusations of hidden finalism that one hears being leveled against men one would have thought offhand to be proof against such a charge, as for example the protagonists of evolutionary theory, and, chief among them, Charles Darwin himself.[13]

In the same context, we can cite the example of Auguste Comte who, hostile on principle to any explanation going beyond laws, as we have just seen, was even more hostile to any finalistic concept. He does not hesitate, however, to define the goal of biology in the following terms: "Given, the organ or organic modification, to find the function or the act, and reciprocally" (*Cours* 3:211 [Martineau 307]). Surely the concept of a *function* to which an organ is in some sense predestined is essentially finalistic.

Actually, then, we see that the finalist and anti-finalist take their stand on common ground and profess a common faith: clearly they both believe that phenomena require an explanation outside laws, beyond laws.

Of course, what we have just established for astronomy and biology is valid for the vast intermediate field of physico-chemical sciences as a whole. Everywhere the search for explanation prevails. Here is just one example among many. In October and November 1911 a "Council of Physics" in Brussels brought together the leading physical scientists of all

Europe. Suffice it to say that France was represented by Henri Poincaré, Brillouin, Madame Curie, Langevin and Perrin, while among the foreigners represented were H. A. Lorentz (who presided over the assembly), Nernst, Planck, Jeans, Rutherford, Kamerlingh Onnes, not to mention others almost as famous. Now, a glance at the proceedings published by the *Council*, in which are reported the papers presented there and the discussions they provoked, suffices to establish that the sole aim of all this work was to search for an actual physical theory, for an assumption concerning how phenomena are produced (a type of assumption so odious to Auguste Comte and indeed so inadmissible according to his conception of science). What is wanted is a hypothesis capable of explaining a whole series of phenomena unquestionably observed by authoritative scientists, phenomena which clearly contradict all the theories that had thus far been formulated. Nothing equals the ardor with which these eminent scientists pursue the search for this conception, unless it is the deep chagrin (the term is not too strong) they manifest in the face of the impossibility of constructing a coherent, plausible representation of what is really happening.

That this is what is at stake and not the search for laws is something no attentive and impartial reader of these pages of the most advanced science could doubt. They have empirical laws, and they also know to what cases each of them must be applied. The unfortunate thing is that if, based on a series of these laws, they have formed a representation of reality that could produce these relationships, this representation is contradicted by what results from other laws valid under other circumstances. Here are a few cogent passages. Einstein notes that Planck's hypothesis (concerning *quanta of energy*), "as simply as it allows Planck's formula to be obtained, nonetheless appears singular and disconcerting when examined more closely." In fact,

> we must picture how the elements of energy move. Since they are so far apart at low temperatures, they must move completely independently. Moreover, if one wants to speak of a simple periodic oscillation of the atoms, a *quantum* must remain connected with the same atom during at least the duration of a half oscillation. If it then passes to another atom, it can only be to a neighboring atom, and naturally according to the laws of chance .... To escape this conclusion one would have to make altogether unlikely assumptions concerning the displacement of *quanta.*

Under these circumstances, Planck's theory "does not truly constitute a theory in the usual sense of the word, at least not a theory that can henceforth be developed coherently." The famous physicist further declares that one "would need to imagine the mechanism that produces an accumulation of radiating energy," and regrets having to concede a property, "even though we do not see the mechanism by which it can be explained." He observes that the

> difficulties encountered by a satisfactory theory for these fundamental phenomena appear at the present time to be insurmountable. Why does an electron in a metal bombarded by Röntgen rays acquire the high kinetic energy observed to be characteristic of secondary cathode rays? Since all the metal is in the field of the Röntgen rays, why do only a small portion of the electrons acquire these cathode ray velocities? How does it happen that the energy is only absorbed at so very few points? In what way do these points differ from the others? We remain without answers in the face of these questions, as we do for many others. ('L'État actuel du problème des chaleurs spécifiques,' *Brussels Conf.* 420–421, 429, 431, 436)

He praises Nernst for having "done much to present these questions in a concrete form," that is, for having tried to conceive an actual representation of reality, and refuses to accept a purely mathematical definition for the probability that is introduced into these phenomena, demanding a "physical definition,"[14] that is, again, committing himself to an actual representation.

We felt it necessary to lay particular stress on these statements of Einstein's, because he is one of the protagonists of the most recent phase of physics: we know, indeed, that this scientist is, with Minkowski, the author of the famous "principle of relativity" (or rather, of the *principles* of relativity, for he successively put forth two somewhat divergent principles) which threatens such a profound upheaval in traditional physics, and also because he played such a major role in the study of the phenomena with which the Council of Brussels concerned itself.

Moreover, it is easy to find comments by other participants at this meeting showing that these scientists think exactly like Einstein on this subject. For example, H. A. Lorentz, the illustrious Dutch physicist, states in his opening presidential address that one "cannot be satisfied with admitting that a molecular oscillator exposed to bombardment by gas atoms can acquire energy only by finite portions of a determined size; we have the right to demand that someone conceive a type of action between the gas molecules and the oscillator that leads to this result," which is evidently to demand a theory as to how these phenomena are produced and a representation. In the course of the discussion he mentions different artifices he had imagined as to how these phenomena are produced, which artifices unfortunately yielded no result. Planck judges that it "would no doubt be desirable if one could give a physical definition of thermodynamic probability valid in all cases," but that it "is not actually possible to find one in the present state of our knowledge," which is why the physical definition must remain undetermined for the time being. Brillouin speaks of a model which reduces the discontinuities of energy to discontinuities of structure (that is to say, again, which replaces the physical observation with a representation meant to explain it. Lord Rayleigh, who did not attend the meetings, explains in a letter that the theory of the elements of energy has unquestionably "already led to interesting consequences, thanks to the skill of those who have applied it," but that it is nevertheless "hard to consider it to be a representation of reality" (*Brussels Conf.* 7, 14, 50, 115–116, 123–124).

It is noteworthy that, in the face of the duly recorded failure of all attempts at explanation, none of those present thought of proclaiming that they should give up efforts of this kind and be content with purely empirical formulas. Planck did observe that two physicists, Larmor and Debye, seemed to want "to take ... a phenomenological stance," but the author of the theory of *quanta* immediately added that he himself found it impossible to stop at that point (*Brussels Conf.* 100), and those present seemed to agree completely with this point of view: they continued to speak of theory, actual representation, and the way phenomena are produced, as if the purely phenomenalistic viewpoint had never been suggested.

It is true that Planck, although he rejected the opinions of Larmor and Debye, felt obliged to explain his position in such a way as not to fly in the face of positivistic orthodoxy. It is, he said, "of the utmost importance to seek possible relations between the *quantum* of action and other physical constants, in order to fix and, at the same time, to broaden its meaning." It is indeed certain that any physical theory, however unsuited to the observations, has an enormous importance from the standpoint of the development of science – even of a purely lawlike science limited to the prediction of facts. A science stripped of theory would appear in some sense entirely finished, static, while true science, we feel, must be in flux, evolve, progress. Nevertheless, considerations of this kind appear clearly inadequate to explain the attitude of passionate curiosity on the part of the participants in the Council of Brussels. To speak only of Einstein, what motivates his *whys* and *how-can-it-bes*? How can one explain the constant intervention of the image, the physical model, and the fervor with which he demands it? And what could the accusation of unlikelihood possibly mean if it were not a question of an actual hypothesis about how phenomena are produced, about what is *really going on*? In the domain of empirical laws, everything is equally probable. It would seem on the whole to require a singularly astute interpretation to fit all this into the framework of positivism. The truth is that if there had been one true positivist present, he would surely have risen to his feet at the very first words and vigorously protested: You are wasting your time, you are pursuing a chimera, or rather you are behaving unscientifically, since you are manifestly looking for a hypothesis as to what is happening in space, as to what really lies behind phenomena, whereas you should be confining yourselves to seeking laws and to formulating only assumptions concerning laws.

Could it be argued that we are dealing here with a turn of mind peculiar to scientists, a sort of pathological tendency instilled in them by the particular kind of occupation to which they devote themselves, as opposed to the ordinary mortal? It must be noted that Auguste Comte himself was of the opposite opinion, believing that the tendency to seek beyond the law, to inquire into causes, was the attribute of an intelligence untutored in science, while the true scientist reacted against such an inclination.[15] But it is easy to convince ourselves how right the founder of positivism was in the first part of his assertion (if not, as we have seen, in the second). We need only talk with an educated man who has only a superficial scientific background to be immediately confronted with questions such as "What is electricity really?" or objections like "But they don't know what electricity actually is!" Now, in one form or another, what he is asking for is obviously not a body of laws (which the interlocutor presumes are already known) but a theory he would like to be framed in common sense terms, or at least in terms of mechanical theory. We can also directly pose the question of the positivistic conception of science, asking such an interlocutor to grant that any magical procedure whatsoever, if it were consistently followed with the right results, would be just as good a basis for establishing a law as any physical experiment. That follows strictly from Auguste Comte's definition, since it is understood that one must abstain from all research bearing on "how phenomena are produced." But our interlocutor will most certainly not allow this point of view, and we are almost sure to end up with the objection: "I can't imagine *how* that happened, whereas for any experiment in physics I see perfectly that the action of a material substance, a force, or something of the sort, was able to produce this change." Furthermore, we can ascertain that people who believe in the efficacy of magical procedures certainly make the most of the successful results of previous experiments, but never fail to support this argument by a theory, claiming to explain how, by the action of an agent or a mysterious force, whose power the ordinary man mistakenly fails to take seriously and which he is, moreover, incapable of setting in motion, the observed phenomena must inevitably have been produced.

This is because in fact the predominant role of explanation in science is not measured solely by the zeal with which we seek it, but also by the profound inner satisfaction we experience when we believe we have attained it, or at least attained something more or less resembling such an explanation. Anyone can try the experiment directly on himself by taking up the study of a mechanical theory he did not previously know; if the theory is more or less in agreement with the facts, he will surely have the feeling that he understands the *why* of the facts in question. Great scientists, without any antipositivist bias, merely concerned with defining their method more accurately, have observed the same thing: "If I can make a mechanical model," says Kelvin, "I can understand. As long as I cannot make a mathematical model ... I cannot understand,"[16] and the whole evolution of science confirms this claim. It suffices to open any textbook to any chapter to see that science is in fact full of these explanatory theories, and there is no need to delve very deeply into the history of scientific thought to establish that it has always been thus, that there has never been anything truly worthy of the name science that remotely resembles the positivistic scheme. Thus the existence of explanatory science is a *fact*, one would almost be tempted to say a brute fact, so clearly does its concrete nature stand out in the face of the puny artifices that have been used to try to fit it into the framework of the positivist's theory, or at least to minimize its importance. Comte, however, as we know, was perfectly candid on this point, at least in the passages where he sets forth his doctrine in its most complete form: anything dealing with the way in which a phenomenon is formed, that is to say, any explanatory theory, must be ruthlessly driven out of science. Yet another thinker, whose name we are unaccustomed to seeing associated with that of the founder of positivism (although his attitude toward science was, as we shall see in the course of this work, in many ways analogous) rejected outright the whole of explanatory science: we are referring to Hegel. Doubtless this philosopher does not mean to condemn all explanation in general; on the contrary, he attempted to realize a complete system of such explanations, but a system conceived in a spirit completely different from the one that directs the efforts of science in this domain, efforts that to him appear entirely vain and absurd, so much so that he considers that there is nothing there worth saving. It clearly requires great audacity and an unlimited confidence in the certainty of one's deductions to fly in the face of the indubitable fact of explanatory science, as these two philosophers have done.

As far as Comte is concerned, however, one must also add that upon occasion he has – perhaps submitting less to the rigorous logic of his preconceived ideas than to his powerful scientific instinct – expressed himself less trenchantly; his position in such cases amounts to one that recognizes a certain utility in representative theories, insofar as they are designed to connect phenomena temporarily, where a lawlike connection is still momentarily lacking, and thus prepare for the establishment of this connection, that is, the discovery of new laws, which remains, in the last analysis, the unique, the true goal of science. Therefore, being unable to disregard entirely the irresistible tendency of our mind to go beyond observations resulting from a pure and simple generalization of experimental data, he seeks to explain it by our inclination to prepare the way for future progress. He even goes so far as to declare that

> whatever one may say about it, absolute empiricism would not only be totally sterile, but even altogether impossible, for our understanding can not act without some doctrine, false or true, vague or precise, which may concentrate and stimulate its efforts, and afford ground for enough speculative continuity to sustain our mental activity. (*Cours* 4:471 [cf. Martineau 525])

But he still maintains that at bottom it is the *positivistic* conception alone, that is, the establishment of a purely lawlike relationship, that fully satisfies – or at least ought to satisfy – our minds, explanations intervening only where the positive conceptions are lacking:

> This need to set out the facts in an order we can easily understand (which is the proper object of all scientific theories) is so inherent to the way we organize that, if we did not succeed in satisfying it with positive conceptions, we would inevitably return to the theological and metaphysical explanations to which it originally gave rise.

Comte's followers naturally preferred to hold to the second of his stances, which is less paradoxical in appearance than the first. They doubtless continued to develop their former hypotheses and constantly to create new ones, and, as always, they subordinated everything in science to research for and establishment of these hypotheses, but they did all they could to conceal the fact, to treat, or at least appear to treat, the figurative conceptions as something secondary, as a sort of appendix they regretted not being able to do without for the time being and needed to remove as quickly as possible. Thus Rankine, in an authoritative theoretical work, asked that in formulating theories we apply ourselves, insofar as possible, to extending the process of abstraction. In that way we would arrive at a "body of principles," created solely by induction, which "will be free from the uncertainty which must always attach even to those mechanical hypotheses whose consequences are most fully confirmed by experiment." Yet Rankine does not mean entirely to rule out the hypothetical method in theories; he recognizes that "in almost every branch of molecular physics it may be held, that a hypothetical theory is necessary," but he sees it as only a "preliminary step to reduce the expression of the phenomena to simplicity and order, before it is possible to make any progress in framing an abstractive theory."[17] It has even been suggested that this is only a particular need of certain individual minds, others (obviously superior) not finding it necessary to have recourse to it. It was the illustrious Clerk Maxwell who expressed himself in these terms,[18] and one could surely find no more striking proof of the truly prodigious influence that the powerful mind of Auguste Comte exercised on ensuing generations than the fact that this opinion comes from a man who devoted most of his efforts to establishing mechanical theories and whose principal work, the *Treatise on Electricity and Magnetism*, is wholly dominated by a preoccupation with the mechanical explanation of electrical phenomena.

The whole series of considerations to which we have just called attention seems to lead to an inevitable conclusion: No more for the scientist than for the man of common sense does law suffice to explain the phenomenon. It plays an immense role in science, to be sure, because it allows prediction and consequently action. But it does not satisfy the mind, which looks beyond it for an *explanation* of the phenomenon.

If we now return to what we established in Chapter 1 on the subject of science's creation of new *things*, it becomes clear that what impels science to proceed in this manner is precisely this need for explanation. Why cannot the physicist, in speaking of the expansion, etc., of a steel bar, simply consider it in the form provided by common sense? Evidently because the phenomenon of expansion would then be inexplicable, whereas it seems to be explained if we suppose the bar to be composed of particles separated by intervals which are assumed to increase in size when the bar expands.

Thus these two powerful tendencies, one that posits a world of ontological entities as the substratum of phenomena, and another that pursues the explanation of these phenomena, are combined and intertwined in science. They are so intertwined that it is almost impossible to speak adequately of the manifestations of the one without touching upon the domain of the other. It seems to us to go without saying that the true explanation is at the same time a real explanation, in terms of what underlies the phenomenon, what *is*. Only the inhabitants of an insane asylum, Hartmann rightly says, could attempt physical explanations using scientifically unreal concepts.[19]

We shall see in a later chapter (Ch. 16, p. 453) that it is possible to understand these two tendencies as having at least partially arisen from the same source and that certainly even our ontological tendency – the fact that sensation is, instantaneously, transformed into perception – is itself ultimately only a product of the need to explain these sensations. For the time being, however, it is preferable that we suppose the two tendencies to be not only distinct, but also opposing, as they are in a certain sense, since, as we shall see in Book Two, science, in carrying its explanations as far as they can go, finally destroys the ontological reality it originally found indispensable.

NOTES

1. *Novum Organon*, I, 81, in *The Works of Lord Bacon* (London: William Ball, 1837), 2:444 [*The New Organon and Related Writings*, ed. Fulton H. Anderson, trans. James Spedding, Robert Leslie Ellis and Douglas Denon Heath (New York: Bobbs-Merrill, 1960), p. 78]. Cf., for example, *De Augmentis Scientiarum*, Bk. 2, Ch. 2; Bk. 3, Ch. 6; Bk. 7, Ch. 1, in *Works* (London, 1837), 2:315–316, 2:341, 2:388 [*The Works of Francis Bacon*, ed. Spedding, Ellis and Heath (Boston: Taggard and Thompson, 1843), 8:415, 8:517–518, 9:193–194]. Cf. also *Redargutio philosophiarum*, *Works*, ed. Basil Montagu (London: William Pickering, 1829), 11:465, 466, 474.
2. *Elements of Philosophy*, *The English Works of Thomas Hobbes*, ed. Molesworth (London, 1839), 1:7. Cf. *Leviathan* (London: Dent, n.d.), p. 363 [Ch. 96, ¶ 1].
3. Comte, *Cours* 2:20, 1:51 [Martineau 135, 40]. Cf. 6:618 [Martineau 803]: "to see in order to foresee."
4. *The Republic*, Bk. 7, 527A-B [Paul Shorey trans.]. Cf. John Burnet, *L'Aurore de la philosophie grecque*, trans. Reymond (Paris: Payot, 1919, p. 11 [*Early Greek Philosophy*, 2nd ed. (London: Adam and Charles Black, 1908), p. 11]. Burnet, moreover, rightly points out that the distinctive trait of Greek thought is the powerful curiosity with which this people was endowed, a curiosity which permitted it to collect and use the small bits of knowledge the barbarians had

acquired (*L'Aurore* 28–29 [*Early Greek Philo.* 28–29]).

5. *Metaphysics*, Bk. 1, Chs. 1–2, $980^{a}22$, $982^{b}12$, $981^{b}20$ [W. D. Ross trans.].
6. *Essais* (Paris: Flammarion, 1908), 4:187 [*The Complete Essays of Montaigne*, trans. Donald M. Frame (Stanford: Stanford Univ. Press, 1958), p. 815].
7. Federigo Enriques, 'La critique des principes et son rôle dans le développement des mathématiques,' *Scientia* **12** (1912) Supplément: 78.
8. Percival Lowell, 'The Atmosphere of Mars,' *Scientia* **19** (1916) 19.
9. George Cuvier, *Leçons d'anatomie comparée*, Paris: Baudouin, 1805 (Year 14 [of the First French Republic]), 1:x [*Lectures on Comparative Anatomy*, trans. William Ross (London: T. N. Longman and O. Rees, 1802), 1:xxviii].
10. Henri Poincaré, *La science et l'hypothèse* (Paris, Flammarion, n.d.), p. 258 [*Science and Hypothesis*, trans. George Bruce Halsted (New York: The Science Press, 1905), p. 155]; 'Sur la valeur objective de la science,' *Rev. de méta.* **10** (1902) 266.
11. Edmond Goblot, *Traité de logique* (Paris: Armand Colin, 1918), p. 22.
12. *Méd. expér.* 140 [Greene 89]. The finalists, not without some justification, exult in these concessions by the creator of modern physiology. Cf., for example, Hans Driesch, *Der Vitalismus als Geschichte und als Lehre* (Leipzig: Johann Ambrosius Barth, 1905), p. 121 [*The History and Theory of Vitalism*, trans. C. K. Ogden (London: Macmillan, 1914), pp. 134–135].
13. Cf. Georges Bohn, 'Idées nouvelles sur l'adaptation et l'évolution, Deuxième Partie: Conception physico-chimique de l'évolution,' *Scientia* **18** (1915) 86, and Henri Piéron, 'Les Instincts nuisibles à l'espèce devant les théories transformistes,' *Scientia* **9** (1911) 203. Similar accusations have been made against Ernst Haeckel; cf. Lásár Róth, *Schelling und Spencer*, Berner Studien zur Philosophie und ihrer Geschichte Vol. 19 (Bern: C. Sturzenegger, 1901), p. 33.
14. *Brussels Conf.* 438. Cf. also p. 115: A purely mathematical definition of probability is "shocking," for the result would be that "Boltzmann's equation has no physical content."
15. Cf., for example, *Cours* 2:169: "We obviously cannot know what this interaction of the stars and this weight of the terrestrial bodies ultimately are .... Only minds entirely foreign to scientific studies are able to be concerned with this today"; *Cours* 2:268 [Martineau 200]: "All good minds today recognize that our study of nature is restricted to the analysis of phenomena in order to discover their *laws* ... and can have nothing to do with ... the mode of their production." It is clear that in both these passages Comte appeals to scientific opinion, which he opposes to that of the crowd.
16. [*Lectures on Molecular Dynamics and the Wave Theory of Light*, in *Kelvin's Baltimore Lectures and Modern Theoretical Physics* (Cambridge: The MIT Press, 1987), Lecture 20, p. 206.]
17. William John Macquorn Rankine, 'Outlines of the Science of Energetics,' *The Edinburgh New Philosophical Journal*, New Series, **2** (July-Oct. 1855) 125.
18. James Clerk Maxwell, *Scientific Papers* (Cambridge: University Press, 1890; reprint New York: Dover, 1952), 2:219–220.
19. Eduard von Hartmann, *Das Grundproblem der Erkenntnistheorie* (Leipzig: Wilhelm Friedrich, 1889), p. 22.

BOOK TWO

THE EXPLANATORY PROCESS

CHAPTER 3

DEDUCTION

What the scientist is seeking beyond law is, of course, often designated by the term *cause*, which in this sense becomes almost synonymous with the term *explanation*: when one knows the cause or causes of a phenomenon, the phenomenon will be explained and the mind will declare itself satisfied.

"The mind of man," says Claude Bernard," cannot conceive an effect without a cause, so that the sight of a phenomenon always awakens an idea of causation." "All human knowledge," adds the great biologist, "is limited to working back from observed effects to their cause" (*Méd. expér.* 54 [Greene 33]).

Plato declares, "Everything that becomes or is created must of necessity be created by some cause, for without a cause nothing can be created" (*Timaeus* 28A [Jowett trans.]), and Aristotle similarly states, "Nature makes nothing which is purposeless or doomed to frustration" (*On the Heavens* $291^{b}12$ [W. K. C. Guthrie trans.]).

Can we relate this search for causes to the meaning of the verb *to explain* arrived at in Chapter 1?

Let us note first of all that a cause, in what might be called the most vulgar sense (we shall see later in what way this sense is inaccurate, or at least incomplete), precedes, in time, the phenomenon it is to explain, that is, its *effect*. Let us further note that the term "cause" is frequently replaced by the term *reason*. Spinoza, for example, consistently makes the substitution,[1] and Leibniz is entirely in agreement with him on this point. But Leibniz sharpens the meaning of the term *reason* by adding qualifiers: he speaks of *determining* or *sufficient* reason.

These two adjectives obviously complement one another. To what, indeed, is reason to suffice? Apparently to determine, to produce the phenomenon. "I hold that to be a sufficient Cause," says Hobbes, "to which nothing is wanting that is needful to the producing of the Effect,"[2] and Leibniz himself speaks of

> the great principle, commonly but little employed, which holds that *nothing takes place without sufficient reason*, that is to say that nothing happens without its being possible for one who has enough knowledge of things to give a reason sufficient to

determine why it is thus and not otherwise.[3]

Thus cause and effect must be mutually entailed, mutually implied. "The knowledge of an effect depends on and involves the knowledge of a cause," says Spinoza (*Ethics*, Pt. 1, Axiom 4).

But as soon as there is a *reason*, then it can be a question only of an operation of our reasoning. Thus the reason for a phenomenon must be of such a nature that it is sufficient to determine this phenomenon in our reason. In other words, we must be able, by the cause or reason, through reasoning alone, to infer the phenomenon. This is what is called deduction. Cause can therefore be defined as the starting point for a deduction whose outcome will be the phenomenon. And it is indeed an everyday experience that once a deduction of this sort has been accomplished, our reason declares itself to be satisfied – except that it then demands the reasons for the cause and so on, in an indefinite regress.

If we now reconsider our etymological meaning of the verb "to explicate" [see Ch. 1, note 1 above] in terms of these new considerations, we can do no better than recall the passage from Bossuet and the particular image it suggests [p. 9 above]. However, in order to cast more light upon the essential aspect of the image, let us forget for a moment that it concerns a development starting from a seed (we shall have occasion to return later to this part of the statement) and think only of the leaves and flowers that are unfolding. If these leaves and flowers, which seemed to appear so suddenly on the tree, were already there earlier, completely formed, but simply folded up, then it would have been possible for an attentive observer, by examining the buds, by unfolding (or *explicating*) their contents, to see these branches and leaves at a time when the natural process did not yet show any trace of them, and consequently their subsequent appearance would have no longer held any mystery for him; it would have been explained.

Thus to explain a phenomenon certainly means, as an intermediate sense, to make manifest what was hidden; but it must be understood that this is applied both to the phenomenon considered to be the effect and to its cause or causes. Moreover, one need only consider the actual situation a bit more closely to realize why it must be so. What we consider, in the first place, to be a phenomenon, what particularly seems to us to need to be explained, is a change, a modification in time; it is the fact that there were a *before* and an *after* which could be distinguished from one another. "The thing," says the mathematician Riemann, "would remain what it was, if nothing else were added to it. That is what creates the impulse to seek, for every change, a cause."[4] Explanation consists in showing that given all the antecedents, what followed from them could be inferred from them by deduction, was simply their *logical consequence*. Indeed, in order to satisfy us, the process must be purely rational, while an empirical law contains elements that are foreign to our reason, that reason could have conceived to be other than they are. Apparently it cannot do so, without denying its own nature, in the case of the elements we for this reason call rational, and the term *sufficient cause* or *sufficient reason* implies the conviction that it is possible to conceive a phenomenon as something that, according to our reason, could not fail to be produced, whose opposite would seem absurd to us, as if it implied a disagreement, not only with the facts (which is the case for an empirical statement), but also with essential elements of our reason. "It is manifest," says Hobbes, in a passage directly following his definition of sufficient cause [p. 47 above], "that whatsoever is produced, is produced necessarily: for whatsoever is produced hath a sufficient Cause to produce it, or else it had not been."[5] Thus what we seek, when we speak of understanding a phenomenon, is to understand it as necessary, to "show that it necessarily depends on necessary judgments," as Lalande says so well precisely with regard to the term *expliquer*. "It is the term *must*," declares a contemporary biologist, "that distinguishes what is causal from what simply refers to an empirical function."[6]

Obviously this is a characteristic and, moreover, very essential trait of human reason. "It is not in the nature of reason to regard things as contingent, but as necessary," states Spinoza (*Ethics*, Pt. 2, Prop. 44).

The term "cause" can, of course, be defined quite differently. Let us immediately set aside the interpretation reducing it to a concept containing nothing more than the idea of succession. That actually amounts to confusing cause with law. This is what was done by Berkeley and Hume in particular, and later by Taine, Helmholtz, Hannequin and Ostwald (cf. IR 1–2 [Loewenberg 17–18]). Taken literally, such a theory would lead to a sort of extreme positivism, affirming that it is inconceivable for us to seek anything at all beyond laws, whereas even Comte admits that such research exists, merely finding it blameworthy. Moreover, it is hardly necessary to stress, after what we came to see in Chapter 2, how little this confusion between law and cause conforms to the mind that truly animates science.

On the other hand, it can be pointed out that the term "cause," in the sense in which we usually use it, is the concept of something which is a source of change and that it thus implies an active element, as seen in the expression *efficient cause*. That is no doubt correct. But even the most superficial observation suffices to establish that this concept is not the one used by science. In fact, the efficient cause can be clearly conceived only as deriving from a will, which must be aiming toward a goal. Now science, at least as we understand it at the present time, sets aside all finality, or at least tolerates it only provisionally (cf. pp. 194 ff. below). It manifestly seeks to understand nature as the product of blind necessity, resulting from the motion of inert matter. Thus causal explanation in its scientific sense consists precisely in eliminating everything referring to the concept of *efficient cause*.

Finally, we can stress the undeniable fact that we speak of cause principally under the category of existence, while the concept of reason is reserved for the category of knowledge. Thus one could say that Spinoza, Hobbes and Leibniz were mistaken and that the confusion between *cause* and *reason* is the initial error of a certain philosophy: reason alone allows us to draw a consequence, while from a cause, on the contrary, we cannot deduce the effect.[7]

Certainly if we consider reality, this point of view can only appear completely justified. For no phenomenon we shall ever observe shall we truly be able to deduce the consequent from the antecedent; yet one can confidently affirm that never in the entire history of human knowledge has any explanation been furnished that exactly followed the pattern we have just sketched, since the most foolhardy deductions of the medieval Aristotelians, like the most perfectly mechanistic theories of modern physicists, contain purely empirical elements. While this is true, it proves only that on this point, as on many others, humanity pursues an inaccessible and transcendent goal; it is not a proof that such a pursuit does not exist. In fact, it is easy to see that the empirical elements of these deductions appear there only provisionally, so to speak; one accepts them for the time being, because one cannot do otherwise, because one is aware that the problem is so arduous that even the slightest progress toward a solution must be considered a triumph of reason. But behind this acceptance there is always a mental reservation, namely the belief that the empirical rule can later be eliminated, replaced by a rational deduction. We still remain convinced deep down inside that, as Goblot put it so well, "wherever experience and deduction reveal to us a constant order, there is surely a logical necessity yet to be discovered," for "we cannot be content with *de facto* truths, we must have *de jure* truths."[8]

No one felt this more deeply than Georges Cuvier, and his testimony is particularly convincing, not only because it involves one of the greatest names of all time in science, but because it deals with biology, a science where rational deductions have thus far had little success. Moreover, it is clear that Cuvier, though conversant with all the properly philosophical learning of his epoch (as is demonstrated by the way he cites Kant and even – to refute them – the works of the German "philosophers of nature") was on this occasion guided not by any theoretical consideration, but by a simple desire to clarify the methods of his science on a point he rightly considered extremely important. Indeed, it is in examining the role played by the bodily organs in the different classes of animals that Cuvier arrives at the following affirmation: "It is on this mutual dependance of the functions, and the aid they reciprocally yield to one another, that the laws which determine the relations of their organs are founded – laws which have their origin in a necessity equal to that of metaphysical or mathematical laws."

What we have here is not an offhand remark by the great scientist, but a deep conviction, one of the foundations of his whole work. He notes that

> an animal ... which can only digest flesh, must, to preserve its species, have the power of discovering its prey, of pursuing it, of seizing it, of overcoming it, and tearing it in pieces .... Agreeably to this necessity, a sharp tooth, fitted for cutting flesh, is never coexistent in the same species, with a foot covered with horn, which can only support the animal, but with which it cannot grasp any thing; hence the law by which all hoofed animals are herbivorous; and also those still more detailed laws which are but corollaries of the first, that hoofs indicate dentes molares, with flat crowns, a very long alimentary canal, a capacious or multiplied stomach, and several other relations of the same kind.

These are laws that

> may be said to be reduced by reasoning from the knowledge we have of the reciprocal uses and functions of each organ. Observation having confirmed these laws, we are authorised to follow an opposite course under other circumstances; when, therefore, we observe constant relations of form, between certain organs, we may conclude that they exercise some influence on one another.[9]

Elsewhere in the same work he adds that

> As we clearly see the causes of this relation between the organs of these two functions [here he is dealing on the one hand with the means of respiration and on the other with the manner in which the movement of the nutritive fluid is accomplished], we are authorized to presume that other relations equally constant, which exist between them,

have also their foundation in causes of the same kind, though they are not so evident to us. [Meyerson's brackets]

As an example of these constant relationships for which we do not know the reason, Cuvier cites the fact that, among animals with double circulation, those who receive air directly through the lungs have their two arterial trunks close together and provided with muscular ventricles joined in a single mass, while those who breathe through gills have the two trunks separated (*Leçons d'anatomie comparée* 1:48–49 [Ross 1:49–50]).

In a later work designed to sum up the most important results of his scientific work for the general public, that is, in the famous *Discourse on the Revolutions of the Surface of the Globe, and the Changes thereby Produced in the Animal Kingdom*, Cuvier again stresses these views most energetically and clearly. "There are," he says, "a great number of cases in which our theoretical knowledge of these relations of forms is not sufficient to guide us, unless assisted by observation." Indeed, one can understand the reasons for the correlation between the diverse traits characteristic of hoofed animals in general; however,

when we proceed to consider the different orders or subdivisions of the class of hoofed animals, ... the reasons upon which these particular conditions or rules of conformation are founded become less evident. We can easily conceive, in general, the necessity of a more complicated system of digestive organs in those species which have less perfect masticatory systems .... But I doubt whether it would have been discovered, independently of actual observation, that ruminant animals should all have cloven hoofs, and that they should be the only animals having that particular conformation; that the ruminant animals only should be provided with horns on their foreheads; that those among them which have sharp tusks, or canine teeth, should want horns, etc.

As all these relative conformations are constant and regular, we may be assured that they depend upon some sufficient cause; and, since we are not acquainted with that cause, we must here supply the defect of theory by observation, and in this way lay down empirical rules on the subject, which are almost as certain as those deduced from rational principles, especially if established upon careful and repeated observation. Hence, any one who observes merely the print of a cloven hoof, may conclude that it has been left by a ruminant animal.

No doubt

it is quite impossible to assign reasons for these relations; but we are certain that they are not produced by mere chance, because, whenever a cloven-hoofed animal has any resemblance in the arrangement of its teeth to the animals we now speak of, it has the resemblance to them also in the arrangement of its feet.[10]

All these considerations serve, moreover, to support the following statement which, like the first passage we cited, and even more precisely, if that is possible, brings the biological sciences closer to mathematics:

In the same manner as the equation of a curve regulates all its other properties; and, as in regard to any particular curve, all its properties may be ascertained by assuming each separate property as the foundation of a particular equation; in the same manner, a claw, a shoulderblade, a condyle, a leg or arm bone, or any other bone separately considered, enables us to discover the description of teeth to which they have belonged; and so also reciprocally we may determine the forms of the other bones from the teeth. Thus, commencing our investigation by a careful survey of any one bone by itself, a person who is sufficiently master of the laws of organic structure, may, as it were, reconstruct the whole animal to which that bone had belonged. *(Discours sur les révolutions* 102 [Kerr 95])

Has Cuvier gone too far in making this last statement? So Blainville,[11] only a generation later, believed he could demonstrate, and science does seem to have decided in his favor against his great predecessor.[12] But from our present perspective, whether Blainville was right or wrong is irrelevant. Indeed, supposing that Cuvier had been mistaken from one end of his work to the other and there was nothing worth saving from his system (which is certainly not the case), that would simply prove he had misjudged the constancy of the relationships he studied; it would not prove (which is the only point that concerns us here) that he was wrong in claiming that the existence of constant relations leads us to assume the existence of an internal connection. On the contrary, the fact that he inferred such a link on the basis of what might today be considered insufficient evidence would, rather, be yet another proof of the vigor of the tendency he was obeying, a tendency able to lead astray a mind of his caliber to the point of deluding him about the results of his research.

Now, there can be no doubt as to what this tendency is. For Cuvier, biology is not at all a purely descriptive science. It must, on the contrary, seek actual explanations, necessary relationships. Empirical laws are insufficient; observation merely prepares the way for theory, which thus constitutes the only true goal of scientific knowledge. This theory must be rational, it must search for relations capable of being demonstrated like those characterizing the properties of geometric figures. Its method can thus be only deductive; therein lies its true essence.

Closer examination of Cuvier's texts reveals a difficulty, however. What interests us here is the purely causal method of deduction. But that is not what Cuvier is thinking about when he speaks of the relations between the various animal organs. On the contrary, what obviously guides him is, above all, the idea of purpose. If we accept Cuvier's ideas on the necessary rationality of science, must we therefore also accept his finalistic concepts, that is to say, must we take deduction, which is the goal of science, to be composed of elements taken indiscriminately, so to speak, from considerations sometimes of cause and sometimes of purpose?

We have already dealt with the concept of purpose and its role in the biological sciences and have acknowledged that it does indeed play the role of an explanatory principle, of a factor serving to make purely empirical relations rational (p. 35). It is a question to which we shall have occasion to return later. For the time being, let us merely note that Cuvier himself certainly meant to limit his teleological considerations to the sciences of organisms. He has left us a most remarkable work, the *History of the Progress of the Natural Sciences from 1789 to the Present*, where, in a magnificent tableau, he embraces all the sciences of his epoch (the term *natural sciences* for Cuvier also including the *physical sciences*). Now, when he treats physics, chemistry, etc., Cuvier nowhere introduces the slightest hint of finalistic or teleological considerations and handles his subject entirely in the spirit of post-Galilean and post-Cartesian science. He finds the phenomenon of impact to constitute the only true explanation of all the phenomena observed in the physical sciences (cf. the passages cited below, p. 58), and one sees clearly, by the way in which he speaks of these explanations yet to come, that (predictably) he considered the rules he had formulated concerning the rationality of the biological sciences to be even more clearly applicable to the physical sciences. In this domain, however, there could no longer be any question of finality, and rationality could therefore be achieved only by means of the concept of cause.

"The most essential part of the causal relationship," says Riehl in a remarkable though apparently little remarked work,[13] "is not its temporal, but its logical content. What causality brings to our attention is the possibility of connecting events separated in time, by means of a conclusion." Furthermore, the logical content is so much more important than all the rest that the way in which cause and effect behave with respect to time appears indifferent, as it were. At one time there was much discussion as to whether a cause always had to precede its effect in time or whether they could be simultaneous. The truth is that we use the term "cause" indiscriminately in both cases, that in particular we use it quite well to characterize the relation between two properties both of which are taken to be inherent in matter, and thus permanent, as, for example, when we say that sulfur's affinity for oxygen is the cause of its combustibility. There is no real impropriety of language here, but only the observation that one of the properties of sulfur can *be deduced* from the other.

It is nonetheless remarkable that language (and not only ordinary language, but philosophical language itself) tends to establish a close similarity of terms, and sometimes a sort of confusion, or, if one prefers, a veritable identification between temporal and logical relationships. Thus the "consequent" and the "consequence" seem to be separated only by a nuance, and we speak of the "logical priority" of a concept.[14] As for the term *consequence* [French: *suite*], we use it indiscriminately in both senses. Lalande's *Dictionary*, after designating as the first meaning of the word: "*what follows another thing, what succeeds it*," notes as a different meaning: "*what follows from something else, what results from it, logical consequence* [*conséquence*]," and observes that it can come to mean (as in a phrase from Leibniz concerning the "natural consequence of things") the "*relation of logical or causal dependence*" itself.[15] We shall have occasion to return later to this peculiarity of terminology.

The close connection established by the human mind between the search for causes and ontological conceptions is obvious. We have already mentioned (p. 21) that we generally designate as theory any conception, figurative or abstract, from which phenomena can be deduced, and we have indicated at the beginning of this chapter (p. 47) that the terms *cause* and *explanation* are really synonymous. Moreover, nothing could be easier than to convince oneself directly of this connection. For example, the most characteristic form of the question that activates the search for causes is the one beginning with *why*? Now, this is the same sort of question to which, as we have seen, the figurative hypotheses of the sciences are responding. They have thus been imagined in order to satisfy our causal tendency, our inclination to understand reality as necessary or reasonable. If the physicist (as portrayed on p. 26) dissolves the solid iron bar into vibrating molecules, it is so that he can then *deduce* from this reality the phenomena of expansion, elasticity, etc. Similarly, if the chemist assumes that the element sulfur exists in a substance, it is so that he can deduce how the substance will react. Furthermore, it is just as easy to see that common sense ontology proves to be no exception in this respect. Why do I need only turn my head to see the table alternately appear and disappear? It is because the table is a *thing*, because it exists outside me, in space, and because the movements of my head, in space, can therefore regulate its appearances.

The peculiar nature of figurative theories or hypotheses gives rise to a phenomenon that has always greatly astonished scholars and philosophers, namely, the extreme weakness exhibited by their foundations, or at least by what one would judge, at first glance, to have to play this role. To take a well-known example, everyone is familiar with the dominant place occupied in contemporary chemistry, and more particularly in organic chemistry, by the concept of valence: it is truly, as Ladenburg affirms, "one of the most important principles" of this science. Now the same scientist feels compelled to admit that

> the subject of valency, quite apart from any mathematical basis (which is at present altogether wanting), must still be called a very anomalous and uncertain one, and that there is no existing conception of it which is capable of dealing in a logical manner with the whole domain of chemistry.

It should be noted that this chemist is not at all opposed to the theory; on the contrary, he has contributed greatly to developing some of its consequences and has attempted to alleviate some of its grave defects (particularly by means of his famous "prism" formula for benzene, designed to meet the objection based on the fact that the alternation of single and double bonds does not create cases of isomerism). He concludes, moreover, that scientists will have to continue, in spite of everything, to be satisfied with this idea of valence.[16]

That is certainly the opinion of the immense majority of contemporary theoretical chemists, as is seen not only from the general tenor of the publications, but also from the progress valence theory is making in annexing new domains. The historical account of one of its conquests is to be found in the fine book by Urbain and Sénéchal, which represents science's latest word in an area that is particularly difficult and interesting because it began quite recently and is still rapidly evolving. The authors extol the merits of Alfred Werner, who succeeded in constructing "a rational system of structural formulas" in the field of inorganic complexes, bringing to it "a needed clarity."

> Before Werner had fashioned the system, which rests on a concept of valence peculiar to electrolytes ..., the chemistry of cobalt ammines and of analogous derivatives of platinum, rhodium, chrome, etc., interested only a limited number of specialists; no other chemist concerned himself with them.

But "at the present time we are witnessing a complete about-face. This part of chemistry now appears as clear to us as it previously appeared obscure." As a result, "in spite of its imperfections," the theory of valence "remains precious, because it throws a bright light on the chemistry of metastable complexes, whose richness in derivatives and in diverse series would otherwise be a hopeless confusion."

Furthermore, the authors are clearly aware of the nature of the "imperfections" of the theory, since they state that "the notion of valence seems artificial" and that "the rigor necessary to raise it to the level of a scientific principle is certainly lacking." What is particularly interesting to note is that as the domain of valence has been extended, the intrinsic difficulties of this concept have also been considerably increased. The chemist accustomed to the formulas of organic chemistry can only be violently shocked when first confronted with the formulas of Werner and his school, where oxygen appears to be trivalent and nitrogen quadrivalent (water, like ammonia, having a "coordination number" of *one*).[17]

Werner himself recognizes this situation. In fact, after having quoted Kekulé's words, "Atomicity is consequently the fundamental property of atoms, a property that is as constant and unchanging as atomic weight itself," the author declares that this concept of the founder of "structural chemistry" later has had to be abandoned. "Thereby," continues Werner, "the concept of valence has lost the simplicity of its original form .... The principal importance of the concept of valence, which is to furnish a solid basis for the establishment of structural formulas, thus vanishes." One is obliged to assume that valence varies according to combination with this or that particular element, according to temperature and pressure, and probably also according to other physical factors (such as the solvent in which the reaction takes place). "That is why contemporary chemistry's concept of valence is so lacking in precise form."[18]

Of course, chemists had excellent reasons to accept all these anomalies (we shall go into this at greater length, pp. 228 ff.), but we can see clearly that the light brought by the theory, according to the informed judgment of Urbain and Sénéchal, to a formerly obscure and chaotic part of science certainly did not come from the particular clarity of the underlying concept.

In other cases, we cannot point to any such explicit acknowledgments on the part of the theoretical scientists themselves, but it can easily be seen directly that the situation is analogous. For example, throughout the ages (or at least, to be more precise, until the twentieth century, that is, until the rise of the electric theories, but of course excluding that period of the Middle Ages when the Peripatetic explanation by matter and form was held in high esteem) the ideal scientific explanation was deemed to be explanation in terms of impact. "Once we abandon the phenomenon of impact," says Cuvier, who, here again, admirably comprehends and summarizes the foundations of the scientific *credo* of his time, "we no longer have any clear idea of the relation of cause to effect," and he regretfully observes that "no matter how general each of them [the physical theories] has become, they are still a long way from being reduced to the laws of impact, which alone will be able to change them into true explanations" (*Histoire* 1:2 [Meyerson's brackets]). Now, nothing is more certain than the fact that the phenomenon of impact itself is by nature quite inexplicable, and furthermore, this situation had been sufficiently brought to light well before the epoch when Cuvier was writing, both by philosophy and by science itself. The philosophic discussion is linked to the name of Hume in particular,[19] and he, as is well known, aimed at the very heart of the problem that concerns us here. He in fact seeks to demonstrate that the consequent cannot, by any effort of pure thought, be deduced from the antecedent, and to this end he attacks the phenomenon that, as we have said, passed for the prototype of the rational, namely, impact. Hume points out that

> the first time a man saw the communication of motion by impulse, as by the shock of two billiard balls, he could not pronounce that the one event was *connected*: but only that it was *conjoined* with the other. After he has observed several instances of this nature, he then pronounces them to be connected. What alteration has happened to give rise to this new idea of connexion? Nothing but that he now feels these events to be connected in his imagination, and can readily foretell the existence of one from the appearance of the other.[20]

In other words, the consequences of mechanical impact can be described by means of a law but can never be deduced by means of a purely rational process.

Scientists, for their part, ever since the second half of the seventeenth century had taken infinite pains to achieve a rational explanation of impact; this was the real goal of the various corpuscular, dynamic, and hybrid theories that were propounded, and the resounding debates on this subject, particularly in the eighteenth century, established beyond question that all these attempts had utterly failed (cf. IR 63–75 [Loewenberg 68–77]). However, this situation in no way prevented physicists from behaving as if the phenomenon of impact had been entirely explained and presented a "clear idea of the relations of cause to effect"; they obviously continued to consider it *rational*, seeking to reduce other physical phenomena to it and considering them in their turn to be explained if the reduction seemed successful.

But it is perhaps even more significant that Hume's demonstration, as completely convincing as it seems to be, was incapable of convincing all the philosophers. Thus, for example, Riehl, Avenarius and Hermann Cohen claimed to draw from the principle of the conservation of energy the conclusion that as soon as there was "constancy" of what they considered to be the essential characteristic of the phenomenon, there had to be a rational connection between the antecedent and the consequent, and that, by this fact, the demonstration itself was found to be null and void.[21] Now, it hardly bears pointing out that this attributes to the principle a significance it cannot at all sustain. Something is undoubtedly conserved, but it is impossible to give any verbal definition of this something, this "energy," though its mathematical expression is perfectly well-defined. And it is completely inaccurate, in spite of certain terms currently used by physicists (which philosophers err in sometimes taking too literally) that the energy recovered at the end of a change is *worth as much* as the energy observed at the beginning. Energy, while supposedly remaining "constant," *dissipates* continually and, in the conditions under which we are forced to operate, only a small part of this dissipated energy (of thermal energy, for example) can be transformed back into a higher level of energy (into kinetic energy on our scale, for example). It is thus not correct that the consideration of energy, in this sense, truly determines the essential aspect of the phenomenon. The consideration of entropy is, in virtue of Carnot's principle, much more important. Now, entropy is not constant, but constantly increases, and this continual increase constitutes the true source of becoming, of change in the world.

On the other hand, it is quite remarkable that eminent thinkers have so misunderstood the essential teaching of Hume. It gives us an idea of the power of the philosophical instinct that moves human thought to affirm the deducibility, the rationality of the external world, and the attitude of the physicists on this question of impact thus ceases to surprise us.

Similarly, we shall not be surprised to see that science frequently uses more or less fictitious entities, created *ad hoc*, in order to explain palpable and familiar phenomena. *Obscurum per obscurius*, as the expression goes, and one could even say "light by means of darkness," if one defines these terms as positivism does, recognizing as legitimate only the reduction of a less familiar phenomenon to a more familiar one. The explanation of heat, which we know by direct sensation, in terms of caloric fluid, an entity whose fictitious nature is unquestionable for us today, was evidently an explanation of this type. But analogous observations can be made concerning contemporary physics. The tendency to explain heat and light by means of the ether, a mysterious fluid with contradictory properties, falls under the same category. For the atoms themselves, until recent discoveries allowed us to come so close to them, the situation was largely analogous: the proofs by means of which we deduced their existence were neither very direct nor very conclusive. But the mere fact that a certain number of phenomena could be explained by means of this assumption was enough to allow the majority of physicists and particularly chemists to believe in it absolutely; this is what made it possible for philosophers, such as Stallo, who had too much confidence in the positivistic theory of science, or for scientists who, despite their talent in a limited area of science, lacked a true scientific instinct from the standpoint of general principles, to attack atomism in general, by declaring it unscientific.[22]

The situation is perhaps even more apparent in the case of the most recent form taken by physical hypotheses: the electrical hypothesis of matter. Indeed, what could be stranger, from the standpoint of the *familiar* phenomenon, than this conception which reduces the whole body of things to being only manifestations, varied in appearance, of a single fundamental process, the electrical process? To explain heat and light, which I sense immediately, indeed to explain even the mechanical action I am able to perform directly with my bodily organs, by electricity, which I cannot perceive directly, whose effects I can experience only as mechanical – does that not seem contradictory? And is it not just as contradictory to try to reduce this mechanical phenomenon, which gives me at least the illusion of comprehensibility, to another that is posited from the very beginning as inexplicable, since it is declared to be ultimate and since even the most extravagant imagination could not claim that it is *rational*, that it conforms to the deep-seated demands of our understanding? However, there is no doubt at all that this theory did not in the least offend the instinctive predispositions of physicists, who, on the contrary, welcomed it with unmistakable favor; at present one can confidently affirm that the whole of science is filled with the electrical theory, to the point that all the hypotheses one encounters are more or less bound up with it. Thus it is quite clear that in this case the authority, the explanatory force of the theory cannot possibly originate in the fundamental fact, which remains a mysterious, unfathomable X.

The tendency to create fictitious entities for the purpose of explanation is so strongly rooted in us that it was necessary to put us on guard against it by a special declaration. That, to our mind, is the true sense (at least in the scientific domain) of the famous "Ockham's razor," which forbids us "to multiply entities beyond necessity." There is no doubt that the epoch of "substantial forms" particularly needed a restraint of this kind, just as it is certain that the restraint has proved largely inoperative. Molière too, so many centuries after Ockham, made fun of the *soporific virtue*[23] of the physiologists of his time. But Molière, with the sure touch of genius, has put his finger on a case of extreme and flagrant abuse. The soporific virtue, in fact, was of use only in the particular case for which it had been invented. If, on the contrary, this obviously fictitious entity could have been used to deduce several phenomena, it is not certain that the hypothesis would have been as reprehensible. That is why the theory of a caloric fluid, although it assumes the existence of an entity quite as fictitious as the *soporific virtue*, was an excellent scientific theory. As we know, it actually explained quite satisfactorily an enormous number of known facts and thus furnished a completely acceptable representation of reality, as well as constituting a first-rate instrument for research. With its help Black had formulated the theories of latent heat and specific heat, and Sadi Carnot also used it in establishing his principle. Granted, it had great difficulty accounting for a certain number of striking observations, such as the experiments of Davy and Rumford, and even for certain facts of vulgar experience (in each case involving the production of heat by friction). But those who are surprised at this and on these grounds blame the great physicists of the first half of the nineteenth century who, like Biot, set themselves up as fervent champions of the theory, thus show that, exactly like those who criticize Aristotle and his followers in the Middle Ages because they made do with a theory that explained projectile motion very badly, they do not understand the true role of physical theories.[24] Theories are indispensable to our intellect because we cannot live without an image of reality. And since reality is partially irrational in nature, one can be certain in advance that no conceivable representation of it will ever be completely adequate. As a matter of fact, *every* theory of any generality that has ever held sway in science has had to leave aside some facts it could not account for and others that directly contradicted it. Skeptics need only refer to the Council of Brussels discussed above; they will discover how the undeniable and immediately verifiable fact that a silver spoon does not shine in the dark contradicts all modern theories on radiation, which nevertheless form the basis for contemporary physics (*Brussels Conf.* 12, 16).

Another hypothesis, which fell into disuse half a century before that of caloric fluid, presents perhaps an even closer analogy to Molière's soporific virtue. Phlogiston is, in fact, nothing more than the *combustibility virtue*. Nevertheless it too was an admirable scientific theory, as is unfailingly recognized by anyone who takes the trouble to examine at all closely the science of the epoch. It is surely paradoxical to want to belittle the merit of Lavoisier, one of the most genuinely great men of science of all time, and only extreme chauvinism could have motivated such an attitude in Ostwald. In any case, the phlogiston theorists have considerable merit. It is true that they intentionally ignored the facts suggesting an increase in weight due to combustion, facts which were already quite generally known at the time their doctrine arose and which Jean Rey in particular had laid out with great perspicacity, but in so doing they were simply making use of the privilege inherent in any theory, a privilege from which all theories have profited more or less extensively and which no scientific theory of any generality has ever been able to do without (and, let us add, probably never will). They based their work on the seemingly obvious principle that similar properties indicated the presence of analogous components. The extent to which this conception dominated the whole of chemistry at that time is seen in the striking example offered by the history of *acidum pinque*. The fact that Black had decisively established that lime lost weight as it was transformed into quicklime in no way prevented universal acceptance (even initially by Lavoisier) of the theory according to which a hypothetical entity, this *acidum pinque*, was presumed to have entered into it. This was due to the fact that as a result of the transformation lime acquired well-defined properties and that, according to prevailing ideas, such a change could only result from the involvement of a quality-bearing principle (cf. IR 378 ff. [Loewenberg 332]). Modern chemistry has totally renounced this position, but it has only done so one step at a time, and the fact that attitudes have changed must not prevent us from recognizing how natural the former viewpoint was, how true to the innermost tendencies of the true scientific spirit, and how, as a result of its abandonment under constraint and force, the true explanation of chemical phenomena, of the properties that appear and disappear in reactions, became more difficult. Indeed, phlogiston – combustibility – *passed* from one body to another, and this primordial fact, which could be demonstrated by many conclusive experiments and thus provided an invaluable support to qualitative theories of nature in general, seemed important enough to ignore a few observations which appeared to be sporadic, paradoxical facts. Certainly, as time went on and discoveries enlarged the field of knowledge, new facts rebelled more and more against the framework of the theory. But its reputation was such, one had such a feeling of being on the right track, that even the authors of the most astonishing and memorable of these discoveries – Priestley, Cavendish, Scheele – at once took pains to formulate them in such a way as to show that they conformed to the hypothesis of phlogiston, explaining through auxiliary hypotheses everything that seemed to escape it, and indeed resolutely leaving these considerations in the shadow, if necessary, as if they were devoid of theoretical importance. This fact, the powerful hold the theory of phlogiston had on men as able as the above-named chemists, who must certainly rank among the great "discoverers" of all ages and all sciences, would in itself seem sufficient, even without the intrinsic reasons we have set forth, to demonstrate the great scientific value of the theory. It is equally significant that Black did not abandon the theory until 1791, that is to say, many years after posterity had judged it completely untenable on the basis of the work of Lavoisier and his disciples (see Appendix 2). Likewise, the circumstances under which it was established and maintained evidently prove (just like the history of caloric fluid and of many other hypotheses) that the concern for explanation and the tenacious will to extend its domain at any cost so far outweigh any other consideration in the march of science that the truths that initially seemed the most plausible, the most well-established facts, are set aside, intentionally forgotten as it were, upon the appearance of a more comprehensive theory allowing a much greater number of phenomena to be reduced to a system, to be connected by deduction.

At the time Lavoisier appeared, then, there were in chemistry, on the one hand a significant number of facts interconnected by a theoretical conception considered beyond doubt, and on the other a considerable and, above all, a growing number of observations that were difficult to fit into this theory, or even directly contradicted it. But it required a mind of a quality even beyond that of a Cavendish, a Scheele, a Priestley – as great as these men of science were – to bring together these apparently disconnected facts and produce a new explanatory conception. The real resolution could not be better characterized than it was by Cuvier shortly after it was achieved: "The new theory," said the great biologist,

> is only a bond felicitously bringing together particular facts recognized at different times by different men .... But it is precisely the creation of this bond that constitutes the incontestable glory of Lavoisier. Before him, the specific phenomena of chemistry could be compared to a sort of labyrinth whose deep and tortuous alleys had almost all been travelled by many hard-working men; but their junctions, their connections to one another and to the whole, could be perceived only by a genius able to raise himself above the construction and take in the whole plan with the eye of an eagle. *(Histoire* 1:69–70)

Thus, appearances to the contrary, scientific theories do not draw their prestige from the solidity of a supposedly fundamental observation or assumption. In the case of impact, the relationship between cause and effect is no more clear than in any other phenomenon, yet this fact has not prevented mechanical explanations from dominating physical science as a whole. In the same way, the contradictory nature of the ethereal fluid and the weakness (until quite recently) of the demonstrations in favor of the existence of atoms have in no way stopped the development of theories in these two areas, just as the fictitious nature of caloric fluid or phlogiston was no obstacle to the dominance exercised by the two concepts over the best-tempered scientific minds. For phlogiston the situation is particularly clear (which is why we have treated this case in somewhat more detail): to us, separated by several generations from those who lived under the domination of the hypothesis, this entity seems so fictitious that we can hardly conceive that anyone believed in its existence, can hardly even understand the nature of this "element," this "negative oxygen," as it was sometimes called, nor what role it was presumed to play in a given circumstance.

But if the weight of theories, their "explanatory power," does not come from these fundamental facts, then the only other explanation is that their source lies in the process that characterizes them, namely, in deduction itself. And this is indeed the case: we *want* phenomena to be explicable and, as we have seen, that can be achieved only by establishing a logical connection between the antecedent and the consequent. It is for this reason that any operation leading, or only seeming to lead to this explanation, this comprehension, immediately seems to assume a particular prestige. It also explains the fact that a theory never disappears unless confronted with another theory. We noted something analogous above, in speaking of the way in which causality and finality in the biological sciences come together and take possession of one another, so to speak, leaving no room between them for pure lawfulness. The phenomenon is still more pronounced in the realm of the physical sciences, in the case of the battle between different causal conceptions. The history of phlogiston theory clearly illustrates this, and here again Cuvier had a very acute perception of the way things actually came about. Enumerating the many fine discoveries that preceded the birth of the new chemistry, he said: "But these experiments, while bringing home the insufficiency of the phlogiston theory, did not immediately yield a better one" (*Histoire* 65). A theory never dies of natural causes, as a result of its constitutional weakness, or from old age, that is, solely because facts have subsequently been discovered that do not fit into its fundamental assumptions: like the priest of Nemi, it must always be assassinated by its successor. It is dethroned only if there is another theory ready to inherit the throne, and so long as this heir does not come forth, one puts up with all the failings, one finds a way to live with all the difficulties.

Thus, to speak of figurative theory is to speak of deduction. But it is obvious that the converse is not true, since, as we have seen, in addition to these conceptions, science knows abstract theories founded on principles (see p. 21 above). In fact, science also uses deduction to connect its propositions directly to one another without the intervention of figurative representations. For example, through the Newtonian discovery, Kepler's laws are deduced from the law of universal gravitation, the same law governing the phenomena of free fall on earth, and by virtue of the works of Maxwell and Hertz, the laws of optics are henceforth deduced from those governing electrical phenomena. Each step along this path constitutes scientific progress, since by reducing the number of propositions we need to retain, it produces an economy of intellectual effort. This use of deduction fits into the framework of the positivistic theory of science, which consequently considers it the only legitimate use: deduction must start from one law and end at another – since a hypothesis, if it is not itself an assumption relative to a law, must be, as we have seen, strictly banished, or at least considered a mere temporary stopgap. Moreover, deduction in no way modifies the character of the laws, which remain propositions founded solely in experience. Thus there could be no question in this system of any logical necessity at all.

However, the inclination compelling human reason to conceive the external world as necessary is so strong that this tendency sometimes intrudes, surreptitiously as it were, into the most orthodox positivistic presentations. That goes back a long way. Indeed, Sophie Germain, after quoting the famous passage where d'Alembert declares that "the Universe, for someone who could embrace it from a single point of view, would be only a unique fact, one great truth," goes on to say:

Let us add that, following our inmost conviction, this unique fact must be necessary. And indeed we seek the essence or necessity of each thing and these two expressions are equivalent, for, when we know the essence, we see that the being to which it belongs could not not be, nor could it be different from what it is.[25]

Obviously the concept of logical necessity is established here with all possible clarity. And yet the author shows herself, on the other hand, to be imbued with the conviction that hypotheses such as those constructed by Descartes – in which, "supported by a small number of certainties, the man of genius has tried to make up for what he lacked in positive observations by allowing gratuitous assumptions" – ended up establishing "purely fantastic relations" between things; but that was "the mark of a time that had just ended" and "the efforts of the human mind then changed direction completely." As a matter of fact, "until then, one had always sought the causes of phenomena. We then began to consider them in themselves. Instead of the *why*, we wanted to know the *how* of each thing." Thus "we must finally bring the different branches of our knowledge to a harmony they formerly owed solely to our imagination" *(Considérations générales* 145, 150, 154, 230). This is quite straightforward positivism, and it is understandable that Comte and his disciples recognized Sophie Germain as one of their spiritual ancestors.

It could no doubt be argued that, precisely because she was a precursor, the famous mathematician could not clearly envisage all the consequences entailed in her way of thinking. Here then is a quite recent example. On the subject of hypotheses and laws, Goblot generally expresses himself in a manner entirely conforming to the positivistic epistemology. "The hypothesis is an anticipation of the law. It is law itself arbitrarily constructed by the mind." As to the idea of cause, it is "obscure and multivocal." Scientists have taken little interest in it, because it is of little importance for them. "The precise, univocal and clear idea of *law* is the only one operative in inductive reasoning."[26] Now Goblot, as we saw earlier (p. 50), is convinced that the constant order throughout nature is only the outer manifestation of an inner logical order and that furthermore the human mind is not satisfied with *de facto* truths but demands *de jure* truths. Consequently, laws must be recognized as such directly, without recourse to hypotheses concerning the manner in which phenomena are produced. That is certainly impossible at the present time, but will be able to be accomplished in a kind of ideal future.

If all laws were known, the world of experience would be completely illuminated .... One cannot see what would remain to be discovered. The laws themselves would no longer be only constant relations, nor even empirically necessary relations: they would without doubt be logically necessary relations. (*Traité de logique*, 331)

Such a conception is obviously altogether foreign to true positivism.[27] Of course, in practice the two theories seem to blend into a single program, since they both seek nothing but laws and endeavor to formulate them so as to embrace as many phenomena as possible, to deduce them from more and more general propositions: the hope of ultimately establishing the logical necessity of the laws, being a long-range view, seems to remain entirely inoperative. But there is the essential difference that while the two doctrines set forth the same prescriptions, it is not for the same reason: the positivist is guided only by the consideration of the efficacy of the effort, whereas for Goblot the deduction exhibits an intrinsic value from the standpoint of thought, since all deductions are in fact only fragments, links detached from a great chain and, what is more, destined ultimately to be reintegrated into this chain, which will embrace the totality of the phenomena of the universe and show them all to be necessary, in conformity with reason. Thus deduction no longer takes place uniquely with an eye to action, it also aims at explanation. Furthermore, this consequence was spelled out by Goblot himself. Experimental analysis, he says, "has as its goal not only a mastery of the powers of nature by the human will, but also to make nature intelligible; it means to subject nature not only to man's will, but also to his intelligence" (*Traité de logique*, 285–286).

At first glance, Goblot's theory seems to effect a sort of synthesis, or at least a rather felicitous compromise. Indeed, it ends up sustaining the positivistic program while at the same time giving it a natural basis more consonant with our actual psychology, since it does not overlook the thirst for knowledge we all experience within ourselves, the thirst which is the driving force of science. It is therefore not surprising that this way of understanding the role of deduction seems plausible. As a result, we believe that conceptions of this sort lie behind many statements, especially by scientists, who to all appearances are completely faithful to the epistemology of Comte and Mach. It is Goblot, however, who must be credited with making them explicit.

We shall return to his theory in Book Four (pp. 423 ff.), at which time we shall see what its true philosophical foundation is and what difficulties it encounters in this regard. What must be noted here is that by the very fact of adopting the positivistic position, the theory disregards, as does the theory of positivism proper, the ontological nature of science: the obvious fact that science starts with common sense and that, as we have established (cf. pp. 26 ff.), having destroyed this metaphysics, it is immediately obliged to invent another, that it can destroy one ontology only by means of another ontology, since it stands to reason it cannot deduce something real from an unreal concept, that, on the contrary, the element of reality, the ontological character is actually greatly intensified in the entities created by science if one compares them to common sense objects. On these grounds, the theory thus shows itself – just like the doctrine of positivism and for the same reasons – to be in disagreement with the true process of scientific thought.

NOTES

1. Cf., for example, at the very beginning of the *Ethics*, "Of everything whatsoever a cause or reason must be assigned, either for its existence, or for its non-existence" (Pt. 1, Prop. 11, 'Another Proof').
2. Thomas Hobbes, *Of Liberty and Necessity*, in *The Moral and Political Works* (London: n. p., 1750), p. 484.
3. Leibniz, *Opera* 715 [*Principles of Nature and Grace*, § 7, Parkinson 199].
4. Bernhard Riemann, *Gesammelte mathematische Werke und Wissenschaftlicher Nachlass*, 2nd ed. (Leipzig: B. J. Teubner, 1892), p. 522.
5. Hegel has similarly stressed the fact that "this thinking study of things" implies the need "of showing the necessity of its facts" (*Enc.*, *Logik*, 6:3; cf. also 6:14 [Wallace 3–4; cf. 15]).
6. Hans Driesch, *Naturbegriffe und Naturурteile* (Leipzig: Wilhelm Engelmann, 1904), p. 42.
7. T. H. Green has shown how essential this distinction between cause and reason is from the standpoint of positivism (*The Logic of J. S. Mill*, *Works*, London: Longmans, Green, 1911, 2:300). Indeed, cause, becoming the empirical antecedent, is thus subordinated to law. Green, who is a Hegelian (and in fact one of the fathers of Anglo-American neo-Hegelianism), not only does not accept this point of view, but declares that "the absolute antithesis between the relation of reason and consequence and that of cause and effect is part of the false antithesis between thought and reality" (2:302).
8. Edmond Goblot, 'Le Concept et l'idée,' *Scientia* **11** (1912) 105 and 'Sur le syllogisme de la première figure,' *Rev. de méta.* **17** (1909) 359. Cf. also his *Essai sur la classification des sciences* (Paris: Félix Alcan, 1898), pp. 23, 32, 49, 50, and *Traité de logique* (Paris: Armand Colin, 1918), pp. 107, 196.

9. Georges Cuvier, *Leçons d'anatomie comparée*, Paris: Baudouin, Years 8–14 [of the First French Republic: 1800–1805], 1:47, 55, 57 [*Lectures on Comparative Anatomy*, trans. William Ross (London: T. N. Longman and O. Rees, 1802), 1:47–48, 55–57].
10. *Discours sur les révolutions de la surface du globe et sur les changements qu'elles ont produits dans le règne animal*, 6th ed. (Paris: Edmond D'Ocagne, 1830), pp. 102, 105, 107 [*Essay on the Theory of the Earth*, trans. Robert Kerr, 3rd ed. (Edinburgh: William Blackwood, 1817; reprint New York: Arno Press, 1978), pp. 95, 97–98, 100].
11. Henri-Marie Ducrotay de Blainville, *Ostéographie des mammifères récents et fossiles* (Paris: J. B. Baillière et fils, 1839), 1:A:36–39.
12. See B. Petronievics, 'La loi de l'évolution non-corrélative,' *Rev. gén. sci.* **30** (1919) 240–242.
13. A. Riehl, 'Causalität und Identität,' *Vierteljahrsschrift für wissenschaftliche Philosophie* [Leipzig] **1** (1877) 373.
14. One can see for example that Edmond Goblot, in his recent *Traité de logique* (Paris: Armand Colin, 1918), constantly uses the terms *prior*, *posterior*, *consequent*, etc., in their logical sense without adding any qualifiers (see, for example, p. 192), which, moreover, given the nature of his work, can lead to no confusion. We might add that the philosopher himself calls attention to this unusual identification and formulates a theory intended to explain it (cf. p. 90 below).
15. André Lalande, article on 'Suite,' 'Vocabulaire technique et critique de la philosophie,' Fascicule No. 19, *Bull. soc. fr. phil.*, 17th year, March-April 1917, published in May 1919, p. 120 [André Lalande, *Vocabulaire technique et critique de la philosophie* (Paris: Presses Universitaires de France, 1980), p. 1066]. Furthermore, it can be seen that there are a whole series of linguistic phenomena obviously deriving from an analogous tendency. Thus the word *puisque* [since], etymologically formed to signify a temporal relation is, on the contrary, one of the words the language uses to indicate most clearly the relation of reason to consequence. The term *rapport* [Eng.: relation] itself has come in ordinary language to take on the precise meaning of *cause*: *Il n'est pas venu, rapport à la maladie de sa fille* [He didn't come, due to (literally: "related to") his daughter's illness]. And no sooner are two sentences placed one after the other than the causal link is established: *He has such good connections, he will succeed*, or *He's sick, he can't come*. We owe these observations to a very interesting paper by Brunot presented to the French Philosophical Society, 3 June 1920, on 'The Essential Vice of the Grammatical Methods of Analysis' [apparently not published as such; see Brunot, 'Les formes du langage et les formes de la pensée,' *Bull. Soc. fr. phil.* **16** (1921) 106–112]. Let us note that the order in which the two clauses are placed in the compound sentence does not affect the meaning: *He will succeed, he has such good connections*, and *He can't come, he's sick* do not mean anything different from the original compound sentences. This remark is not without interest from the standpoint of the theory attempting to explain the confusion (cf. below, Ch. 4, p. 90).
16. Albert Ladenburg, *Histoire du développement de la chimie depuis Lavoisier*

jusqu'à nos jours, trans. Corvisy, 2nd ed. (Paris: A. Hermann et fils, 1911), p. 298 [*Lectures in the History of the Development of Chemistry since the Time of Lavoisier*, trans. Leonard Dobbin (Edinburgh: The Alembic Club, 1900), p. 309].

17. Georges Urbain and A. Sénéchal, *Introduction à la chimie des complexes* (Paris: A. Hermann et fils, 1913), pp. 50–53. It might be useful to dwell briefly on these divergencies, for as the theory becomes more established in science, scientists are apt to be less aware of them. Thus Urbain and Sénéchal's statement that Werner's theories were "closely modeled on those of organic chemistry" (p. 321) could lead us astray. The appraisal is correct, if one thinks of the way Werner utilizes the symmetry proper to the octahedron, since this part of the hypothesis is indeed strictly *modeled* on that of Le Bel and Van't Hoff's tetrahedron (on this subject, see p. 227 below). But with regard to the fundamental concept of the theory, that of valence, there is not agreement, but clear contradiction, a contradiction that Werner then strives to overcome with the aid of an auxiliary hypothesis.
18. Alfred Werner, *Neuere Anschauungen auf dem Gebieteder anorganischen Chemie* (Brunswick: Friedrich Vieweg & Sohn, 1913), pp. 18–19, 24, 26. Cf. also p. 36: "We thus arrive at the conclusion that the restricted theory of valence does not permit us to deduce useful representations concerning the formation and structure of these combinations, which is why we find ourselves forced to deduce their structure on the basis of their properties without taking into account the ordinary conceptions of valence."
19. See Appendix 1 on the precursors of Hume.
20. David Hume, *Essais philosophiques sur l'entendement humain*, trans. Renouvier and Pillon (Paris, 1878), p. 469 [*An Enquiry Concerning Human Understanding*, Sect. 7, Pt. 2, *Enquiries concerning Human Understanding and Concerning the Principles of Morals*, ed. L. A. Selby-Bigge, 3rd ed. (Oxford: Clarendon Press, 1975), pp. 75–76].
21. Harald Höffding, *Der Totalitätsbegriff* (Leipzig: O. R. Reisland, 1917), p. 78.
22. Remarkably enough, Auguste Comte judged correctly in this case and declared atomic theory to be a "good hypothesis" (*Cours* 6:64 [erroneous citation]).
23. [*The Imaginary Invalid*, Interlude 3]
24. Well before Black, physicists' attention had been drawn to the phenomena of the production of heat by friction; they were explained at that time almost as we explain them today, by the increased motion of the particles (cf. Georgii Ernesti Stahlii, *Fundamenta chymiae dogmatico-rationalis et experimentalis*, Norimbergae [Nuremberg]: impensis B. G. M. Endteri filiarum, & Vid. B.J.A. Engelbrechti, 1732, p. 19).
25. Sophie Germain, *Considérations générales sur l'état des sciences et des lettres aux différentes époques de leur culture*, *Oeuvres philosophiques* (Paris: Paul Ritti, 1878), pp. 158–159. The text reads "*ni être, n'être pas différent* [neither be, not be different]," which is nonsense and an obvious printer's error (the edition is full of them). Moreover, the original edition (Paris: Lachevardière, 1833) gives the correct reading instead ["*ni n'être pas ni être différent* (neither not be nor be different)"] (p. 57). [Nor is the 1927 edition of *De l'Explication dans les sciences* free of errors. We have corrected as many as possible, usually silently. In this

case, what the 1879 (not 1878, as printed here, though correctly identified as 1879 elsewhere) edition actually says is "*ni être ni n'être pas différent* (neither be nor not be different)"]. Sophie Germain seems to have borrowed this notion in part from Laplace; see Appendix 3.

26. Edmond Goblot, *Traité de logique* (Paris: Armand Colin, 1918), pp. 291, 295–296. Cf. also his *Essai sur la classification des sciences* (Paris: Félix Alcan, 1898), p. 47: "Causality ... is a transitory notion, which rational science endeavors to eliminate and whose role is so unimportant in the theoretical explanation of phenomena that scientists have not even felt the need to clarify its equivocal and obscure meaning."
27. It can be pointed out in this context that Comte himself, even while presenting the role of reasoning in the sense we described on p. 65, and insisting on the fact that it tends to be substituted for direct observation, is nevertheless so far from granting deduction a dominant influence that he explicitly protests against all "vicious exaggeration" in this area and declares that "the number of really irreducible laws is necessarily much greater than would be suggested" by the "dangerous illusions founded on a false understanding of our mental powers and of scientific difficulties" (*Cours*, 6:601).

CHAPTER 4

THE RATIONALITY OF THE REAL

Underlying science's pursuit of explanation there is obviously a postulate: it is the affirmation that nature is explicable, in other words, that the way it behaves is in conformity with the paths followed by our reason. That is an assumption human thought has formulated from the dawn of its evolution. Anaxagoras and Hermotimus before him, Aristotle tells us, proclaimed that "reason was present – as in animals, so throughout nature – as the cause of order and all arrangement" throughout the world *(Metaphysics* 984^{b}14–19 [W. D. Ross trans.]). However, in the affirmation of this mysterious and (as we shall see later) quite imprecise agreement, there seems to be a thesis our understanding is reluctant to call upon, at least directly, in its immediate lines of argument, and from which it would attempt rather to free itself. The prestige of the positivistic conception of science – undoubtedly a completely theoretical prestige, we saw in the preceding chapter, but nevertheless a very real one, as can easily be seen by the study of epistemological works as well as science books – certainly rests in large part on the vague feeling that by dispensing with *hypotheses*, *metaphysics*, one would have no need to appeal to the agreement between nature and mind. This sentiment is justified in part, but only in part, and it is interesting, from the standpoint of the legitimacy in science of theoretical deductions in general, to examine this important question somewhat more closely.

The first thing we notice is that, in a certain sense, the affirmation of this agreement is indispensable for all science. This is what is proclaimed in the above-mentioned assertion by Hermotimus, since the rule of intelligence in nature appears to be necessary for regularity, even for lawfulness itself. That immediately follows, moreover, from this simple reflection: in order to reason about nature, we must obviously assume that it is to some extent adequate to our reason. Furthermore, let us allow for a moment that it really is possible to construct a science in conformity with the positivistic ideal, composed of laws and assumptions concerning laws, without any theory of reality; surely even such a body of doctrine will include interpolations and extrapolations. No one has ever tried to measure everything directly; such a task we feel to be as impossible as it is idle, and it is obvious that the vast majority of the facts, the data we use in the laboratory as well as in our everyday life, are and will always be data that have been interpolated, deduced. Because of this, there is no doubt that, even though we believe we are adhering to pure experience, we are continually constrained to go beyond its limits. As Auguste Comte himself so admirably put it, science is essentially designed to dispense with all direct observation, insofar as the different phenomena allow, by permitting the greatest number of results to be deduced from the smallest number of immediate data. "Is not this the real use," he adds, "both in speculation and in action, of the laws which we discover among real phenomena?" Likewise, in another passage he declares that "the positive spirit is for ever enlarging the logical province at the expense of the experimental, by substituting the prevision of phenomena more and more for the direct exploration of them" (*Cours* 1:99, 6:600–601 [Martineau 55, 799]). Of course for him the term *rational* does not have the same meaning we assign it, in conformity, moreover, with common usage; it does not mean *reducible to elements originating only in our reason.* But Comte is right in the sense that even explanation by law, which is all he envisages, and which is not a true explanation, but simply a step toward explanation (which can indeed occur only once the relation of law has been established), cannot be accomplished except by an operation of our reason; this operation would be meaningless if we did not suppose that there is *conformity between reason and nature* on this point, that nature, for example, in the intervals between experimental observations is as *continuous* as our understanding, by instinct one might say, posits it to be.

However, it is not that we attribute more regularity to nature than it actually has, as is sometimes suggested; on the contrary, we are innately convinced that nature, down to its deepest manifestations, is ineluctably governed by rigorous laws. Nevertheless Comte does not seem to be of this opinion. Did the author of positivism really imagine that there existed in nature entirely arbitrary phenomena exempt from all law? One hesitates to credit it, yet his attitude on this point seems rather surprising. "Natural laws, the true object of our research," he declares in the final volume of his *Cours*, "can never remain strictly compatible with too detailed an investigation" (6:637–638 [cf. Martineau 809]). At first glance one might believe that he is only expressing the seemingly obvious truth that laws, being the expression of the temporary state of our knowledge, must necessarily yield to others as science advances. But that would be to misconstrue his position.

Earlier we called the belief in the metaphysical existence of laws an *excessive stress* in the positivistic conception [p. 11 above]: what we had in mind was the doctrine as it seems to be professed at the present time by most of the qualified men who call themselves its supporters. When it comes to the opinions of Auguste Comte himself, however, the situation is much less clear. For Comte, indeed, the laws that are discovered, if they achieve a certain degree of generality (like Mariotte's law, for example), are to stand forever. Any subsequent research attempting to shake them, or simply to modify or elucidate their content, is deemed utterly otiose and must be strictly forbidden. That is a theme to which Comte returned time and time again and on the subject of which he expressed himself most forcefully. Piling up terms of reprobation, he declared procedures utilizing overly precise measuring instruments to be "incoherent or sterile," proceeding from an "always vain ... and seriously disturbing curiosity," from a "childish curiosity stimulated by greedy ambition." He loudly protested "the abuse of microscopic research and the exaggerated merit still too often accorded to such a dubious method of investigation," and did not hesitate to invoke the secular arm of "the true speculative regime" of the future against the "active disorganization" which seemed to threaten the system of positive knowledge as a result of these attempts *(Cours* 3:369–370, 6:637–639; cf. IR 6 ff. [Loewenberg 20 ff.]). It is curious to note – the coincidence is obviously not purely fortuitous – that Comte, in support of his strange distrust of optical instruments overly perfected for his taste, could have appealed to the testimony of the greatest of his spiritual ancestors. The only kind of microscope Bacon knew was the single lens, and he speaks highly of the services it renders and especially of those that can be expected of it. On the other hand, he finds the telescope, which had just been invented, to be an extremely questionable means of investigation (*Novum Organum* II, 39). Of course, Bacon had excuses that cannot be claimed for Comte. To demonstrate how far Comte's attitude toward the microscope diverges from everything professed by science, all we need is the following quotation by two contemporary laboratory physicists who, without any preoccupation with theory, state a simple matter of fact. "Of all the instruments of physics," say Aimé Cotton and Henri Mouton, "the one that has rendered the most services in all branches of science taken as a whole is probably the microscope."[1]

Obviously, if we are to be forbidden ever to touch the laws in question, it is because these laws must present the definitive formula of the true relations between things, which is to say that they must constitute ontological entities. But if, on the other hand, we are to abstain from overly detailed research for fear of shaking these laws, it is because we cannot arrive at other more precise ones, and because beneath this ordered nature, whose true laws can be known, there is only a pile of chaotic facts entirely beyond the reach of any rule, at least of any rules we can know.

To be sure, that is a surprising statement. However, it is not an isolated example in the history of human thought. Plato, whose conception of science was in many respects at the opposite pole from that of positivism – since, contrary to the renunciation of any attempt at explanation basic to the Comtian doctrine, he believed firmly in the fundamental rationality, the explicability of the universe – took a similar position. For him, indeed, the regularity of nature, its lawfulness, was only a corollary of this rationality, so much so that where the latter ceased, the former could not subsist; because of this, truly inexplicable, nonrational facts necessarily had to be simply fortuitous, eluding all knowledge (cf. Zeller, *Phil. der Griechen* 2_1:543 [Alleyne 227–228]). Aristotle, on this question as on many others, in spite of the divergencies of principle separating him from his teacher, seems to have simply followed in the latter's footsteps, and we discover in him the same assumption of the existence of facts governed only by chance (cf. Zeller, *Phil. der Griechen* 2_2:336, 428, 650 [Costelloe 1:364–365, 465, 2:179–183]). But that part of the doctrine, despite the great authority of the names of its authors, seems to have had only limited success, even in antiquity; the diametrically opposed doctrine put forth by the Stoics, which insisted, on the contrary, on absolute determinism for all the phenomena of the universe, including those of the will (of which we shall speak a little later, p. 89), quickly prevailed.

From another angle, Comte's doctrine might be compared to certain thoroughly modern conceptions of science. Indeed, we know that a certain number of perfectly valid propositions appear to the modern scientist not only as not being inherent in nature, in things themselves – for that goes without saying, and Comte's assumption on this point is seen to be simply extravagant, utterly contrary to the true spirit of science which is, as we have recognized, nominalistic or at least conceptualistic – but as characterizing only the world on our scale. Underneath this world of visible masses, the *molar* world, there lies (as we have seen in the discussions at the Council of Brussels) a quite different world, or rather there lie several worlds, first the molecular world and then the atomic or subatomic ones. Since our usual laws were deduced only by observation of the molar world, they are applicable only to the phenomena of the molar world. Their existence is due to the fact that the number of elementary particles called into play by each phenomenon in this world is excessively large, and thus regularities are established by the simple play of the laws of chance. These regularities are therefore purely *statistical*, in the same way that in a large country almost the same number of deaths occur each year, although each individual death is due to multiple and quite diverse causes, causes that statistics can permit itself to ignore (except in the case of a general cataclysm, of course) while still formulating sufficiently precise rules. Does that not seem to be a brilliant confirmation of Comte's theory: a world of strict rules superimposed on another ruled by chance alone?

It is nothing of the kind, however, as is easy to demonstrate. To pass from molecular to molar phenomena, we appeal to the laws of chance, basing our work on the calculation of probabilities, because we are ignorant of the way molecular, atomic, subatomic, etc., phenomena actually fit together, and because this process allows us to disregard our ignorance, to eliminate (if we may use a mathematical image) this unknown from our calculations, by committing only errors that – so long as we are dealing with molar phenomena, and consequently gross phenomena – remain imperceptible. But we remain nonetheless convinced that the underlying molecular, etc., phenomena themselves also obey a perfectly strict lawfulness. And the proof is that, contrary to what Auguste Comte commanded, we never stop probing further and further into these phenomena; we apply ourselves to inventing more and more detailed methods for this purpose and we have perfected our research instruments – in particular the microscope, so odious to the founder of positivism, as we have just seen – well beyond the limits assigned to its use in the first half of the nineteenth century. The phenomena dealt with by the Council of Brussels (cf. Ch. 2, pp. 35 ff.) belonged without exception to the submolecular and subatomic worlds. The scientists who came together in this Areopagus did not succeed in constructing an explanatory theory for these facts, but, as we have pointed out, the debates themselves leave no room for doubt that they implicitly acknowledged there must be one, and that only the perspicacity of the physicists had thus far been wanting. As to the laws governing these phenomena, they claimed to know them, and it is hardly an exaggeration to state that, on the very points where this knowledge was imperfect, their conviction that such laws existed was not touched by the slightest doubt.

Well before there was any question of using the concept of probability in physics, Sophie Germain, in explaining that a result founded on probability "attests to our ignorance," immediately added that "if one knew perfectly" the different circumstances of the phenomenon, "one would know that the event is inevitable or that it is impossible, and this impossibility would be evident for the instant immediately preceding the one for which the realization of the same event would be just as evident."[2] Here, we see, the use of probability is allied with the strictest determinism. That is just as clear for the data of human statistics we just mentioned. To calculate the probability of an eventuality such as the death of an individual (in view of determining a life insurance premium, for example) not only does not preclude considering this eventuality as strictly conditioned by particular causes, but determinists have been known to base demonstrations of their doctrine on the regularity of statistical results.

In order better to explain how statistical law, founded on chance, is nevertheless compatible with the determinism of the underlying molecular phenomena, it will perhaps be useful to look at an example. Consider a liquid of known temperature. This is a precise datum capable of entering into any number of propositions, that is, capable of serving as the basis for a great number of predictions which will all be verified with great exactitude. It is, however, only a statistical indication expressing the average of the molecular thermal movements in the liquid. If we observe a drop of this liquid under a good microscope, we are quite likely to discover, if not the molecular motion itself, at least movements of tiny particles directly caused by the impact of the molecules: this is Brownian motion. Now, in the case of one isolated particle set in motion by this movement, one can no longer speak of temperature; indeed, we are no longer dealing with an average, since we have the individual particle before us. Nevertheless, the movements of this particle appear to be entirely determined by the impacts it receives; that is in fact the very assumption on which Perrin's calculations are based.

Thus these conceptions concerning the role of chance and probability can lend no support to Comte's thesis, and his assumptions as to the way laws are capable of being destroyed by the study of overly minute phenomena exhibit only a purely superficial resemblance to recent ideas of what is occurring in the molecular and submolecular worlds. Must we then attribute to this thinker, who elsewhere shows himself to be so imbued with the spirit of modern science, frankly indeterministic views of the same sort as those of Plato and Aristotle? The following passages, in any case, seem to suggest otherwise. Speaking of the role of mathematics, Comte declares that "all of organic physics and probably also the most complicated parts of inorganic physics are by nature necessarily inaccessible to our mathematical analysis, in virtue of the extreme numerical variability of the corresponding phenomena," and a little farther he adds that these phenomena nevertheless "are submitted to mathematical laws," but that "we are condemned to remain forever in ignorance of them because they are too complicated" (*Cours* 1:114–117). It is therefore at least highly probable that he saw these overly minute phenomena, to which he so strictly forbade science access, as fitting precisely into the category of those that, though being perfectly determined, were governed by laws unfathomable by human understanding. What the basis was for Comte's beliefs, which run so directly counter to what science has been professing since Galileo and Descartes, is certainly quite difficult to determine. On the other hand, we clearly perceive the motives behind them, motives that have no connection with the properly epistemological theories of positivism: these opinions are actually inspired by sociological considerations, namely, by the need for an unshakable authority (cf. Bk. 3, Ch. 13, p. 355 below). Moreover, Comte himself observed elsewhere that "the most terrible sensation we are capable of, is that we experience when any phenomenon seems to arise in violation of the familiar laws of nature" (*Cours* 1:52 [Martineau 40]). Now this sensation can arise only from the fact that – in spite of Plato, Aristotle and even Comte himself – we find the idea of phenomena that would actually escape the dominance of lawfulness (unless they are governed by the free will of acting beings) inconceivable, and we have seen that the whole attitude of the modern scientist (as revealed by the discussions of the Council of Brussels, for example) conspires to demonstrate this.

But if contemporary science sees the universe as strictly governed by laws, it nevertheless does not believe (as we saw on pp. 11 ff.) that it truly knows these laws. Thus it seems that here we are limited to just the general affirmation of the perfect *lawfulness* of nature.

This proposition is in fact sufficient, provided we understand that, by the very fact of affirming this lawfulness, we implicitly stipulate that it must be of a particular nature, namely, so constituted that it can be revealed to us. For, finally, if we do not know *the* laws of nature, we have still succeeded in formulating *some* laws, and furthermore everyone knows that we have not even waited for science in order to do so: to live, to act, we had to anticipate, and we could do so only by supposing that we could know some of the rules the phenomena obey – even if they obey them only roughly, or only appear to obey them. That postulate is not even peculiar to human intelligence. "Foresee or you will be eaten" is, according to Fouillée's apt expression,[3] a principle that must be obeyed by any organism, no matter how low an organism we imagine it to be – unless, of course, we imagine it, as Descartes did, to be formed in the manner of a mere mechanism, without sensations or acts of will.

That we are in fact dealing here with a particular condition going beyond what would be a purely abstract lawfulness – conceived in itself or, if one prefers, perceptible only to an intelligence of a higher order, such as the *intellectus angelicus* of the Scholastics – can be shown by considering, for example, the laws of planetary motion as revealed by Kepler and Newton. Are these really the rules the celestial bodies obey? We have no way of knowing, or rather we know that now and henceforth, due to the fact that we have been able to determine the relations between the phenomena somewhat more precisely, several of these laws no longer seem to us to be anything more than merely approximate rules (cf. p. 17 above).

There nonetheless remains the primordial fact that these laws were able to be formulated and that, just as they are, they allow us to make predictions of truly prodigious exactitude. Now one needs only a rudimentary acquaintance with celestial mechanics to realize that the discovery of these laws is primarily due to one specific circumstance, namely, the particular way in which the motions in our planetary system happen to be arranged. If the trajectories of the planets came much closer to each other than they actually do, the perturbations would be incomparably stronger and the trajectories would eventually have assumed such irregular shapes that no Kepler would have been able to recognize in them the basic outline of an ellipse. But especially if the masses of the planets were less insignificant in relation to that of the sun, or if we lived on a stellar body belonging to a system made up of three bodies of almost equal mass, the complication of the movements would be such that the discovery of the laws governing them would have been delayed indefinitely, indeed even rendered virtually impossible for an intelligence of the same order as our own.

To realize this, one need only consider the genesis of Kepler's discoveries. This great scientist happens to have been one of the few who have made a point of informing us, at least in part, of the paths followed by their intellect. Thus we can see that what was at the root of all his discoveries is a sort of panmathematicism or very general Pythagoreanism: the firm belief that the celestial motions had to be arranged according to very simple mathematical proportions. In his first work, the *Mysterium cosmographicum*, he believes he can establish, starting from the astrological notion of "trigons," that the distances of the planets correspond to the five regular polyhedra inscribed in a sphere, and in the *Harmonice mundi* he seeks to apply to astronomy the proportions of the musical intervals, entirely on the model of the theories customarily attributed to Pythagoras himself.[4] Bailly, in his *History of Modern Astronomy*, judges these attempts quite harshly. "In all these harmonic relations," he says, "there is not a single true relation; in a host of ideas there is not a single truth. He [Kepler] became human again after having shown himself to be a spirit of light" (2:120 [Meyerson's brackets]). But Delambre, speaking of the same hypotheses, assesses that "all things considered, one could say, on the contrary, that Kepler always showed himself in the same light" and that if his way of proceeding must be called *folly*, as Bailly claims, "this folly created the glory of Kepler, leading him to the discovery of his immortal laws."[5] It is, indeed, by trying all sorts of proportions, guided no doubt sometimes by simple analogies and sometimes by more precise conceptions, such as that of *conservation* (cf. Ch. 17, p. 504, below), that he finally found the true relationships. But once again, none of that would have made sense without his unshakable faith in the simplicity of the relations he was seeking. It so happened that Kepler, with the keen eye of genius, had chosen a particularly propitious field in which to apply his ideas; one could even say that nature, in this regard, surpassed his hopes, for he had begun by trying an ovoid path for the trajectories of the planets, and it is only after the fact that he realized it was an ellipse – which is a simpler curve than the ovoid – that actually corresponded to the observations. But assuming the complications of which we spoke above, he clearly would have arrived at no real result by that route.

Needless to say, we are not claiming here that these arrangements exhibit a deep-seated finalism, that they are, for example, the consequence of a decree of Providence essentially committed to arranging the planetary world in such a way that its inner workings could be recognized by man. On the contrary, nothing stands in the way of our admitting that the particular state of the solar system is the consequence of an evolution whose laws we may one day discover. Still, it must be admitted that this is a particular state, as can be seen from the simple mathematical consideration (for we are of course concerned here with the simplicity of mathematical figures and calculations) that the *simple* case is generally a unique solution, and hence infinitely less probable than the more complicated case.

If we now turn from the contemplation of the solar system to the contemplation of nature as a whole, we shall see that it has analogous characteristics. All phenomena are undoubtedly interconnected, but not inextricably so. On the contrary, they seem to be mysteriously arranged in sorts of series, in such a way that our intelligence, and even the most rudimentary intelligence among animal organisms, can easily extract from it what is necessary to fabricate rules and predictions.

Indeed, these rules, as we know, can only be concerned with general concepts, *genera*; nature must therefore be constituted in such a way that we can arrive at such concepts. Thus – and that is obviously only one particular aspect of the question – we can formulate predictions only if phenomena repeat themselves. Now no phenomenon in nature can reproduce itself exactly; there will always be circumstances that differentiate it from the one we observed previously, even if it is only the change of place due to the continual movement of the earth or of the solar system as a whole, or the position of the stars in the sky, which can never be the same at two consecutive moments. If the configuration of the starry sky had a predominant influence on the course of phenomena, animals and primitive men would be incapable of anticipating anything in life, and consequently incapable of living; and if place had any such influence, we ourselves, who lack any means of determining whether we are changing place, would have the same problem. But such is not the case, and furthermore we sense it immediately, if imprecisely. We of course know that one phenomenon is tied to all the others, that it is the consequence of the state of the universe at the preceding moment, the whole universe being, in Leibniz's phrase, "all of one piece, like an ocean" (*Théodicée*, *Opera* 506 [Huggard 128]). But we feel at the same time that there exists a sort of hierarchy among all the conditions that influence phenomena, things being arranged in such a way that only very few of these conditions exert a really significant action which must be taken into account for a first approximation, the action of the others then being able to be understood as leading to "perturbations" of the principal action. For example, it is certain that no gas strictly follows Mariotte's law; but a certain number of them, including the one we continually have close at hand, namely, atmospheric air, nevertheless follow it rather closely and within limits broad enough that the law was able to be formulated and that later the study of "anomalies" (a study Comte, as we have seen, intended to bar) could serve as a basis for the even more fertile generalizations of kinetic theory.

The powerful mind of Montaigne had already recognized this peculiar trait of our reality, a trait difficult to describe precisely, it is true, but nevertheless quite characteristic. "As no event and no shape is entirely like another," he says in Chapter 13 of Book 3 of the *Essays*, "so none is entirely different from another. An ingenious mixture on the part of nature. If our faces were not similar, we could not distinguish man from beast; if they were not dissimilar, we could not distinguish man from man."[6] Arthur Balfour has found a particularly felicitous expression to designate the special quality of this "ingenious mixture on the part of nature": he calls it her "'fibrous' structure."[7] In fact, these series of coherent phenomena of which we have just spoken do resemble the fibers of an organic tissue; these fibers are certainly not entirely independent of the tissue, to which they strongly adhere at both ends, but between these two points their attachments are slight and few in number, so that the fiber can, at least roughly, be isolated. Of course, just as in the earlier case of the planetary system, it is not at all necessary that we associate an idea of finality with the concept of this arrangement. On the contrary, nothing prevents us from hoping that if we eventually penetrated more deeply into the mystery of things, we would discover there, along with the solution of many other mysteries, the reasons for this particular *structure*. In any case the very fact of this structure is immediately obvious to any observer willing to take notice.

Furthermore, this fact explains and to some extent justifies the affirmation of the ontological existence of laws. If we cannot actually assume that *our* laws govern phenomena, we must nevertheless admit that in the phenomena there exists (or, to speak with Bertrand Russell, *subsists*, *subexists*) something that corresponds not only to *lawfulness*, but also to *our* laws.

Thus even the scientist most determined to limit his study to laws and assumptions concerning laws (according to the precept of Auguste Comte) is obliged to stipulate an agreement between nature and our intellect, an agreement concerning first the intrinsic lawfulness of nature (which he takes to be unlimited) and secondly the capacity of this agreement to be conceived by the human intellect. Moreover Comte himself, as we have seen, not only implicitly posited the first part of this proposition, but went beyond it, since he assumed that our intellect was capable of discovering the true laws of nature. He did not speak to the point suggested in the second half of the proposition, but it is not impossible that a vague yet real sense of such a *structure* is what moved him to formulate the strange proposition to the effect that science was to renounce absolutely certain kinds of research – which he declared of no use at all to man and, what is more, doomed to sterility – such as research into the physical composition of the stars. It is well-known that science, only a few years after the publication of the *Cours de philosophie positive*, proved his prediction to be blatantly false. Through the results of spectral analysis, not only are we now well-acquainted with the chemical composition of the visible celestial bodies, but these results have also led to quite important observations on the earth itself, the utility of which could consequently not be denied by the most orthodox Comtian. For example, helium (as its name would suggest) was discovered in the sun before it was discovered on earth, and there can be no doubt that this substance plays a major role in the modern conceptions of radioactive transmutation of the elements, since it is presumed to be a part of all atoms without exception (or at least of all those subject to such transmutations). Still another element, called *nebulium*, has recently been discovered in the nebulas but has so far not been found on earth. Comte sought to justify his interdiction by pointing out that there

> exists, in all categories of our research and concerning all important relations, a constant and necessary harmony between the study of our true intellectual needs and the actual scope, present or future, of our real knowledge. This harmony ... derives simply from the following obvious necessity: we need to know only what can more or less directly act upon us; and, conversely, by the very fact that such an influence exists, it sooner or later becomes a sure means of knowledge for us. (*Cours* 2:11 [cf. Martineau 133])

It is clear that, taken literally, this reasoning cannot be defended: all parts of the perceived universe can and even must, directly or indirectly, act upon us, and a part having no possible relation with us would not be something "that we did not need to know," as Comte puts it, but something that we did not know, whose existence we have not imagined, in a word, something nonexistent. But if one concentrates on the expression *necessary harmony*, one cannot help thinking that what he had in the back of his mind at that moment was perhaps the idea of this peculiar structure of nature, which allows us to formulate laws, as well as the conviction that, in order to do so, we must first be able to disregard remote events. His mistake, then, would be to have overgeneralized, to have transformed into an essential and permanent condition of all scientific research what is in actuality only the condition for a first approximation, for the attempt to isolate the *fiber*; subsequent efforts must aim, on the contrary, at reattaching the severed connections to the tissue as a whole, that is, at rectifying the proposition so as to take account, insofar as possible, if not of all conditions (which surpasses human intelligence), at least of the greatest number, in order to come closer to the real, in which this phenomenon is only an integral part of the great All.

What is more, to understand where the scientist actually stands on this primordial question of the presumed necessary agreement of reason and reality, one need only observe him impartially, not at a moment when he takes himself to be doing philosophy of science, but when he is, simply and instinctively, *doing science*. It is easy to see that in passing from strictly "positive" considerations (that is, those having to do with relations, with assumptions concerning laws) to hypotheses concerning being, he in no way has the impression, as the positivistic theory requires, that he is changing fields, that he is leaving the domain of science and entering that of metaphysics; most of the time he is not even aware that his attitude toward reality has changed, that he is henceforth appealing to an agreement between reason and reality which he was not previously assuming.

There can be no doubt on this subject insofar as scientists of the past are concerned. Duhem, whose great authority in these matters is reinforced here by the nature of his personal beliefs, which are diametrically opposed to such a position, writes: "There is no doubt that several of the geniuses to whom we owe modern physics have built their theories in the hope of giving an explanation of natural phenomena, and that some even have believed they had gotten hold of this explanation."[8] He also remarks that the great scientific theories – and in particular the doctrines of the Peripatetics, the atomists, Descartes and Boscovich – were entirely dominated by metaphysical concepts and were thus only extensions of philosophical systems (*La Theorie physique* 11 ff. [Wiener 10 ff.]), an obvious proof that they both had the same goal, namely, the explanation of reality. But even in glancing through the works of those who currently use these hypothetical conceptions, including the most prudent among them, one senses that they attribute quite a different degree of reality to hypotheses than they do to a purely mathematical concept: the reader will have been amply persuaded of this by our quotations from the debates at the Council of Brussels.

Of course, explicit affirmations of reality have become a bit less frequent of late: Comte's and Mach's anathemas undoubtedly have a hand in this, and we shall also see in our Book Four (pp. 375, 400 ff.) that in fact, in the course of the last few generations, there has been a significant change in the true attitude of science on this subject, a change in which positivism was not without influence. But the most obvious reason probably lies in the fact that scientific hypotheses themselves are in the process of undergoing a profound change, of "changing their spots," so to speak, as a result of the establishment of the electrical theory of matter, which we discussed in the preceding chapter (p. 60). This does not prevent scientists, as soon as they bring atoms and ether into play, from implicitly reasoning as if these were not concepts, but real things, indeed even the only real things, since they are supposed to explain all reality.[9] Far from limiting science to laws, or from considering hypotheses as temporary surrogates for future laws, scientists manifestly and constantly subordinate the latter to the former. Duhem provides us with excellent examples of this subordination.[10] Thus, when optics classifies phenomena of the prism and the rainbow in the same category, while Newton's rings are classified with Young's and Fresnel's interference fringes, or when biology treats the swim-bladder of fish as homologous with the mammalian lung, both these sciences are obeying purely theoretical considerations, hypothetical conceptions. And even the most flagrant anomaly to be found in the application of a law (for example, Gouy's phenomenon with regard to the impossibility of perpetual motion) appears to be *explained* as soon as theory can account for it.

This is also the reason for the unquestionable fact – though from the positivistic point of view it can only appear to be an inexplicable and thoroughly reprehensible anomaly – that if we give way to reflecting on the *true laws* of nature (in the sense that, for Comte, Mariotte's law had to be such a law), we shall not be able to wean our thought from considerations concerning the true being of things. Indeed, we all know that physical laws generally present themselves as bearing a sort of qualitative coefficient, undoubtedly hard to define but nevertheless quite real; these are not, or are only rarely, physicochemical laws in general, but rather laws of this or that particular field of science. Certainly closer and closer relations have been established between these various fields, and science has made great efforts to reduce all phenomena to a single model – but this work is far from finished. Excluding biological phenomena, where this sort of work can be said to have hardly begun, we see that there remain three clearly distinct major divisions in contemporary science: mechanics, electricity (including everything concerning the study of magnetism, light and radiant heat), and finally chemistry. Let us also eliminate chemistry; its specificity does in fact rest on the observation that there are indecomposable ultimate substances, elements, which are qualitatively different. Now, this is an assumption that our inmost feeling accepts only with distaste; furthermore, it has been challenged by a whole series of observations and considerations (cf. p. 229 below), and this situation does not allow us to think of chemical phenomena as constituting the true foundation of things. But the same is not true for mechanical and electrical phenomena. Either of them is capable of appearing to us as *ultimate*, as constituting the real texture of things and consequently as necessarily serving to explain all other phenomena. Until quite recently, physicists did not doubt that this role had to belong to mechanical phenomena, but, as we mentioned on page 60, the situation has changed in the last few years and at present scientific opinion is for the most part inclined to consider electrical phenomena as fundamental. However, the older conception still numbers many authoritative supporters and physics finds itself in this respect more or less in a state of transition, or, perhaps better, in a state of struggle between two opposing currents. Now that very circumstance allows us to observe the remarkable fact that laws are in what might be called a state of subserviency with regard to theories, in the sense that our knowledge of these laws, far from being independent of our knowledge of being (as positivism supposes), seems to us to be subsequent to it. Indeed, if we suppose that nature is ultimately mechanical, its actual laws will all have to be laws of mechanics, while if the primary phenomena are presumed to be electrical, all known laws will have to be reduced to the laws of electricity. If Comte affirms (after Blainville) that "all natural effects ... might be referred to laws of extension or laws of motion" (*Cours* 1:106 [Martineau 56]), it is because at that moment, unmindful of his own principle requiring strict disregard for anything concerning the *foundation* of phenomena, and carried away by the deep-seated scientific beliefs of his time, he implicitly assumes that all phenomena can be, at bottom, only mechanical.

That these are not abstract considerations, applicable only to the somewhat remote question of "primary" phenomena, can be seen by examining what happened in one small area of the physical sciences where, by a unique privilege, the phenomena dealing with motion seem almost independent of those of the rest of the universe. It is probably obvious that we are referring to the astronomy of our solar system, which, such as we now understand it in virtue of Newton's principle, seems to offer a representation – very simplified, it goes without saying – of what our conception of the universe in general might be when we have succeeded in resolving it into a mechanism. Some have even been willing to see here the factor responsible for the tendency toward mechanical explanation in science today: since planetary astronomy had quickly achieved a development that seemed perfection itself and had struck us with its elegant order, one might have gotten the idea that all the other sciences ought to resemble it. We hope that the reader, in the course of this work, will become convinced that the causes of this tendency are much more deep-seated, and that if the desire to imitate the order of the solar system in all of physics may have played a role in its evolution, that role could only have been a secondary one. But, in the present case, the analogy is valuable. Indeed, one need only glance over the history of planetary astronomy to realize that it was almost impossible to formulate laws of any precision concerning the apparent movements of these stars on the celestial sphere, so long as it was not known what these movements in space "really" were. No matter how many spheres were amassed in the case of the system of Eudoxus, or epicycles in the case of Ptolemy's system, they could arrive at only very rough approximations, inadequate even considering the quite significant margin of error characteristic of their imperfect measuring devices. It is clearly impossible to imagine that anything at all even remotely resembling Kepler's laws could have been established without the Copernican concept of heliocentric motion first having been acquired.

But of course at the present time we can no longer nourish the hope that science will reduce sensible reality to a mechanism, nor, in general, to any rational system, and consequently the laws we formulate can never pretend to the dignity of true laws of nature; they will never be anything more than the image – probably very remote from the original – of this fundamental lawfulness that governs nature, an ephemeral image dependent both on the nature of our mind and on the changing state of our knowledge. However, it is not necessary to invoke this conviction, which has of course become current only rather recently, to understand the real attitude of the physicist toward the laws he formulates. One need only observe that with respect to this or that particular field to which these laws apply, the physicist clearly feels that he has *not yet* succeeded in reducing the phenomena to a rational whole (whatever his opinion may be as to the future possibility of such a reduction). As a consequence, he can only consider these laws to be approximate, and he always seeks to interpolate cautiously. His method can be compared to that of the mathematician who constructs his curves by means of infinitely small straight lines or who, by determining their radius of curvature at a particular point, seems to resolve them into arcs of a circle. As a matter of fact, however, he does neither, and what the inventers of infinitesimal calculus have created is precisely procedures allowing the use of devices of this sort without diverging from reality in the process.

To sum up what he have just seen about the presuppositions underlying all science, we can say that those formulated by Auguste Comte as well as by contemporary positivists are not essentially different from those one is obliged to make in acknowledging the true role of theoretical science. Between them there is only a difference of degree: while Comte and the present-day positivists are not completely of one mind on this subject, they still assign much stricter limits to the sort of preestablished harmony between our reason and the external world than science actually does in constructing its theories. It could be said that in this sense the nominalism of modern science is less complete than it might seem at first glance. Bertrand Russell points out that since "general and *a priori* truths must have the same objectivity, the same independence from the mind as that possessed by the particular facts of the physical world," it follows that

> logic and mathematics force us ... to admit a kind of realism in the scholastic sense, that is, to admit that there is a world of *universals* and truths that do not bear directly on this or that particular existence. This world of universals must subsist, although it cannot exist in the same sense in which particular data exist.[11]

Nothing can be more true, in fact. No scientist doubts that, no matter how far we push mathematical deductions on the one hand and physical research on the other, the former will remain entirely consonant with the latter and will continue to furnish the framework for it. This must be because there is something mathematical in things, because from this point of view there is perfect agreement between our sensations and our understanding. But the agreement appears just as complete to us insofar as the *lawfulness* of phenomena is concerned. Moreover, there is also a much less precise but nonetheless real agreement that makes this lawfulness perceptible to our understanding – which, as we have seen, entails the affirmation that reality has a *structure* and that, in particular, there must exist in it something that corresponds in some way to our concept of genus. From all these points of view, then, science today is actually realistic in the medieval sense of the term: it believes in the existence, in things, of something that is manifestly a concept of our reason.[12]

But, once again, the very fact that throughout the ages man has reasoned about nature proves that he has always assumed it to be perfectly rational and consequently deducible. As a matter of fact, it could be argued that this faith was not complete, since all our scientific deductions are thoroughly blended with purely empirical elements. But we have also seen that these elements stand only as toothing stones:[13] in the last analysis, as Cuvier said, any really valid empirical rule must hide a rational relation. Our real goal, then, is the complete deduction of nature. By seeking a cause for every phenomenon, man implicitly affirms that he believes nature to be entirely explicable. That is why all definitions of sufficient cause or reason (cf. p. 47 above) already imply this postulate.

But no one insisted more vigorously and eloquently on the universality of this necessary rational connection between all the parts of the great All than the ancient Stoics. Alexander of Aphrodisias sums up their doctrine in these words:

Our adversaries teach that this world is one, that it contains within itself everything that exists, that it is governed by a living, intelligent and rational nature, and that all beings reside there subject to eternal laws, which take place in series, as in a chain, so that what happens first becomes the cause of what happens later. In this way all things are linked together, and nothing happens in the world without something else following from it and being connected to it as to its cause, nor can anything that follows be detached from what precedes it, since it is impossible not to consider it a consequence of what precedes and a result that is tightly bound to it. In short, all that happens has for consequence another thing, which necessarily depends on it, as on its cause, in the same way that all that happens has as an antecedent another thing to which it is connected as to its cause. Nothing in the world actually is or happens without cause, because nothing is separated and isolated from all that precedes it. Just imagine! The world would be divided, disconnected, and would no longer remain one single world directed according to one single disposition and one single economy if some motion were produced in it without a cause. Now it would be to introduce such a movement into the world if one supposed that all that happens does not previously have its cause, from which follows necessarily all that is and all that happens. If our adversaries are to be believed, it is as impossible for anything to be without cause as it is for something to be made from nothing. And it is from infinity to infinity that, according to them, this regime of the universe evidently and imperturbably unfolds.[14]

The most important point to retain from this admirable passage is that these philosophers posited not only the universal determination of phenomena, but also the fact that their connection must appear necessary to us. It is indeed an affirmation of the complete rationality of the universe, and certainly this postulate, which is the very basis for our reason, was never expressed more perfectly. The moderns have added nothing to it, and the well-known statements of Spinoza on this subject – "Nothing in the universe is contingent, but all things are conditioned to exist and operate in a particular manner by the necessity of the divine nature," or "Things could not have been brought into being by God in any manner or in any order different from that which has in fact obtained" *(Ethics*, Pt. 1, Props. 29, 33) – seem to be, apart from the theological or pseudotheological form, mere summaries of the rendition of Alexander of Aphrodisias. To give another example, the famous passage from the Introduction to the *Philosophical Essay on Probabilities*[15] by Laplace, where he stresses the necessary connection of all the phenomena of the universe (see Appendix 3), likewise seems to be a sort of paraphrase of these declarations.

The same postulate turns out to be the determining cause of the similarity between the terms used to characterize the logical and the temporal relations of phenomena, a similarity that sometimes goes so far as to confuse these two very different concepts, as we showed on p. 54. Although this point has perhaps not attracted as much attention on the part of logicians in general as such a strange anomaly seems to merit, Goblot[16] was rightly struck by it. After noting that "'logical anteriority' is not an anteriority at all," he continues: "What explains this metaphor is that our discursive thought sees itself obliged to admit the consequence *after* it has admitted the principle: the nontemporal order of logical dependence prescribes that thought follow the temporal order of its discursive assertions." Thus it is simply because, when we set forth a deduction in any form, we customarily enunciate the reason before the consequence that we tend to use for the latter a term that clearly tends to confuse it with what is posterior in time. Can one really believe that this purely external circumstance is sufficient to explain a phenomenon so abnormal, so clearly marked and moreover so general, since as Goblot observes, it is more or less reproduced in all languages – especially given that, as is easy to see, the circumstance in question is in no way a truly essential characteristic of the logical relation? Indeed, it cannot be claimed that the notion of consequence always presents itself to our mind later than the notion of its reason. That is undoubtedly the case, as Goblot says, for formal discursive reasoning; there we certainly state the logical antecedent before its consequent. But that is not the point where the thought process began (and Goblot, whose theory generally tends to minimize the importance of the syllogism in reasoning, knows it better than anyone). On the contrary, it began by searching and, in the great majority of cases, by searching for the *explanation* of a phenomenon. Of course, that is not a rule without exceptions, and Goblot is right in repudiating the claim that induction "always and necessarily aims at discovering the cause of a given fact" (*Traité de logique* 290). It certainly can happen that we seek the effects of a given cause; thus a chemist who has succeeded in synthesizing a new compound may try to determine whether it can be used as a therapeutic agent, a dye, etc. But such cases are, if not actually rare, at least much rarer than those in which research begins with the fact it means to explain. As Plato said, science is born of wonder, and Riemann's statement reaches the same conclusion. Goblot himself remarks that the error he has pointed out is quite common: it undoubtedly results from the fact that the process in question – which tends to go back from a phenomenon to its cause – is the path our reasoning habitually follows, instinctively as it were, the opposite process being the prerogative of a more reflective and mature scientific thought.

Therefore, since the logical consequent actually presents itself to our mind before its antecedent, it becomes altogether bizarre that these particular terms should have been chosen to designate them.

The true reason for the anomaly is more profound; it results, not from an external circumstance, but from the very essence of the two concepts and from their intimate connection: if language tends to identify them, it is because they are actually identified in our thought; it is because these two relations, which form the basis for our concept of the external world, are one and the same. In virtue of our irrefragable conviction of the essential rationality of the world, we think, we must irresistibly think that every relation of succession reveals – and at the same time hides – a relation of logical dependence. The anterior *cannot* be anything but the cause of the posterior; it *must* contain the explanation for it. That is not immediately apparent, but if we only search, look closely, we shall see this link appear; one has only to develop thoroughly, to *explicate* all that the preceding state implies, for the state that succeeds it to be *explained*.

Finally, one more proof that man has always believed nature to be explicable is the fact that, from his first steps in this domain, he has tried to divine it, or at least to divine its essential texture, while considering anything that did not allow itself to be so divined as secondary, as having to be disentangled later, indeed even as purely "accidental," as unworthy of the attention of the true thinker. We saw above (pp. 75 ff.) how, for Plato and for Aristotle, a part of the phenomena of nature escapes all rationality, and no doubt their precursors professed similar opinions. The fact remains that these philosophers are really trying to deduce nature. For the most part, the point of departure for their "global deductions" today appears rather bizarre; even when we succeed in more or less understanding the physical (or perhaps meteorological) observation upon which they were based, we never tire of marveling that anyone could have considered it sufficiently important to deduce the entire universe from it. But we must realize that this is really only one more example – and a striking one, because we are dealing with very ancient and remote things – of the eternal weakness of any starting point for an explanatory theory claiming to embrace a significant part of nature; and these theories, we know, wanted to embrace it all. The essential thing, once more, is not where one starts, but the deduction itself. And furthermore, as soon as one is firmly convinced that nature is entirely comprehensible, does it really make any difference whether one starts with one phenomenon or another? Provided that it is correctly analyzed, we should attain, through it, the true creative principle or principles, and from then on all the rest should unfold by reasoning: as Cuvier tells us, whether it is the tooth, the shoulder-blade or the condyle that one holds, one will still rediscover the same entire skeleton, just as the geometer recovers the curve with all its properties equally well from any one of these properties.

The same is also true for the theory that for so many centuries appeared, in the eyes of the world, to sum up all the work of Greek thought, namely, the theory of Aristotle. Infinitely more complete in many respects than the Ionian cosmogonies, since the Stagirite built on ground prepared by the criticism of Heraclitus, the Sophists and Socrates, it also presents an attempt at global deduction of nature. How this deduction actually works, by what means the phenomena, with the aid of the concepts of matter and form, are constituted, has been covered in sufficient detail by the textbooks that we can dispense with setting it out here. Let us merely point out that deduction dominates the whole system. Everything must be reduced to the syllogism, and Aristotle knows no scientific demonstration except by syllogism, such demonstration, as Zeller aptly puts it, being for him a conclusion resulting from premises that are themselves necessary *(Phil. der Griechen* 2_2:232 [Costelloe 1:243]). So pronounced is this that it has been possible to say that Aristotle's science was not a physics but a logic. That is in fact the impression it gives to a man schooled in modern science. But it is clear that, for the master of Peripateticism, as well as for his followers in antiquity and in the Middle Ages, the two are identical, since nature cannot be anything but logical.[17]

It must be noted, to be sure, that there was something more than a purely logical apparatus in Peripatetic science. Since it began with the four primary qualities, it tended to develop a veritable qualitative physics. It thus brought together two distinct scientific tendencies, which could sometimes be seen to conflict, as in the discussions of *mixtures* (cf. IR 371 [Loewenberg 327–328]). But on the other hand this double nature was a source of considerable strength for the doctrine, often allowing it to adapt better to circumstances: alchemy, in particular, with its qualitative elements – although these qualities are different from those of Aristotle's four elements – certainly grew out of this qualitative conception of nature. But for the moment we shall leave that aspect of the system entirely aside, in order to deal exclusively with its logical aspect.

When confronted with the starting point of the theory, we experience almost the same astonishment, or at least an astonishment of the same kind as that the Ionian systems inspired in us. It is true that here we perceive more clearly what determined the choice of the four elementary substances (borrowed, of course, from Empedocles): they were certainly above all the expression of what we today call the states of aggregation of matter. But how could Aristotle attribute such predominant importance to just these sorts of phenomena? And if it is true (as indeed seems probable) that the choice of the four fundamental qualities which create these substances and thus appear to be elements of elements, as it were, was determined by a profound conviction of the primacy of the sense of touch, on which the establishment of these qualities depends, how could they have been satisfied with these primitive experiences? Certainly the observation as to the relative unimportance of the starting point of theories explains why it was possible for so many centuries to close one's eyes to the weakness of these fundamental observations. But just as for the Ionian systems, something more was involved. If they in some sense disdained new experience, hastening rather, by logical analysis of the most frequent, the most familiar phenomena, to go up to the principles, in order to come down again immediately to complete explanations of nature as a whole, it is because they were convinced that this was the most appropriate method for knowing nature. They believed they could succeed rapidly, and even, so to speak, immediately, in knowing the true essence of things and in defining it exactly. Once this was done, the task of the physicist, as described by an ancient Peripatetic, was to be limited to "demonstrating each of his propositions, by drawing them from the essence of the bodies, from their potency, from what best suits their perfection, from their generation, from their transformation."[18] Thus all the physicist could need had to be deducible from the definition itself, by means of a few rational principles, and new experience *could not* provide anything new.

To be sure, that state of mind seems quite far removed from our own. It is, however, not impossible to find a parallel in a very recent epoch. Hegel, as we shall see later, undertook a task, if not identical to the one that the Ionians or Aristotle had in view, at least quite similar, in the sense that, although he did not claim to deduce nature in its entirety, he nevertheless believed he could recreate everything essential in it through his metaphysics. Now, one of the most respected Hegelians of our time, McTaggart, in explaining how the master established the foundations of his logic (which serves as the point of departure for his *philosophy of nature*, as we shall see), says, "What is required, therefore, is not so much the collection of a large mass of experience to work on, but the close and careful scrutiny of some part, however small. The whole chain of categories is implied in any and every phenomenon" (*Studies* 19–20). There is no doubt that Hegel would have subscribed fully to this statement, and we can see how close it is to the principles that guided the Ionians.

Nevertheless, for the man schooled in modern science, confronting a specific phenomenon seems to involve an almost inconceivable gamble or an arbitrary choice. For example, it is said that when Father Scheiner informed his provincial of his discovery of sunspots, the latter replied that this could not be. "I have read my Aristotle several times from beginning to end," he reportedly said, "and I can assure you that I found nothing of the kind in it. Go, my son, set your mind at rest and be assured that what you take for spots in the sun are the fault of your glasses or your eyes." Montucla, who reports this anecdote, believes it to be a tale fabricated by someone who wanted to poke fun at the Peripatetics.[19] Possibly so, and that would be altogether in the spirit of the polemics of that epoch, as can be seen by the example of Galileo. Galileo, who is not only one of the most powerful geniuses in the history of science, but also an admirable writer and a most formidable polemicist, very adroitly manages to illuminate this weakness of the system of his adversaries who, he says, are convinced that research must be carried out, not in the world, not in nature, but in texts, that all that is involved is merely confronting texts: "because men of that sort believe that philosophy is some book like the *Aeneid* or the *Odyssey*." And he adds (he is writing to Kepler): "Oh how I wish that we could have one hearty laugh together!"[20] It is true that in giving the argument this striking form, by which scientific knowledge is equated to erudition pure and simple, that is, to a knowledge having fixed limits and incapable of indefinite progress, Galileo lends it a quite preposterous appearance. But even supposing that among his adversaries there were men professing opinions of this sort, they certainly do not form the majority, and their opinions were not derived from the foundations of the reigning doctrine. What they believed rather is that physics, although differing widely from mathematics as far as form is concerned (since it was to take its format exclusively from logic), nevertheless had to resemble it, in that – since they were both rational sciences – they could both, as they progressed, simply add, as Duhem would have it, new verities to other verities, equally incontestable, already previously established. Progress in mathematics (and no doubt Galileo's adversaries believed in the possibility of such progress just as much as he did himself) had to leave standing the geometry of Euclid. It was certainly not contradictory to suppose that the same could be true in physics. Experience, since it intervened at the most only to verify deductions resulting from seemingly incontestable principles, was thus a redundancy, that is, basically useless, true progress having to come chiefly from progress in deduction. Or else, if experience could lay claim to some utility, it is solely insofar as it stimulated that deduction, in other words insofar as its results fitted into the more or less extended framework of the prevailing theory. If, on the other hand, experience claimed to contradict prevailing theory, one could confidently disregard it; it could only be an error which would surely be explained sooner or later, and to think of overturning such a complete and firmly established body of doctrines because of trifles of this kind would have been madness.[21]

Furthermore, the divergence here between the Peripatetics and their opponents was perhaps not so great as might at first appear. Many of the latter believed, just as firmly as the partisans of Aristotle, in the rationality of nature. The ancient atomists were certainly of this number, and it is also in this sense that the doctrine of universal mechanism was understood by the great majority of physicists from the Renaissance on. What separated the two camps was not the fact of deduction, but its form. The Peripatetics mean to use logical deduction exclusively, while their opponents, who constantly refer to spatial images, thus have recourse to mathematical deduction. Towards the end of his life Galileo writes,

I truly think that the book of philosophy is the book of nature, which is perpetually open before our eyes but, because it is written in letters different from those of our alphabet, cannot be read by everyone. The letters in this book are triangles, squares, circles, spheres, cones, pyramids and other mathematical figures quite appropriate for this kind of reading.[22]

Are there really two fundamentally different processes in these two kinds of deduction, or can they ultimately be reduced to a single pattern? This is the weighty question of the relationship between logic and mathematics, and we know how keenly it has been discussed during the last few years and what valuable works it has elicited, in Italy with Peano, Padoa and their disciples, as well as in France and England with Bertrand Russell, the late lamented Couturat and so many others. Henri Poincaré's profound observations on mathematical reasoning[23] speak to the same question. For Poincaré, deduction as practiced by the mathematical sciences is entirely distinct from the syllogism and can be reduced to the type "demonstration by recurrence." His theory was actively combatted by Goblot, who, although he agrees with Poincaré that it is impossible to reduce mathematical deduction to the syllogism, denies the importance of the role of demonstration by recurrence and believes he can establish that what governs the process of any true deductive reasoning in general (and of mathematical reasoning in particular) is a "constructive activity" of the mind, leading to a "logical discovery."[24] These are very important questions which we have done our best to work around; later, in dealing with the concept of identity, we shall have occasion to examine more closely one particular area of this difficult domain. Let us only note here that for our purposes both kinds of deduction behave similarly, seem to exercise the same authority on the mind, and create the same mental attitudes, or at least very analogous ones. That is why we earlier took the liberty of invoking the example of mathematics in order to clarify the reasoning process of those who used a system of logical deduction in science, where this particular use strikes us as an anomaly.

Let us also note here how easily the two kinds of deduction seem to be substituted for one another: the Aristotelian system, which constitutes the most perfect application of exclusively logical deduction and can therefore be considered a *panlogism*, descends directly and immediately from the Platonic system, many of whose dominant characteristics it conserves, despite the fact that in many respects Platonism is the most complete expression of *panmathematicism*.[25] For Plato the foundation of nature – what we today call by the Kantian term the *thing in itself* – is mathematical and only mathematical. All reality is composed uniquely of geometric figures.[26] And of course everything is also rational, or at least everything that is truly real; what is not rational is in some sense outside the universe, that is, it is not knowable, being governed solely by chance (cf. p. 75 above).

During the Renaissance, Peripateticism undergoes an eclipse and mathematicism returns to favor. Descartes reduces it to a system: for him mathematics is at the center of everything and must be all sufficient. Consequently, it is clear, everything can only be rational, and it must be possible to arrive at physics through deduction – this time mathematical – starting with a few principles. Thus Peripatetic panlogism and Platonic and modern panmathematicism and panmechanism come together in this faith in the complete rationality and, therefore, in the deducibility of nature.

Even before Descartes, of course, Bacon had proclaimed the coming of a quite different science, based on experience, and in which deduction in any form could play only a completely subordinate role. But his doctrine actually exerted very little influence on the course of science, either in the Renaissance or during the succeeding centuries (cf. IR 447 [Loewenberg 390–391]). The antideductive bias of the illustrious chancellor was altogether too great a departure from the innermost tendencies of the period and, we might add (although he deserves immense credit for having attempted this reaction, certainly justifiable in part), also a departure from the true tendencies of science and of the human mind. Descartes, no doubt like many of his contemporaries, does not even seem to have understood very well what the doctrine meant. He chiefly saw in Bacon the vigorous fighter, waging a furious combat against the common adversary, Peripatetic science; but beyond that, he so misunderstood the full significance of the theory that he believed he was in perfect agreement with its author.[27]

It is without question Descartes who was the real legislator of modern science. Of course, no matter how deep and ineffaceable an imprint was left by this most powerful of all minds, it would be an exaggeration to say that the physicist today continues to draw his inspiration without reservation from the views of the author of the *Principles*; yet the foundations of the modern scientific credo strongly resemble them in many respects. We are no longer convinced that everything in nature is mathematical, but we wish it were, because the mathematical is still the only thing we find to be rational, and it is only through mathematical deduction that we seek to understand nature and explain it. In discussing the *method applied in general physics*, Bouasse asserts that

what makes the method of physics original is ... the importance assumed by deductive reasoning .... Within its domain physics seeks to reconstruct the world, to deduce it by purely syllogistic means from a general principle once established. No one denies that this is and has always been the avowed goal of physicists.[28]

Going on to speak of an early branch of physics, catoptrics, the author adds:

I seem to minimize the role of experience in a strange way; once the principle is discovered, experience no longer plays a role except to verify the geometric deductions: in the specific case its role was consequently useless. As strange as that may seem, it is quite true .... (*De la Méthode* 126 [1909 ed. 78])

Bouasse rightly considers these observations to be valid not for physics alone, but for all the sciences. Physics is merely "of all the sciences of nature, the one whose method has been fixed for the longest time, while so many others are still groping to find a method." For that reason, "it can be taken as the prototype of a perfected experimental science. In fact, all the others strive to resemble it."

Bouasse is equally correct in saying that this trait is not peculiar to contemporary science but is valid for science throughout the ages. Science has never varied in its search for theory and rationality, and it undoubtedly never will, given that we are dealing with a primordial need of the human mind; as we saw, the physicists at the Council of Brussels proved to be just as ardent in the pursuit of this goal as any of their predecessors (*De la Méthode* 124 [1909 ed. 76].

NOTES

1. Aimé Cotton and Henri Mouton, *Les Ultramicroscopes et les objets ultramicros-*

copiques (Paris: Masson, 1906), p. l.

2. Sophie Germain, *Considérations générales sur l'état des sciences et des lettres aux différentes époques de leur culture, Oeuvres philosophiques* (Paris: Paul Ritti, 1878), p. 161.
3. Alfred Fouillée, 'Les Origines de notre structure intellectuelle et cérébrale,' *Rev. phil.* **32** (1891) 576.
4. Jean-Sylvain Bailly, *Histoire de l'astronomie moderne* (Paris: de Bure, 1785), 2:6 ff.
5. Jean B. J. Delambre, 'Kepler,' *Biographie universelle ancienne et moderne*, ed. Michaud (Paris: Mme C. Desplaces, 1843) 21:527 ff.
6. Michel de Montaigne, *Essais* (Paris: Flammarion, 1908), 4:194 [*The Complete Essays of Montaigne*, trans. Donald M. Frame (Stanford: Stanford Univ. Press, 1958), p. 819].
7. Arthur James Balfour, *L'Idée de Dieu et l'esprit humain*, trans. J. L. Bertrand (Paris: Bossard, 1916), p. 242 [*Theism and Humanism* (New York: Hodder & Stoughton, 1915), p. 202].
8. Pierre Duhem, *La Theorie physique* (Paris: Chevalier & Rivière, 1906), p. 46 [*The Aim and Structure of Physical Theory*, trans. Philip P. Wiener (Princeton: Princeton Univ. Press, 1954), p. 31]. Max Planck is still more emphatic, stating that all great physicists have "believed in the reality of their representation of the world" (*Die Einheit des physikalischen Weltbildes*, Leipzig: S. Hirzel, 1909, p. 36). Wilhelm Wundt compares the theories of "economy" and "convention" to the legal fictions abounding in the history of law; they are attempts to establish the genesis of knowledge independently of all its real history; even the most stubborn supporter of these conceptions is forced to admit that the principles of science were not really created in that way (*Die Prinzipien der mechanischen Naturlehre*, Stuttgart: Ferdinand Enke, 1910, pp. vii-viii).
9. Max Planck explicitly asserts that atoms and electrons are as real as the heavenly bodies or the objects around us and that contemporary physicists speak the language of realism and not the language of Mach (*Die Einheit* 33–34, 37). Henri Poincaré similarly states (*Science et méthode*, Paris: Flammarion, 1908, p. 186 [*Science and Method*, trans. Francis Maitland (London: Thomas Nelson and Sons, n.d.), p. 172]) that in the physical sciences the term "existence" does not have the same sense as in mathematics: "it no longer signifies absence of contradiction, but objective existence." We saw above (p. 24) that Cournot had already clearly recognized the true attitude of science concerning mechanism.
10. Pierre Duhem, *La Théorie physique* 33, 35 [Wiener 23–24, 25]. In a recent short treatise, Max Planck pointed out how general this process is and how characteristic of the true evolution of science. Thus we now classify acoustics with mechanics, and magnetism and optics with electrodynamics. What was formerly called the *physics of heat* has been divided up, radiant heat being classified with optics (and electrodynamics), while the rest is treated under the headings of mechanics and kinetic theory.
11. Bertrand Russell, 'L'Importance philosophique de la logistique,' *Rev. de méta.* **19** (1911) 289–291. Russell's text reads *universels* [instead of *universaux*], but we do not feel we are being unfaithful to his thought in substituting the term

more often used in this context. [This does indeed seem to be merely a question of usage, with no technical distinctions at stake.]

12. The close kinship between the modern conception of lawfulness and the Platonic theory of ideas has been set forth quite well by Giovanni Vailati (*Scritti*, Leipzig: J. A Barth / Florence: Successori B. Seeber, 1911, p. 676), who stressed in particular that this theory arose out of the need to create a point of support against philosophic doctrines that, by insisting on the mutability and corruptibility of things, seemed to destroy all possibility of any stability at all in nature.
13. [Stones projecting from the wall of a structure designed to accommodate future construction, to be attached by interlocking stones.]
14. Jean Félix Nourrisson, *De la liberté et du hasard, Essai sur Alexandre d'Aphrodisias*, followed by *Traité du Destin et du libre pouvoir aux empereurs* (Paris: Didier, 1870), p. 260.
15. [Pierre-Simon Laplace, Introduction, *Théorie analytique des probabilités, Oeuvres* (Paris: Gauthier-Villars, 1886), 7:vi-vii (*A Philosophical Essay on Probabilities*, trans. Frederick Wilson Truscott and Frederick Lincoln Emory (New York: Dover, 1951), pp. 3–4)].
16. Edmond Goblot, *Traité de logique* (Paris: Armand Colin, 1918), p. 19.
17. Harald Höffding (*La Pensée humaine*, trans. Jacques de Coussanges, Paris: Félix Alcan, 1911, p. 144) correctly observes that for Kant the categories are only the form of our thought, whereas Aristotle understood them directly, as predicates of existence.
18. Geminus, as preserved by Simplicius. See Pierre Duhem, *Le Système du monde: histoire des doctrines cosmologiques de Platon à Copernic* (Paris: A. Hermann et fils, 1913), 2:77.
19. Jean-Etienne Montucla, *Histoire des mathématiques* (Paris: Ch. Ant. Jombert, 1758), 2:227.
20. Galileo, Letter to Kepler of 19 Aug. 1610, *Opere*, Edizione Nationale (Florence: G. Barbèra, 1890–1909), 10:421–423 [Oliver Lodge, *Pioneers of Science* (London: Macmillan, 1919), p. 106].
21. It should likewise be noted that another aspect of this theory, which strikes us as quite surprising, must have seemed much less paradoxical to the fine minds of this period, namely its excessive admiration for the writings of the ancients. In this respect, as a matter of fact, the innovators of the Renaissance did not differ essentially from their adversaries. The foremost idea of the Renaissance, an idea that is, moreover, accurate and, above all, *salutary*, was more or less as follows: the ancients so far surpassed the generations that came after them that the best we can do is to return to their doctrine pure and simple, freeing it from all the deformations brought by the barbaric or "Gothic" centuries that preceded us. When someone contested the authority of Aristotle and Ptolemy, it was in the name of other ancients: Plato, Democritus, Archimedes, Aristarchus of Samos. Revolt against the authority of the ancients was in general an exceptional attitude on the part of a few isolated thinkers such as Giordano Bruno, Francis Bacon (though he did occasionally invoke the ancients, such as the atomists for example) or Paracelsus, and their authority weakened only gradually, when further study of the ancients showed them to have disagreed among themselves,

and thus one learned to imitate their methods of research rather than trusting blindly in their doctrines.

22. Galileo, Letter to Liceti (1641), *Opere Complete*, ed. d'Albèri (Florence: Società editrice fiorentina, 1842–1856) 7:355.
23. Henri Poincaré, *La Science et l'hypothèse* (Paris: Flammarion, n.d.), pp. 11 ff. [*Science and Hypothesis*, trans. George Bruce Halsted (New York: The Science Press, 1905), pp. 5 ff.].
24. Edmond Goblot, *Traité de logique* (Paris: Armand Colin, 1918), pp. 165, 256 ff., 271.
25. Cf. Ch. 14 (pp. 363 ff.) for the precise meaning we give the first syllable of these terms *panlogism* and *panmathematicism*. For panmathematicism considered as *panalgebrism* or *pangeometrism*, cf. Ch. 15, p. 410.
26. Plato's position on *panmathematicism* has sometimes been interpreted quite differently. We are not at all competent to take a stand in the debate and are content (limiting ourself to contemporary philosophers) to invoke the authority of Brunschvicg (*Les Étapes de la philosophie mathématique*, Paris: Félix Alcan, 1912, pp. 43 ff.), Léon Robin (*Études sur la signification et la place de la physique dans la philosophie de Platon*, Paris: Félix Alcan, 1919, passim, esp. pp. 38, 49, 60, 64, 74) and Burnet (*Greek Philosophy*, Part I, London: Macmillan, 1914, pp. 312 ff.).
27. Cf. Gaston Milhaud, 'Descartes et Bacon,' *Scientia* **21** (1917) 188–189.
28. Henri Bouasse, 'Physique générale,' in *De la Méthode dans les sciences*, 1st series, 2nd ed. (Paris: Félix Alcan, 1910), p. 124 [1909 ed., p. 76].

CHAPTER 5

IDENTITY AND IDENTIFICATION

Let us return once again to the image suggested by Bossuet's phrase, to the leaves and flowers that unfold. What is the source of the satisfaction the mind derives from this *explication* (the term taken this time in both senses)? It is, as we saw, the fact that the leaves, which we first believed to have appeared suddenly, have been revealed to us as having been preformed, preexistent. We have explained the phenomenon, the change, by deducing the consequent from the antecedent, by showing that the consequent was necessarily what it was, *could not* be different from what it was, because it was already implicitly contained in the antecedent. All it did was explicate itself, unfold, but it is the sole fact of the unfolding that makes all the difference between the antecedent and the consequent; beyond that, there was nothing changed, nothing created, nothing destroyed, things remained what they were. And it is obviously the fact of having been able to show this, of having been able to reduce the striking but purely apparent change to an underlying real identity, that is the source of the feeling of satisfaction which the explanation gives us.

That should come as no surprise. Let us recall Riemann's statement: we seek a cause only because there is change. Thus the best way to deal with it (and indeed the only way, as we shall show later) is to show that the change does not exist.

Moreover, we are dealing with reasoning, and the concept of identity of course plays a predominant role in all reasoning. Although this primordial truth has certainly been felt by mankind throughout the ages, Leibniz seems to be the first to have clearly expressed the indubitable fact that identity constitutes the ideal of all rational thought.[1] He never tires of reaffirming the fundamental proposition that "in necessary truths there is a demonstration or reduction to identical truths."[2] However, the truth that only the identical really conforms to the demands of our reason was already implicitly contained in the Eleatic conception of the sensible world. Indeed, if for Parmenides the world appears in the form of a sphere, the same in all its parts and immobile, that is to say, without diversity in time and space, it is because only such a world would seem rational to us.

There is certainly a connection between this Eleatic conception of being and Plato's conception of the role of the Same and the Other as substances composing the soul as well as the external world. Only the Same produces rational knowledge.[3]

A century after Leibniz, Condillac quite forcefully expresses analogous ideas: "I agree," he says in the *Langue des calculs*, "that in this language, as in all others, one makes only identical propositions whenever the propositions are true," and in his *Logic* he says that the "evidence of reason consists uniquely in identity."[4] Quite recently, a respected philosopher has strongly insisted on the importance of these considerations. "Logic and mathematics," says Höffding, "demand the sameness of thought in its most rigorous and ideal form – as absolute identity – for only in that way does it become possible to conclude and calculate with rigor," elsewhere observing that "identity is the measure of thought, the condition for any formation of concept, judgment and conclusion," and that this is "our highest and clearest principle of thought, the principle of principles."[5]

Furthermore, a mere glance at any mathematics book convinces us how right Leibniz was. What most generally characterizes a proposition is that the different terms of which it is composed are connected by an equal-sign, while a demonstration generally consists only of a series of equations, one after the other.

However, there is undeniably a paradoxical side to this observation. The true principle of identity, $A = A$, has the appearance of a mere tautology, or in Kant's terms, a purely analytic proposition. At most, it appears to be useful for simplifying the terms of a statement by eliminating what is recognized to be identical in the concepts we mean to connect, almost as we do in canceling terms that are repeated on both sides of the equal-sign in an equation. Now we clearly need something else here; we need a synthetic proposition. How does the principle of identity manage to play this role? How can it be an instrument for furthering our thought?

The truth is that the principle of identity, as we apply it in our arguments, is never purely analytic. No scientist or philosopher, and probably no man of sound mind, ever entertained the idea of enouncing $A = A$ in the sense of a perfect tautology, for he could in fact only repeat it indefinitely without advancing his thought in the least.

Let us examine how the concept of identity is applied in some very simple cases in mathematics. Suppose we are in the process of demonstrating the equality of two triangles and we state that *the line AB is equal to itself*. Why is it necessary to formulate this apparent tautology? It has to be because there were circumstances, however slight, that could induce me to consider a different idea: the line *AB* bordered one of the triangles on its right side and the other on its left side; but it is a line, and I must remind myself that, as such, it has no thickness, that therefore the same length must be involved. Or else the two triangles whose equality is to be demonstrated were positioned differently; what one must do is recall that they *nevertheless* have the same base, and so on. Here again are two points C and C' which we want to demonstrate must coincide, must actually form only a single identical point; because this point was reached by two different routes, one might think at first glance that there were in fact two distinct points, and therefore it was necessary to demonstrate the contrary. Or to take another example, no one (except in the case of a standard device, for example adding the same quantity to both sides of an equation) would write $a^2 - b^2 = a^2 - b^2$; but might write $(a + b)(a - b) = a^2 - b^2$. Granted this is what we call an *identical* equation, but the identity is not immediately obvious, at least for a beginner in mathematics, because what is written on the two sides of the equal-sign is not identical in point of fact: on the one side there is a a product and on the other a sum. The equal-sign simply means that things are equal in certain respects or would be by certain conventions. For example, if we replace a and b by natural numbers and perform the operations indicated by the signs, the numerical result will be the same on both sides; or if we manipulate the product $(a + b)(a - b)$ according to established rules, we shall succeed in actually transforming it into $a^2 - b^2$. But only then will there truly be identity – identity that we refrain from writing, for it would serve no purpose to do so.

Thus A = A is never really a true tautology. If we found it necessary to state the formula, it is because there were reasons which could lead us to think that there was no identity, that two different things were involved and not one and the same thing. In our thought, A = A is always followed by a sort of implied qualification beginning with "although ..." or "in spite of the fact that ...." There must be *something*, some sort of circumstance, differentiating the second A from the first, and what the proposition affirms is that this circumstance is irrelevant from the point of view that interests us at the moment.

Hegel vigorously insisted on this important observation. "This proposition in its positive expression A = A," he says in the last part of his *Science of Logic*, "is, in the first instance, nothing more than the expression of an empty tautology. It has therefore been rightly remarked that this law of thought has no content and leads no further." But "truth is complete only in the unity of identity with difference," and the principle of identity is actually "not merely of analytic but of synthetic nature." It contains "more than is meant by [it], to wit, ... absolute difference itself" (*Wiss. der Logik*, 4:33–37 [Miller 413–416]). The *Logic* of the *Encyclopedia* defines this thought in greater detail. "The propositional form [A = A] itself contradicts it: for a proposition always promises a distinction between subject and predicate; while the present one does not fulfil what its form requires." Hegel deduces from this that identity must not be understood "as abstract Identity, to the exclusion of all Difference," later adding: "No doubt the notion, and the idea too, are identical with themselves: but identical only in so far as they at the same time involve distinction." By formulating an identity, even in its most rudimentary and rigorous form, we at the same time suppose or posit diversity, for otherwise the proposition would serve no purpose and "we could not but own [thinking] to be a most futile and tedious business" (*Enc., Logik*, 6:230–232 [Wallace 213–215]).[6] Thus our mind demands "in the case of difference ... identity, and in the case of identity ... difference" (*Enc., Logik*, 6:237 [Wallace 219]) and "the elements distinguished are ... at the same time declared to be identical with one another and with the whole" (6:317 [289]), so that, for example, cause and effect, "if they are distinct, are also identical" (6:305 [278]). There is obviously a contradiction there, but it is a *necessary* contradiction: "The Other, the negative, contradiction, disunity, therefore also belongs to the nature of mind." This is why "Ordinary logic is ... in error in supposing that mind completely excludes contradiction from itself. On the contrary, all consciousness contains a unity and a dividedness, hence a contradiction" (*Phil. des Geistes*, 7_2:25–26 [Miller 15–16]).

However, Hegel – as all those who have even a superficial knowledge of his thought will have guessed – did not arrive at this observation through an analysis of mathematical deduction. Indeed, one of the most prominent features of the physiognomy of this powerful thinker, whose singularities are so strongly pronounced, is how little his thinking was influenced by mathematics (although he knew it well and even seems to have studied it with great zeal). This trait strongly distinguishes him from Kant, of whom it has rightly been said that he was a "philosophic Newtonian." Hegel's position, on this point as on so many others, seems nevertheless to be quite closely connected to that of his great predecessor;

but it is connected by the theory of the antinomies. For Kant, as we know, these antinomies are four in number; they are, moreover, notions concerning cosmological conceptions. Hegel, while he greatly appreciates this theory of the antinomies, which he considers to constitute "one of the most important steps in the progress of Modern Philosophy," finds it insufficient and extends its limits enormously. "The Antinomies," he says, "are not confined to the four special objects taken from Cosmology: they appear in *all* objects of every kind, in all conceptions, notions and Ideas. To be aware of this and to know objects in this property of theirs, makes a vital part in a philosophical theory" (*Enc.*, *Logik*, 6:102–103 [Wallace 98–99).[7]

It is clear that since Hegel's thought is entirely general it must be applicable to mathematical concepts as well. Such an application, like any introduction of mathematics into reasoning, offers the great advantage of making thought more precise. But we must immediately warn the reader that in so interpreting Hegel's position, we are straying somewhat from his own interpretation. In fact, not only did Hegel not start with mathematical considerations, but he explicitly stated that his observations on the process of reasoning, such as we have just summarized them, did not generally apply to mathematical reasoning.[8] We shall return to this subject a little later, at which time we shall see what was at the source of this anomaly in the thought of the philosopher. For the moment, and in order better to grasp the development of that thought, let us not hesitate to return, somewhat in spite of Hegel himself, to the mathematical arguments we had advanced.

If we had to demonstrate the identity $(a + b)(a - b) = a^2 - b^2$, it is because the two terms of the equation had not initially seemed identical to us. Similarly the points C and C' did not appear to coincide until after the demonstration, and finally, to realize that the line AB was equal to itself, we had to pass through the *in spite of*, which, as we saw, accompanies any application of the principle of identity. Thus the two terms to be identified, at a given moment, seemed to us to be distinguished from one another by some feature or another – to be *diverse* – without which, quite simply, we would have had only a single term and would not have taken the trouble to formulate our proposition, which would then have been only a veritable tautology devoid of interest. Now, since these terms were certainly only concepts, things of thought (the figure serving only to sustain the faltering imagination), they were consequently not identical at that moment. How then could they become identical? Solely because we decided to disregard what we had initially seen in them as different. But that did not make the element of diversity disappear; it still exists and in reality renders the proposition contradictory.

Thus the contradiction underlying the simplest geometric proposition is inseparable from my thought. Indeed, I can reason only by applying the concept of identity, by identifying things, concepts that at first seemed diverse. I am therefore, always and everywhere, condemned to affirm *simultaneously*, as Hegel rightly said, the identity and the diversity of these things and these concepts. If I wish to reason, if I mean for my thought to progress, I have no other choice than – while remaining aware of the diversity and keeping this notion clearly in mind (it is the process by which this inherent contradiction is exposed that Hegel, invoking the name of Plato, calls *dialectic*[9]) – to *go beyond* it.[10]

What we have just come to see with respect to the proposition customarily called "the principle of identity" makes us understand how it can play the role of a synthetic proposition, can serve to further our thought. It does not express the identity already preexisting in our mind, already recognized, which would be otiose: it makes us recognize it, it introduces it where we were not aware of its existence. And it accomplishes this by making us see that, out of the thousand characteristics we could observe concerning a given concept, only a small number actually mattered from the standpoint of the reasoning in which we were engaged, the others being completely irrelevant and consequently able to be eliminated, ignored, treated as if they did not exist.

The reader will have understood that we are dealing here with an altogether fundamental point in the theory of reasoning. We are therefore allowing ourself to dwell on it at more length and to cite in particular an example of somewhat less rudimentary geometric reasoning. We choose that offered by the proof of the Pythagorean theorem concerning the square of the hypotenuse, and we beg the reader's indulgence for confronting him once again with this ancient figure which reminds us all of our first steps in the domain of science. It is the most venerable of figures, not only because of the role the theorem and its demonstration played in the development of Greek geometry, mother of all the sciences which constitute the glory of the human mind, but also and above all precisely by the fact that for more than two thousand years all intelligences that have opened themselves to the scientific understanding of the world have used it to begin their climb. It thus reminds us that no matter how immense may be the knowledge acquired by civilization, accumulated through the genius of great men, through the labor of countless generations, we can nevertheless really profit from it only if we make ourselves worthy by our individual labor, in obeying these words of Goethe: "Whatever you inherit from your late / Forebears, see that it is possessed" (*Faust*, lines 682–683 [Charles E. Passage trans.]).

Let us recall, then, what I am sure few of our readers have forgotten, that after having drawn the right triangle as well as the three squares, one draws the perpendicular *BKJ* and then two auxiliary straight lines *DC* and *BH*. One then begins by demonstrating that the two triangles *ADC* and *ABH* are equal. These two figures, as the demonstration establishes, are actually *identical*, the sole difference being the positions they occupy in the total figure. From the equality of the two triangles one proceeds to the equality of the square *ABED* and the rectangle *AKJH*. Here the two figures are manifestly dissimilar.

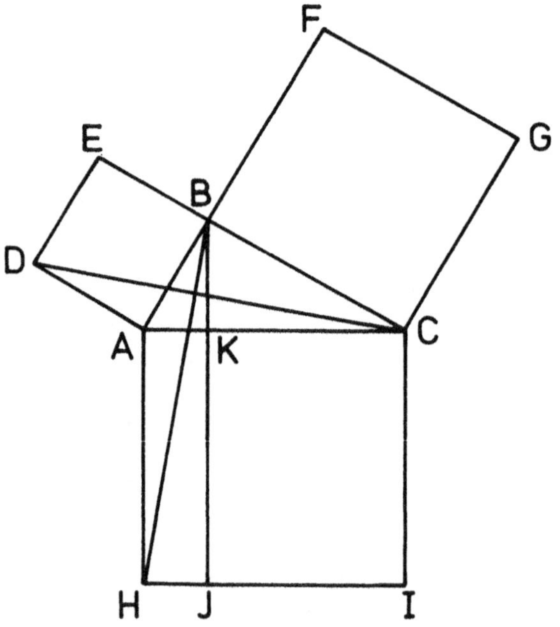

But one shows that they nevertheless have the same area. This is done by establishing that in each case the figure has an area twice that of a given triangle, and that the two triangles involved are precisely the ones whose identity was previously established. The operation is repeated on the other side, and we are finally convinced, by glancing at the figure, that the two rectangles put side by side indeed form the square constructed on the hypotenuse.

Thus one has actually proceeded from equality to equality, each time setting aside – consciously ignoring, declaring irrelevant to the argument – the diversity that stood in the way of the identification. Was it irrelevant from the standpoint of our attention, that is, irrelevant to the conception as our mind first formed it? Certainly not, for in that case the work of the demonstration would have been for naught, and we all remember that this was not the case, that, on the contrary, the demonstration required a quite significant stretching of our intelligence. At the very point where perfect identity is established, in the equality between the two triangles, it is not all smooth sailing, if we may be permitted this expression. Of course, we are all fully convinced in advance, even prior to any geometry, that displacement in space can in no way affect identity, that position is perfectly irrelevant to identity. However, although the two triangles are in reality only one and the same triangle which has simply been rotated 90° around point *A*, such is the disparity of this position, so little analogy is there between the two triangles from the standpoint of their role in the figure as a whole, that as I write these lines some fifty years later, I still recall perfectly my astonishment at that demonstration of identity, and the difficulty I had at first in finding the straight lines to draw, a difficulty that was obviously only an expression of the *unexpected*, and thus surprising, nature of the figures. Many readers will probably have similar memories.

Between the squares and the rectangles there can no longer be any question of true identity, and our equation of one with the other must obviously be based on the understanding that shape is irrelevant insofar as area is concerned. Moreover, here again, we had to be *shown* the relation between the triangle and the respective rectangular figure. This is because the situation is not altogether the same in the two cases; otherwise the two rectangular figures would be alike. The trick consists precisely in fitting the same triangle, in an analogous way, in one case to a rectangle and in the other to a square, because each time one uses a different side of the triangle as a base, which of course is why the two rectangular figures must be at the same time different in shape and equal in area. And if we finally reached the point where one glance at the figure was enough to show that the two rectangles added to one another formed the square constructed on the hypotenuse, we still needed this glance or the memory of the way in which we had divided the square into these rectangles before beginning the proof. It is thus that we came to write $AC^2 = AB^2 + BC^2$, to relate, to connect by an equal-sign what had initially seemed absolutely different to us, namely, two squares on one side and a single one on the other. But of course this equal-sign includes a restriction: it refers only to areas. From all other points of view, and in particular those concerning shape and position, the differences subsist.

The role of the concept of identity is altogether analogous in algebraic reasoning. Taking up again our earlier example, we discover that in order to prove that

$$(a + b)(a - b) = a^2 - b^2$$

we first use a theorem on the multiplication of algebraic sums, writing:

$$(a + b)(a - b) = a^2 + ab - ab - b^2.$$

It goes without saying that in writing this equation we do not mean to claim that the terms on the two sides of the equal-sign are identical in all respects, which they obviously are not, since on the left there is a product and on the right a sum of four products. What is being affirmed is (as we have seen) that there is identity if we adopt certain conventions, for example that the result will be identical if the letters *a* and *b* are replaced by any numbers at all. In other words, here again we declare that the quite apparent divergences between the two terms have no effect from the relevant point of view and can thus be entirely ignored. We then see that in the algebraic sum of four terms one and the same term *ab* appears first as positive and then as negative. Now it has previously been established that a number is not modified by the fact that another is added to it and subtracted from it at the same time. Thus *+ab* and *-ab cancel*, that is to say that, always assuming the same conventions, it is immaterial whether one writes them or omits them.

These are obviously very elementary proofs, and more than one reader may have smiled at our earnest efforts to expose their inner workings. But it is clear that *all* mathematical demonstrations are of this type. Indeed, all mathematical deduction is composed of a series or (to use an expression that Henri Poincaré applied to his schema of reasoning by recurrence) a *cascade* of equations,[11] and each time we write an equal-sign it goes without saying that what is placed to the right and to the left of it cannot really be identical, that it cannot be the actual repetition of the same formula – for then reason would not progress – but that what we have is an identity that will be recognized if we consider things from a particular point of view. The whole art of mathematical reasoning consists precisely in choosing what can be ignored, eliminated – either in virtue of unchanging mathematical conventions, as for example the convention recognizing the invariability of a geometric figure transported in space, or in virtue of previously demonstrated *identities* – so that what remains will provide us with a proposition bringing us closer to what we are trying to demonstrate. The larger the step we take, that is, the more unexpected the proposition at which we finally arrive, the more felicitous, the more *elegant* the reasoning will be. But we must immediately add the proviso that the rhythm of this progress must to some extent be adapted to the intelligence, if not of the common man, at least of the common mathematician. If, on the other hand, the great *inventor* or *discoverer*, in whose mind relations are revealed directly, so to speak, fails to break down his giant steps sufficiently to make it easier for those that follow him to attain his heights, he runs the risk of remaining misunderstood for a long time or even of being called an "abrupt genius" when people finally recognize his merit, as happened for that unfortunate child who was one of the most authentic mathematical geniuses of the nineteenth century: Évariste Galois. Hegel, who, as we shall see in a later chapter (pp. 281 ff.), professed rather strange ideas concerning mathematics, strongly protested against the fact that in the course of a geometric proof one sometimes proceeds to a construction that at first does not appear to be *necessary* – that is, to be indicated by the very content of the theorem – or, as he formulates it,

> does not arise from the nature of the theorem; ... the injunction to draw just these lines, an infinite number of others being equally possible, is blindly acquiesced in .... Later on this design then comes out too, and is therefore merely external in character.[12]

Hegel, toward whom nature seems to have been rather parsimonious in bestowing the gift for mathematics, yet who had made a considerable effort (no doubt a bit belatedly) to assimilate this learning, must have experienced, when confronted with rather elementary reasoning, the same feeling of "abruptness" that (if we may compare very small things to very great ones) the mathematician feels in the presence of the reasoning of Galois. Furthermore, we noticed the same thing above with respect to the auxiliary lines in the Pythagorean proof. The proof is so surprising that it seems a little artificial: at first the student has the impression of witnessing a sort of feat of prestidigitation that is marvelous but rather hard to bring off successfully. In this sense, therefore, Hegel's feeling is well-founded, and there is no better way to confirm it than by observing that Leibniz, certainly one of the greatest mathematicians of all time, expressed a similar opinion, accusing geometricians of "extorting the reader's assent" and "constraining the mind rather than enlightening it."[13]

One could to some extent diminish this feeling of surprise by proceeding more analytically, for example by making clear from the outset of the demonstration where one is going, and never losing sight of the fact that one's goal is to establish a relation between the different areas, showing how one can search for various figures that might be appropriate for this purpose. Although that would make the demonstration more "psychological," it would certainly be to the detriment of that synthetic elegance which makes the Pythagorean theorem a precious monument to the ingenious Hellenic mind. One might also consider that it would be unprofitable to try to encumber it in this way and would be better to substitute a different procedure, as recent texts sometimes do. But it must be understood that, no matter what one does, one will not completely avoid the element of surprise, the impression of artifice, in mathematical deductions. Indeed, this surprise is only the expression of the fact that, proceeding from equality to equality, one has nevertheless managed to bring together, to connect with an equal-sign concepts that were previously quite far apart. In other words, it is the expression of what this deduction contains that is new, not present in our consciousness. Now this new, this previously unconscious element, is exactly what we are seeking; it is in order to have it revealed to us that we do mathematics, and mathematics is valuable only insofar as it does reveal this element to us, or, to use Kant's terminology, insofar as it reveals itself to be truly synthetic. Poincaré has said that in introducing this concept of *synthetic a priori judgments* Kant only baptized the difficulty offered here by the explanation of mathematical reasoning, instead of resolving it (*La Science et l'hypothèse* 10 [Halsted 5]). That is largely correct, but Kant should nevertheless be given credit for stating the givens of the problem more precisely. The fact that these propositions are *a priori* shows that they must have previously been part of the content of our intellect in one way or another, since we were in fact able to call them forth without having recourse to external observations, to actual experiences. It is clear, however, that they were present only implicitly: the student to whom we are going to explain the multiplication of $(a + b)$ by $(a - b)$ knows very well what a sum, a difference, a multiplication or a square is, yet he still does not know that by multiplying the sum of two numbers by their difference he will obtain the difference between the squares of the two numbers. Likewise, the student for whom we are going to prove the Pythagorean theorem knows the definition of a right triangle perfectly well, but is not aware that this proposition is deduced from it: if one asked him to construct a right triangle such that the sum of the sides exceeded the square of the hypotenuse by one-tenth, he would not find the statement of the problem at all absurd and would no doubt make numerous attempts and measurements before convincing himself that the task was impossible. To require that the revelation produced by mathematical deduction "arise from the nature of the theorem," as Hegel would have it, is to require, given the fundamental simplicity of mathematical concepts, that there be no revelation at all. Hegel himself realized this to some extent, for in his system he was careful to put mathematics in a separate category, to attribute to it a particular kind of reasoning entirely distinct from that undertaken in the other branches of knowledge. In short, in spite of many incursions into this domain, he did not attempt to reform mathematics as he did the physical sciences; everything leads us to believe, on the contrary, that when he taught mathematics (as he did on several occasions), he did so more or less in the traditional manner, without too much admixture of deduction by concept (cf. Ch. 11, p. 283, below).

To sum up what we have just said about the way our intellect applies the concept of the identical, let us say that the most important thing to recognize is the active role of this same intellect. For of course our reason could not act as it does if something corresponding to this notion of identity did not preexist in the concepts it manipulates in this way; the situation therefore is altogether analogous to the one we observed in the preceding chapter (p. 81) concerning the existence of *genus*. But since the rational element does preexist, our reason intentionally – although the intention often remains unconscious – increases the importance of this rational element to the point of causing the whole of the concept in question to be resorbed into it, since reason affirms that this concept is identical to another even though it knows the two are distinct. Reason thus exercises a veritable constraint on concepts, it imposes identity on them or makes them *assume* it – if we may grant the word the meaning it has in the expression *assumed name*.

It is certain that the way in which what we conventionally call the principle of identity is stated, and even the name with which the principle is embellished, do not entirely correspond to its true role, which explains the erroneous assertion that it is only a tautology. For this reason, Stanley Jevons, in his logical system, felt compelled to formulate, alongside the principle of identity, a different proposition that he calls the principle of the *substitution of similars*, which he considers to be applicable in much the same way we have explained the role of the concept of identity in mathematical demonstration.[14]

But as a matter of fact, as Hegel taught us, the principle of identity as tautology does not and cannot play any role in our reasoning process; even when we say we are dealing with identicals, we are *always* dealing only with similars, which we make identical by momentarily setting aside what diversifies them. And since, moreover, Jevons's principle clearly depends on the concept of identity, it might be preferable to choose a name that would emphasize both the active role of the proposition and the way in which it is connected with the tautological statement of the principle of identity, whose role it must actually assume. We propose the term *schema* or *process of identification*, with the understanding that the final three syllables of the word *identification* have their full meaning, as they do for example in the word *simplification*. Identification here will thus mean not only the act by which we recognize the identical where it exists, but also the act by which we reduce to identity what at first appeared to us not to be identical.

Let us note, moreover, that if all mathematical demonstration necessitates continuous application of the process of identification, the concept that constitutes the starting point for such a demonstration also could only have been formed through the operation of the same process. Indeed, what we intended to demonstrate by the Pythagorean theorem is not something applicable to a particular right triangle; it is a proposition concerning *any* right triangle in general. Logicians have devoted much study to the processes by which our reason comes to form these general concepts. But here we need only note that to form the idea of a right triangle in general, we must set aside what characterizes this or that particular figure, as for example its dimensions or the size of its angles. We declare these particulars to be negligible from the standpoint of the operation at hand. Thus, here again, we have essentially reduced something diverse to something identical. And the same is true for any mathematical demonstration, because all of them necessarily bring general concepts into play.

If we now recall what we came to see (Ch. 1, pp. 13 ff.]) concerning the concepts underlying the laws of physics, it becomes evident that these concepts in their turn are necessarily the product of an analogous operation. As we saw, a proposition dealing with sulfur presupposes the framing of the concept of the *genus* sulfur, an operation that can be accomplished only by paying attention solely to what the different samples found commercially have in common, disregarding what makes them different. Consequently, there too we proceeded by identification.

However, it is important to note that, although the mental process is essentially the same in all three cases, it is not applied under the same conditions. What we are doing in each case is, to be sure, making identity out of diversity in order to reason, but the stage of the reasoning process where this operation occurs is not the same. When we form the concept of a genus, either in mathematics or in the physical or natural sciences, it is because we have been struck by a resemblance. It is this resemblance that makes us conceive the abstract entity we call a *triangle*, *sulfur* or *man*, and, in the same way, the resemblance between certain phenomena gives rise to the concept of *electrical phenomenon*. At the moment when we form these concepts, we undoubtedly know we are going to reason about them, but we do not yet know what lines the reasoning will take. In other words, here the process precedes reasoning properly speaking. Therefore, despite the fact that the identification was made for the sake of a process of reasoning, it does not seem any less spontaneous or natural for that.

The same is not true for mathematical deduction or demonstration. Granted, at each particular step in the demonstration – for example at the moment when we have grasped what two figures have in common – by the very fact of stating this, by connecting these two concepts with an equal-sign, we create a *genus*, and thus it could seem that we have proceeded entirely according to the path followed by reason in the operation just discussed. But this would be to lose sight of the fact that what we have here is only a step in the demonstration: obviously the demonstration, considered as a whole, does not admit such an interpretation. The student before whose eyes one constructs the figure of the right triangle surrounded by the three squares certainly notices no relation of similarity between the sum of the areas of two of these squares and the area of the third. For him this relation initially results simply from a decree: it is the statement of the theorem, and the very fact that this statement is followed by a demonstration proves that what is involved is not something that can be immediately perceived, but something that must be established, that it is necessary to look for similarities, identities, in order to arrive at the desired identity by linking them together. Thus here it is the process of reasoning that must bring us to a recognition of the similarity, of the possibility of applying the schema of identification, since the identity does not force itself on our attention all by itself. So little does it force itself upon us that, even in the case of the intermediate steps, the student at first has difficulty discovering the geometric figures whose areas he must connect with the equal-sign, and even after these figures have been discovered, he is slow to recognize that they are equal.

It is there obviously – it is in that particular application of the schema by which it truly forces our intellect to accept the identity – that we must seek the reason why the mind sometimes resists demonstration in its mathematical form. Hegel, precisely because he was not particularly gifted in mathematics, is not a bad judge on this point. What he finds blameworthy, as we saw, is the fact that demonstration does not arise from the nature of the theorem and that the constructions to which we have recourse do not at first seem necessary, do not seem to follow from the concept itself. What does this mean, and what would need to be done to satisfy him? We should simply confine ourselves to proceeding solely by means of similarities suggested by the concept and the figure. It is the process our reason follows in framing the concept of genus, and there is no doubt that this process appears, in its very essence, more spontaneous, more natural. But one also notices how foreign it is to the true spirit of mathematical demonstration, which does of course progress by means of the same process of identification, but on the condition of directing it, not allowing it to follow the natural bent of the mind, but leading it to choose and to find what will be able to take it closer to the predetermined goal.

Does the process of identification, which is so essential from the standpoint of mathematical reasoning, as we have just seen, also have an application in the reasoning of the physical sciences beyond what is necessary for the formation of the concept of genus itself? To answer this question we have only to recall what we earlier came to see concerning deduction and rationality. We saw in fact that physics intends to connect the antecedent and the consequent by a rational link, by demonstrating that the consequent is the *necessary* consequence of the antecedent. Now, as Leibniz tells us, necessary truths must be reducible to identical truths.

Therefore we must be able to demonstrate that the consequent is identical to the antecedent – except, of course, for a few specific details which differentiate them, but which we shall declare negligible (as we did for mathematical demonstration and the framing of the concept of genus). In other words, if a phenomenon is to seem to be explained, to be rational, there must be equality between cause and effect.

This is what Leibniz clearly saw and formulated as lucidly as one could wish. He declares that "the integral effect can reproduce the entire cause or its counterpart" and that "the whole effect is always equivalent to its full cause."[15] He finds this proposition so obvious that he uses it for a *reductio ad absurdum*: "It would follow," he writes, in an attempt to demonstrate that a proposition is erroneous, "that the cause could not be entirely reestablished, nor substituted for its effect, which, as one easily grasps, is entirely contrary to the habits of nature and to the reasons of things" (*Mathematische Schriften* 6:206).

The unique importance of the role played by the concept of identity in causal reasoning did not escape Riehl's attention. Indeed his work bears the title *Causality and Identity*, and the passage we cited (p. 54) is immediately followed by the following considerations:

> By that determination [of causality insofar as it establishes a logical connection between events separated in time] one finds, at the same time, the transition between the subjective meaning of causality and its objective meaning. The transition is accomplished by means of the principle of identity, which governs our conclusions in general and is consequently valid for causal conclusions as well .... The prior condition for a causal conclusion is an equation between the antecedent and the consequent .... Exactly in the same measure that they are identical, there is the possibility of connecting them by a conclusion, and consequently it is also in the same measure that one can establish a causal connection. We explain a change if and to the extent that we succeed in reducing it to an immutable being or to an identical sequence (for example to a uniform rectilinear motion). For this reason we can also define causality as the application of the principle of identity to time or, more precisely, to the succession of events.[16]

Thus the schema of identification applied to the temporal phenomenon takes a particularly regular form, in the sense that the implicit "although" is always the same: things have remained what they were, in spite of the fact that time has elapsed between the two observations. They have persisted in time.

However, they certainly seem to us to have changed; otherwise we would not have gone to the trouble of seeking an explanation. What then could have caused the diversity? What circumstance – an unimportant one, it goes without saying, this being the postulate implied by identification – could have been modified?

The objects of the external world, forming the whole of our perceptions, are subject to only two sorts of entirely general conditions, namely, conditions of time and of space. The causal postulate consists in denying, in eliminating the influence of time. All we have left, therefore, is space. Thus, what may have changed is the arrangement in space, and the most perfect explanation will consist in showing that what existed before has subsisted after, that nothing has been created and nothing has been lost, that as the result of the phenomenon no change has occurred – except insofar as spatial configuration is concerned. The most perfect explanation of a change can only be its reduction to a spatial function.

We now understand better why the explanation of the appearance of the leaves on which Bossuet's image is based is so satisfying for the mind. And we also see how it comes about that, as we pointed out at the beginning of this chapter, the two meanings of the verb *to explicate* come together and almost merge in this image. It is because that is truly a model explanation: ultimately all explanations must conform to this type.

That this is actually the case in the physical sciences, we have shown by numerous examples in a previous work (IR, Chs. 2–5; cf. also Chs. 7 and 17 below). The whole of mechanical theory is obviously simply a system designed to reduce reality to a collection of unmodifiable parts producing all change by their displacement alone. That is clearly seen from the very origin of this conception in the ancient world, with Democritus and Lucretius, and in spite of the enormous mass of scientific knowledge introduced between that time and our own (we can say without exaggeration that whatever the intellectual and artistic development of the Greeks in the age of Democritus may have been, their real physical knowledge did not go beyond that of quite primitive peoples), modern scientists think absolutely like the ancients insofar as these principles are concerned: they would like to be able to explain the totality of phenomena by figurative constructions in space and – as the debates at the Council of Brussels demonstrate – consider it to be a failure, an impediment to the development of science, when things cannot, by some sort of artifice, be so arranged. Likewise the principles of conservation derive their authority primarily from the fact that they tend to favor the idea that certain concepts (considered for that very reason as somehow assuming particular importance, dignity), such as velocity, mass and energy, persist through all change, are conserved and merely change place. We also showed that the explanatory role of displacement is not limited to the mechanistic or atomistic form of science, which is its present form (as it was also that of at least some of the ancient physicists), but is generally applicable to any physics that follows a truly scientific path; in particular the Peripatetic science of the Middle Ages depended on it, insofar as it escaped its properly logical framework and attempted to constitute a qualitative physics and chemistry. That is clearly revealed in the role played by the "substantial qualities" as they move about (IR, Ch. 10, pp. 366–396 [Loewenberg 323–346]); in the present work we have encountered a rather striking example of this in phlogiston theory and in the way it explained a whole series of chemical phenomena by the displacement of the *combustibility virtue* (see pp. 62 ff. above).

Thus, it is important to note, all true scientific explanation of a temporal phenomenon fundamentally rests on the permanence of something, of any concept whatever. Sometimes, in the case of very general concepts, science has felt obliged to formulate this permanence explicitly: these then are the principles of conservation properly speaking. At other times the permanence is stated more or less implicitly, or even taken entirely as understood, but in such cases a little effort suffices to pick it out. Thus chemistry before Lavoisier never explicitly declared that the quality of combustibility was to be considered indestructible, but it is clear that in positing the existence of phlogiston, which was only a hypostasis of this quality, and in treating this phlogiston as a substance that passed from one compound to another without ever disappearing, it in fact implied the notion of conservation. Finally, it also happens that the theory only incompletely defines what it is that actually must be conserved, and here it is contemporary science that offers us a typical example, by its formula of the conservation of matter. The principle is generally stated in such a way that it refers only to the constancy of weight in a chemical reaction. Now it is easy to see that this is not the only thing that contemporary chemical theory stipulates must be conserved. As a matter of fact, the whole of this vast construction implies in almost every detail – in the theorems of physical or general chemistry as well as in the formulas of chemistry proper – that elements are permanent, at least in ordinary chemical reactions (that is, excluding, for example, radioactive transformations): sulfur must remain sulfur; hydrogen, hydrogen. Undoubtedly we would be at a loss to identify clearly what it is in the element sulfur that is conserved when it enters into a compound: that is obviously why we do not state this part of the principle. But we certainly assume that something *must* be conserved: it is for this reason we call sulfur an element and sulfur dioxide a compound and write the formula of this gas as SO_2.

Our statement above concerning the relation between principles of conservation and explanatory theories can be turned around: any statement of conservation tends to give rise to an explanatory theory. That is why when confronted with anything which is said to be conserved and which is at first, of course, only a scientific abstraction, such as heat or combustibility, we feel a sort of irresistible need to hypostasize it ontologically, to transform it into a *being*.

That is not a new discovery for us, but only a somewhat different and, especially, a somewhat more precise form of the tendency we discussed in Chapter 3 (pp. 59 ff.), a tendency that impels our understanding to create fictitious entities – if these entities seem to be able to serve in the explanation of phenomena – and against which Ockham's apothegm is a useful but too often insufficient barrier. Examples of hypostases of this sort abound. For example, heat, because it can be claimed to be conserved, becomes *caloric*, and combustibility for analogous reasons becomes *phlogiston*. These are examples from the past, but it is clear that modern science is animated by the same spirit. The Newtonian simplification of Kepler's formulas, by showing that they could be deduced from the postulation of a single mathematical expression remaining invariable in time, immediately gives rise to the concept of *force*, this force being without question, given the way it is presented in the textbooks, an ontological entity; and even energy, in spite of the patent fact that it is only an integral and that it is utterly impossible (whatever the texts may say) to furnish a verbal definition of it (cf. IR 317 ff. [Loewenberg 280 ff.]), manifestly tends to be transformed into such an entity. That tendency, of course, reveals itself less in true science than in the somewhat murky domain that borders it; but however little importance one may wish to attribute to the philosophic energetics of Ostwald, the birth of a conception in which energy appears to be a veritable *thing-in-itself* and even the only true thing-in-itself (cf. IR 396 ff. [Loewenberg 346]) is nonetheless quite typical from this point of view.

It is obviously also the close kinship between explanatory theories and principles of conservation, the explanatory character of the latter – that is to say the fact that they seem to us to conform to the exigencies of our reason, to constitute a step on the way toward the rationalization of reality which is the true goal of all science – which makes them seem to assume a particular dignity much greater than that of simple empirical laws: they enjoy, as it were, the double authority of laws and theories, since the agreement of reason and nature seems to be revealed in them. That is why we are satisfied with insufficient proof in cases of this kind. And what is much more, we have the tendency, as soon as conservation is involved, to go on to observations which seem to contradict the proposition and which we consequently strive to explain by means of auxiliary hypotheses or similar devices of all kinds. What happened with regard to Black's caloric and the efforts made to "explain away" the very obvious objections arising from the production of heat by friction is a case in point. We spoke of it on the subject of theories (p. 61); but that is precisely because, as we have just seen, in the domain of conservation, theory and law come together and merge to a certain extent. In that particular case, it is evident that by defending the theory of caloric, one was at the same time maintaining the underlying affirmation of the conservation of heat. Furthermore, science today is undoubtedly still of the same mind. A palpable proof was furnished quite recently by the discussions that followed the discovery of the phenomena of radioactivity. Indeed, it is obvious that in their immediate sense the observed phenomena appear to contradict the principle of the conservation of energy, because we see an energy appear without being able to detect the disappearance of any other. Now, all those who speculated about these phenomena clearly started from the implicit postulate – which they did not even feel the need to state in the immense majority of cases, it seemed so much a *matter of course*, so *natural* – that the energy whose appearance was observed *could not* be created *ex nihilo*, that it could only be the transformation of an energy that existed previously even though we were not able to perceive it, this in spite of the fact that the principle of the conservation of energy can in no way be considered unassailable from the standpoint of the demonstrations on which it is based, nor even to have a very firm experimental basis.

Outside of the physical sciences proper, in the domain of biology, a striking example is offered by the theories of the preformation or encasement of germs, which for a long time enjoyed considerable favor. According to these conceptions, every organism, with all its characteristic traits, was to be found enclosed, *preformed*, in its germ, which was presumed already to contain, simply reduced in size but otherwise complete, the germs of all the beings that this primitive organism, its descendants and the descendants of its descendants would bring forth in the future, no matter how distant. "The researches of the moderns has taught us, and it is approved by reason," says Leibniz,

that the living things whose organs we know, that is to say plants and animals, do not come from putrefaction or chaos as the ancients believed, but from *pre-formed* seeds, and consequently from the transformation of pre-existing living things. There are little animals in the seeds of the large ones, which by means of conception assume a new vesture, which they appropriate, and which enables them to be nourished and to grow, so as to pass on to a wider stage, and propagate the large animal .... And what has just been said of large animals occurs also in the generation and death of these spermatic animals themselves; that is to say, they have grown from other smaller spermatic animals, in comparison with which they can be reckoned large; for everything in nature proceeds *ad infinitum*. Thus not only souls but animals also are ingenerable and imperishable: they are only developed, enveloped, re-clad, stripped, transformed .... (*Opera* 715–716 [*Principles of Nature and Grace*, Sect. 6, Parkinson 198–199])

These lines date from 1714, but already, nearly twenty years earlier, Leibniz had written:

... the transformations of MM. Swammerdam, Malpighi, and Leeuwenhoek, who are among the best observers of our day, have come to my assistance and have made me admit more readily that the animal and every other organized substance does not begin when we think, but that its apparent generation is only a development, a kind of increase. (*Opera* 125 [*New System, and Explanation of the New System*, Parkinson 118])

As we see, Leibniz appeals to results obtained by the biologists of his time. Indeed, when he wrote these lines, "preformationist" conceptions were already widespread among them. As early as the beginning of the seventeenth century, Harvey had pointed out the primordial importance of the egg in the generation of animals; it may well be that the famous formula generally attributed to him – *omne vivum ex ovo* – is not to be found in his writings, but he certainly said things very close to it. And if he did not absolutely exclude all possibility of spontaneous generation (which, like the other form of generation, would be presumed simply to pass through the egg[17]), it is nonetheless true that what he did say was preparing the way for such a conclusion, which biologists were quick to formulate.[18] Swammerdam, however, seems to be the actual author of the theory, as Leibniz suggests. Soon afterward, the illustrious Malpighi, who of course discovered the capillaries and thus confirmed an important aspect of Michael Servetus's and Harvey's work on the circulation of the blood, believes preformation can be confirmed by precise observations of the egg. In his work *On the Formation of the Young Animal in the Egg*, he entitles one chapter: "The tissues of the young animal preexist in the egg and have a more remote origin, just as takes place in the eggs [seeds] of plants."[19] Leeuwenhoek's microscopic discoveries, in particular those of blood globules and spermatozoids, especially capture the imagination of his contemporaries: one school, that of the "spermists" claims henceforth that the animal is preformed in the sperm, while the "ovists" follow the teachings of Harvey and locate the preformation in the egg.

But it is undoubtedly Bonnet and Albrecht von Haller who gave the theory its most absolute expression. Bonnet states not only that it "is demonstrated that the chicken exists in the egg prior to fertilization," but that there "is no true generation in nature"; we "improperly give the name generation to the beginning of a development which makes visible to us what we could not previously see." And he concludes: "All the parts of the universe are thus contemporary. The Efficient Will realized in a single act all that could be realized" (Driesch 47 [Ogden 50–51]). This is what Albrecht von Haller translates into the concise formula which reveals so well the fundamental tendencies of the system: "There is no becoming."[20] Bonnet and Haller are ovists, but the spermists are no less robust in their faith in preformation. Some go so far as to imagine spermatozoids with human heads (Rádl 181).

Certainly biologists today would take care not to fall into such excesses. But they are moved by tendencies that come very close to those of their ancestors, as attentive observers have discerned. Le Dantec,[21] F. Houssay[22] and Appuhn,[23] whatever differences they may have, are in perfect agreement on this point, and Prenant, after having reviewed the modern theories of heredity and the more or less figurative elements they take to persist – such as Darwin's gemmules, De Vries's pangenes, Spencer's physiological units, Nägeli's micelles, Weismann's biophores, etc. – observes that "here we have almost returned to the time of the quarrels between ovists and spermists."[24] Nor does Jacques Loeb, the illustrious American biologist, fear the idea of preformation. In treating the conceptions of the "new embryology," the author of the discovery of artificial fertilization uses the term on several occasions and shows what, according to our present-day ideas, could be found *preformed* in the egg and the ovule (*La Dynamique* 349–350 [cf. *Dynamics* 194–197]). Another American biologist, Bateson, in commenting on the assumptions underlying the Mendelian conceptions, which, as we know, have many adherents among modern biologists, states that the great difficulty encountered by these theories is in the origin of dominant traits. There are some who manage to conceive that these traits have existed since the dawn of life, which constitutes "a truly extraordinary rehabilitation" of the old theories of preformation.[25] Finally E. S. Russell, while admitting that the vulgar theory (in another sentence he calls it "almost grotesque") of preformation has become indefensible as a result of progress in biological research, nevertheless maintains that "the idea of the explanation upon which this theory is based continues to exist today with undiminished force." He adds by way of example that "few theories have had more influence on recent biology than Weismann's theory of heredity and development, which is out and out preformationistic." Moreover, the English biologist is of the opinion that if one hypothetically supposes the substances that transmit heredity in the egg to be similar in nature to the *hormones* of Bayliss and Starling, "the major objection against the preformationistic theory disappears" (*ibid.* [erroneous citation]).

It is perhaps even more striking in the present context to observe what a hold the idea of the preformation of the organism has on the minds of men who are not professional biologists. Here briefly is how Maurice Maeterlinck expresses himself in a recent work, the tenor of which is already sufficiently suggested by the title, *Heredity and Preexistence*:

> We do not know in what way those who will be born of us, down through the generations, already live in us, but it is certain that they do so. Whatever the number of our descendants down through the ages, whatever transformations they are forced to undergo by the elements, the climates, the territories and the centuries, they will keep intact, through all vicissitudes, the life principle they have drawn from us. They did not get it from anywhere else or they would not be what they are. They actually did issue from us, and they could not have done so had they not first been there.[26]

What then does our understanding find so attractive about these conceptions? E. S. Russell is doing no more than elaborating Albrecht von Haller's famous formula when he declares that the fundamental reason preformation pleases us is that "the idea of an active creative organism is repugnant to the intelligence, and that we try by all means in our power to substitute for this some other conception."[27] Indeed, what nature shows us in organisms is beings that are constantly changing. Now all change, as Riemann tells us, forces us to seek a cause, in other words, seems enigmatic to us until a cause has been found. Since, on the other hand, according to this mathematician, everything must remain what it is if nothing else is added to it, it follows that in a truly closed system any change whatsoever is entirely impossible. But in certain respects an organism seems to present this kind of system. It is a clearly delimited "individual" – we have direct knowledge of this through our consciousness of ourselves. No doubt this individual is related to what surrounds him – he takes nourishment, he receives sensations and he reacts – and these relations can certainly serve to explain certain of his peculiarities. E. S. Russell quite rightly pointed out that the Darwinian theory (as opposed on this point to Lamarckism) is nothing but an attempt to seek out the factors transforming the species in the surroundings of the organism rather than in the organism itself. But for the most striking trait of the living being, namely the minute exactitude with which it reproduces its procreator, the surroundings can clearly be of no use to us. Consequently we are greatly perplexed. What is the source of this novelty that the organism continually presents to us, since it cannot come from the outside? The simplest explanation (and, it must be granted, the only valid one, that is, if the phenomenon must allow an actual explanation) is to say: But there is nothing new! Everything that seems new was always there, in the organism itself or in the one from which it descends. It issued from that organism, therefore it was already there, as Maeterlinck says so admirably, and there is hardly need to point out that the famous poet's formula is only the strict application of the assimilation between the temporal and the logical relationships discussed on pages 55 and 90, which is so clearly exemplified by the indiscriminate use of the term *consequence* [*suite*] in either sense. Thus by unceasingly tying the consequent to the antecedent, one comes to see that everything was preformed, that everything preexisted. All creation must be simultaneous, as Bonnet said.

Another indication that here we are indeed dealing with an eternal and unchanging tendency of the human mind is the very terms by which we customarily describe the changes brought about in organisms by time. When we speak of *development* or *evolution* (the second word being equivalent to the first, since *volvere* means *to roll up* [and *evolvere* therefore means *to unroll*]) we are obviously alluding to the very image employed by Bossuet, an image that is clearly preformationistic: these parts were in some way found in the seed, that is, they preexisted; all they did was grow, develop, unfold [explicate themselves].[28]

Thus we can sharpen what we said above on the subject of scientific explanation. The satisfaction we derive from it does indeed have its source in the deductive process; but this deduction (at least insofar as the physical sciences as we know them today are concerned) must be accomplished through displacement, that is, by means of a spatial function. And if we find explanation by displacement the most perfect kind of explanation, it is because it is the only explanation that is real.

Of course that also applies to the ancestors of this science in antiquity, the atomists. Leucippus and Democritus, says Aristotle, know only three differences, which are the cause of all phenomena: form, order and position (Aristotle, *Metaphysics* $985^{b}14$). There is no doubt that the word *form* is used here in the spatial sense; they are therefore purely spatial relations. Descartes takes up this tradition with unparalleled vigor. Of a body undergoing combustion, he says,

> Posit 'fire' in the wood, posit 'heat' in the wood, and make the wood 'burn' as much as you please. If you do not suppose in addition that some of its parts are moved or detached from their neighbors, I cannot imagine that it would undergo any alteration or change.[29]

What does he mean here? The phenomenon of combustion, in its external appearance, was certainly as familiar to him as it is to us, and he knew what modifications it produces in bodies. How can he claim not to be able to imagine it? It is because he means to imagine the real phenomenon, the one behind what is apparent, and this, for him, can only be rational. Now displacement alone constitutes a rational phenomenon, and consequently, if there is no displacement, there must be no real change, which is to say that nothing could have happened.

Has science subsequently strayed from the path pointed out to it by the man who was unquestionably its principal inspiration in modern times? There is ample evidence that seems to prove the contrary: the proposition that phenomena must be explained by matter and motion is accepted as a current principle. Now matter itself – as we saw in examining the starting point for various theories – seems to us to be something mysterious, something for which we seek an explanation. If this were not so, it would be incomprehensible that anyone should have attempted to reduce it to atoms – which are sometimes termed material, but which are certainly anything but matter, given the strange properties that characterize them, such as their indivisibility and their absolute elasticity.[30] It is even less comprehensible that anyone should have wanted to constitute matter by means of Kelvin's rings or Helmholtz's singular points, that is to reduce it to ether, with its contradictory properties, and finally to explain it by electricity, that is to say, by something fundamentally inexplicable. The reason must be that, of the two terms we have just mentioned – matter and motion – the first can offer us no adequate help from the point of view of explanation, given that explanatory force is exclusively lodged in the second.

Furthermore, no one was more explicit on this point than Auguste Comte on those occasions when, as sometimes happened, he lost sight of his doctrine and was content to state the scientific credo of his time. Let us recall the passage we cited above (p. 86) according to which "all natural effects can be understood simply as necessary results, either of the laws of extension, or of the laws of motion." Here obviously there is only space and motion, in the form of a spatial function.

That is clearly only a particular consequence of the conclusion at which we had arrived in Chapter 3 (p. 64) concerning the role of deduction. Since deduction is the effective factor in all explanation, its essential element must therefore be revealed by that through which the consequent is attached to the antecedent and at the same time differentiated from it. Now what changes, we have just seen, can only be spatial.

In the preceding pages we have broken down as far as possible the operations of our reason that accompany the search for a cause. We have attempted to show, through examples selected from the actual science of the present and the past, that there is nothing fictitious about the intermediate stages through which, according to our theory, these operations pass. The reader may have noted that in our examples the most disparate epochs are to some extent intermingled. That demonstrates the fundamental unity of the human mind, which always remains essentially the same, so that science throughout the ages, despite appearances to the contrary, retains a powerful fundamental solidarity. But it also makes us see that the logical order we believe we have established has not been the historical order. That is easily understandable. It is because these operations are of a great simplicity and because they follow one another so rapidly in the individual consciousness itself that they seem to trigger each other immediately, so that we remain unaware of the intermediate steps. Thus the etymological explanation of the term *to explicate*, as we have derived it from Bossuet's statement, leads directly, as we have seen, to the concept of explanation through the identity of the antecedent and the consequent, and the image suggested by this statement is an example of a preformationist hypothesis, a very advanced form of causal hypothesis.

We deemed it advisable to dwell on these observations in order to avert a misunderstanding that could possibly arise from the exposition of our position. In the preceding pages, we have pointed out the importance of the role of mathematical deduction, which dominates contemporary science, and have related how, since Galileo and Descartes, it has come to be substituted for syllogistic deduction, which had been the basis for the science of the Middle Ages. Does that not seem to support the oft expressed assertion that the entire form of science as we know it is due to the influence of mathematics?

The mathematical and the physical sciences have certainly become more and more interdependent in the course of the last few centuries. It is well-known that mechanical considerations made an important contribution to the birth of the infinitesimal calculus and that since that time many advances in mathematics were the direct result of problems that physics posed for the science of those who calculate. But the influence is still more obvious in the opposite direction. Everyone is aware that the actual form given the law in modern physics is that of integral calculus, and there is no doubt that if mathematics today underwent a development anything like the one produced by the creation of infinitesimal calculus, physics in its turn would almost immediately make an immense leap forward.

Thus the assertion we mentioned above is certainly for the most part justified. Is it entirely justified? To be more precise, is it the influence of mathematics which is responsible for the mechanistic and atomistic form of modern science?

Obviously the examination of contemporary science, so saturated with mathematics, can teach us very little on this subject, and we must appeal to the science of the past. Now one cannot read the exposition of a Greek atomistic system, such as that of Democritus by way of Aristotle's refutations, or that of Epicurus in *De rerum natura*, without being struck, on the one hand by how much this science resembles our own (we have already stressed this point in Chapter 4, p. 96), and on the other by the total absence of anything resembling mathematical calculation. Of course, since we possess no writings of Democritus, there is a remote possibility that the mathematical parts had more or less fortuitously disappeared from the résumés that have come down to us. However, it appears highly unlikely that Aristotle, who, as we are well aware, consistently treats the atomists with marked consideration, should have passed over such an important feature in complete silence. Moreover, this lack of mathematics properly speaking in the system becomes all the more significant given that Democritus was principally a mathematician; that is in fact one of the few specific details we know about this great thinker.[31] As to *De rerum natura*, there cannot be the slightest doubt. In this unique masterpiece the system is laid out with incomparable lucidity and attention to detail, and we see clearly that it contains next to no mathematics and that properly mathematical deduction had no appeal for the author.[32] Here, however, we must make one thing clear. To explain the diversity of substances, the atomists stipulated the motion, the displacement of their corpuscles, and also a diversity of shapes (cf. Ch. 8, p. 216 below). In both cases, this amounted to using a spatial function, to which, as we have shown, they had no alternative (cf. p. 118, above). Now the science of space is geometry. If, then, this branch of knowledge is considered to include all observation, even the most primitive, concerning, for example, the properties of spatial figures – such as the fact that the tetrahedron ends in sharp points or that small spheres slide easily against one another – it is certain that *De rerum natura* is overflowing with such "mathematics." If, however, as simple reason seems to dictate, we reserve the term mathematics for an actual body of doctrine created by deduction, and if we acknowledge that these quasi-intuitive observations must date back to an early state of our intelligence, a state where the beginnings of philosophy and of all the sciences appear intermingled, so to speak, and whose expression is found in this common sense world we all perceive when we open our eyes in the morning, the absence of true mathematics in the atomistic systems of antiquity becomes striking.

We are thus forced to conclude that the ancients arrived at mechanism, not because they were impelled by the requirements of mathematical deduction, but directly through consideration of the identity of the antecedent and the consequent.

Aristotle's testimony fully confirms this. According to him, the atomic doctrine of Democritus developed from the doctrine of the Eleatics, and in particular from that of Parmenides, for whom the universe was a sphere, immutable in space and time, imperishable and incapable of undergoing change. It is in order to explain how being, though itself permanent, can engender a world where change, generation and destruction seem to prevail that Leucippus multiplied this being, positing elements "infinite in number and invisible owing to the minuteness of their bulk" [*On Generation and Corruption* 325^{a}30, Harold H. Joachim translation]. That circumstance is extremely important from the standpoint of the theory of science in general, and we shall return to it in our Book Four (Ch. 17, pp. 498 ff.). What we must stress here is that it would be a mistake to call upon mathematics in order to explain the role of mechanistic atomism in science, as did Hannequin, among others (cf. IR 101 ff. [Loewenberg 94]). Certainly, before Leucippus and

Democritus, the Pythagoreans had formulated a sort of mathematical atomism, which presupposed a discontinuous space composed of points, and it is possible that these concepts were not without influence on the atomistic physicists, as Aristotle himself indicates in the passages we have mentioned. But this influence could only have been wholly indirect, for it is evident from the way in which Democritus in particular insists on the existence of the void, necessary for the motion of the atoms, that he supposes this void, that is to say space, to be continuous, geometric, and not discontinuous, arithmetic, as the Pythagoreans do, and on this point Lucretius is entirely of the same opinion (*De rerum nat.* I, 330 ff., 420–421).

That does not, of course, exclude the possibility that, as far back as this primitive stage, inquirers could have more or less vaguely conceived the idea that by resolutely stripping matter of its qualitative attributes and reducing it to spatial quantity (as any true mechanism does), one made possible the use of mathematical deduction and consequently also a rationalization of reality, or at least a partial one. But it was rather (unless one takes into account the profound connection between these two sorts of considerations, which is precisely what we have sought to establish) a kind of coincidence, and true mathematics did not begin to exercise its powerful influence on science until much later than this early epoch.

Thus the essential form of our science appears to us to be shaped above all by the concern to explain what changes by what persists. It is in obedience to this irresistible tendency that we seek the sufficient reason for the phenomenon, and the principle of sufficient reason is therefore only a form of the process of identification. It is, through the identification of the antecedent and the consequent, the application of this process to the succession of phenomena in time.

If we now return to the kind of considerations we have developed in this chapter (pp. 115 ff.) and ask ourself what the particular conditions are under which this schema is applied here, we shall easily see that deduction in the physical sciences resembles that of mathematics in that, just like the latter, it does not follow our mind's natural bent by progressing through similarities that impose themselves on our attention, but instead applies itself to seeking out these similarities. Only it pushes this process to its limit, constraining us to connect, by identifications, things we initially took to be highly dissimilar. For, finally, as different in shape as the space circumscribed by two squares may be from that of a single square, we nevertheless understand, as soon as we have grasped the meaning of the term *area*, that there can be equality in this case. But how can there be equality between the metal sodium and the gas chlorine on the one hand and crystals of salt on the other? And yet we unhesitatingly write the familiar chemical equation. Likewise, what similarity is there, for the immediate understanding, between a mass of calm water in a high reservoir and an electric current? But here again we succeed in convincing ourselves that there is a similarity in a sense that seems very essential to us, since we are dealing with energies in both cases and since these energies can be equated. In order to make us discover these identities which seem, to any unbiased mind, to be terribly *far-fetched* in the literal sense of the term, all that was necessary was that the phenomena succeed one another. As a matter of fact, chlorine and sodium, if put together, will form salt and can, inversely, be produced from this substance, while on the other hand it seems impossible to break either of them down, at least by ordinary laboratory methods, which fact makes us consider them to be elements. They *must* then be contained in salt; it is impossible for them not to be there in some way, and although this substance has totally different properties from either of its components, we nevertheless call it sodium chloride and write its formula as NaCl. Similarly the fact that water, by falling from a height and thereby activating a turbine and dynamos, could produce an electric current suffices to make us declare that the water reservoir contains energy, that it is a reservoir of energy, energy which is of course qualified as *potential* (because it does not immediately reveal itself), but which we nevertheless assimilate with that of the turbine and the electric current.

What is perhaps even more noteworthy here is the fact that, contrary to what takes place in mathematical deduction, the search for identity in physics is not usually preceded by the decree that explicitly orders it, which we call the statement of the theorem. That could not be more easily explained: there is no need to proclaim something about which all men are in close agreement, nor to call for obedience to what constitutes an irresistible inclination of any normally constituted mind. The student, when shown the three squares of the Pythagorean theorem for the first time, has no preconceived idea of how these areas could possibly be related: initially, he would not be at all shocked if one stated a relation quite different from the actual one. The theorem must thus be clearly stated so that he knows in what direction the process of identification will move. But as soon as we perceive a phenomenon, a change, we expect it to be explained, to be made rational. This time, then, the process of identification works spontaneously, so to speak – and, if necessary, in spite of ourselves. That is why we are not aware of it and must analyze scientific procedures to find it at the foundation of our reasoning.

But the role of the process of identification in the physical sciences is not limited solely to what has to do with change. If we but ask ourselves, we realize that we demand an explanation not only of what changes, but also of what persists. Even Aristotle, so circumspect when dealing with cause, sometimes obeys this tendency. Criticizing the position of Democritus, who explained the agitation of atoms by pointing out that it had always existed, he states:

> But it is a wrong assumption to suppose universally that we have an adequate first principle in virtue of the fact that something always is so or always happens so. Thus Democritus reduces the causes that explain nature to the fact that things happened in the past in the same way as they happen now; but he does not think fit to seek for a first principle to explain this eternal state. (*Physics* $252^{a}32$–$252^{b}1$)[33]

One could, of course, assume that in this case the Stagirite's opposition is largely motivated by the fact that it is a question of an agitation, that is, of a motion. Indeed, since the ancients are unacquainted with the concept of inertia, the concept of a body *in a state of motion* was infinitely less familiar to them than it is to us, and, particularly for Aristotle, motion is consistently assimilated with change. Nevertheless it would seem to be a somewhat forced interpretation in the present case, because here the agitation is certainly assimilated with a state (as shown by the expression *eternal state*). But let us consider an additional citation that we feel leaves no room for doubt:

> We suppose ourselves to possess unqualified scientific knowledge of a thing, as opposed to knowing it in the accidental way in which the sophist knows, when we think that we know the cause on which the fact depends, as the cause of that fact and of no other, and, further, that the fact could not be other than it is. (*Post. An.* $71^{b}7$–12 [G. R. G. Mure trans.])

Here it is clearly a question of cause, while all becoming, all change, seems to be excluded.

Leibniz has expressed this way of thinking quite clearly: "There is a reason even for eternal things. If it should be supposed that the world has existed from eternity, and that there have been only globules in it, a reason must be given why there should be globules rather than cubes."[34]

And he is obviously following the same line of thought when he protests against attributing arbitrary qualities to substances. "Thus," he says,

in the order of nature (miracles apart) God does not arbitrarily give to substances such and such qualities indifferently, and he never gives them any but those which are natural to them, that is to say qualities which can be derived from their nature as explicable modifications.[35]

But all this was virtually contained in his statement of the principle of sufficient reason itself: "Nothing ever comes to pass without there being a cause or at least a reason determining it, that is, something to give an *a priori* reason why it is existent rather than non-existent, and in this wise rather than in any other."[36]

Indeed, if the first part of the statement seems to refer solely to becoming, in the second part the philosopher, by a sort of mental leap, obviously goes further, since he asks the reason for that which "is existent."

We have only to refer back to what we set forth above to realize how right Leibniz was. Physics, as we noted, seeks to explain not only phenomena, but matter itself (p. 126). Now matter is, essentially, what exists and what we must even consider to be eternal. Thus what we are really seeking is the very reason for eternal things, as Leibniz said. Moreover, there are abundant examples of this sort of concern in science. For example, when Cuvier set out to establish a relation of dependency between the characteristic traits of animal structure (Ch. 3, pp. 51 ff.), he certainly did not conceive of them as having any relation of succession at all; this thought was all the further from his mind because for him, as we know, a species was something absolutely stable, appearing with all its particularities and disappearing in the same way. And chemistry is also obeying the same eternal tendency of the human mind when it expresses astonishment at the diversity of substances, which astonishment, according to the informed opinion of Job, constitutes the starting point for that whole science.[37]

One need only open any textbook to see that the goal of chemical theory is to establish a rational link between the various properties of a substance, and it is quite clear that even where it cannot affirm anything to this effect, the existence of such a link remains a true article of faith.

As a result, there obviously tends to be established, in the area of causal relations, a genuine confusion of two nevertheless quite distinct orders of ideas, which immediately suggests that, in spite of this distinction, there must be a close connection between them. This can easily be seen.

The goal of all explanation is, in the most general sense, to make us understand the world – which we initially perceive only as a *de facto* truth and consequently an accidental truth – as something necessary, a truth of reason. Thus, in the representation of reality sought by science, everything must be rational. Consequently, we must justify even the starting point of deduction before the tribunal of sufficient reason.

It is evidently in attempting such a justification that philosophers and scientists have often, without hesitation and almost without transition, enlarged the concept of causality and sought the causes of things understood to be permanent. But the intimate relation that links the two problems can also be discovered by a somewhat different route.

With Riemann, we noted that the problem of causality properly speaking arises due to the fact that things change, and we have seen that theories attempting to explain this change basically end up equating the antecedent and the consequent, declaring that nothing was created and nothing lost, that everything has persisted – in other words, denying this very change. Thus the determination of the sufficient reason for diversity in time consists in the fact that we submit this diversity to a process by which we endeavor to reduce it to an identity.

The same is true of the sufficient reason for what does not become, but is. There too, what seems to us to need to be explained is the fact that there is diversity, and once again we can explain the diversity only by a process of identification. Earlier we asked: Why do things change in time? And the answer, the explanation, consisted in declaring that the change is merely apparent, that it does not really exist, since the consequent is, at bottom, identical with the antecedent. We now ask: Why does what we perceive in space appear to be diverse? And, if we are to explain this diversity, we can follow no other way but that which consists in denying it, in claiming that the astonishing variety we think we are observing is only apparent, that it hides a fundamental identity, all the different kinds of matter that occupy space being essentially only one and the same matter.

This is the concept of the unity of matter, and a quick glance through science and its history suffices to convince us that this concept has constantly made its influence felt in all theories of physical reality. The ancients, atomists as well as Peripatetics, consider it to be a truth requiring no demonstration, and the Middle Ages follows in their footsteps. For Descartes too "all matter in the whole universe is of one and the same kind" (*Principes* II, 23), and everyone who comes after him seems to agree on this point. When the concept of qualitatively diverse elements is established as a result of the profound transformation chemical theories undergo in the second half of the eighteenth century, chemists still only conditionally accept this given imposed on them by experience. They never cease searching for relations between these elements, which clearly shows that they do not consider them truly ultimate, that, on the contrary, the multiplicity of this given strikes them as an anomaly. In their heart of hearts, as we know by the authoritative testimony of Berthelot, they always retain the hope of going past what they consider a temporary limit.[38]

This is because, as Hannequin rightly saw, the oneness of matter is the secret postulate of all atomism.[39]

But it is not enough to reduce the diversity of different kinds of matter to unity. It is also necessary that the existence of this unique matter itself conform to the requirements of the principle of sufficient reason. Let us recall Leibniz's injunction: if the world is composed of globules, we must be able to show why there are globules and not cubes. Now, under these conditions, what properties can we attribute to this unique matter? The reply is as simple as, at first sight, it seems paradoxical. Indeed, it is evident that no physical property can possibly seem to us to be really motivated by sufficient reason, that on the contrary any quality with which we try to endow nature will inevitably appear to us to be occult, because only spatial properties prove to fit the needs of our mind, to be really necessary. Ultimately, therefore, truly rational matter can only be space.

That this is really the spirit that animates science can be demonstrated by studying the different theories of matter and their evolution in history. First of all, there can be no doubt about the true legislator of modern science, Descartes. For Descartes matter and space are indistinguishable, there is no space apart from matter and, reciprocally, matter is stripped of all non-spatial properties, so that it is obviously only hypostasized space.

Modern science does not proceed with the same somewhat shocking candor, but it is easy to see that by means of a detour involving a sort of hypocrisy (unconscious to be sure) it has the same objective in view. In fact, after seeking to unify matter, it then takes pains to construct this matter out of a "universal medium," ether (at least insofar as one does not mean to abolish the concept completely, by immediately substituting for it, as we saw above, this definitive and formidable X of electricity), and finally it does its utmost to allow this ether to be to some extent confounded with space. It is perhaps this last circumstance that is most characteristic, most revelatory of the power of the irresistible current that sweeps human thought along in this domain and makes it a sort of unwitting accomplice of deceptions at the same time complicated and naive. What could be more astonishing, indeed, than to observe that this ether, which must be endowed with all sorts of bizarre and more or less contradictory properties – whose density, for example, according to the expert testimony of Sir J. J. Thomson and Sir Oliver Lodge, must be supposed "immensely greater than that of any known substance"[40] – is only another name for the void, since, as Maxwell noted, the properties of the ether are those of the void,[41] and because, according to Nernst, the hypothesis of ether is only the "theory of the void."[42] Thus the ether is actually a hypostasis of space, as Helmholtz admitted in so many words, and as Kant before him had recognized in the case of the caloric fluid that in the physics of his time played a role analogous to that of our ether.[43]

The paradoxical aspect of this situation sometimes strikes the scientists themselves. When, without any preoccupation with method, but with a simple concern for clarifying the meaning of a theory, they run up against a conception in which the contrast is revealed at all tangibly, they are apt to be surprised. This is what happens to Whittaker, for example, when he observes that the ether under consideration in a theory is "simply space endowed with certain dynamical properties."[44] Similarly, Duhem judges that he has sufficiently refuted certain theories by saying that they tend to reduce matter to space,[45] and Henri Poincaré loudly protests against the demands of some of the adherents of pure mechanism, who try to reduce everything to a matter "having nothing but purely geometric qualities,"[46] this matter, we hardly need point out, obviously being simply space. Castelnuovo goes even further, recognizing that this phenomenon is not exceptional, but general. The practice nevertheless appears blameworthy to him, and he seeks its cause in an opposition between the old spirit and the new: we make a concession to the new spirit by attributing the motion of matter to the ether, "but then we attribute to the ether properties that reestablish in it the system of absolute rest we thought we had gotten rid of."[47] The reader will have been convinced, or at least we hope he will, that the cause lies much deeper, namely, the necessity of identifying space and the ether.

Thus, diversity in space is unquestionably an enigma for us, a grounds for astonishment if not identical, at least very similar to that we discover in the case of diversity in time. As a consequence we cannot escape the conclusion that if our reasoning is correct, the goal of explanations and theories is really to replace the infinitely diverse world around us by identity in time and space, which clearly can only be space itself.

NOTES

1. Harald Höffding, *Der Totalitätsbegriff* (Leipzig: O. R. Reisland, 1917), p. 14. Cf. also his *La Pensée humaine*, trans. Jacques de Coussanges (Paris: Félix Alcan, 1911), pp. 73, 172. Höffding has also pointed out that among the Greeks there appears, "formulated with the energy of thought" characteristic of them, "the principle of identity, the absolute will to obtain the oneness of thought with itself ... the absolute ideal to which thought tries to assimilate all knowledge" (*La Pensée humaine* 124).
2. Leibniz, *Opera* 83. Cf. *Opuscules*: "*Absolute necessaria propositio est quæ resolvi potest in identicas*" and "*pervenitur ad demonstrationem seu identitatem*" (p. 17). "*Alioqui veritas daretur quæ non posset probari a priori, seu quæ non resolveretur in identicas, quod est contra naturam veritatis, quæ semper, vel expresse vel implicite, identica est*" (p. 518) [Parkinson: "An *absolutely necessary* proposition is one which can be resolved into identical propositions" and "one ... arrives at a demonstration or an identity" (pp. 96–97). "Otherwise there would be a truth which could not be proved *a priori*, i.e. which is not analyzed into identities; and this is contrary to the nature of truth, which is always, either expressly or implicitly, identical" (p. 88)]. Cf. also *Opuscules* 18 [Parkinson 97–98], 374, 387, 388, 513 [Parkinson 7] and Louis Couturat, *La Logique de Leibniz* (Paris: F. Alcan, 1901), p. 210.
3. Plato, *Timaeus* 37A: "And because she is composed of the same and the other and of the [intermediate] essence, these three, and is divided and united in due proportion" [Jowett trans.; Meyerson's brackets]. For "the intermediate essence," see the earlier passage at 35A, a translation of which can be found in Léon Robin's admirable work on *La Place de la physique dans la philosophie de Platon* (Paris: Félix Alcan, 1919), p. 55, note 1. Cf. also *Timaeus* 28A.
4. Étienne Bonnot de Condillac, *La Langue des calculs* (Paris: C. Houel, Year 6 [of the First French Republic: 1798]), p. 60, and *Logique*, *Oeuvres* 22:177 [*Philosophical Writings of Etienne Bonnot, Abbé de Condillac*, trans. Franklin Philip (Hillsdale, N. J.: Lawrence Erlbaum Associates, 1982), p. 415].
5. Höffding, *La Pensée humaine* 22, 276 [erroneous citation].
6. The *Logic* figures in the monumental edition of the works of Hegel published through the efforts of his disciples (Gans, Henning, Michelet, etc.) in two different forms, first in Vols. 3, 4 and 5 under the title of *Wissenschaft der Logik*, and later in a shorter form in Vol. 6, as the first part of the *Encyclopädie der philosophischen Wissenschaften im Grundrisse*. The first part of the *Wissenschaft der Logik* (Vol. 3 of the *Werke*), for which Hegel himself had prepared a new edition before his death, and the *Logik* of the *Ençyclopädie*, put together from notes taken in his courses by his disciples, date in their present form from the same epoch, that immediately preceding the death of the author. The two other

volumes of the *Wissenschaft* (Vols. 4 and 5 of the *Werke*) are, on the contrary, the reproduction of the first edition of this work, which appeared in 1814 and 1816. We shall cite the two works as *Wissenschaft der Logik* (the first volume according to the Stuttgart edition, 1832, and the two others according to the Berlin edition, 1834) and as *Encyclopädie, Logik*. [The 1832–1840 Berlin edition of the *Werke* is cited here in all cases, however; see Bibliographic Abbreviations, p. xxviii.] In the second part of the *Wissenschaft der Logik* (5:287 [Miller 793]), Hegel says that "abstract identity, which alone analytic cognition knows as its own," is "essentially the identity of distinct terms." This is obviously (apart from the difference of nomenclature concerning abstract identity) the same conception as in the later work.

7. Similarly, *Wiss. der Logik*, 3:217 [Miller 191]: "But profounder insight into the antinomial, or more truly into the dialectical nature of concrete reason demonstrates *any* Notion whatever to be a unity of opposed moments" [adding "concrete" to conform to Meyerson's usage; see Bibliographic Abbreviations, page xxviii above]. Also *Phänomenologie*, 2:16 [Baillie 82]: "The beginning, the principle, or the Absolute, as at first or immediately expressed, is merely the universal .... Even the mere transition to a proposition, is a form of mediation, contains a process towards another state [*ein Anderswerden*] from which we must return once more." Furthermore this is the reason that "a so-called fundamental proposition or first principle of philosophy, even if it is true, is yet none the less false just because and in so far as it is merely a fundamental proposition, merely a first principle" (2:19 [Baillie 85; Meyerson's brackets]).
8. On this point the Hegelians have remained faithful to the spirit of their master, and thus McTaggart, for example, rejects with some indignation Eduard von Hartmann's objection to the dialectical method, pointing out that his reasoning is founded in mathematics (*Studies* 94–95).
9. This definition is not at all intended to exhaust the meaning of the term in the Hegelian doctrine. It is only one aspect of this concept which plays so important a role in Hegelianism, but it is the one that interests us here. For a different aspect of the same notion, cf., among others, Wallace, *Prolegomena* 287.
10. Hegel uses the term *aufheben*, and Boutroux, in his admirable exposition of Hegel's doctrine in the course of the discussion of René Berthelot's thesis 'Sur la nécessité, la finalité et la liberté chez Hegel,' pointed out the capital importance of this term in the Hegelian philosophy (*Bull. Soc. fr. phil.* **7** [1907] 142; cf. 162 ff. for Berthelot's observations on the same subject). Hegel, who is particularly fond of (sometimes fanciful) etymologies, and even puns – he is, according to the keen observation of Eugenio Rignano ('Les diverses mentalités logiques,' *Scientia* **22** [1917] 123), in his two capacities as a German and as a metaphysician, essentially an *auditory* personality – insists on the double meaning of the word in question, since it means both "to keep" and "to abolish" (*Enc., Logik*, 6:191 [Wallace 180]; see also p. 167 [159], where Hegel seems to say that it is the former philosophies which contain the later ones *aufgehoben* within them – but that is undoubtedly a simple printer's inversion, as anyone knowing anything about Hegelian thought will recognize [the inversion is set right in the Wallace translation]). On word plays in Hegel, cf. also William

Wallace, *Prolegomena* 11. Trendelenburg (*Log. Untersuch.* 1:89) believes that Hegelian dialectic's attempt to go back to Plato is entirely unjustified, and that the master to whom it could properly appeal in antiquity would more likely be Proclus.

11. [E.g., *La science et l'hypothèse* (Paris, Flammarion, n.d.), p. 20 (*Science and Hypothesis*, trans. George Bruce Halsted, New York: The Science Press, 1905, p. 11).]
12. Hegel, *Phänomenologie*, 2:33–34 [Baillie 102]. On the context of this passage, see pp. 281 ff. below.
13. Leibniz, *Opuscules* 33: "*Nam Geometræ accurate quidem sua demonstrant, sed animum cogunt magis quam illustrant, in quo quidem admirationem sibi majorem pariunt, dum invito Lectori assensum extorquent, eumque arte improvisa circumveniunt, sed memoriæ atque ingenio Lectoris non satis consulunt, quia rationes causasque naturales conclusionum quodam modo occulunt, ut non facile agnoscatur modus, quo sua inventa obtinuere.*"
14. William Stanley Jevons, *Logic*, Science Primers (New York: American Book Company, n.d.), p. 75. Cf. also his *The Principles of Science* (London: Macmillan and Co., 1892), pp. 5, 9 [, 17]. For information concerning Jevons's predecessors in this line of thought, see *Principles* xvi ff. and 21.
15. Leibniz, *Mathematische Schriften*, ed. Gerhardt, *Gesammelte Werke* (Halle: H. W. Schmidt, 1860), 6:439, and *Opera* 716 [Parkinson 200].
16. A. Riehl, 'Causalität und Identität,' Vierteljahrsschrift für wissenschaftliche Philosophie [Leipzig] **1** (1877) 373–374 [Meyerson's brackets]. We mentioned above how little notice Riehl's work has received. To some extent this is explained by the fact that not only is its scientific content rather poor (the reference to uniform rectilinear motion, which is quite unclear, as the reader may have noticed, is among his better efforts; the sentences dealing with mechanism, p. 383, are still less clear), but also by the fact that the very concept of causality is complicated and muddled by considerations which contradict the principal definition. For example, the author seems to accept the definition according to which causality is the tendency to reduce the extraordinary to the familiar (p. 379). As we shall see in Chapter 15, what is involved is a more general phenomenon, which is by this fact invested with a special significance.
17. Emanuel Rádl, *Geschichte der biologische Theorien in der Neuzeit*, 2nd ed. (Leipzig: W. Engelmann, 1913), pp. 135, 138.
18. Hans Driesch, *Der Vitalismus als Geschichte und als Lehre* (Leipzig: Johann Ambrosius Barth, 1905), pp. 46–47 [*The History and Theory of Vitalism*, trans. C. K. Ogden (London: Macmillan, 1914), pp. 49–50].
19. Marcello Malpighi, 'De Formatione pulli in ovo,' *Opera omnia* (Lugdun [Leyden]: Apud Petrum Vander Aa, 1687), 2:53 [Meyerson's brackets].
20. Louis Couturat, 'Compte rendu critique du IIe Congrès de philosophie – Genève, Séances de Logique et philosophie des sciences,' *Rev. de Méta.* **12** (1904) 1059.
21. Félix Le Dantec, 'Les Néo-Darwiniens et l'hérédité des caractères acquis,' *Rev. phil.* **47** (1899) 1–41.
22. Frédéric Houssay, 'Les Théories atomiques en biologie,' *Congrès international de philosophie de 1900, Bibliothèque du Congrès international de philosophie*

III: Logique et Histoire de Sciences (Paris: 1901), pp. 595–607.

23. Couturat, 'Comte rendu' 1059–1061.
24. A. Prenant, 'Questions de biologie cellulaire,' *Scientia* **9** (1911) 480–481, 487.
25. E. S. Russell, 'Le Problème des espèces et de leur origine,' *Scientia* **18** (1915) 426.
26. *Le Monde nouveau*, No. 1, 20 March 1919 [p. 11], as cited in the *Mercure de France* of 1 May 1919, p. 132. Before Maeterlinck, Ernest Renan was captivated by the idea of the preformation of the organism, which he connected to speculations on hyperspace. He supposed, in fact, that "'modern geometry's reflections concerning space having more than three dimensions may have some connection with reality,' thanks to the resources they might provide for making clear how 'the successive generations are contained within one another'" (Georges Sorel, 'Vues sur les problèmes de la philosophie,' *Rev. de méta.* **18** [1910] 610).
27. E. S. Russell, 'The Influence of the Theory of Evolution on Morphology,' *Scientia* **20** (1916) 355.
28. In Nicholas of Cusa who, as we said in Chapter 1, p. 9, employs the term *explicatio* in a usage close to the etymological sense, this term appears to be a synonym of *evolutio* (Rudolf Eucken, *Geschichte der philosophischen Terminologie im Umriss*, Leipzig: Veit, 1879, pp. 82, 187).
29. Descartes, *Oeuvres* 11:7 [*Le Monde, ou Traité de la lumière*, trans. Michael Sean Mahoney (New York: Abaris Books, Inc., 1979), p. 9]. Pascal, of course, judges that Descartes erred in trying to find the actual movements underlying phenomena and to "assemble the machine," for that "is quite ridiculous" as well as "pointless, uncertain, and arduous." Nevertheless "in general terms one must say: 'That is the result of figure and motion,' because it is true" (*Pensées* 361 [Krailsheimer 52]).
30. Maxwell laid great stress on the difference between the properties we are obliged to impute to the particles and those of the visible bodies which they constitute. See Lawrence J. Henderson, *The Order of Nature* (Cambridge: Harvard Univ. Press, 1917), Appendix, p. 219.
31. It seems to have been Democritus who determined the volume of the pyramid and the cone, without providing a proof of these theorems (cf. Max Simon, *Geschichte der Mathematik im Altertum in Verbindung mit antiker Kulturgeschichte*, Berlin: B. Cassirer, 1909, p. 181. Cf. also p. 190 and Gino Loria, 'L'infiniment grand et l'infiniment petit d'après les mathématiciens de l'antiquité,' *Scientia* **18** [1915] Supplément: 230). According to Paul Tannery, it is quite likely that Democritus demonstrated the impossibility of constructing a cone by means of superimposed circles (*Rev. phil.* **20** [1885] 396). This would thus be an additional proof that he began with the idea of the continuity of space and, consequently, rejected anything resembling a geometric atomism. Gino Loria (*Le Scienze esatte nell' antica Grecia*, Modena: Società Tipografica, 1893, Bk. I, p. 63) expresses doubts as to Tannery's interpretation, although he allows that the problem of the cone may have been suggested to Democritus by his atomistic conceptions.
32. That is especially obvious concerning the discussion of certain problems of astronomy, such as those having to do with the motion of the sun and the stars or

the phases of the moon (V, 510–769), for which the scientists of antiquity had sometimes provided accurate, mathematically incontestable solutions. Even when Lucretius gives an account of these explanations, he does so hesitantly, and places them beside others that are erroneous.

33. [We have used the R. P. Hardie and R. K. Gaye translation, with the exception of the final word, at which point we have substituted "eternal state" for "'always,'" following here the French translation by Barthélemy Saint-Hilaire which Meyerson is using. This specific wording plays a key role in Meyerson's argument.]
34. Couturat, 'Sur la métaphysique de Leibniz,' *Rev. de méta.* **10** (1902) 3. Cf. Leibniz, *Opuscules* 519 [Parkinson 88].
35. Leibniz, *Opera* 203 [Parkinson 168–169]. St. Thomas was of exactly the opposite opinion, since, in speaking of the fact that heavy bodies tend toward the center, he forbade asking the reason for this phenomenon: there is no explaining natures (Antoine D. Sertillanges, *Saint Thomas d'Aquin*, Paris: Félix Alcan, 1910, 1:126). Leibniz was thus not in error when, precisely on the subject of Newtonian gravitation, he protested against the occult qualities of the scholastics (cf. IR 514 [Loewenberg 447–448]). But it is doubtful that St. Thomas found this part of his doctrine in his master Maïmonides, since the fact that the latter means to grant God only negative attributes, as we shall see (p. 153 below), seems to indicate a contrary tendency.
36. Leibniz, *Théodicée, Opera* 515 [Huggard 147]. Cf. also *Opuscules* 25: "*Principium omnis ratiocinationis primarium est, nihil esse aut fieri, quin ratio reddi possit, saltem ab omniscio, cur sic potius quam non sit, aut cur sit potius quam aliter.*" Cf. *Opuscules* 11, 402 [Parkinson 94], 553. In comparing these statements among themselves as well as to the one we cited on p. 117, we see that they do not all express with equal clarity the need for seeking the cause of what persists. However, the *nihil esse aut fieri* of the present note, as well as the earlier statement concerning the *globules*, shows that there was indeed no hesitation in Leibniz's mind on this point.
37. André Job, 'Les Progrès des théories chimiques,' *Bull. Soc. fr. phil.* **13** (1913) 47.
38. Marcellin Berthelot, *Les Origines de l'alchimie* (Paris: George Steinheil, 1885), p. 289. On this deep-seated tendency of chemistry, see below, Ch. 7, pp. 217 ff.
39. Arthur Hannequin, *Essai critique sur l'hypothèse des atomes dans la science contemporaine* (Paris: G. Masson, 1895), p. 166.
40. Oliver Lodge, 'The Aether of Space,' *Nature* **79** (1909) 324.
41. Maxwell, *Scientific Papers* (Cambridge: University Press, 1890; reprint New York: Dover, n.d.), 2:323.
42. Walther Nernst, 'Sur quelques nouveaux problèmes de la théorie de la chaleur,' *Scientia* **10** (1911) 292.
43. Kant, *Vom Uebergange von den metaphysischen Anfangsgründen der Naturwissenschaft zur Physik* (Frankfurt: Moritz Schauenburg, 1888), pp. 111, 119, 121. Among contemporary philosophers James Ward in particular has clearly recognized this identification of ether and space (*Naturalism and Agnosticism*, London: A. and C. Black, 1899, 1:132 ff.).
44. Edmund T. Whittaker, *A History of the Theories of Aether and Electricity*

(London: Longmans, Green, and Co., 1910), p. 420.

45. Pierre Duhem, *L'Evolution de la mécanique* (Paris: A. Joanin, 1903), pp. 177–179 [*The Evolution of Mechanics*, trans. Michael Cole (Alphen aan den Rijn: Sijthoff & Noordhoff, 1980), pp. 94–95].
46. Henri Poincaré, *Electricité et optique* (Paris: Georges Carré et C. Naud, 1901), p. iv.
47. G. Castelnuovo, 'Le Principe de la relativité et les phénomènes optiques,' *Scientia* **9** (1911) Supplément: 51–52.

CHAPTER 6

THE IRRATIONAL

To be sure, the conclusion we reached in the preceding chapter appears altogether paradoxical at first glance. Is it possible that science, incontestably the dominant force in the life of our times, the one whose inspirations modern man is proud to follow, ultimately seeks to explain everything by Space, even though the latter seems to us to be something inert, a form devoid of content? Is it believable that *this* is what science is?

We must first make an important qualification. Yes, *this* is indeed what science is – but that is not all there is to science. Science, obeying the tendency to explain which is characteristic of the human mind, does in fact seem to wish to reduce everything to what is rational. But on the other hand, infinitely careful to keep in contact with nature, it recognizes the limitations nature imposes on its efforts. There thus arises a concept the nature and importance of which have perhaps not always been clearly recognized, but which has nonetheless exercised a profound influence on the evolution of science: it is the concept of the *irrational*, i. e., that which, among the elements science is led to use, appears destined, by its very nature, to resist any ulterior reduction to purely rational elements.

Are there such elements in mathematics itself? Let us first consider the question of *diversity*, to which we referred in the preceding chapter (pp. 134 ff.). Can diversity, as mathematics knows it, be qualified as irrational? That would seem to be a simple question of definition. It is certain – we see it clearly in the very process of mathematical reasoning as we analyzed it above (pp. 103 ff.), as well as by the Eleatic conception which dissolves all diversity into an indistinct whole – that diversity, whatever it may be, is fundamentally distasteful to our reason, which seeks to impose identity upon it. On the other hand, it is just as evident that the existence of this diversity is the very condition for the functioning of reason, since reason can act only on the diverse.

But setting aside this element of diversity, which we have seen to be inherent in any thought whatsoever, a glance at geometry suffices to show that this science conceals a given, an element our reason cannot draw from within itself and for which it consequently cannot *give a reason*. This given, it goes almost without saying, has to do with space, with the fact that it has three dimensions and allows the Euclidean postulate. Why indeed does space have a specific number of dimensions? Why is there not a fourth, or an infinite or a fractional number, and so forth? And why are things arranged in conformity with the Euclidean system and not with those of Lobachevsky or Bolyai? The mere fact that it is possible to believe in the existence of a fourth dimension (as even practicing scientists have done, among them, of course, the astronomer Zöllner) and that Lobachevsky, Riemann and Helmholtz claim to have verified the validity of the Euclidean postulate by astronomical measurements (see Appendix 21), clearly proves that we are dealing here with a particular structure of our space, a structure our reason is obliged to accept as a fact, that is, with a true irrational.[1] Indeed, a sound appreciation of this situation would seem to be what lies behind the claim that geometry is substantialistic in origin (cf. Ch. 1, p. 25 above).

It can easily be seen that the physical sciences recognize the existence of a whole series of these regions where all attempt at explanation seems barred, or, if one prefers, doomed to certain failure. These regions make up what might be called physical *irrationals*, and the irrational first to be recognized is doubtless that constituted by sensation. Democritus already declares that "by convention are sweet and bitter, hot and cold, by convention is colour; in truth are atoms and the void."[2]

The ancient atomists consistently maintained this teaching. "Do not imagine that colour is the only quality that is denied to the atoms," says Lucretius. "They are also wholly devoid of warmth and cold and scorching heat; they are barren of sound and starved of savour, and emit no inherent odour from their bodies" (Lucretius, *De rerum nat.* II, 842–846; cf. also II, 737–738, 797–800, 808–809).

The attitude of Democritus, as the ancients already understood perfectly, according to the testimony of Sextus Empiricus and Diogenes Laertius, was tantamount to "abolishing the qualities";[3] in that respect, as a contemporary philosopher has rightly pointed out, Democritus's position, by its purely rational starting point, was more consistent than Locke's with the position adopted by Galileo and Descartes and by all of modern science along with them.[4] This position was formulated, in all its rigor, by Hobbes.

All which qualities called Sensible, are in the object that causeth them, but so many several motions of the matter, by which it presseth our organs diversely. Neither in us that are pressed, are they anything else, but divers motions; (for motion, produceth nothing but motion). But their appearance to us is Fancy, the same waking, that dreaming. (*Leviathan* [London: Dent, n.d.], p. 3)

His contemporary Pascal remarks that such a way of looking at things – a direct and ineluctable consequence of the Cartesian philosophy according to which everything was to be explained by figure and movement – is difficult for our reason to accept.

When they say that heat is merely the movement of certain globules and light the *conatus recedendi* [centrifugal force] that we feel, we are amazed. What! is pleasure nothing but a ballet of spirits? We had such a different conception of it, and these feelings seem so far removed from those other ones, which, we say, are the same as those with which we are comparing them! The feeling of fire, the warmth which affects us in quite a different way from touch, the reception of sound and light, all seem mysterious to us. And yet it is as straightforward as throwing a stone. It is true that the smallness of the spirits entering the pores touches other nerves, but they are still nerves. (*Pensées* 497–498 [Krailsheimer 244; Krailsheimer's brackets])

Leibniz has given a picturesque and arresting form to this position: "We are moreover obliged to confess," says he in his *Monadology*,

that perception and that which depends on it cannot be explained mechanically, that is to say by figures and motions. Suppose that there were a machine so constructed as to produce thought, feeling, and perception, we could imagine it increased in size while retaining the same proportions, so that one could enter as one might a mill. On going inside we should only see the parts impinging upon one another; we should not see anything which would explain a perception. (*Opera* 706 [Parkinson 181])

Thus we see here an initial limit – and clearly a definitive one – to our desire to understand nature, to conceive it as structured in conformity with the needs of our reason, as rational. This limit has been clearly recognized by science. To be sure, we do not find physicists expressing it as explicitly as Hobbes or Leibniz. But this is because there is no need to do so, since the very premises on which the whole of science stands implies a sufficiently clear attitude on this question. As soon as one declares that matter and motion constitute the unique essence of all phenomena, one precludes all explanation of the true quality, the *quid proprium* of sensation. As Bergson correctly points out, "it is ... of the essence of materialism to assert the perfect relativity of sensible qualities,"[5] and it is easy to see that, as a matter of fact, modern science proceeds as if there could be no doubt on this point. Whether we are concerned with the optics of Descartes, with that of Newton or Fresnel, or, finally, with that of contemporary scientists, for whom light is an electrical phenomenon, it is certain that the theories will disclose no trace of an attempt to deduce what is specific in our sensation of the color red;

the part of science which seems to be designed to deal particularly with sensation – physiological optics – resolutely leaves aside anything resembling an explanation of the transition between motion and sensation. To the physicist, such an attitude seems so natural that he cannot imagine any other. As a consequence, any attempt at theory that includes sensation itself he finds absurd, or at least pointless, doomed to sterility; he does not even consent to discuss it, contemptuously ruling it out in advance, so to speak. That is what so irritated Goethe, whose *Farbenlehre*, in spite of Hegel's and Schopenhauer's support, could never attain the honor of a serious refutation, even in Germany, where these men were so influential. Nor did anything come of all the complaints by philosophers about the fact that science, in its explanatory theories, obviously left out something very essential which is an integral part of our conception of the external world. Says Bradley,

> The sensible life, the warmth and colour, the odour and the tones, without these Nature is a mere intellectual fiction. The primary qualities are a construction demanded by science, but, while divorced from the secondary, they have no life as facts. Science has a Hades from which it returns to interpret the world, but the inhabitants of its Hades are merely shades.[6]

Scientists themselves sometimes seem surprised at the image of the universe their theories would impose on our understanding. Henri Poincaré humorously expressed this state of mind when he said that universal mechanism winds up supposing that a superior intelligence – God – would have something like the same sensation contemplating the world that we have in watching a billiard game.[7]

Science nevertheless goes its merry way; in the nineteenth century it even stiffened its position or, if one prefers, strengthened its attitude on this question. First, as a result of progress in the sciences of the organism, the problem of sensation had become more evident, so that we can report a few quite explicit statements by scientists. "Everyone knows," says Cuvier, "that the production of a perception, or the action of external bodies on the self which results in a sensation or an image, is an eternally incomprehensible problem, and that on this point there exists a gap between the physical and the moral sciences that all the efforts of our mind will never be able to fill."[8] "Physiologists," declares Alexandre Herzen, who, as we know, was himself a renowned physiologist, "could ... study the nerves and the brain for centuries without ever managing to form the slightest idea of what a sensation is ... if they did not subjectively experience these states of consciousness themselves."[9]

These generalities go no further than Leibniz's famous passage on the *mill*. But now science is prompted to go more deeply into the matter. In about 1830, the physiologist Johannes Müller formulates the doctrine of "the specific energy of the sensory nerves," which asserts that the particular quality of a sensation depends not on the action of the external cause, but on that of the transmitting organ, the nerve. For example, we can excite the optic nerve in various ways, first normally by what we term light, but then also by shock or mechanical pressure or by electrical action; under these quite different circumstances we shall always experience the same kind of sensations, namely, sensations of light. Quite recently the theory has been developed further, in the sense that it seems to have been established that in general the nerves are capable of transmitting to us only sensations of a single kind, so that when we appear to experience, simultaneously and by the same organs, impressions different in nature, as occurs particularly for cutaneous impressions, these impressions actually are located in perfectly distinct parts of our epidermis. For example, there are assumed to be four specific cutaneous senses: contact, cold, heat, pain, each with special peripheral organs, particular paths of conduction, distinct centers, etc.[10]

As far as his fundamental conception is concerned, however, Müller had been preceded by the philosophers. We do not know Democritus's opinion on the genesis of the qualitative diversity of our sensations, but later atomists seem to have adopted the position of Empedocles, according to which the receiving organ had only a passive role, merely exercising a choice among the impressions offered to it. Only those whose size and shape fit those of the pores opening into the organs can affect them; the others, being too large or too small, either cannot get in or else slip through the pores without touching them.[11] This, as a matter of fact, is the preferred theory of Lucretius (*De rerum nat.* II, 679–685; VI, 985 ff.). However, there are other passages where he seems to suggest that the body producing a specific sensation in us contains nothing that is peculiar to that sensation and that this element must thus belong to the action of our organs (VI, 960 ff.). That may be the source of the altogether remarkable ideas Montaigne developed on the subject. Indeed, Montaigne begins by observing, like Lucretius, that the same objects can create quite different sensations in different organisms.

That things do not lodge in us in their own form and essence, or make their entry into us by their own power and authority, we see clearly enough. Because, if that were so, we should receive them in the same way: wine would be the same in the mouth of a sick man as in the mouth of a healthy man; he who has chapped or numb fingers would find the same hardness in the wood or iron he handles as does another.

But he then becomes more explicit: "The sick lend bitterness to sweets, whereby it is evident that we do not receive things as they are, but in one way and another, according to what we are and what they seem to us." Thus, after observing that even in nature things can be strangely transformed – "The moisture that the root of a tree sucks up becomes trunk, leaf, and fruit; and the air, being but one, by being applied to a trumpet is diversified into a thousand kinds of sounds" – he comes to ask himself: "Is it our senses, I say, which likewise fashion these subjects out of various qualities, or do they really have them so?" And he finally concludes: "Now, since our condition accommodates things to itself and transforms them according to itself, we no longer know what things are in truth; for nothing comes to us except falsified and altered by our senses."[12] Obviously, and without speaking of the deep properly metaphysical content of these lines – which foreshadows a large part of the evolution of philosophy in the centuries that followed – the idea that the true quality of sensation belongs exclusively to the subject is expressed with all the clarity one could wish. Thus we must not be surprised – especially given Montaigne's great influence on European thought as a whole – to see reappear from time to time this conception to which the nineteenth century was to give its definitive form. The same form is already found almost complete in Hobbes, who immediately following the passage we quoted above (p. 144), supports his statement that "their appearance [that of the qualities] to us is Fancy, the same waking, that dreaming" with the fact that "pressing, rubbing, or striking the Eye, makes us fancy a light" *(Leviathan* [London: Dent, n.d.], p. 3 [Meyerson's brackets]).

Müller's contribution, therefore, is actually reduced to the fact that he systematized the concept and stressed its importance. In addition, he initially had to defend it against numerous adversaries, for the interpretation of the fundamental fact on which Hobbes had relied does seem to have been contested, particularly in Germany; scientists, and in particular physicians, had formed the opinion that what was involved in this case was an actual production of light. That had even been the starting point for Müller's research: he had been called upon to give expert testimony concerning the assertions of a witness who claimed to have recognized a malefactor, in total darkness, by the light flashing from his eye following a blow the aforesaid malefactor had struck him. The other experts had generally found this assertion quite plausible, and Müller's debates with his colleagues led him to go more deeply into the question.[13] But this concept of the role of the sensory nerves was so consistent with the very principles of explanatory science that it could not fail to triumph rapidly.

One factor that no doubt contributed greatly to making its triumph complete is that, as physical theories progressed, it became more and more obvious that there could be no parallelism between the ways our different sense organs interpret the external phenomenon. One need only reflect on the absolute disparity between impressions of light and those of sound, as is indicated, for example, by the fact that a mixture of colors never forms anything except a single shade, while a group of sounds forms a chord, although in both cases the external phenomenon is considered to be a series of vibrations. Similarly, it was recognized that the phenomena we directly perceive form only a small part of those of the same nature which the external world has to offer: thus the narrow visible spectrum is actually flanked on both sides by considerable extensions, indicating the existence of rays to which our eye remains insensitive. Moreover, in order to transform these vibrations into light and sound, the eye and the ear use a total number of intervals extraordinarily different in range, the eye scarcely a sixth and the ear approximately ten octaves. Likewise one must admit, as Tyndall pointed out, that the intensity of our sensation varies quite differently from the energy of the vibratory motion involved.[14] Particularly striking discoveries have quite recently been made in this area concerning the maximum intensity of the sensation of light. It had been generally assumed (a viewpoint tacitly implied by many accounts in the classic texts) that the output of light increases indefinitely as the temperature rises. Now this is not so. The yield peaks at about 6000°, beyond which point it diminishes rapidly. Thus, using a stellar pyrometer, Nordmann found a temperature of 13,300° for the star Algol; however, in proportion to its total radiation, this star emits two times less light than the sun, whose temperature is only about 6000°. There seems to be a correlation between the temperature of the sun and the region of the light spectrum where our retina reaches its maximum sensitivity, a correlation that would obviously be the result of an adaptation of our visual organ, enabling it to use the sun's light as advantageously as possible.[15]

But science had already taken a new and highly important step in this direction toward the middle of the last century. Thanks to the work of

Ampère and Melloni, it was established that the impressions we receive through the different sense organs can be only one and the same external motion: for example, the same vibrations can be sensed by our eye as light and by our cutaneous organs as heat (cf. IR 330 [Loewenberg 293]). One need hardly point out that here again there was nothing that did not agree perfectly with the premises of theoretical science. Therefore scientists were perfectly willing to accept the data in question.

The same was not true for philosophers. To be sure, Montaigne, Hobbes and Leibniz loudly proclaimed the principle of the irrationality of sensation, as we have seen. But there has been no lack of contrary currents; and it is even a former Hegelian strongly tinged with materialism, D. F. Strauss, who declared scarcely fifty years ago, in defiance of Leibniz's demonstration, that he did not find it at all established that sensation was inexplicable from the scientific point of view, and that only time would tell.[16]

However, it is chiefly the statements concerning the last-mentioned discoveries – those establishing the identity, outside our organs, of the phenomena of light, heat, etc. – that elicited protests from highly respected philosophers, such as Lotze in Germany, and Boutroux and Bergson in France. These thinkers advanced more or less precarious theories. Lotze, for example, theorized that the qualities were actually inherent in the things themselves, that they could act on us only by movements, which movements then recreated these qualities in us, much as the telephone receiver reproduces the original sound, although the sound has traveled along the line in a quite different form. For Boutroux, as for Bergson, the movement, which we take to be simple is, on the contrary, complex, and our organs somehow draw from it diverse elements that already preexist there (cf. IR 333 ff. [Loewenberg 295 ff.]). None of these conceptions has had the slightest effect on the course of science.

Should we be surprised at this opposition? On the contrary, it seems to us that the sentiment behind it is not too hard to discern. It is, in fact, quite simply the deep and indestructible faith in the explicability, the rationality of nature. At bottom, can philosophy as a whole be anything else than an attempt to establish this rationality, or at least to get as close to it as possible? And consequently, if we accept the existence in nature of an element radically irreducible with respect to our reason, inexplicable, irrational, is this not tantamount, in the vivid words of a contemporary English philosopher, to a sort of *suicide* of this very reason? Modern philosophy, unlike ancient philosophy, has had to submit to this harsh extremity, as Burnet points out,[17] but its resistance to observations of this kind, when they come from without, is only too natural.

In sharp contrast to science's attitude toward the irrationality of sensation was its reaction to another discovery of the same kind, namely, that having to do with transitive action. Not only has science not excluded this action from the domain of explanatory theories, but one might say it has made it the foundation for these theories, since, as we have seen, all mechanistic explanation is ultimately based on impact. But we have also noted (Ch. 3, pp. 58 ff.) that science has been led to admit, for the sake of peace and quiet as it were, that there was no possible process for explaining, for enabling reason to conceive what happened at the moment when two masses were assumed to act upon one another. It went even further. For when it was in fact established that, even supposing this mechanical action to be entirely explicable, one was powerless to use it to explain duly observed phenomena, whose laws were known and which were considered important (such as electrical phenomena), science did not hesitate to turn the theory completely around, by reducing the mechanical phenomenon to the electrical (Ch. 3, p. 60). As a matter of fact, this curious about-face signifies quite simply that science dropped once and for all the idea of explaining transitive action, which it recognized as an irrational element. And it is surely quite remarkable (this is only a slightly different aspect of the reasoning we presented in our previous chapters) that it came to this realization, not through the demonstrations, however convincing, of philosophers such as Hume, nor even by considering the futility of the efforts of Huygens and Leibniz, Newton and Boscovich, Kelvin and Hertz and so many others, but through the simple concern for extending the domain of deduction.

Nevertheless, and under whatever form science finally accepted this notion, it is clear that it is by its very nature aprioristic, exactly as in the case of sensation. That results from the deduction of Hume, who, moreover, as we have seen, was not without predecessors. One cannot even say that, in this domain of transitive action, science has spelled out these notions of aprioristic thought as it had done in the domain of sensation: it has added nothing to Hume's formula, which it indeed accepts more implicitly than explicitly.

For two other irrationals, on the contrary (if we may be permitted for the moment to *enumerate* them in this fashion, leaving until later in the present work our explanation of what is to be made of this classification), the role of science has been much more active. Of course, their existence was able to be deduced by pure reasoning (and we shall see that that has actually been accomplished), but it could only be in the form of rather vague notions; progress in experimental science was needed in order to give them body and life, in order to endow them with a definite and truly convincing form. These two irrationals are those deduced from the existence of diversity in time and space. The reader will not be surprised to find these two diversities coupled in this way: we observed (Ch. 5, p. 118) how closely connected the two problems are and noted that philosophers had frequently passed, one might say without transition, from the first to the second. This is what Newton did in his turn in a passage where he affirms the irrationality of this double diversity. Newton's argument, which is found at the end of his *Principia*, is purely aprioristic. That fact will surprise no one except those who, on the strength of the *hypotheses non fingo*, have become accustomed to seeing this great man as the prototype of the *positivistic* scientist, distrusting all apriorism and basing his arguments strictly on experimental data. We have already shown where he really stood; and certainly Hegel was right to praise him for not having limited his program in this way, but for having quite often devoted himself, like any scientist worthy of the name, to pure reason.

Newton, then, having arrived at the last page of his work and looking back over its general outlines with a single glance, is led to pose the question of the deducibility of nature. No doubt the question presented itself to his understanding with all the more precision because a conception that prevailed over many minds, among them the best of his epoch – namely Cartesianism – claimed to have achieved precisely this global deduction. Therefore he obviously has Descartes and his disciples in mind when he affirms that "Blind metaphysical necessity, which is certainly the same always and everywhere, could produce no diversity of things," which diversity is found "suited to different times and places."[18] In other words, this double diversity in time and space cannot be deduced *a priori*; it is capable, in short, of no complete explanation: it is essentially irrational.

It is not clear that Newton was at all influenced by the opinions of earlier thinkers, and no doubt this most powerful of minds arrived quite independently at this profound conclusion. But anyone more or less familiar with the course of human thought will not be surprised to see that the thought of the great Englishman does not exist in isolation. Basically, the image of Parmenides' sphere, showing that what is undifferentiated in time and space cannot be deduced, made to conform to reason, already depends on an analogous point of view. The same can be said for the demonstration of God's existence that the Karaite Jew Jeshua ben Juda borrows (toward the end of the eleventh century) from the Arabic *kalam*: the atoms, which are uncreated, are indifferent to the place and time in which they are found; therefore the fact that they are found in a given place and time cannot be deduced from the very fact of their existence and must consequently have a special cause, which can only be the will of God. Or consider this claim by Maimonides: the only true attributes of God – the being whose existence is demonstrated *a priori* – are necessarily negative attributes;[19] in other words, anything particular, or diverse, cannot be deduced and appears irrational.

Also related to the same current of ideas is Gersonides' conception referred to in Chapter 1. Indeed, whether or not it is a theological monstrosity, it is certainly nothing less than a monstrosity from the philosophic standpoint, for it is quite obvious what Gersonides meant to say: God knows only true science, which can treat only the universal. If one takes into account the fact that for the Jewish philosopher, as for the Middle Ages as a whole, only what is deducible can be part of true science, the statement becomes: only the universal in nature is deducible – which is, in effect, to define deduction's outermost limits, that is, once again, to affirm that everything truly diverse is independent of reason.

Nevertheless, it must be understood that what these very general arguments provide is actually only a brief guideline. They warn us that in this domain, that of diversity in time and space, not everything can be made comprehensible or, to continue to make use of our own nomenclature, that it must contain irrationality; but it in no way follows that everything in it must be irrational. Indeed, if this were so, explanatory science would have no grasp of phenomena of this sort. Now, these phenomena, as we know, really form the whole domain of explanatory science. For example, to begin by speaking only of diversity in time, we saw with Riemann that the need for explanation arises from change, from the fact that there is a difference between the antecedent and the consequent. Therefore, if in fact everything in this change were irrational in nature, there would exist no scientific explanation of any sort, for no one would be able to imagine that the whole of enlightened humanity had for so many centuries devoted itself to such a futile exercise, the very illusion of explanation becoming impossible. But we hardly need labor the point;

the existence of explanatory science is a patent, undeniable fact. Thus, Hegel to the contrary (cf. pp. 281 ff. below), scientific explanation has been able to deal productively with this primordial problem of temporal diversity, which is to say, in a certain measure at least, to resolve it.

We know how science sets about accomplishing this task. Its principal instrument is mechanism. By affirming that everything must be reduced to different arrangements of eternally immutable parts, it maintains the permanence of being proclaimed by Parmenides, while at the same time "saving" the diversity of appearances.

Can this solution be complete? If it were, Newton's deduction would have led us astray and it would be possible to create diversity by means of the undifferentiated, to deduce diversity *a priori*. That is clearly absurd and thus, even supposing the mechanistic program to have been entirely successful, *it is not possible*, in reality, for these endeavors to have entirely achieved their goal.

The nature of the obstacle standing here in the way of our understanding of phenomena was not specified until the nineteenth century with what was perhaps the most memorable, the most scientifically productive discovery witnessed by this remarkably fertile century – that of Sadi Carnot. In the final analysis, what any explanation seeks is identity between the antecedent and the consequent. Now we all have a powerful immediate feeling that what there is between them is not identity, but diversity, that today is not exactly the same as yesterday and that tomorrow cannot be entirely assimilated with today, that time marches on, that phenomena follow a definite course in time, having a beginning, a middle and an end. The more complex a phenomenon is, the clearer that seems to us: for the phenomena of organic nature, the most complicated of all, the idea of reversing their course does not even occur to us. Who can imagine a world where men regurgitate their food, reconstituted, through their mouths and children go back to their mothers' wombs? For simpler phenomena, however, such a return does not seem as absurd to us, and finally, for a particular class of them, the phenomena of "pure" or "rational" mechanics, we explicitly stipulate the possibility of such a return, their "reversibility." There can be no doubt at the present time that this is an entirely artificial conception, that rational mechanics is only an abstraction, constructed, as its very name would indicate, to meet the demands of our reason; yet rational mechanics is what comes to mind when one supposes that nature is reducible to matter and motion. Therefore, before Carnot's discovery, and in spite of the just mentioned feeling of the march of time, physical phenomena are generally presumed to be reversible.

Leibniz, as we saw (Ch. 5, p. 117), explicitly formulated this postulate of reversibility, by stating that it must be possible to reproduce the cause from its effect. Carnot's major contribution was to show that the prototype of irreversible phenomena is an extremely simple phenomenon, namely, the transfer of heat, passing from a body of a higher temperature to another of a lower temperature. As a matter of fact, we all sense immediately that in this case return is impossible, at least by the same direct route, that heat will never flow naturally from a less warm body to a warmer one, and that a difference of temperature will never be spontaneously produced in a system of two bodies having the same temperature. Granted, that is only a fact of experience, but it is an experience so general that our thought is incapable of disregarding it: the fact of *heat transfer* certainly forms an integral part of our concept of heat. Furthermore, *it can only be* a fact of experience. For what is truly aprioristic, rational, in science must, as the meaning of the term indicates, conform to the needs of our reason. Now reason, which expresses itself through the principle of causality, demands that everything be conserved, be permanent, whereas Carnot's principle stipulates a continuous change in the same direction. Therefore this proposition contains nothing that caters to the innermost tendencies of our reason, it does not share at all in that *plausibility* (as we have put it)[20] that distinguishes the other very general propositions that can be brought together under the heading of principles of conservation. Moreover, the history of these various principles fully confirms this claim (see IR 298 ff. [Loewenberg 266 ff.]).

Carnot's principle is so implausible, the human race is so disinclined to believe in continuous change in one and the same direction, that it has always made great efforts to free itself from such a conception. Since the universe invariably suggested the idea of an incessant change, it was theorized that these successive states, granting that they were not identical, had to be *equivalent* and, after running through a cycle, had to return to their original state. This is the conception of the serpent *Ouroborus* (who bites its tail) and the *Great Year*, which is to be found in many cosmogonies and the latest repercussions of which are still to be found in the quite modern theories of Rankine, Spencer, Haeckel, and Arrhenius.[21] But Carnot's principle actually puts an end to any attempt to return to identity through the circuitous device of cyclical change; he teaches us in effect that the successive states of a system cannot be equivalent, that there is something essentially differentiating them, namely, energy, which, although it is conserved, although it remains constant in certain respects, nevertheless loses in quality, is constantly *dissipated.* This point is perhaps still not very well understood at the present time, as is shown by the perpetual resumption of the attempts we have just mentioned – attempts that are, once again, perfectly explained by the enduring and powerful force of the causal tendency.

Yet the resistance – doomed before it starts – to Carnot's principle and its consequences is not the most remarkable manifestation of the causal tendency in this domain; even more remarkable is the action by which science has come to explain the principle itself, to make it rational to a certain extent, by furnishing a mechanical theory for it, based on the concept of *probability* or, as they put it, on *statistical* conceptions. This theory, due largely to the efforts of Maxwell, Boltzmann and Gibbs, starts, like any mechanical conception, from the argument that it is possible to produce perceptible changes by modifying the order in which the elementary particles are classed; however, the theory in question has the particular feature of appealing to the very large number of these particles, which allows the laws of probability to come into play. The following example will make clear what we mean.

Consider two containers filled with the same gas at different temperatures. According to the kinetic theory, that means that the average speed of the molecules in each container is different; but of course we are dealing only with averages, around which the speeds of the molecules in the two containers *oscillate*: the speed of one particular molecule at a given moment can be quite different. Now let us put the containers in thermal contact. For that it is not necessary that the gases be able to mix; it is sufficient that the partition separating them become permeable to heat. The result, as we know, is that a complete temperature equilibrium between the two containers will be established more or less rapidly. Here is how the theory explains what has happened. The warmer gas molecules, having a greater average kinetic energy, communicate this energy, by impact, to the molecules of the wall, which in turn (the wall being supposed permeable to thermal motion) transmit it to the cooler gas. This process necessarily continues until the average speeds on both sides of the wall are equalized, that is, once again, until the two gases have the same temperature, at which time it of course stops, in spite of the fact that the impacts continue, each gas receiving on the average as much molecular kinetic energy from the other as it loses to it.

If we now consider the two containers taken together as a single system from the standpoint of heat – as it effectively is from the moment the wall becomes permeable to heat – we realize that at that very moment the distribution of the particles is *improbable*, since one part of the space contains particles with a more rapid average motion and the other contains particles with a less rapid motion, the two clearly separated by a plane of demarcation. Eventually, as the heat is communicated, the distribution tends to become less and less improbable, until finally, when the temperature has equalized, the average speed is the same everywhere, oscillating around a single average, the differences being only those consistent with the laws of chance; the distribution has become as probable as it can be.

In order to grasp the nature of this process more firmly, it may be useful to have recourse to a very simplified image, replacing motion, which distinguishes the particles in the kinetic hypothesis, by another property, color for example.

Let us then imagine a rectangular box having more or less the shape of a double cube. At the point where the two cubes touch, it is divided in two by a thin wall which can be inserted and removed at will. We put it in place and pour into each compartment an equal number of marbles, round and smooth enough to slide easily against one another and, moreover, identical in all respects except that those in the right compartment are white and those in the left one, black; there will of course be enough of them for the laws of probability to come into play, let us say several thousand. Having put them in place, we remove the wall and begin by shaking the box more or less vigorously a certain number of times. It is obvious that with each shake the marbles, clearly separated at the beginning of the operation, will tend to mix more and more; with each shake their original altogether improbable distribution – for if we had poured them into the box pell-mell, without there being a partition cutting it in half, it is highly improbable that, all by themselves, they would have turned out to be arranged in the prescribed manner with all the whites to the right and all the blacks to the left – will become more and more *probable*.

It is in this respect that our image resembles that of kinetic theory for phenomena obeying Carnot's principle – that is to say, in reality, for all phenomena except, of course, those occurring on the molecular scale. A body which is not thermally insulated from those around it but has a higher temperature constitutes a group of molecules of improbable distribution, equivalent to that which arranged the white marbles to the right and the black ones to the left in our box. But this improbability tends to diminish with time, just as that of the distribution in the box did when we shook it, until it finally disappears entirely – as will also occur in the box after it has received a great many sufficiently vigorous shakes. The shakes – the reader will no doubt have realized – are indispensable for our purposes; the particles of kinetic theory, because they are in motion, have a tendency to mix spontaneously, whereas those of our image are inert, because we have replaced motion by color.

Let us now return to reality, as represented by kinetic theory (the image of the box will serve us again later when we come to see the true significance of the principle), and consider, instead of a caloric phenomenon, a mechanical phenomenon, not as it is schematized by rational mechanics but as it is presented to us by physics.

Here is a body which is moving. By that fact, all its component particles have a common velocity with respect to those of the medium in which the movement takes place (excepting, of course, the velocities they both may have in virtue of their molecular movements). Here again we have an improbable distribution, in the sense that we recognized earlier. But, as a result of friction, the velocity of the body in motion tends to diminish, that is, part of its kinetic energy is transformed into the kinetic energy of the particles of the medium it drags along, while another part directly becomes molecular motion either of the body itself or of the medium, that is, becomes heat. The first part, the motion communicated to the particles of the medium, moreover, is also transformed into heat, due to the fact that these particles quickly tend to communicate their movement to neighboring particles and that each of these communications of motion conditions the transformation of a part of the molar motion into molecular motion. Finally, this heat produced by friction tends in its turn to be dissipated. We thus eventually end up with a group of bodies at rest with respect to one another and having the same temperature, that is to say a group in which the velocities of the particles are distributed in a *probable* way, like the marbles in the box after it has been shaken.

In this way each phenomenon that occurs in the world (always excepting molecular phenomena) plays a role analogous to a shake of our box; with each phenomenon the probability of the distribution increases. And this increase is clearly what determines the direction in which these phenomena occur; it is the mainspring of becoming, it is the reason we have before our eyes a continually changing spectacle in the world around us. Once again it must be understood that we are speaking of the world on the human scale, for if we observe through a microscope a cut made in an ore sample millions of years old and find trapped there a small amount of liquid in which suspended particles are swimming, we see them animated with molecular motion, Brownian motion, which has endured for those millions of years without changing or dissipating. But for molar motion, the rule appears to have no exceptions: everything happens in one and the same direction, with no possible turning around. To suppose otherwise is to suppose the possibility of a world of reversed phenomena.

Certainly the hypotheses of cyclical change discussed above, of which Arrhenius's theory is the most recent and best-developed, do not mean to appeal to any such reversal: they would not have us digest before we have eaten. For them, on the contrary, in the world around us, that is, not only in the terrestrial world, but also on the sun and in the immense majority of the stellar bodies, phenomena would proceed in the customary way and energy would continue to dissipate and be dispersed. But at some time, by chance, in some star or other, as the result of a cataclysm, the opposite event would take place, that is, energy would reconcentrate itself all at once, after which events would resume their course and energy would slowly begin to dissipate again, creating the innumerable phenomena that we know.[22]

Henri Poincaré has expressed the objections to the famous Swedish chemist's hypothesis in scientific language (see Appendix 4). But we believe that it is possible – using precisely the image with which we sought to illustrate how kinetic theory explains the continuous change imposed by Carnot's principle – to show why, once this theoretical conception has been accepted, any cyclical return becomes inadmissible, unimaginable.

In effect, what is being asked of us is quite simply to imagine that after being thoroughly mixed by a large number of successive shakes, the black and white marbles, as a result of a particular shake, could again find themselves distributed as they were at the beginning of the operation, the whites to the right and the blacks to the left, with a vertical plane separating them, as if we had just at that very instant removed the partition. Obviously we could effect this rearrangement ourselves by taking out the marbles and replacing them one by one. But that is the work of a conscious agent. Similarly, Maxwell's famous *demon*, who could open or close at will a molecule-sized aperture between two containers filled with a gaseous mass of uniform temperature, could sort out the molecules moving more rapidly from those moving more slowly. In this way he would manage to separate the gas into two masses of different temperature. But the demon too is an intelligent agent. What we are being asked to believe here, on the contrary, is that the separation could be brought about by an unconscious agent, a blind force of nature, not acting with an end in view – that is, in our image, by a single shake of the whole box.

To be sure, that is not impossible, strictly speaking. Everything about this distribution is only a matter of probability, and the eventuality envisaged can thus also be no more than extremely improbable. But we have a strong feeling that already in the case of our box the improbability is enormous. Moreover, it obviously increases with the number of elements involved: it is well known that if a single element is added to n others, the number of possible permutations is multiplied by a factor of $(n + 1)$. Thus in the universe, where the number of elementary particles appears to us as an extraordinarily large figure, the improbability of a return to the previous state is measured by a number of an even higher order than the order of the number of these particles itself. This observation is not without relevance, for many arguments in this domain seem to be implicitly based on the argument that the improbability of a return makes no difference, since there is infinite time for it to be brought about. But that is simply a mental exercise based on the supposition of the existence of a finite world in infinite time. If, on the contrary, the limits are allowed to increase at the same time for both of them, there is no doubt that the improbability of a return (that is, in short, the time necessary to bring it about) will increase at a much higher rate than the increase of the number of elementary particles. In our everyday life – our every act attests to it – we consider this improbability as being equivalent to the certainty of the contrary. A mason who, according to Perrin's excellent example, waited for the brick he needed to be lifted to the scaffolding by Brownian motion would quite properly be considered mad.

It is a fact that the hypothetical process occurring in distant stars by means of which Arrhenius intends to reestablish the course of events so that everything can begin again – whatever objections may be advanced – is far from appearing as extravagant to us. But that is only a consequence of this fundamental realization of the irrationality of Carnot's proposition. Indeed, no matter how strong a conviction we have of the order in which phenomena must occur, the conviction nevertheless contains no *a priori* element and is only a generalized experience; that is why the conviction only really comes to bear under circumstances not too different from those in which the experience was formed. On the other hand, when someone speaks to us of the celestial expanses and of forces whose action is little known or totally unknown, our imagination is liable to falter as to the direction in which phenomena will proceed. But we have only to return to kinetic theory and to its conception of increasingly probable distribution (which is what we did by calling upon the image of the box) to understand that events *must* have a specific direction and that there can be no turning back, even of a cyclical nature.

Obviously the believers in "eternal return" will always be able to fall back on the claim that all this is valid only for our limited world, while "in the world at large, ... quite other conditions obtain," as Haeckel said.[23] We shall see later what this way of avoiding the problem really means.

Thus it is impossible to escape from the grip of Carnot's principle by the old cosmogonic device that the Greeks called the *Great Year*, and therefore the irrational remains intact: we are forced to believe in an unending evolution, always in the same direction, and to suppose that we are located in a particular phase of this process. There is certainly something here that resists our reason, which will always be inclined to wonder why, since the world has existed for an infinitely long time, we have not yet reached the final state, Clausius's "heat death" (Arrhenius, *L'Evolution des mondes* iv). Of course we can console ourselves somewhat with the reflection that these are difficulties on the order of the "cosmogonic antinomies" set forth so well by Kant, difficulties which loom up each time the infinity of time and space is involved; and that in the particular case, if the final stage has not yet arrived, it is because the effect attributed to the infinity of time was no doubt counterbalanced by the effect of the spatial infinity of the universe, clearly acting in the opposite direction. Nevertheless, our imagination and our reason can obtain only very limited satisfaction from this quarter.

It is easy to see, however, that at bottom we are dealing here with something very general, to wit, the irremediable distaste our reason feels when confronted with any *given*, with anything that by its very nature seems to escape rational deduction. Why don't we live in the time of King George III of England? asks McTaggart (*Studies* 162). This is obviously another form of the very question posed by Arrhenius, but here we grasp more clearly that what disturbs reason is the realization that we find ourselves at a *particular* moment of a development that we are nevertheless obliged to consider continuous. Pascal stated the enigma in all its generality:

Why have limits been set upon my knowledge, my height, my life, making it a hundred rather than a thousand years? For what reason did nature make it so, and choose this rather than that mean from the whole of infinity, when there is no more reason to choose one rather than another, as none is more attractive than another? *(Pensées* 428 [Krailsheimer 87])

It is simply the realization that all these givens are irrationals, or that ultimately there must be irrationals; that even supposing we managed to deduce a certain number of them – that is, of course, to deduce them from other givens or, if one prefers, to deduce the givens in part from each other – we shall obviously not succeed in deducing them all.

The persistence and the definitive nature of the irrational underlying the concept of continuous change stand out perhaps even more precisely from a supposition necessarily entailed by kinetic theory, namely that of an improbable initial state. Indeed, given that things change because they tend to arrange themselves more and more in conformity with a probable distribution, it follows that they must have been distributed in an entirely improbable way at the beginning of time (no matter what meaning we attach to the expression). This initial distribution constitutes a precise irrational given. As a matter of fact, we could escape it only by assuming that this improbable state grew out of a more probable state, which would be to have recourse to eternal return, as we did earlier to escape the necessity of "heat death"; and we have just been persuaded that this is an impracticable way out.

This circumstance, however, must not keep us from recognizing what an enormous step explanatory science made toward the rationalization of the external world by the statistical theory of continuous change. Granted, mechanism explained change from its very inception – that is the very purpose for which the human mind constructed the theory. But these explanations had never sought to do anything more than make us understand change as *possible*. Now statistical theory goes further, making us understand it as *necessary*, as required by the very fact of the existence of a diverse world, that is, one constituted in opposition to the needs of our reason. On this account, therefore, change itself is rationalized, up to a point – and it is hardly necessary to point out that it is precisely due to this introduction of rational elements into the domain governed by Carnot's principle that we ourselves were able to *reason* just now about this principle and to point out the difficulties encountered by the supposition of cyclical change.

The problem of diversity in space, which is to say that of explaining the properties of substances, constitutes the principal task of chemistry (Ch. 5, pp. 133 ff.). We shall have occasion to return to this subject, at which time we shall see how this science tackled the problem. Let us only note in passing that the existence of a true clearly delimited and definitive irrational has not yet been recorded in chemistry. On the other hand, chemistry has, by another route, arrived at a precise notion of this sort which does enter into the same domain of spatial diversity: it is the data on the absolute size of molecules.

These discoveries (which we already had occasion to mention in connection with the triumph of kinetics over thermodynamics, Ch. 1, pp. 22 ff.) came as something of a surprise. Of course atomism is as old as science, and, particularly in chemistry, it had, since the beginning of the nineteenth century with Dalton, Avogadro and Ampère, become so prevalent that one can without fear of contradiction say that it was an integral part of science itself. Indeed, what formed its backbone, as it were, and at least partially transformed it into a body of rational doctrine – namely, chemical formulas – was totally imbued with atomism. In the late nineteenth century it would have been a truly impossible task to set forth a chapter of this science, and especially of organic chemistry with its countless derivatives, without introducing atoms and molecules. Thus Perrin was certainly correct when he summed up the situation by saying that actually "for a long time chemists had not seriously doubted a reality supported by so many confirmations."[24] Nevertheless, no doubt largely due to the perennially powerful influence of Auguste Comte's ideas, chemists have often denied in their theoretical statements the principles they invariably followed in practice. We saw (Ch. 2, p. 42) that this was true of the great physicist Maxwell. But many chemists went further still: they rejected, as if it were some sort of insult, the suggestion that they were capable of believing in the *reality* of these atoms and molecules, although they never ceased talking about them. For example, Henri Sainte-Claire Deville, in discussing the question of the duplication of certain atomic weights, which was a topic of the day, says: "At bottom, all these questions assume importance only in the eyes of those who admit and believe at the same time that atoms have an absolute weight."[25] Apparently he found the latter concept so preposterous that he believed he could use it as a *reductio ad absurdum*.

It seems reasonably unlikely that this great chemist, if he had lived thirty or forty years later, would have maintained the same attitude in the face of the closer and closer union of chemistry and atomism. However, it is at that time, and on the very eve of the discoveries which were going to confer upon it a veritable consecration, that atomism suffered extremely violent attacks on the part of a renowned chemist. Obviously we have in mind Ostwald, whose resounding campaigns were at first undertaken ostensibly, according to a competent critic, as a sort of reaction against the too rigorously materialistic conceptions of certain theorists;[26] but we must add that they quickly turned into propaganda campaigns in favor of the "energetistic" position of the author, who was thus combatting what he considered an illegitimate ontology only on behalf of another ontology, namely his own. Certainly many chemists disapproved of these attacks,[27] which have, moreover, remained without the slightest influence on the actual course of science: this is the epoch that saw the rise of the work summed up in Urbain's and Sénéchal's book, work which unquestionably springs directly from the atomistic conception. Nevertheless, the mere fact that Ostwald's writings appeared and were taken seriously by scientific opinion, at least for a time, seems rather significant as an indication of the lack of prestige of the atomic theories. It is just as remarkable that so little attention had been paid to the fact that by applying kinetic theory to well-known data developed by Clausius, Maxwell and Van der Waals it was possible to calculate a first approximation of the absolute number of molecules in a volume of gas ("Avogadro's number"). Physicists criticized this calculation because it required multiple hypotheses and had difficulty believing that the procedure enabled them to arrive at "molecular reality" – to use Jean Perrin's expression ('Les Preuves,' *Idées modernes* 5). Similarly, the quite convincing demonstration by which Gouy established the true nature of Brownian motion in 1888 initially created very little stir. However, little by little, scientific opinion began to be roused, especially when the atomistic conceptions received support from an unexpected quarter: electrical theories. As early as 1881 Helmholtz had expressed the opinion that electricity might exhibit an atomic structure,[28] but the suggestion at first fell on deaf ears. It is only much later that a whole series of discoveries, particularly Millikan's famous experiment (in which one sees, by direct observation of a droplet suspended in a gas, that the electrical charge passes discontinuously from one value to another), made this viewpoint compelling (Perrin, 'Les Preuves,' *Idées modernes* 46). From then on, obviously, the atomistic position in general acquired new prestige. We have seen that Lucien Poincaré, with great perspicacity, pointed to this shift in scientific opinion as early as 1898, his evidence being all the more impartial because he found this a regrettable turn of events (cf. Ch. 1, p. 22). He was not alone in his opinion, and even after Einstein and Smoluchowski had almost simultaneously developed the quantitative theory of phenomena (in 1905 and 1906), when a first cinematographic verification undertaken by Victor Henri gave negative results, "the physicists most attached to kinetic theory" were surprisingly ready to assume, as Perrin notes, that the calculations in question had to hide some unjustified hypothesis ('Les Preuves,' *Idées modernes* 30). But the resistance, quite explicable in terms of longstanding habits formed largely, as we have said, in response to the great prestige of positivistic ideas, did not last long in the face of the proofs that accumulated from that time on. At the present time, Perrin, who might be considered an interested witness because of his considerable role in the revolution that has just taken place, is not alone in proclaiming that "it is becoming ... difficult to deny the objective reality of molecules" ('Les Preuves,' *Idées modernes* 21). Bouty, while expressing doubt as to whether these hypotheses "are a definitive, rigorous expression of reality," nevertheless concedes that "at any rate, they furnish a good approximation of it."[29] Edmond Bauer states that "the molecular constitution of matter can no longer be doubted,"[30] and Henri Poincaré, so loath to exaggerate the value of theories, notes in one of his last writings that "the old mechanical and atomic hypotheses have, during recent years, become so plausible that they have ceased to seem like hypotheses; atoms are no longer just a convenient fiction. It seems almost as if we could see them, now that we know how to count them."[31]

What are the implications of this testimony from the point of view of the existence of the irrational? We have seen (Ch. 5, pp. 135 ff.) that any theory of matter ultimately ends up identifying it insofar as possible with space, the identification generally being carried out (at least in contemporary science) by successive stages. For example, in current kinetic theory, we do presuppose discrete corpuscular atoms, but since we then take them to be composed of subatoms or electrons, the latter being in their turn conceived as "singular points" in the ether, the continuity of the ether, which seemed to be broken by the assumption of the atom, is reestablished. As Bergson astutely perceived, the essential feature of the explanations of matter by the ether is the elimination of "that discontinuity which our senses perceived on the surface."[32] Thus there are two successive operations involved, which must contribute to the same goal yet necessarily go in opposite directions, the first consisting in diversifying atoms, and the second in making this diversity disappear. And it is obvious that the less thoroughly the first operation has been carried out, the easier the second will be to accomplish, in other words, that the less the atoms and the corpuscles have been differentiated from the surrounding medium, the more easily they will be resolved into the undifferentiated ether. From this standpoint, the vagueness characterizing atomic theories up until the discoveries of the last few years could not have been more propitious. No doubt the scientists who professed not to believe in "molecular reality" were above all obeying the positivistic injunction to forgo all ontology; but perhaps some of the very same scientists who dealt with these theories felt vaguely that there was an advantage, generally speaking, in not making the entities they created too substantial, so that, having remained shadowy, so to speak, they could later be dissolved into nothingness. The determination of the absolute dimensions of molecules puts an end to this twilight state which invited misapprehensions. It clearly defines the situation by showing that there is a precise *given*, a diversity which is definitive. For even if we then reduce the molecule to atoms and the atoms to subatoms, it will nevertheless remain true that at an average given distance from a material center, there is *something else*, namely space empty of all matter (whatever sense one may give to this word), after which one runs into another center. In other words, the subsequent dissolution of the molecule may well explain to some extent – by the nature of the atoms and subatoms or by properties with which one will endow the ether – the absolute dimensions that the present-day experiments and calculations force us to attribute to the molecules, and these givens will thus no longer appear to us as being *ultimate*. But this explanation will not eliminate the molecular discontinuity that has now been established; it can only add to it new discontinuities, within the molecule itself, just as this molecular discontinuity does not eliminate the discontinuity which visible objects establish in space, but adds itself to it. That is certainly a result that might have been anticipated for the time when molecular dimensions would become known. All science has done is confirm the existence of an irrational whose existence could be deduced a priori, but it has delineated and elucidated the notion in a singular way.

Let us note, however, that the analogy between the two irrationals we have discussed last – namely those that reveal, on the one hand, "the improbable initial state" in the statistical theory of Carnot's principle and, on the other, the absolute dimensions of atoms – although it is real and profound, as is shown by the aprioristic considerations and in particular by the comparison with Newton's line of reasoning, is nevertheless extremely limited. What is analogous is solely the problem in its most abstract form – the existence of *diversity* in time and space – and the instrument by which science went about resolving it, also conceived in its most general form: mechanism. But in all other respects, for example, the exact way in which science proceeded and the scope of the result obtained, there is no resemblance. On this last point in particular, that of the nature of the result, it can easily be seen that the statistical image really embraces the entire domain of this vast concept of temporal change; at least it seems to embrace it, that is, we *can*, if we must, imagine that everything fitting into this category must be reducible to displacements in conformity with the laws of probability – whereas the absolute dimensions of molecules clarify and make more precise only one well-defined aspect of the problem of diversity in space. It should be noted in this regard that the work involved in determining Avogadro's number refers to molecules and not to atoms. In other research it is really the atom that is concerned; but it is the atom as chemistry conceives it, that is, the qualitative atom. Finally, theories like those of Sir J. J. Thomson, Rutherford or Moseley actually tend to reduce the diversity of the chemical atom to the uniformity of the two electricities. But as important as these conceptions may be, it is certain that the explanation of the almost infinite number of properties we are obliged to attribute to the chemical atom is hardly even sketched out. Given the purely empirical way, by simple trial and error, in which science is obliged to proceed on this path, it is of course entirely impossible to foresee to what extent this future effort of explanatory theory will be able to succeed and what obstacles – irrationals – it will come up against. Perhaps the most we are able to conjecture is that there will very likely be grounds for admitting new irrationals. What leads us to envisage this eventuality is the fact that chemistry, in spite of the attempts that have been made throughout the ages, or at least since chemistry has existed as a science, to incorporate it into physics, still unquestionably presents all the characteristics of a distinct science. Certainly such efforts have become especially vigorous of late and have led to impressive results in this or that particular specialty. A whole science has been created, recording extremely interesting observations and results; its name, *physical chemistry*, makes clear enough what it hopes to accomplish. On the other hand, the term *physicochemistry*, which is encountered more and more frequently and which designates the two sciences taken as a whole, might seem to suggest that the merger has already taken place. However, one need only look a bit more closely to discover that there may well be a program in place, but no result, and that actually the specificity of chemical phenomena remains intact. Now this specificity could well hide one or more distinct irrationals. Of course it is not at all impossible that this irrational or these irrationals are linked with one or more others whose existence further investigation will force us to recognize in one branch of physics or another. For example, Max Planck is inclined to allow that the boundary between phenomena that evolve continuously (according to the laws of classical dynamics) and those produced by *quanta* of action is the dividing line between physical and chemical phenomena. These *quanta*, which would undoubtedly constitute a new irrational, would therefore serve to explain the specific phenomena of chemistry.

Entire molecules, atoms and perhaps also free electrons would move according to the laws of classical dynamics; atoms and electrons constrained by a molecular bond would obey the laws of the theory of *quanta*. Physical forces, gravitation, electric or magnetic attraction or repulsion, cohesion would be exerted continuously; chemical forces, on the contrary, by *quanta*.

That is obviously possible, and it would be hard to find a more qualified opinion on the subject than that of the author of the theory of *quanta* himself. One must bear in mind, however, that this hypothesis arose not in connection with chemical phenomena, but with those of black body radiation. Planck, it is true, seeks to establish a connection between the two kinds of conceptions, pointing out that the law of action by *quanta* would be connected with the law according to which, in chemistry, masses "can act only in clearly defined and discontinuously variable proportions," while in physics they act in any quantity whatever.[33] There may indeed be a profound analogy there and the second of these discontinuities may be able to be deduced from the first. But for the moment that is only a pure and simple possibility, since the deduction does not yet seem to have been attempted, and at first glance the link between this hypothesis and those used in current attempts to explain the notion of valence (like those of Abegg or Werner, for example) or to furnish an image of the structure of the chemical atom (as Sir J. J. Thomson undertook to do) does not seem very obvious.[34] But once again Planck's opinion carries great weight and one can therefore only wait and see what the future will bring.

Moreover, even if, as Planck hopes, this irrational should ultimately appear to be demanded by the distinction between physical and chemical phenomena, that is, to be an anomaly whose appearance could to some extent have been anticipated at this particular juncture in science, it would not remain any less true that the notion arose in connection with black body radiation, where nothing, it would seem, suggested the existence of such an anomaly, which is to say that in reality it appeared *unexpectedly*. Brillouin pointed this out when he summed up the results of the discussions at Brussels in a sort of minimum formula "which might appear quite timid to the youngest" of those in attendance. "It seems quite certain," said the eminent physicist, "that from now on we must introduce into our physical and chemical conceptions a discontinuity, an element varying by jumps, of which we had no idea a few years ago" ('Conclusions générales,' *Brussels Conf.* 451). Indeed, this *unexpected* aspect of the new irrational came out quite strongly in the course of the Brussels discussions, and it is easy to see that the perplexity of the participating scientists was principally due to their feeling that scientific explanation had run up against a new obstacle (if not several of them) whose extent cannot even be defined at the present time.

Brillouin's observation is very important, for it proves that the irrational is essentially unpredictable, that it can arise unexpectedly anywhere, even in the phenomena we thought we knew the best and whose theory may have seemed more or less complete, definitive. For example, for a long time the movements of the stellar bodies belonging to our planetary system appeared, in virtue of Newton's discovery, to form a privileged region of science, a region where explanation was complete. To be sure, there were anomalies that could not be completely accounted for, such as that of the motion of Mercury. But it was hoped that future discoveries – and in particular that of an intra-Mercurial planet, which Leverrier sought – would permit these irregularities to be eliminated without having to modify the foundations of the theory. Now, as we know, this research was unsuccessful. On the other hand, Einstein's new theory is said to account completely for the anomaly in question. If the views of this physicist are generally accepted by astronomers, the result will be a profound upheaval in the Newtonian theory, since the very conception of space, which is its essential foundation, will be entirely modified. And who can say whether future observations by means of instruments whose precision has been immensely increased, or even by means of research procedures as yet unknown, will not still later come to destroy the whole edifice. We need only recall the surprise caused by the discovery that it was possible to do research on the chemical composition of the stars (a possibility Auguste Comte had gone out of his way to deny explicitly a short time before) to make us very circumspect in this regard.

At the same time it is important to note that even where we might possibly suspect the existence of an irrational, we are entirely incapable of predicting what form it will take. Let us consider spatial and temporal diversity. There is no question that these two concepts are closely connected, and Newton could guess that this double diversity concealed irrationals. But even the genius of a Newton would have been inadequate – unless he followed exactly the path mapped out by Carnot, Maxwell and Boltzmann – to guess that change would be made rational by means of statistics and that then the improbable initial state would emerge as the irreducible element.

This is why, for example, science, as we remarked earlier (p. 163) teaches us nothing about a chemical irrational (or irrationals, since there could be several). No doubt its existence is extremely likely, but in no way is it certain. As a matter of fact, less than a century ago one could have made a similar assumption concerning the phenomena of light, which then appeared to have nothing in common with electrical phenomena except the fact of being able to be treated by the theory of central forces. Much later it took the genius of a Maxwell to surmise that the apparent dissimilarity hid an identity that did not encounter any irrational.

It is all the more impossible to say what form the chemical irrational will take when science defines it more precisely. Will we be able to reduce all properties of compounds to those of the elements, that is, will we succeed in establishing for the elements a conception such that, through it and through the position of the elementary particles in space, all chemical reactions as well as all the physical phenomena exhibited by the bodies in question will be explained? We do not know, and all we can say is that it cannot be claimed at the present time that there is something inaccessible to our reason here; on the contrary, we can perfectly well allow that, particularly by endowing elementary particles with more or less complex properties, it may be possible to succeed in conceiving all the rest to be rational.

As a consequence, all of the irrational will be concentrated, so to speak, in the elements. That would be altogether logical, since the true element, that which must remain undecomposable, indestructible, uncreatable, is by definition an irrational, something that reason is condemned to acknowledge as an eternally recalcitrant given. In this sense Schelling was right when he said that the chemical elements "are nothing else than just so many refuges of your ignorance" (*Ideen*, I, 2:27; cf. 2:295–297 [Harris 21; cf. 233–236]); but long before him and at a time when Lavoisier's reform was still in limbo, Bailly wrote that "the elements of the world are the last retrenchments of nature."[35]

Obviously, if we assume that physicochemistry will really succeed in carrying out its program, it will be necessary for these retrenchments to be taken by storm, that is, for it to be recognized that the chemical elements are not true elements. Will the discontinuities that served as the basis for the concept of *quanta* be sufficient to the task (which, as we have seen, does not appear very likely) or (as is more plausible) will they play at least some role in this process of rationalization? Will there be general satisfaction with the concept of discontinuity as the form of the irrational element, or will it be necessary to resort to concepts of a kind not yet even contemplated? It certainly seems that we can say nothing on this subject, or, to be more precise, can make only a quite general and, as it were, negative prediction. As a matter of fact, it appears highly unlikely that all the properties we currently attribute to the elements will be recognized as irreducible, as ultimate: speculations on the interrelations of the elements stand in opposition to it. No matter how large a role we attribute to pure hypothesis in his speculations, there is no denying that Mendeleev's periodic table, confirmed and refined by the recent admirable discoveries of Moseley,[36] establishes links between a good many properties of the chemical elements, and Sir J. J. Thomson's theory shows us that these relations are perfectly capable of being explained by assumptions concerning the nature of the atom, that is, by means of the way its parts are arranged in space.[37] Thus the very fact that one recognizes the probable (or even certain) existence of the irrational in a certain class of phenomena in no way signifies that theoretical science must cease trying to explain and rationalize them. This proposition is demonstrated by the history of all the irrationals discovered one after another. To cite only one example, it is clear that insofar as temporal change is concerned (if one disregards Newton's deduction), the fact that thermal phenomena do not conform to the requirements of our reason, which demands conservation, was well established by Carnot, although Maxwell and Boltzmann much later succeeded in introducing kinetic theory into this area and thus partially rationalizing it, thereby giving the irrational a definitive form. But basically, as we have said (p. 153), the very existence of theoretical science as a whole already suffices to demonstrate that this is the case. Indeed, man, in spite of his invincible tendency to believe in rationality, obviously sensed very early that there is irrationality in nature, that nature is not entirely explicable. We saw in Ch. 4 (p. 92) that the boldest speculations of Ionian philosophy surely implied certain mental reservations in this regard. At any rate, Heraclitus's statements on universal change express this sentiment quite clearly and those throughout history who have questioned nature through experiments were thereby implicitly affirming that they were abandoning the idea of deducing it. Yet humanity has not ceased seeking to explain nature, and furthermore its efforts have largely been crowned with success. Our proposition thus is actually a truism. Nevertheless, we shall see a bit later how failure to recognize this obvious truth was able to create misunderstandings (cf. pp. 183 ff.).

In sum, we can make only negative or altogether imprecise pronouncements in this area. We know where complete rationalization is impossible, that is, where the agreement between our reason and external reality comes to an end: those are the irrationals already discovered. But we do not know – and shall never know – where the agreement exists, since we can never be sure that there will be no new irrationals to add to the old ones. That is why we shall never be able really to *deduce* nature, even by taking into consideration all the given and irreducible elements, all the irrationals that we know at a given moment; we shall always need new experiments and these will always pose new problems, causing new contradictions between our theories and our observations to leap out at us, as Duhem puts it.

NOTES

1. We shall see later, p. 388, note 3, that philosophers have attempted to use more or less complicated contrivances to deduce the tridimensionality of space, which also goes to prove that this is not a determination our reason immediately recognizes as its own.
2. Friedrich Wilhelm August Mullach, *Fragmenta philosophorum graecorum* (Paris: Ambrosio Firmin Didot, 1860), p. 357 [G. S. Kirk and J. E. Raven, *The Presocratic Philosophers* (Cambridge: University Press, 1957), p. 422].
3. Cf. Ralph Cudworth, *The True Intellectual System of the Universe* (London: Richard Royston, 1678), p. 8.

4. Paul Natorp, *Forschungen zur Geschichte des Erkenntnisproblems im Alther-thum* (Berlin: W. Hertz, 1884), p. 183.
5. Henri Bergson, *Matière et mémoire* (Paris: Félix Alcan, 1903), p. 66 [*Matter and Memory*, trans. Nancy Margaret Paul and W. Scott Palmer (London: Swan Sonnenschein, 1911), p. 79].
6. F. H. Bradley, *Appearance and Reality* (London: Swan Sonnenschein, 1893), p. 493.
7. Henri Poincaré, *La science et l'hypothèse* (Paris: Flammarion, n.d.), p. 193 [*Science and Hypothesis*, trans. George Bruce Halsted (New York: The Science Press, 1905), p. 116].
8. Baron Georges Cuvier, *Histoire* 1:235. Actually we have a few reservations about citing this passage as evidence of the attitude of the science of the time. As a matter of fact, everything goes to show that the great biologist was well aware of contemporary philosophic thought, even of German thought – he quotes Kant and even, quite frequently (to refute them), the *Naturphilosophen*, and the terminology of the passage, with its allusion to the *self* and to *moral sciences* would tend to suggest an extrascientific inspiration. Schelling already noted the influence of German philosophic thought on Cuvier (*Zur Geschichte*, I, 10:200). Moreover, we know that Cuvier, originally from Montbéliard, which until 1792 belonged to the Duke of Württemberg, spent several years as a scholarship student at the *Kazlsschule* of Stuttgart.
9. Alexandre Herzen, *Le Cerveau et l'activité cérébrale* (Lausanne: J. B. Baillière et fils, 1887), pp. 33–34.
10. Ioteyko and Stefanowska, *Psycho-physiologie de la douleur*, reviewed in *Rev. de méta.* **17** (1909) Supplément: 7.
11. See the passages of Theophrastus quoted by John Burnet, *L'Aurore de la philosophie grecque*, trans. Reymond (Paris: Payot, 1919), pp. 281 ff. [*Early Greek Philosophy*, 2nd ed. (London: Adam and Charles Black, 1908), pp. 284 ff.].
12. Montaigne, *Essais* (Paris: Flammarion, n.d. [1908]), Bk. 2, Ch. 12, 2:296, 347, 348–349 [*The Complete Essays of Montaigne*, trans. Donald M. Frame (Stanford: Stanford Univ. Press, 1958), pp. 422, 452, 453–454; we have inserted our own translation for material in the second quotation not included in the Frame translation (p. 452)]. It should be noted that Montaigne does not cite Lucretius on this subject, which fact, given his customary procedures, would seem to indicate that he was not influenced by him, or at least that the influence was only indirect.
13. See Hermann von Helmholtz, *Vorträge und Reden*, 4th ed. (Brunswick: Friedrich Vieweg und Sohn, 1896), 2:220.
14. We have treated these considerations at somewhat more length in IR 331 ff. [Loewenberg 293 ff.].
15. See Nordmann, 'Le Rendement lumineux des corps,' *Scientia* **13** (1913) 477.
16. David Friedrich Strauss, *Gesammelte Schriften* (Bonn: Emil Strauss, 1877), 6:269. Trendelenburg also declares that "the activity of the sensory nerves has *not yet* been reduced to motion" (*Log. Untersuch.* 1:209).
17. John Burnet, *Early Greek Philosophy* (London: Adam and Charles Black, 1892), p. 191. The picturesque expression concerning the "suicide" of philosophy is not

found in the corresponding passage of the French translation (*L'Aurore de la philosophie grecque*, trans. Reymond, Paris: Payot, 1919, p. 207), which is based on a different edition of the original.

18. Newton, *Principia*, 3rd ed. (London: Innys, 1726), p. 529 [*Mathematical Principles of Natural Philosophy*, trans. Andrew Motte and revised by Florian Cajori (Berkeley: Univ. of California Press, 1934), 2:546; we have substituted "diversity of things" for "variety of things" at the end of the first quotation to conform to Meyerson's French translation]: *A caeca necessitate metaphysica, quae utique eadem est semper et ubique, nulla oritur rerum variatio. Tota rerum conditarum pro locis ac temporibus diversitas, ab ideis et voluntate entis necessario existentis solummodo oriri potuit.* The passage is not found in the more often cited second edition (London, 1713), which is reproduced in the Amsterdam edition of 1714; it is thus a reflection of Newton's later years (he was seventy-one in 1713 and eighty-four when the third edition was published). The *General Scholium*, which contains the passage quoted, begins with an account of the difficulties encountered by the vortex theory with respect to planetary motion; there is thus no doubt that Newton was indeed thinking of the Cartesian deduction.
19. Isaac Husik, *A History of Medieval Jewish Philosophy* (New York: Macmillan, 1916), pp. 56, 265.
20. [Meyerson uses the term "plausible" to denote statements "intermediary between the *a priori* and the *a posteriori*." Such statements are instances of general statements which are *a priori*, but, insofar as they are *a priori*, indefinite. Only experience can make them definite, "but in this matter experience plays a peculiar role, in the sense that it is not free," since it must conform to the more general *a priori* constraints involved. Such statements are therefore not strictly *a priori* nor merely *a posteriori*. They are, in this technical sense, *plausible* (IR 159–160; Loewenberg 147–148).]
21. Cf. IR 302–303 [Loewenberg 269–270]. The *Philosophers of Nature* also formulated conceptions of this sort. Cf. for example Schelling (*Weltseele*, I, 2:349–50, 381) on the current which, in the organic as well as in the purely mechanical domain, "turns back into itself" and on the "invisible power" reducing all the phenomena of the world to the eternal circular current. Cf. also *Erster Entwurf*, I, 3:124, and *Transc. Idealismus*, I, 3:490 [Heath 121–122].
22. Svante Arrhenius, *L'Evolution des mondes*, trans. Théophile Seyrig (Paris: Ch. Béranger, 1910), pp. iv, 204.
23. [Ernst Haeckel, *Les Énigmes de l'univers*, trans. Camille Bos (Paris: Schleicher frères, 1902), pp. 283–284 (*The Riddle of the Universe*, trans. Joseph McCabe, New York: Harper & Brothers, 1902, pp. 246–248)]. Cf. Ch. 8, p. 215 below.
24. Jean Perrin, 'Les Preuves de la réalité moléculaire,' *Idées modernes* 1. Smoluchowski similarly attests that since Dalton "chemists have never ceased to think as atomists, not even some twenty years ago when philosophers and physicists (Mach, Ostwald) had inspired a short-lived but powerful movement against it" ('Anzahl und Grösse der Moleküle und Atome,' *Scientia* **13** (1913) 28. On the other hand, Perrin notes that many chemists saw in atomic theory "only a useful tool and expressed reservations, sometimes strictly verbal if truth

be told, concerning the fundamental issues in question" ('Les Preuves de la réalité moléculaire,' *Brussels Conf.* 157).

25. Henri Sainte-Claire Deville, *Leçons sur la dissociation professées devant la Société chimique de Paris, le 18 mars et le 1er avril 1864* (Paris: Ch. Lahure, 1866), p. 354 [erroneous citation]. Kolbe's attacks against Van't Hoff on the subject of the stereochemical conceptions, and his lack of moderation as well, are in the same spirit. Kolbe considers the fact of having sought the position of atoms in space the height of audacity [*Dreistigkeit*], and this way of treating scientific questions seems to him to be "not too far from believing in witches and rapping spirits" [Meyerson's brackets]. See J. H. Van't Hoff, *Dix années dans l'histoire d'une théorie* (Rotterdam: P. M. Bazendijk, 1887), pp. 19–20 [*Chemistry in Space*, trans. J. E. Marsh (Oxford: Clarendon Press, 1891), pp. 17–18].
26. Giuseppe Bruni, review of Wilhelm Ostwald's *Prinzipien der Chemie*, in *Scientia* **4** (1908) 380.
27. Thus Van't Hoff, though he was linked with Ostwald by common campaigns (in favor of ideas belonging rather to the former than to the latter of these two chemists), made a clean break with him when he declared war on atomism. At the Congress of Vienna in 1906, Van't Hoff affirmed that atomism would still render great services (Bruni, 'L'Oeuvre de L. H. Van't Hoff,' *Scientia* **10** [1911] Supplément: 60). On the lack of consistency in Ostwald's system concerning the definition of the elements from the standpoint of the laws of chemical composition, cf. Alfred Werner, *Neuere Anschauungen auf dem Gebiete der anorganischen Chemie* (Brunswick: Friedrich Vieweg & Sohn, 1913), p. 2.
28. Edmund T. Whittaker, *A History of the Theories of Aether and Electricity from the Age of Descartes to the Close of the Nineteenth Century* (London: Longmans, Green, and Co., 1910), p. 397. It should be noted that Whittaker dates the beginning of the renaissance of atomic theory from this statement by Helmholtz (made in a lecture at the Chemical Society of London).
29. E. Bouty, 'La théorie cinétique des gaz, Deuxième partie: Ses progrès et ses difficultés,' *Scientia* **19** (1916) 266.
30. Edmond Bauer, 'Les Quantités élémentaires d'énergie et d'action,' *Idées modernes* 115.
31. Henri Poincaré, 'Les Rapports de le matière et de l'éther,' *Idées modernes* 357 ['The Connection between the Ether and Matter,' *Smithsonian Institution Annual Report, 1912* (Washington, 1913), p. 199].
32. Henri Bergson, *Matière et mémoire* (Paris: Félix Alcan, 1903), p. 223 [*Matter and Memory*, trans. Nancy Margaret Paul and W. Scott Palmer (London: George Allen & Unwin, 1911), p. 266].
33. Max Planck, 'La Loi du rayonnement noir et l'hypothèse des quantités élémentaires d'action,' *Brussels Conf.* 114.
34. Victor Henri, in his research on the spectra of absorption, through which he has succeeded in penetrating so deeply into the domain of the inner structure of the chemical molecule, arrives at the conclusion that "the theory of *quanta* cannot serve as a guide for the study of the chemical effects of radiation" and that consequently "there are grounds for introducing other theoretical considerations

on the structure of the molecules and on intramolecular energy" (*Etudes de photochimie*, Paris: Gauthier-Villars, 1919, p. 215).

35. Jean-Sylvain Bailly, *Histoire de l'astronomie ancienne*, 2nd ed. (Paris: De Bure fils ainé, 1781), p. xi.
36. Moseley, while studying the x-ray spectra of the elements, succeeded in establishing that *the square root of the frequency of a given line of the spectrum is a linear function of the atomic number*. Mendeleev's conception is thus provided with a precise numerical basis, while at the same time the anomalies exhibited by the system are explained. A brief résumé of this work will be found in Adolphe Lepape's Appendix to Frederick Soddy, *Le Radium* (Paris: Félix Alcan, 1919), pp. 337 ff. How tragic that Moseley was killed in the Dardanelles at the age of 28.
37. We shall deal with this theory at more length in Ch. 8, p. 230.

CHAPTER 7

BIOLOGICAL PHENOMENA

Still, even taking into account this very important qualification, the image of science at which we have arrived cannot help astonishing us. It seems to shock us most of all, of course, when we consider the biological sciences. Would science really ever presume to use these same methods of reduction and spatial assimilation to approach, to attack those infinitely particularized beings, at once so changeable and so persistent, so distinct from what surrounds them, in short all that prodigious whole we call life?

However, let us recall Bouasse's very sound and important remark to the effect that "all the sciences of nature" strive to resemble physics[1] (p. 98). This suggests that the aims of the biological sciences may not differ fundamentally from those of the physical sciences.

We must nevertheless admit that the image currently presented by the biological sciences is quite different from the model provided by physicochemistry. The most striking difference is the considerable place all the sciences of the living being accord to finalistic considerations.

We have already touched upon this question in Chapter 2. We came to see that the concept of an end did, as a matter of fact, have a certain explanatory force and that its intervention in science was motivated by the human mind's resistance to the conception of a purely lawlike phenomenon entirely deprived of explanation: that is why, if all causal deduction seems lacking, the finalistic explanation appears capable of filling the gap to some extent. And since the sciences of the organism obviously are still in rather a primitive stage of development at the present time and since cases of actual reduction of biological processes to purely physical ones, indeed even the beginnings of such reductions, are very rare,[2] the *quantitative* prevalence (if we may use this term) of the finalistic considerations should not surprise us. In fact they are so prolific that still today scientists, including highly competent ones, sometimes bind them together into coherent bodies of doctrine embracing a whole class of phenomena, indeed even claiming to embrace the totality of the phenomena of the organism, at least in certain respects. They then form what are called *vitalistic* theories, the designation signifying that, for all the phenomena located within the declared limits of the theory, everything takes place according to particular rules entirely distinct from those valid for the inanimate matter treated by the physical sciences – that the vital processes are "autonomous," as one of the protagonists of the doctrine, Driesch, puts it.[3]

We cannot relegate these conceptions entirely to the past, as is sometimes done. On the contrary, in recent days they appear to have regained considerable strength and a certain favor among biologists, particularly in Germany, as the result of the work of Driesch and others.

A superficial glance at the history of the biological sciences might even suggest that what we have here is not progress of the science in one particular direction, but a struggle between two equivalent principles that dominate by turns. Mechanistic theories of vital phenomena actually have a long history; they abound among the ancient atomists and also among the scientists of the Renaissance; they appear to triumph in the materialistic philosophy of the eighteenth century. And yet vitalism subsists and even seems to flourish again, as we have just noted. As a matter of fact, in the history of biology, vigorous shifts of opinion in either direction frequently seem to have provoked equally violent reactions. For example, the radical mechanism of Boerhaave and Lémery was followed by the equally extreme vitalism of Stahl. Must we conclude that this is nothing more than mere seesawing back and forth? On the contrary, closer examination reveals that such is not the case, that in reality finalism, or vitalism, has consistently lost ground and that its retreat is a direct and ineluctable consequence of progress in the physical sciences. Rádl, one of the most stubborn champions of the vitalistic cause, openly acknowledges as much in a work teeming with extravagances and contradictions, but also concealing, in addition to copious but sometimes unreliable erudition, a few not unoriginal views. From one end of his book to the other, he never stops deploring what he calls "the decline of biology," a decline that he sees as beginning in the Renaissance, or at least immediately after Paracelsus (whom he considers the champion of "absolute vitalism"), and still continuing today.[4] First of all, this view cannot help surprising us, particularly since its author provides no evidence of biology's lofty conquests in earlier epochs or even at the time of Paracelsus. But this is because what Rádl is really lamenting is simply the decline of biology's *prestige* among the sciences. What grieves him is the waning or loss of this prestige, the surrender of biology's position as an entirely independent science (independent above all, of course, of the physical sciences), which upon occasion had even dominated the entire scientific domain. In other words, he identifies biology with the most radical vitalism, which explains why its loss of status seems to him to be conditioned by the very rise of the science as we know it. "The constitution of modern physics was brought about at the expense of biology" (Rádl 1:122); "Caesalpino has succumbed, Galileo's science has triumphed, but at the same time biology has fallen into decline" (1:126); "Galileo's lifelong battle ended up as an extermination campaign against biology" (1:153). The author follows this "decline" through the centuries, lavishing the bitterest (and sometimes, one must add, the most unexpected) criticisms against the eminent minds who have made the science what it is today, whenever these men manifested even the slightest tendency to connect the science of the living being to that of nonorganic matter. Thus seventeenth century biology is a "science of epigones"; Leeuwenhoek is only a "dilettante"; there is "no original thought" in Réaumur; Albrecht von Haller is a thinker "of an extraordinary platitude," for whom vitalism is evidently "too serious" a doctrine, etc. (1:163, 173, 174, 239). Even Leibniz himself is not spared: he is only a "typical representative of an epoch that tended toward universality and genius but remained attached to the petty side of things"; his philosophy "tended disagreeably toward compromises and set aside all that is sincere and radical, truly profound and healthy" (1:220, 222). The author's fierce prejudice, attesting to the sincerity and intransigence of his vitalistic conceptions, can obviously only enhance the value of his testimony here.

Moreover, as more is learned about physicochemical phenomena on the one hand and biological phenomena on the other, the loss in strength of the vitalistic conceptions accelerates and becomes obvious to all; starting more or less with the end of the eighteenth century – the great age of Lavoisier, Volta and Bichat – one has only too many examples to choose from. The theories of Bichat himself stand as a sufficiently convincing example, particularly in light of their subsequent fate in science. Bichat was not at all an extreme vitalist; on the contrary, his work constituted a strong reaction against the animistic school. As Claude Bernard recalls, Legallois was still trying to locate the seat of life, which he placed in the medulla oblongata, while Flourens lodged it in the vital center of that organ.[5] Bichat breaks with these erring ways; although he is opposed to the somewhat too crude mechanism of Boerhaave,[6] he nevertheless declares himself to be the adversary of Stahl[7] as well and combats Barthez's "vital principle," which, he says, is only van Helmont's *archeus* warmed over (*Anatomie*, Bichat's Preface, 1:vii [Hayward 1:vii]).

However, while Bichat, according to Claude Bernard's felicitous expression, "decentralizes" the vital principle (*Phénom. de la vie* 1:7 [Hoff 7]), he still attributes a considerable role to it. He opposes vital properties to physical properties as something essentially different and declares that "application of the physical sciences to physiology was explanation of the phenomena of living bodies by the laws of the inert. Here then is a false principle" (Bichat's Preface, 'Considérations générales' 1:xxx-xxxi [Hayward 1:25]). As it passes into living bodies, "this matter ... becomes possessed, at intervals, of vital properties, which are then united to physical properties" ('Considérations générales' 1:xxxvii [Hayward 1:27]). In particular – and this is the point we wish to stress here – he affirms that the movement of the fluids in the [capillary] canals of plants "is foreign to the physical properties, the vital ones only direct it" ('Considérations générales' 1:x [Hayward 1:3]). This thesis obviously seems extravagant to us today, and surely no contemporary biologist, no matter how firm his vitalistic convictions might be, would dare stand behind it. Now, it must be noted that insofar as the true mechanism of these movements is concerned, we are basically not much farther advanced than Bichat's contemporaries were, and that in any case no fundamental discovery concerning this mechanism has been made since that time. This only makes the general trend of the evolution of science stand out all the more clearly. As Jacques Loeb rightly says, in citing another example of the same sort (but one in which the progress of physics played a considerable role), "the times are gone when physicians and biologists dared to raise the objection – as they did against J. R. Mayer – that our body inherits its heat" (*La Dynamique* 106 [*Dynamics* 53]).

Thus the biological sciences are no exception. On the contrary, just as Bouasse said, whenever possible they seek to imitate physics, to fit into the framework of physicochemistry.

It may help us clarify the nature of the controversy separating vitalists and antivitalists if we make use of the concept to which we devoted the preceding chapter, namely, the concept of the irrational.

What in fact is the position of contemporary vitalists? Not a single one of them would claim that nothing in living bodies takes place in conformity with the rules that govern inanimate bodies. They all recognize, to cite only the best-known examples, that the blood is circulated by the mechanical action of the heart, which acts like a force pump and a suction pump, that the organism does not create the energy it develops, either in the form of mechanical action or of heat, but that this energy is only a transformation of that supplied by food. Nor would any contemporary biologist maintain that the chemical substances encountered in living bodies (leaving their "organization" aside) can be produced only by a special *vital force*. Granted, not all of them have been able to be synthesized *in vitro* so far, but chemists have already created quite a few of them, and for the rest success no longer seems so remote as to warrant apodictic denials.

But one can go even further, it would seem. There are certainly few vitalists who would profess that the physicochemical explanations found thus far are the only ones possible and that no further progress can be expected along these lines. Moreover, there can be no doubt as to the opinion of working biologists: they clearly feel that what has been accomplished is insignificant compared to what can be done and that science is barely on the threshold of important discoveries. As we saw in the passage from Jacques Duclaux (note 2, p. 201), this scientist, so disinclined to look favorably on past conquests, nevertheless does not at all doubt that the future holds great enlightenment for us in this area.

Therefore, the most the vitalists claim is that certain areas, the limits of which they believe can now be identified (such as the one Driesch would attribute to his *entelechy*), remain entirely inaccessible to any attempt at physicochemical explanation.

Let us recall what we came to see in the preceding chapter with regard to chemistry: the specificity of the phenomena embraced by that science seemed to us to be at least a very strong indication that there were one or more irrationals ultimately to be found there. This observation is all the more applicable to vital phenomena. Indeed if one tries to take in the formidable mass of these phenomena with a single glance and considers how they have been classified (after eliminating, of course, everything to do with sensation and action, which must be considered irrationals of another order), one will hardly be able to avoid the impression that some of these categories (as for example the phenomena of sensibility, of assimilation and growth, of heredity, etc.) are characterized by such originality and complexity that it appears very difficult to imagine that they can be entirely reduced to the reactions exhibited by nonorganic matter. As Montaigne said so well:

What a prodigy it is that the drop of seed from which we are produced bears in itself the impressions not only of the bodily form but of the thoughts and inclinations of our fathers! Where does that drop of fluid lodge this infinite number of forms? And how do they convey these resemblances with so heedless and irregular a course that the great-grandson will correspond to his great-grandfather, the nephew to the uncle?[8]

Thus, if this process of reduction ever becomes far enough advanced (which will probably require many centuries), we shall then have, alongside a large number of phenomena perfectly continuous with those of nonorganic matter, other clearly defined and delimited phenomena in which an essential discontinuity with nonorganic matter will have been recognized, where it will have been demonstrated that the living particle behaves quite differently from a nonliving one. The particular property of matter that will reveal itself on this occasion will thus appear as something irrational.

Considered from this point of view, the vitalistic thesis amounts to affirming that it is now permissible to indicate the limits of possibility in this domain: physicochemical explanation will be able to go only so far, and everything beyond that point will forever remain irrational. The antivitalist, on the contrary, supposes that, to use Claude Bernard's formula, "the vital properties are nothing more than complexes of physical properties" (*Phénom. de la vie* 2:477); cf. also 1:32–33 [Hoff 23–24]), and that, as a result, the properties we today consider characteristic of living matter will one day be recognized as conditioned solely by a certain complexity in the structure of that matter. That is why, if he has difficulty reducing a property of organic bodies to known physicochemical properties, he tends to assume that there is a property involved, which has not yet been discovered, to be sure, but which nevertheless belongs to matter in general and not to living matter alone. For example Bosc, in speaking of Driesch's entelechy, whose existence is still rather dubious (to say the least), thinks he can infer that it is "a principle that can be applied to all bodies."[9] This amounts to allowing that all we would need to do is group together in a certain way a given number (thousands or millions, let us say) of molecules of the bodies we call albumins and vital phenomena would appear. It would even be possible to produce groupings of this kind, thereby achieving what has been called "artificial generation" or "the creation of life."

From the standpoint of rationality, which is what interests us here, two eventualities would then be possible.

The first is that these properties of the groupings (only some of these properties, of course, the great majority of them always presumed to be entirely reduced, explained) appear to be without any possible logical connection either with the properties of the elementary parts or with those properties that can be attributed to the "power of grouping" itself. One would then have a given, an irrational (clearly delimited, it goes without saying) showing up at the time a certain grouping of elementary particles occurs.

Or else (this is the second eventuality) one would attribute to the elementary parts of nonorganic matter itself certain properties which would remain inoperative or make themselves felt only very faintly so long as the groupings are relatively uncomplicated (one must of course assume that phenomena would have been discovered to support assumptions of this sort), but would become much more pronounced as soon as the grouping becomes sufficiently complex, at which time they would succeed in conditioning those groups of phenomena – heredity, assimilation, etc. – mentioned above. That is a conception we already see appearing rather clearly in a few contemporary works, such as those of Jagadis Chandre Bose. But then these properties of the elementary particles would certainly themselves appear to be *givens*, *occult*, inexplicable. They would thus still be irrationals, just like diversity in time and space.

These two suppositions are entirely in line with those we had formulated in the preceding chapter about the chemical irrational, when we asked ourself whether all the irrational could be lodged exclusively in the properties of the elementary particles. This is because in both cases we are concerned with one and the same question, or at least with the same categories of our understanding.

The analogy offered by the future role of explanation in the two domains may help us better understand the true meaning of the vitalist thesis. Indeed, what the advocates of the doctrine are chiefly striving to do is to demonstrate that this or that class of phenomena characteristic of organic matter, as for example one of the categories we cited on page 181, seems incapable of being explained in terms of what we know of the way in which nonorganic bodies behave. These demonstrations may be judged more or less convincing depending on the specificity of the phenomena involved. But what is essential is that they in fact fail to achieve their true goal, which is, as we have just said, to establish a barrier against any future attempt at physicochemical explanation, to deny the theories of nonorganic matter any access at all to the domain whose limits have been defined. In the field of chemistry, that would be equivalent to saying that because it does not at the present time seem to us that everything characterizing elementary atoms can be reduced, by a mechanical or even an electrical theory, to a grouping of subatoms of a single type or perhaps of two types (such as positive and negative electrical particles), we must prohibit all research concerning a theory of this sort or even tending only to establish relationships between properties of the elements (as Mendeleev's system does). It is clear, on the contrary, that this would be an entirely unjustified conclusion. By declaring that the specificity of the chemical elements will never be able to be completely explained by the "power of grouping," we are simply affirming that there must be something irrational in this domain; but we by no means claim, we could not claim that everything concerning this domain is irrational, that rationalization can play no role there. Similarly, while it seems quite probable that the phenomena of instinct or of heredity are really distinct classes, that is, have a specificity that will never be able to be completely reduced, it does not at all follow that they cannot be reduced in part, cannot be partially explained by phenomena such as those characterizing nonorganic matter. Science has certainly discovered some extraordinary analogies along these lines in the last few years.

We referred above to the work of Jagadis Chandre Bose, which is all the more remarkable because it deals with the very essence of substances. By actually studying the molecular phenomena produced by the action of electricity on nonorganic and on living matter, this scientist has established that the reactions believed characteristic of the latter can be reproduced in the former, so that it is impossible to draw a line between the phenomenon commonly considered physiological and the merely physical phenomenon.[10]

But there are many examples to choose from here. For instance, work in colloidal chemistry has revealed that the action of anesthetics, which certainly seemed physiological until now, could be explained quite well without the slightest reference to the concept of life: all anesthetics have the common property of being fat soluble and would seem to act solely by modifying the properties of the lipoids, which explains an observation that had always puzzled physiologists, namely, the fact that chemically inert bodies can be powerful anesthetics (Loeb, *La Dynamique* 85–86 [*Dynamics* 40]). Similarly, the action of potent poisons, such as potassium cyanide, appears to have nothing to do with the specificity of organic bodies, since potassium cyanide proves to be "toxic" with respect to a catalysis of hydrogen peroxide (Loeb 61–62 [*Dynamics* 26–27]).

Another category of phenomena that one might think all the more likely to have to be limited to organisms because they seem to involve psychic activity, nevertheless has its counterpart in nonorganic bodies. These are the phenomena of memory, and the nonorganic equivalent has been called *hysteresis*. They are facts whose explanation, or theory, greatly disturbs physicists, precisely because they appear so abnormal from the standpoint of our customary idea of nonorganic bodies; but they are undeniable facts and the physicist studying the elasticity of metals, for example, runs up against them all the time.[11] At the present time, under certain conditions, a nonorganic body is actually presumed to have a *past*, just like a living body; Boltzmann was able to say that metal wire "remembers" (Bouasse, 'Sur les déformations' 127).

Tropism, the controlling action that certain physical agents, principally light, exercise on organisms, functions much like instinct, and we can, without forcing the analogy too much, understand a certain number of instincts as combinations of tropisms. Now tropism surely seems susceptible to physicochemical explanations.[12]

The fundamental and seemingly almost contradictory observation of cellular biology – namely that the content of the cells seems to be in a liquid or at least semiliquid state of aggregation and yet at the same time certainly to be organic – has lost much of its paradoxical aspect since the discovery of liquid crystals, which possess not only the anisotropy characteristic of all crystalline structure, but which also have the ability to grow and whose aggregates seem to show a striking analogy to certain forms observed by biologists, in particular with forms of myelin.[13]

In general, forms that were believed to be peculiar to living matter seem to be able to be imitated, at least to a point, by reactions involving only nonliving substances. These experiments (particularly the latest ones, those of S. Leduc),[14] which result in highly visible similarities, have excited keen interest. Their import has unquestionably been exaggerated, but the analogy may be more than apparent.[15]

There is certainly a real analogy between the division of a sea urchin egg and that produced in a drop of olive oil by a thread saturated with an alkaline solution[16] and in general that part of the life of the organism which somehow appears the most mysterious and essential of all – that is, the fertilization and development of the egg – has been the object of research and discoveries which suggest that physicochemical explanation will be able to play a considerable role there. One need only cite the memorable work of J. Loeb (see *La Fécondation chimique*), soon followed by that of Yves Delage,[17] on artificial fertilization, as well as the studies of a whole school of biologists grouped around Wilhelm Roux,[18] who devote themselves to what they somewhat pretentiously term "mechanics of development."

To some extent it has also been possible, through the use of basically quite simple experimental arrangements, to imitate phenomena that appeared characteristic of the living cell: Rumbler showed that a drop of chloroform suspended in water behaves toward a wax-coated glass filament like an amoeba which swallows a diatom, that it surrounds itself with a sort of construction if provided with ground glass, and that it forms pseudopods by a simple decrease in surface tension (Przibram, *Vitalität* 16 ff.).

Along the same lines, we note that two major problems, animal energy and the chemical synthesis of the substances that compose organic bodies, problems that might legitimately have been considered fundamentally insoluble one or two generations ago, certainly no longer seem so at the present time. As a matter of fact, the earliest successes in these two areas are even older; the first steps in the synthesis of organic bodies date back to 1828, that is, to the famous synthesis of urea by Wöhler. Since that time, indeed, it had become impossible to claim that the chemical substances qualified as *organic* could be produced only within the organism, on the grounds that only the involvement of the vital force allowed this formation. But one still had the expedient of pointing out that nature, life, operated by methods entirely inaccessible to chemists, since the latter made use in their syntheses of potent means, such as elevated temperatures, strong concentrations, violent reagents, most of the time producing only insignificant results, while the organism, without using any of these agents, apparently accomplishes its work with a quantitative yield. Now today we are familiar with genuine syntheses, such as the polymerization of formaldehyde, which are carried out under conditions analogous to those within the organism.[19] And on the other hand, we also know that these reactions whose yield in the organism appeared so paradoxical can be reproduced in the laboratory through the intervention of minute quantities of ferments, of *enzymes* as they are now called, which act by *catalysis*. These ferments are soluble, which proves that we are dealing with true chemical reactions in which the structure of the organism is not a factor. This was not an idle demonstration, for, as a result of Pasteur's work, the opinion tended to prevail that certain reactions previously considered purely chemical, such as alcoholic fermentation or the acidification of alcohol, were, on the contrary, the work of tiny organisms. Pasteur himself defended this opinion in a resounding controversy with Liebig (Loeb, *La Dynamique* 54 [cf. *Dynamics* 21–22]), whereas Claude Bernard, in speaking of the action of yeast, observed that its nature was unknown but that it "must necessarily belong to the physico-chemical order" (*Méd. expér.* 320 [Greene 201]). It is Claude Bernard who was right, and current opinion holds that microorganisms themselves act only by means of soluble substances without organismic structure, which substances they excrete.

What must be considered the point of departure for the discoveries concerning animal energy is Lavoisier's and Laplace's famous work on respiration, to which, according to J. Loeb's expert testimony, "all the really important discoveries in biological chemistry are linked, directly or indirectly" (*La Dynamique* 14 [cf. *Dynamics* 7]). Thus the source of animal heat had been known from that time on. And since it was also known, by the example of heat engines, that mechanical work could be created by means of heat, the fact it is produced by the animal no longer seemed so mysterious. But, just as for chemical synthesis, it seemed that the process used by the organism had nothing in common with that employed by man. As soon as we had learned through Carnot's discovery that a drop in temperature was indispensable for the functioning of a heat engine and that the yield varied in proportion to this drop, the divergence appeared even greater, for there are only very small temperature differences within the animal organism, yet the energy yield of this engine far exceeds that of the best heat engine. But here again recent discoveries have managed to bring, if not a solution, at least the hope of a solution. We do not yet know how the organism goes about transforming chemical energy from food into the mechanical energy of muscles, but we know of analogous transformations being accomplished under conditions essentially no different from those existing within the tissue. Examples include Engelmann's "absorption process," that of d'Arsonval and Imbert using surface tension, that of Quincke involving spreading, and many others (Loeb, *La Dynamique* 107–109 [*Dynamics* 54–55]).[20] Granted, there is the serious difficulty that such procedures most often seem irreversible, or at most (like Engelmann's process) partially reversible, while those taking place in living organisms are obviously entirely reversible.[21] This would seem, however, to be only a matter of degree, and the obstacle certainly does not seem insurmountable.

A recent discovery, which has justifiably excited great interest, clearly shows how sketchy, and one might go so far as to say crude, the notion underlying the vitalistic position, namely that of the vital phenomenon, still is, and how much more clearly delimited it would need to be before it could be included in strictly scientific propositions. We are referring to Jean Nageotte's elegant series of experiments on the grafting of dead tissues.

Organic grafting is a very ancient procedure, and one that has always vividly captured the public imagination. Indeed, one must appreciate that even in its simplest form of a direct living graft the procedure is somewhat paradoxical and shocks some of our instinctive ideas. We feel that we are *individuals* and assume by analogy that all of humanity and even the entire animal world is composed of such individuals; we therefore find it strange that an *organic* part of such an individual can, without passing through the digestive organs and while therefore maintaining its own organic structure, be joined together with a foreign organism. It should be noted, however, that prior to Nageotte no one had thought of grafting anything except living tissue: it seemed self-evident that life, that mysterious state, was indispensable to the success of the operation, and infinite pains had been taken to keep the grafts alive, even if they were kept quite a long time. Nageotte resolutely violates this rule. He *kills* the tissues before grafting them, by plunging them for prolonged periods in solutions of 10% formol, in 90° alcohol, or even in a solution of corrosive sublimate. Now the grafts made with these tissues succeed perfectly, in general even better than living grafts: the new organism reacts less, tolerates them better than it does living grafts. Moreover, there can be no doubt that these dead tissues, which retain their structure perfectly after the graft, become living tissue again; in certain cases they are actually able to hypertrophy, that is, to create new tissue, which is most certainly a privilege of life.[22] The renowned biologist has endeavored to demonstrate more clearly the nature of the mechanism of this death and reviviscence. The tissue in question, which is connective tissue, is "inhabited" by a certain number of cells of which it is some sort of an excretion. These cells are the only truly living part of the tissue and it is because they have difficulty adapting to the new organism that the reactions observed in the case of living grafts are produced. These reactions are much less severe for dead grafts, because it is easier for the organism to get rid of the "cadavers" of dead cells. Assimilation in this case occurs by the immigration of cells from the new organism. However, for dead grafts, this phase seems to us to correspond to a true reviviscence. In a certain sense, it actually is one, provided one is willing to accept the notion that all the *vital* phenomena are not the product of the connective tissue itself, but solely of the cells. As for the connective tissue, its "fundamental substance" (which is its primitive state), the biologist tells us, is quite simply "a coagulum of albumins contained in the internal medium. It is no more living than the coral of the polyparies."[23]

This observation follows directly from the facts. But in a work of more general scope published later Nageotte makes much more far-reaching conjectures which, it must be admitted, appear quite plausible in the light of the established results. They concern in particular what happens within the cells. We know that cells contain a certain number of granules (or mitochondria) and some intergranular substance. Now Nageotte suggests that "the essence of life" may reside solely in these granules, while the intergranular substance may be analogous to the intercellular substance.[24] We see what an insignificant portion this biologist has come to consider as actually alive in an organism we used to believe was alive in its entirety.

By way of summary, let us say that the phenomena called into play by the vitalistic demonstrations seem too complex, that they do not yet seem to have been sufficiently analyzed by science for us to be able to formulate propositions of such a pronounced "negative dogmatism." Given the vigor with which our understanding pursues its eternal task of rationalization, it certainly will not be deterred by a barrier unless the barrier is of one piece, as it were, unless it presents no hint of a gap. In the case of the vital phenomena, if we try to imagine what a perfect demonstration of this kind might be, we arrive at more or less the following image: we would see organic particles behaving differently from nonorganic particles, that is, they would probably not set themselves in motion while the others remained at rest or vice versa (which would be creating energy, according to our present understanding, and we are persuaded that energy is conserved in the organisms as well as elsewhere) but, for example, they might interrupt or retard a motion (which is the kind of action Driesch attributes to his "entelechy")[25] or follow an *ordered* motion, while the motion of the nonorganic particles in an analogous situation would be unordered, would simply obey the laws of chance (which would be the kind of action Maxwell imagined for his "demon" and which would moreover be consistent with Claude Bernard's well-known dictum that life directs forces it does not create).[26]

There is no doubt that such a demonstration would convince us, and it is not difficult to see why. It is because it would involve a *molecular* phenomenon and therefore a supposedly simple one. Molar phenomena, on the other hand, appear to be necessarily complex, resulting as they do from a great number of molecular movements, at least some of which may be able to take place according to the rules of physicochemistry. To say that this is not the case, that we shall never be able to understand the whole collection of things we call vital as broken down into simpler parts or that, conversely, all the parts going to make it up will necessarily be seen as exempt from the rules of nonorganic nature, obviously involves a very great risk. The risk is all the greater because, as is easy to see, such a claim implies a thesis concerning the functioning of our understanding: it claims knowledge of what our understanding will accept, in other words, how far the limits of the conceivable extend. Now, such knowledge can certainly be attained insofar as the most general foundations are concerned, and our own works aim at just such a goal; furthermore we do not pretend to do anything more than identify the principles from which science has drawn its inspiration down through the ages, and so the thesis in question might not seem unreasonable at first glance. But all we claim to do is to identify these foundations, to show what our reason considers to be perfectly rational and how far it finds this concept to extend in nature, or if one prefers, how far it manages to impose this concept on nature. However, as we have seen, our reason is not only aware of something perfectly rational, it is also aware of something imperfectly, partially rational, which it knows quite well contains an irrational element and nevertheless uses as if it were something perfectly rational for the reduction of that which is yet more irrational. But it is impossible to be too cautious in judging this functioning of our reason, its different forms, and its limits. Not that, as has sometimes been said, our reason modifies or evolves in this respect; everything goes to show, on the contrary, that it is totally immutable. But, because we do not see it function, because we are constrained to deduce the rules of its functioning *a posteriori* by means of concrete examples, we are apt to go astray. Moreover, we are all the more likely to go astray when the things with which we are dealing are newer, less analogous to those we knew before, to the cases where we have been accustomed to seeing our reason operate. Now that is the normal situation in science: even though reason always stays the same, it is constantly led to attack new problems, whose form can be so different from those that are familiar to us that it will be impossible for us to predict how our reason will react. For, as we have said, it is extremely difficult to make our reason apply itself effectively for the sake of a dry run (except in very simple cases). Therefore, if we really want to know what decision it will make in a specific case, often the only thing to do is to try to find out how serious and dedicated thinkers have reasoned in the same, or at least in analogous circumstances. But in a case for which we can find no sufficiently close analogy in either the present or the past, there may be no guidelines, and this becomes all the more likely as the case becomes more complicated. That is exactly the situation in which we find ourselves in speaking of what will appear conceivable or inconceivable in the case of vital phenomena: we do not and cannot know what devices scientific reason will set in play in order to reduce them, and thus it is not at all impossible that something that appears radically inconceivable to one generation will no longer make the same impression on another. Consider the following example, which we find quite cogent. In the whole history of human thought there is surely no more well-balanced mind, none less inclined to rash statements, than that of Montaigne. Now Montaigne is not content to point out (as we indicated on p. 181) that the phenomena of heredity in general are so specific that they appear quite difficult to explain; he also declares that in one particular case the explanation will forever be completely inconceivable. He is referring to the bladder stone disorder he had inherited from his father but which had not manifested itself until he was forty-five.

Where was the propensity to this infirmity hatching all this time? ... If anyone will enlighten me about this process, I will believe him about as many other miracles as he wants; provided he does not palm off on me some explanation much more difficult and fantastic than the thing itself.[27]

Obviously Montaigne is advancing a clearly vitalistic position here: he maintains that there is no conceivable explanatory physical theory for this fact. Although we are certainly still far from being able to formulate a theory of this kind at the present time, it is nevertheless certain that the fundamental impossibility of such a theory would seem much less well-established to a contemporary biologist than it did to Montaigne. Indeed, since the formation of the stone is a chemical process, we must ask how biochemistry deals with this sort of phenomena. Now the reader will see below that biochemists have come to believe that each organic individual might well possess a *distinctive* albumin, unlike that of any other individual; it would thus not be surprising if strongly analogous albumins (as those of a father and son would be) should result in similar phenomena under similar circumstances (that is, at a certain age). But even without resorting to such speculations, nothing prevents us from conceiving that, through the act of procreation, there is transmitted the modification of a compound, a ferment, which is present in infinitesimal quantity and yet causes the substances that form the stone to be deposited in the bladder. No matter how hypothetical we find these suppositions, we cannot deny that they are *conceivable*, that there is nothing here "much more difficult and fantastic than the thing itself": a contemporary vitalist would no doubt prefer to choose another example for such a demonstration.

Thus we are forced to the conclusion that the vitalistic claims are somewhat premature, that they will not be able to find a legitimate place in science until research is infinitely more advanced than it is today. Vitalism, which could well be correct, all things considered, in asserting that organic nature conceals something specific, nevertheless seems to be wrong on each particular issue insofar as it claims to identify once and for all where this irreducible specificity lies, and thereby to limit the field of explanatory research.

It seems to be the somewhat vague but very powerful awareness of this situation, much more than any conviction of the extent of the established results, that constitutes the source of antivitalism's vigor and confidence in the future. It knows that it is seeking to satisfy an eternal penchant of the human mind and that consequently no obstacle, no defense, however well organized it might seem, can stop it. As a matter of fact, stopping it would require nothing less than a very clear demonstration of both the irrationality and the fundamental simplicity of the phenomenon that was claimed to be beyond the reach of antivitalism.

The vitalists themselves, although they of course find this tendency reprehensible, are forced to admit that it is the real attitude of the modern biologist. "The reduction of biological processes to the forces of inorganic nature is taken for granted," observes Driesch on the first page of a work written for the express purpose of demonstrating that biology is an "independent basic science,"[28] that is to say, one independent of the physicochemical sciences.

Furthermore it is not difficult to understand that the weakness of vitalistic theories is due precisely to the fact that they are a form of finalism. Indeed, a glance at the history of biological conceptions will suffice to show how eager the human mind is to free itself from the concept of an end whenever anything more or less related to scientific research is concerned. A particularly good example of this is offered by the great controversy that broke out in the last century over the problem of the evolution of species. Until that time it seemed to be taken for granted that only finalism could furnish synthetic views on the genesis of the organism and its parts. Of course Descartes had declared that the organism is only a machine, and eighteenth century materialists had developed this thesis magnificently. Yet even the most superficial observation revealed, and deeper study confirmed, that each organic being constitutes an ensemble marvelously adapted to its environment and mode of existence. Now mechanistic views appeared entirely powerless on this terrain. It is well-known, moreover, that this accord is the basis for the teleological proof of the existence of God, which held a considerable place in human thought for many centuries. Almost on the eve of Lamarck's work, in a milieu much inclined toward materialism, the Abbé Galiani set out this demonstration very forcefully and eloquently.[29] Kant too was of the opinion that considerations of finality were indispensable for explaining the genesis of the organism. "It is absurd," he says, "... to hope that another Newton will arise in the future who shall make comprehensible by us the production of a blade of grass according to natural laws which no design has ordered. We must absolutely deny this insight to men."[30]

It is clear that the situation changed radically beginning with the second half of the last century. Perhaps one could still find biologists today subscribing to Kant's negative postulate in all its rigor, but it is certain that neither general scientific opinion nor that of the educated public would concur. On the contrary, both appear firmly persuaded that what we at first sight take to be a tendency toward a future state can be adequately explained by one form of evolutionary theory or another (Lamarckism or Darwinism in their different nuances, Weismann's theory or analogous conceptions, Mendelian theory and De Vries's theory of mutations, etc.) or a combination of these various theories, or else the intervention of causes not yet considered by biologists.

The depth and the rapidity of this shift in opinion were strikingly illustrated in a recent article in a major English newspaper concerning the publication of the biography of Hooker,[31] the great botanist who was a friend and comrade in arms of Darwin and Huxley. It is indeed in England that the battle was the most intense, since creationist convictions based on an absolute faith in the literal inspiration of the sacred texts were particularly ardent there. The evolutionist conceptions were the object of the most impassioned attacks, and the year after the appearance of *On the*

Origin of Species in particular, one of the highest placed members of the Anglican clergy, the bishop of Oxford, took the occasion of a Congress of the *British Association for the Advancement of the Sciences* held in that city in 1860, to inveigh directly against Darwin and Huxley (the latter being present) in a manner as vehement as it was unusual. Scarcely a generation later, in 1885, a statue of Darwin was dedicated in the Natural History Museum in London. At this solemn ceremony the Anglican church was represented by its highest dignitary, the Archbishop of Canterbury, who in his speech not only lavished all imaginable praise on the author of the book so reviled a short time earlier, but solemnly declared that the evolutionist conception was perfectly consistent with the Bible. Huxley, who naturally was on the platform, could not resist whispering in the ear of a friend, the biologist Judd, "My dear Judd, you and I will be burned some day, because we do not go far enough for these gentlemen."[32]

Now it is no insult to the great protagonists of the theory of evolution – whose efforts cannot be too much admired and whose names humanity will undoubtedly inscribe among those of its most eminent creative geniuses – to point out that the results actually established in this domain remain inconclusive in the immense majority of cases and certainly do not allow anything like a true demonstration at the present time. Furthermore, one need only listen to some of the polemics between the various schools of evolutionary theory to realize that the very facts that are supposed to support the arguments are very rarely beyond dispute. And as to the evolutionary position as a whole, it certainly appears both imprecise and contradictory as soon as one tries to look at it closely, gradually fading into the distance, as it were, as one tries to approach it.

But why did the defenses of finalism, whose position is easily defined with all the clarity one could wish, and which had enjoyed an uncontested right of ownership in this domain for so many centuries, come to yield so rapidly in the face of what hardly seems a formidable enemy? There seems to be only one explanation: the weakness inherent in any finalistic conception. Of course, we *want* a solution and, for lack of any other, resign ourselves to accepting that one. But it is and will always be only a last resort, and the hold it has on our mind is weak: as soon as a causal explanation becomes available, even a far-fetched or confused one, the finalistic explanation immediately gives way. Moreover, what could be more natural? No doubt the causal conception involves many great philosophic difficulties if pushed to its logical consequences. But the idea that the present can be controlled by the future, which does not yet exist and which, if I assume my own free will, may well never exist, offends the understanding even more, especially if one means to forgo properly theological considerations.

Furthermore, it is clear that finality presupposes prescience, which in turn implies consciousness. If I do something to attain some end or other, first I must, as Lucretius says, have had the thought, the anticipation of what I wanted, and this anticipation must have had an image as its object.[33] Of course Lucretius was an antifinalist. For him, "nothing in our bodies was born in order that we might be able to use it, but the thing born creates the use" (*De rerum nat.* IV, 834–835). But even for Aristotle, the prototype of all finalists, "the final cause does not move unless it is known and desired, and thus has a hold only on beings capable of feeling and wishing."[34] Obviously in appeasing my hunger and thirst, in performing a sexual act, I am conscious only of responding to an immediate need, an obscure instinct, even though upon reflection I come to realize that these are acts directed toward the conservation of my person or my species. But in that case I assume that a higher consciousness, Nature, God, knows these ends; otherwise how could it want them? Anthropomorphism is inevitable here. Unless of course I succeed, as evolutionary theory does, in returning to causality by imagining that the only species able to survive were those in which these needs and instincts had been formed and perfected, in which case finality is only apparent and immediately gives way. But if finality is to be fundamental, it cannot arise from unconscious forces. To see this, one need only think about the more or less surreptitious use sometimes made of final causes in physics. If I say that a light ray goes from one point to the other by the shortest path, and if I want to see anything more than an empirical rule in this statement, I attribute to the ray not only the choice of paths to be followed, but also the anticipated knowledge of the result to be obtained. That is surely a view which, to quote Henri Poincaré's apt expression, "has something about it repugnant to the mind,"[35] and from which our imagination will always seek to free itself. It has succeeded, as we know, in this particular case, and the so-called "economy" of nature has been transformed for us into a sort of prodigality, since we suppose that waves would be propagated in all directions if they did not cancel each other out. Furthermore, there is no doubt that we arrive at this concept of an action aiming at an end out of the consideration of the way we ourselves act, or at least think we act. This would seem to have been definitively established by

Spinoza (*Ethics*, Pt. 1, App.).

This origin of the concept explains why, as the author of the *Ethics* immediately deduces, the concept is anthropomorphic from still another point of view. Like the cause, the end must be intelligible; but for the latter, this postulate leads to a value judgment: it must be perceived as a goal worth pursuing. Now, if there is anything in the universe whose interest takes precedence over that of humanity, we are certainly incapable of conceiving it.

> They [men] are bound to estimate the nature of such rulers [the "rulers of the universe"] (having no information on the subject) in accordance with their own nature, and therefore they assert that the gods ordained everything for the use of man ... but in their endeavour to show that nature does nothing in vain, *i.e.*, nothing which is useless to man, they only seem to have demonstrated that nature, the gods, and men are all mad together.

Injurious things, tempests, earthquakes, maladies have been explained by the anger of the gods. "Experience day by day protested and showed by infinite examples, that good and evil fortunes fall to the lot of pious and impious alike; still they would not abandon their inveterate prejudice" (*Ethics*, Pt. 1, App. [Elwes 2:76; Meyerson's brackets]).

We need only attend to the finalists themselves to see how right Spinoza is. To cite a recent example, the geologist de Lapparent judges that the disposition of coal reserves in the depths of the earth's crust "attests to a marvelously wrought design"; all the characteristics of the deposits, the thickness of the layers, the intercalation of sterile masses, the fact that the coal bearing terrain is not too accessible (which prevents waste), seemed to him to concur in this demonstration. But of course the goal of this "well-woven fabric" can be none other than "to prepare the coming of the king of Creation."[36]

Nevertheless, if it was possible to understand the universe in this way in the past, it is much harder to do so at present and is becoming more and more difficult as our knowledge advances. To speak only of organic nature, each species, even if we imagine it to be constituted with a view to an end, appears to know no other end than itself, its own welfare, its conservation, its propagation. Man seems to play no role in it, and his role in nature in general unquestionably appears more and more diminished, to the point that it seems truly insignificant to us at the present time. How can man be persuaded that all that was made for him alone? Was there no way to arrange things better, or at least more simply? It is the venerable opposition between God's infinite goodness and His omnipotence, an opposition that has plagued theologians throughout the ages, culminating, as we know, in Leibniz's conception of a divinity for whom all things are possible but not *compossible* and who therefore, being unable to create a *good* world, had to be content to create the best world it could.

To be sure, religious faith is less strong today than it was in past centuries. But that does not diminish the problem, for here the concept of omnipotence is actually implied by the argument itself. If Nature or Life (which in this context can only be surrogates for the divinity) have enough power over things to insert themselves into the web of their being, how can we imagine that these things nevertheless seem to resist them? We allow – certainly not without difficulty – that Life is content to direct forces it does not create (as Claude Bernard put it). But then it is at least necessary that, in its perfect prescience, it show no weakness in doing so. That is why if we must admit, for example, the concept of instinct as a fundamental concept, exempt from any subsequent attempt at explanation, this instinct necessarily appears as having to be infallible. We can easily see this in examining the works of those who defended the finalistic conception of instinct. Henri Fabre's work, which is only one long and quite brilliant plea in favor of this thesis, brings to the fore precisely this concept of infallibility, as opposed to the painful gropings, the continual errors of the conscious will, and adversaries who, like Rabaud, seek to destroy these proofs of infallibility or even to show us, like Piéron, instincts leading to acts injurious to the species,[37] are attacking the most essential (not to mention the weakest) side of the theory.

In comparing the case of finality to what we saw in the case of causality, we might say that what here appears to be contrary to finality or imperfectly final is analogous to what we termed *irrational* when we were dealing with causal rationality. If what is irrational from the point of view of finality appears to be much more strongly opposed to the fundamental conception that it limits than the irrational strictly speaking, if it actually appears to destroy it, that is clearly only one more aspect of the primordial fact that the concept of an end is somehow less vigorous than the concept of cause, which is why the latter recovers from an attack to which the former succumbs. We can, if we must, imagine a limited rationality, whereas we find a limited finality absurd.

In spite of this weakness inherent in all finalism, it would nevertheless be an error, as we have seen, to say that it is entirely foreign to scientific thought. Bacon, as we know, protested quite vehemently against the admission of anything resembling a final cause into science. Final causes, according to him, can serve no purpose in science, breeding "waste and solitude in that track."[38] It is obvious, on the contrary, that even if one takes a strictly antivitalistic stance, finalistic deductions can render great service by allowing phenomena to be grouped in terms of a particular viewpoint, thus sometimes paving the way for future attempts at reduction. The most stubborn antivitalist can quite readily admit, for example, that it has been advantageous to class an almost infinite diversity of phenomena under the headings of *heredity* or *instinct* until a more thorough analysis can be made. Supposing that this classification is not definitive, that certain facts currently included under these headings are later recognized as belonging to quite different categories, or even that the specificity of these facts is one day completely abolished, it is certain that the provisional classification will have been necessary, since the kinship of these facts – even if only apparent – is undeniable and science can start only with appearances of this type. Similarly, it may well be that Driesch's *entelechy* – which certainly seems (at least to a layman) a bit too *crude* for a definitive irrational[39] – can render considerable service by grouping phenomena whose analogy is not immediately obvious. It is up to the biologists alone to judge it, to say whether this concept can be of real use to them, whether they find Driesch's arguments sufficiently conclusive for his concept to be admitted into science. Or rather, it is no doubt the march of scientific progress itself (much more than the discussions, as useful as they may be for clarifying the respective positions of the adversaries) that will take care of this decision: Driesch's position will remain defensible so long as there appears to be no possibility of a physicochemical explanation for any of the phenomena he places under the jurisdiction of his "entelechy." But the day that any remotely plausible theory as to the physical means of production can be formulated for any part whatsoever of the group of phenomena this biologist would embrace by his ambitious conception, the staunchest defenders of the conception will certainly hasten to abandon the compromised position and transfer their assumptions elsewhere.

The same is true for other vitalistic concepts accepted much earlier, whose existence – no doubt largely for that reason – seems to us to be better warranted than that of "entelechy." We could return to our earlier enumeration and examine from this point of view all the categories of vital phenomena that explanatory science has begun to attack. Let us limit ourself to the concept of instinct. It is clear that if science had to limit itself to studying it "as is," it would by that very fact be affirming precisely its existence as an irreducible concept, more or less in the manner of Henri Fabre. This would thus amount to declaring that within the limits set by the definition of this concept science must forbid any search for an explanation. Therefore the biologist is correct in affirming with Piéron that the "hypothesis of reducibility appears more fertile," given that it "stimulates research and generates progress," for if "reduction to simpler reflex mechanisms has not been accomplished, that does not mean it never can be" and that "negative facts, unsuccessful attempts, are worthless."[40] That is exactly what Claude Bernard meant when he proclaimed that

> we must therefore get used to the idea that ... we must always seek to exclude life entirely from our explanations of physiological phenomena as a whole. Life is nothing but a word which means ignorance, and when we characterize a phenomenon as vital, it amounts to saying that we do not know its immediate cause or its conditions. (*Méd. expér.* 319 [Greene 201])[41]

But Spinoza had already pointed out that by appealing to final causes, one "takes refuge in the will of God – in other words, the sanctuary of ignorance." Indeed, "when they survey the frame of the human body, they are amazed; and being ignorant of the causes of so great a work of art, conclude that it has been fashioned, not mechanically, but by divine and supernatural skill ..." (*Ethics*, Pt. 1, App. [Elwes 2:78]).

Thus any finalistic conception in science seems to be simply tolerated until it can be replaced by a causal deduction. Nevertheless we would be wrong to try to take Bacon's precept literally and to chase finalism from science altogether. As a matter of fact, since our propensity for rationality is irrepressible, it would be futile to put obstacles in its way: where our mind cannot completely satisfy this propensity by causal deductions, it enters naturally and spontaneously into the path of finalistic deduction, which may be less satisfactory, but still makes things to some extent rational. Everywhere causal explanation seems unable to penetrate as yet, the researcher will necessarily be led to advocate explanations of the sort Cuvier had in mind, that is, mixtures of strictly causal considerations and finalistic considerations.[42] If Bacon could completely misunderstand this very clear situation, it is because his mind was preoccupied with the image of a purely empirical science. But, as we have seen, such a science is something chimerical, something mankind has never known and surely never will.

Bacon's vituperations nonetheless contain a goodly share of truth. First of all, he was entirely justified in criticizing the science of the preceding epoch, which had made such ill use of the finalistic point of view. But he was also perfectly right when he affirmed that from a strictly logical standpoint, everything explained by an *end* is by that very fact exempted from the search for a cause.[43] This situation strikes us much less today, precisely because of our feeling of the inherent weakness of the finalistic position; we know for a fact that whatever limits anyone tries to set to the search for a cause, resistance will disappear as soon as there is the least hope of seeing this search crowned with success. But the situation was very different at a time when faith was still strong and the teleological argument could invoke the powerful support of theology.

More than a half century later, Spinoza, writing in a country which at that time was probably one of the least tyrannical in Europe from the point of view of religious opinions in spite of the coup d'état of the Prince of Orange a short time before, observes that

> anyone who seeks for the true causes of miracles, and strives to understand natural phenomena as an intelligent being, and not to gaze at them like a fool, is set down and denounced as an impious heretic by those, whom the masses adore as the interpreters of nature and the gods. Such persons know that, with the removal of ignorance, the wonder which forms their only available means for proving and preserving their authority would vanish also. (*Ethics*, Pt. 1, App. [Elwes 2:78–79])

Danger from this quarter is no longer very significant, it would seem, and science, by permitting finality to penetrate into that part of the biological domain where causality has not yet been able to establish itself – we have seen, moreover, that it would be futile to try to oppose this penetration – has little to fear from future offensives by partisans of the finalistic conceptions. At any rate, except for this part of science – which will almost certainly continue to shrink as science progresses – all the rest of the domain is and will remain vested in causal explanation.

Now, as we hardly need point out, in all its parts where physicochemical explanation is winning or trying to win acceptance, biology resembles or strives to resemble the physical sciences, in accordance with Bouasse's formula. They clearly have the same goal and procedures and, as a result, what we have seen or shall see to be valid for the latter will also be valid for the former.

Furthermore we have seen in the course of this work that wherever matter seems to allow it, modern biology, exactly like the biology of the past, does not hesitate to resort to explanations of a very advanced causal type: the conceptions connected with the theories of preformation provide us with an outstanding example of this (Ch. 5, pp. 121 ff.).

Let us look at some direct testimony which speaks to the same point. "If humanity lasts long enough," says a contemporary antivitalist, "the time will surely come when scientists will find a mechanical explanation for all phenomena."[44] There is no doubt that the biologist in question had in mind the phenomena of living matter in particular when he performed this act of faith. Let us set aside for the moment the difficulties standing in the way of the realization of this prophecy even insofar as purely physical phenomena are concerned (difficulties that contemporary physicists consider prohibitive, as we know) and let us also ignore possible obstacles (in light of the probable existence of new irrationals) to the reduction of the phenomena of life to pure and simple physicochemistry. Let us take Bouasse's formula for what it no doubt is at bottom, namely, the statement of the ideal but actually inaccessible goal of our reason with respect to nature. We shall then have only to recall that mechanism, explanation by matter and motion, is in reality reduced to explanation by the second term alone, the first being essentially inexplicable. Now motion is undoubtedly explanatory only because it is a spatial function.

Let us now look at the testimony of a vitalist who is if possible even more direct. In seeking precisely to defend the domain of living matter against the encroachment of explanations drawn from the physical sciences, Driesch declares that mechanism (which he acknowledges, moreover, to be perennial and fully necessary) reduces all problems in some degree to problems of geometry.[45] That is indeed its goal and at the same time the true goal of all explanation in the domain of science as a whole. Everywhere there is the same search for identity, either between antecedent and consequent or between two coexisting entities, and everywhere the same recourse to spatial construction in order to equate and diversify them at the same time. In the whole immense field of science, there is and can be no true explanation except by space and the properties of space.

NOTES

1. Henri Bouasse, 'Physique générale,' in *De la Méthode dans les sciences*, 1st series, 2nd ed. (Paris: Félix Alcan, 1910), p. 124 [1909 ed., p. 76].
2. Jacques Duclaux began his fine book on *La Chimie de la matière vivante* (3rd

ed., Paris: Félix Alcan, 1910) with this frank statement: "The only really scientific way to treat the chemistry of living matter would be to write below the title, 'Nothing is known,' and put off the rest until a second edition, which could be published twenty or fifty years from now" (p. i).

3. Hans Driesch, *Naturbegriffe und Natururteile* (Leipzig: Wilhelm Engelmann, 1904), pp. 112 ff. Cf. also his *The Science and Philosophy of the Organism* (Aberdeen: printed for the University, 1908), 1:143.
4. Emanuel Rádl, *Geschichte der biologischen Theorien in der Neuzeit*, 2nd ed. (Leipzig: W. Engelmann, 1913), 1:83, 140, 166, 270.
5. Claude Bernard, *Leçons sur les phénomènes de la vie communs aux animaux et aux végétaux* (Paris: J. B. Baillière et fils, 1878) 1:8 [*Lectures on the Phenomena of Life common to Animals and Plants*, Vol. 1, trans. Hebbel E. Hoff and Roger and Lucienne Guillemin (Springfield, Ill.: Charles Thomas, 1974), p. 8]. Hereafter *Phénom. de la vie.*
6. Some of his most resounding theories are summed up in Claude Bernard, *Phénom. de la vie* 2:433–434 [only Vol. 1 is included in the Hoff, Guillemin, Guillemin translation].
7. Xavier Bichat, *Anatomie générale appliquée à la physiologie et à la médecine*, *Oeuvres* (Paris, 1832), 1:vii [*General Anatomy applied to Physiology and Medicine*, trans. George Hayward (Boston: Richardson and Lord, 1822), 1:vii]. Hereafter *Anatomie.*
8. Montaigne, *Essais* (Paris: Flammarion, 1908), 3:182 [*The Complete Essays of Montaigne*, trans. Donald M. Frame (Stanford: Stanford U. Press, 1958), p. 578].
9. F. Bosc, 'De l'inutilité du vitalisme,' *Rev. philo.* **76** (1913) 375.
10. Jagadis Chandre Bose, 'De la généralité des phénomènes moléculaires produits par l'électricité sur la matière inorganique et sur la matière vivante,' *Rapports présentés au Congrès international de physique* (Paris: Gauthier-Villars, 1900) 3:584–585.
11. Henri Bouasse, 'Sur les déformations des solides,' *Revue générale des sciences pures et appliquées* **15** (1904) 121 ff. Cf. also his 'Développement historique des théories de la physique,' *Scientia* **7** (1910) 293.
12. Jacques Loeb, *La Dynamique* 10, 212 ff., 290, 311 [cf. *Dynamics* 5–6, 118 ff., 158, 175; citation of 311 / Eng. 175 erroneous]. Driesch himself concedes that the fact that all tropisms are subject to Weber's law, which resembles the rules governing the action of masses in chemistry, seems to demonstrate that "something chemical is connected with tropisms" and that we "may assume hypothetically that true simple reflexes are machine-like in every respect" (*The Science and Philosophy of the Organism*, Aberdeen: printed for the University, 1908, 2:9, 12).
13. O. Lehmann, 'Scheinbar lebende fliessende Kristalle, künstliche Zellen und Muskeln,' *Scientia* **4** (1908) 293 ff.
14. Stéphane Leduc, 'Les Lois de la biogénèse,' *Revue scientifique*, 5th Series, **5** (1906) 225–229, 265–268. Cf. Hans Przibram, *Vitalität*, *Experimental-zoologie* (Leipzig and Vienna: Franz Deuticke, 1913), **4**:13–14.
15. Cf. Loeb, *La Dynamique* 80 ff. [*Dynamics* 38 ff.] on the work of Traube. Loeb does express reservations, however, and is "not at all inclined to see artificial

organisms in the morphological imitations of cells and bacteria by means of inorganic precipitates" (*La Fécondation chimique*, trans. Anna Drzewina (Paris: Mercure de France, 1911), p. 339). Prenant, while judging that the physical theories proposed in this domain encounter "insurmountable difficulties," nevertheless acknowledges that "anyone who knows the specters or phantoms produced by various physical forces, magnetic tracings for example, and in another connection has before his eyes the complete and perfect image of the mitotic division of a cell, is struck by their resemblance and hazards a physical explanation." Prenant finds Leduc's efforts "striking" and concludes that "the general features of resemblance [in the case of mitosis] are such that they prohibit us from appealing to a mysterious energy, a vital energy, distinct in nature from known physical energies" ('Les théories physiques de la mitose,' *Scientia* **13** (1913) 380–391 [Meyerson's brackets]). Cf. also Lehmann, 'Scheinbar lebende fliessende Kristalle, künstliche Zellen und Muskeln,' *Scientia* **4** (1908) 293.

16. Loeb, *La Fécondation chimique* 27–28 [cf. *Artifical Parthenogenesis and Fertilization* (Chicago: University of Chicago Press, 1913), p. 21].
17. Yves Delage and Marie Goldsmith, *La Parthénogénèse naturelle et expérimentale* (Paris: Ernest Flammarion, 1913).
18. On this subject see Driesch, *Der Vitalismus als Geschichte und als Lehre* (Leipzig: Johann Ambrosius Barth, 1905), pp. 155 ff. [*The History and Theory of Vitalism*, trans. C. K. Ogden (London: Macmillan, 1914), pp. 171 ff.].
19. Furthermore, Bäyer believes that it is in fact by means of this reaction that synthesis is carried out in the organism. See Loeb, *La Dynamique* 208 [*Dynamics* 114].
20. Cf. A. Betse, 'Neuere Vorstellungen über die Natur der bio-elektrischen Ströme,' *Scientia* **8** (1910) 70 ff., and O. Lehmann, 'Scheinbar lebende fliessende Kristalle, künstliche Zellen und Muskeln,' *Scientia* **4** (1908) 297.
21. See Filippo Bottazzi review of books by H. Bechhold, T. Brailsford Robertson and M. H. Fischer, *Scientia* **12** (1912) 276.
22. On this hypertrophy see Jean Nageotte, 'Reviviscence de greffes ...,' *Bull. Soc. biol.*, 24 Nov. 1918, p. 892 [citation unverified].
23. Jean Nageotte, 'Les substances conjonctives sont des coagulums albuminoïdes du milieu intérieur,' *Comptes rendus des séances et mémoires de la Société de Biologie* [**79** (1916) 833], 21 Oct. 1916, p. 1 of the offprint.
24. Nageotte, 'La matière organisée et la vie,' *Scientia* [**24** (1918) 434–436], Dec. 1918, pp. 9-11 of the offprint.
25. Driesch, *The Science and Philosophy of the Organism* (Aberdeen: printed for the University, 1908), 2:150.
26. Furthermore, this concept has been taken up by Sir Oliver Lodge. Cf. George Henslow, 'Ecology considered as bearing upon the Evolution of Plants,' *Scientia* **13** (1913) 205.
27. Montaigne, *Essais* (Paris: Flammarion, 1908), 3:183 [*The Complete Essays of Montaigne*, trans. Donald M. Frame (Stanford, California, 1958), p. 579].
28. Driesch, *Die Biologie als selbständige Grundwissenschaft* (Leipzig: W. Engelmann, 1893), p. 1.

29. André Morellet, *Mémoires inédits de l'abbé Morellet de l'academie Française, sur le dix-huitième siécle et sur la révolution*, 2nd ed. (Paris: Ladvocat, 1822) 1:135 ff. Cf. IR 354 [Loewenberg 311]. Many theologians seem to have been well aware of the weakness of this position, however: Pascal notes that "no canonical author has ever used nature to prove God," which, he adds, "is very noteworthy." Moreover, Pascal himself never intends to use it. To tell unbelievers "that they have only to look at the least thing around them and they will see in it God plainly revealed ... is giving them cause to think that the proofs of our religion are indeed feeble, and reason and experience tell me that nothing is more likely to bring it into contempt in their eyes" (*Pensées* 445–446 [Krailsheimer 179, 263–4]).
30. Kant, *Critique du jugement*, trans. Jules-Romain Barni (Paris: Ladrange, 1846), 2:77, § 74 [*Critique of Judgment*, trans. J. H. Bernard (New York: Hafner), 1951, p. 248, § 75].
31. Leonard Huxley, *Life and Letters of Sir Joseph Dalton Hooker* (London: John Murray, 1918).
32. *The Times Literary Supplement*, 18 July 1918, p. 334.
33. Lucretius, *De rerum nat.* IV, 883–885: *neque enim facere incipit ullam / rem quisquam, quam mens providit, quid velit ante: / id quod providet, illius rei constat imago.*
34. Thomas Henri Martin, 'Mémoire sur les hypothèses astronomiques d'Eudoxe, de Callippe, d'Aristote et de leur école,' *Mémoires de l'Institut National de France, Académie des Inscriptions* **30** (1881): Pt. 1, p. 255. Cf. Aristotle, *On the Soul* 433ª13–21.
35. Henri Poincaré, *La science et l'hypothèse* (Paris: Flammarion, n.d.), p. 154 [*Science and Hypothesis*, trans. George Bruce Halsted (New York: The Science Press, 1905), p. 93]. Jacques Loeb says that there "can be no economy in work except where there is memory and, as a consequence, reason; blind forces do not spare the means" (*La Dynamique* 224). But Descartes, on the subject of action at a distance, had already protested against an assumption that appeared to him to endow material particles with reason, to the point of making them "truly divine, so that they can know without any intermediary what is happening in very remote places and act upon them there" (*Oeuvres* 4:306 [erroneous citation]).
36. Albert-Auguste Cochon de Lapparent, *Science et apologétique* (Paris: Bloud, 1905), pp. 191–211.
37. Henri Piéron, 'Les Instincts nuisibles à l'espèce devant les théories transformistes,' *Scientia* **9** (1911) 201.
38. Bacon, *De Augmentis Scientiarum*, Bk. 3, Ch. 4, *The Works of Lord Bacon* (London: William Ball, 1837), 2:339 [*Of the Dignity and Advancement of Learning*, *The Works of Francis Bacon*, ed. Spedding, Ellis and Heath (Boston: Taggard and Thompson, 1843), 8:510].
39. Driesch explicitly posits it as such, declaring this concept "autonomous" and "irreducible" (*The Science and Philosophy of the Organism*, Aberdeen: printed for the University, 1908, 1:144, 228). See also 2:249, where Driesch asserts that no chemical substance is possible as a basis for entelechy.
40. Henri Piéron, 'La Notion d'"instinct,"' *Bull Soc. fr. phil.* **14** (1914) 327 ff.

41. It should be noted that Schelling declared in almost the same words that "the vital force was conceived solely as a stopgap [*Nothbehelf*] for our ignorance" and that it is "an authentic product of lazy reason" (*Erster Entwurf*, I, 3:80 [Meyerson's brackets]).
42. Lawrence J. Henderson's interesting book *The Order of Nature* (Cambridge: Harvard Univ. Press, 1917) contains a number of these considerations, judiciously chosen and clearly set forth, with neither a finalistic nor an antifinalistic slant.
43. Höffding rightfully stresses this somewhat negative characteristic of vitalism (*Der Totalitätsbegriff*, Leipzig: O. R. Reisland, 1917, p. 85).
44. Pierre Delbet, 'Sciences médicales,' in Henri Bouasse et al., *De la Méthode dans les sciences*, 1st series, 2nd ed. (Paris, 1910), p. 249 [1909 ed., p. 201].
45. Driesch, *The Science and Philosophy of the Organism* (Aberdeen: printed for the University, 1908), 2:208.

CHAPTER 8

FORMS OF SPATIAL EXPLANATION

In an earlier work we attempted to show directly, by examining present and past theories, that the physical sciences actually do conform to the pattern at which we arrived in Chapter 5, that is, that their explanations ultimately rest on identification in time and in space. But we are forced to admit that a demonstration of this sort remains and will always remain incomplete. No matter how zealous one is, one will never be able to cover more than a small part of the immense body of scientific knowledge, and it will always be possible for the reader to suspect that the examples were chosen and presented with too much erudition or ingenuity to be really convincing, at least with respect to a thesis which, on the face of it, appears so extraordinary. We shall not attempt to take up that demonstration again here, but perhaps a few considerations concerning the way science uses space and its functions in its explanations can help make our thesis a bit more plausible.

A. The simplest and most general use is that suggested by the term *displacement* itself: the explanation rests on the fact that something (having remained identical to itself, which, as we know, is the basis for all explanation) is presumed to have changed place. Because space is by nature uniform, undifferentiated, and because this uniformity is destroyed only by the material objects found in it, displacement must therefore be defined by these objects. In the simplest case there will be something moving from one object, from one material body, to another. Of course, since we are dealing with explanation, with figurative theory, that is, with the substitution of a different reality for what we perceive or think we perceive, for the reality suggested to us by the common sense conception, there can be no question of grasping this displacement directly; we must infer it indirectly, as a result of this or that phenomenon or group of phenomena. That is why what changes place can be understood as immaterial. In the past, as we know, physics often used and even abused this schema; what changed place was above all considered to be a quality-bearing principle, such as caloric or phlogiston. Certainly modern physicists are much less accommodating in this respect. The caloric fluid no longer exists for us; where it was presumed to have changed place, we assume that movement is being communicated. As for phlogiston, we now claim, for phenomena where it supposedly moved from one body to another, that something material (especially oxygen) changes place, but in the opposite direction. At the temperature of our ordinary reactions, oxygen is a colorless gas, and so it goes without saying that we cannot *see* it move. Furthermore, in reactions where it passes from one compound to another, we quite often do not even admit that oxygen gas, molecular oxygen like that contained in our cylinders, was really formed as an intermediate product and then entered into the reaction as such; on the contrary, we suppose that oxygen acted in the nascent state, as atomic oxygen – a hypothetical being *par excellence*. That does not prevent us from being quite convinced that material oxygen really was transferred in the reactions in question – a whole series of observed circumstances calls for this conclusion, principally the considerations of weight admirably brought out by Lavoisier. But – and we cannot emphasize this fact too much, because it casts a vivid light on the true nature of scientific truths – the claims of the phlogiston theorists were also based on observations. Not believing in the principle of the conservation of weight, or at least admitting implicitly that it allowed exceptions which could be more or less explained, they inferred the displacement of phlogiston by the fact that a specific property had passed from one body to another.

Has contemporary science entirely given up the use of immaterial principles in this sense? One would hesitate to make such a claim. Indeed, if it is eventually agreed that electricity is not reducible to mechanical motion, but that on the contrary all other phenomena must ultimately be reduced to electrical phenomena, we wonder how this primitive being, the foundation of all reality, could possibly be understood if not as a sort of immaterial principle. No matter how different it is said to be from the earlier *fluids*, particularly Franklin's electric fluid, it will necessarily retain many points in common with them, chiefly the ability to create phenomena by changing place. But even setting aside this most recent phase and limiting oneself to the representation of the universe according to purely mechanical theory, it is clear that *motion* itself is such a principle. As a matter of fact, motion must pass from one body to another. And if we directly perceive the fact of the passage in this case, for example when one billiard ball bumps into another, the *how* of the passage nevertheless remains completely mysterious. We are thus forced to endow matter with occult properties, such as impenetrability or elasticity, or with forces no less enigmatic.

Because of the fact that the body or the principle presumed to change place may not be directly perceived, it is not absolutely necessary that we know both where it started from and where it went. Of course, it is inconceivable that we should be ignorant of both of them at once, for in that case we would have had no reason for framing the theory. However, it is possible, in the extreme case, that we know only the starting point or the finishing point, with the other end of the chain remaining obscure. In such a case we say that the body in question has "dissipated into space" or "comes from the depths of space." The ancient atomists, as we can see in Lucretius, used this means of explanation extensively. In *De rerum natura*, it is constantly a question of particles from faraway spaces which cause terrestrial phenomena by their impact. Moreover, it is not difficult to see why such conceptions appeared plausible. Since there was no known means of following gases in their peregrinations (Empedocles's famous water-clock experiment[1] having remained completely isolated in this respect), these substances seemed endowed with no more than a sort of semimateriality, at least as they were commonly understood. As late as Van Helmont, who is considered to be the creator of the chemical conception of gas, this *gass* appears to be something halfway between true bodies and immaterial principles (such as his "*blass*," the life principle). Now it is a fact of common experience that a body dissipates its smoke in the atmosphere as it burns and that water evaporates into the air, while in return the very tangible manifestations of rain, snow and hail come from the atmosphere. What then could be more natural than to appeal to surrounding space for other phenomena as well? That has become harder for us, because we are better able to follow matter in its transformations. However, the situation may have changed less than we would at first be inclined to think. This is because in addition to the air, we also have the ether, theoretically filling the depths of space, and of course this ether must act upon terrestrial phenomena (indeed, it is for this reason that we invent it). Now we no doubt believe we can also follow these different ways in which the ether acts (which are what we call forms of energy), but only to a certain extent; we certainly feel that there must be some of these forms that elude us. It is on a conception of this type (although it was framed in the middle of the eighteenth century, before there was any question of our ether) that Le Sage based his theory of gravitation, which explained Newtonian attraction by the action of "ultramundane corpuscles." This hypothesis certainly must not have contained anything that could shock the modern physicist, since Maxwell declared it the only consistent theory of gravitation ever formulated.[2] It finally succumbed only in the face of the difficulty of making it agree with firmly established laws of physics such as the principle of the conservation of energy, etc. (cf. IR 79–80 [Loewenberg 80]). Still more recently, at the time of the discovery of radioactive bodies, when people were wondering what could cause the formidable and mysterious energy they constantly radiate (we spoke of these discussions on p. 121), the hypothesis was advanced that it could be the result of radiations throughout space that our instruments were incapable of detecting and that the bodies in question somehow were able to capture in order to disperse them later.[3] Is there any need to emphasize the kinship of this conception with those of Le Sage and Lucretius? In each case there is the same utilization of space to explain terrestrial phenomena; we know the arrival point of what is happening, and we place the starting point – uncontrollable – in unlimited space. Due to circumstances unnecessary to retrace here, the hypothesis of radiation capture had to yield to that of atomic disintegration, which has been generally maintained since that time. But the very fact that such a hypothesis could arise and be readily accepted by many qualified physicists clearly shows that it would be a mistake to consider the conceptions of the past completely out of date.

B. A much more precise but also more restricted use of the spatial function is suggested to us by "the explication" of the leaves. The leaves have changed neither in size nor in structure; they have changed shape, but so slightly that the identity of the two states does not seem affected in any way. Moreover, we are accustomed to this kind of change: we can fold and unfold a fabric and even fold the leaves back up again almost as they were before and convince ourselves that this is a highly *reversible* modification, which can be done and redone in both directions without leaving any trace in the intrinsic nature of the object that has undergone it.

C. If we now refer to the first part of Bossuet's phrase and consider that we must start with the seed of the plant, we see that this process will not suffice. Supposing that Bossuet to some extent had in mind the preformationist theories so popular in his time, it will be necessary not only to fold and refold the plant in a thousand different ways, but also to reduce its size in very great proportions if this plant with all its parts is to fit into the seed. Consequently the preformationist theories of the organism show us that our understanding does not consider change in size to be a modification that really disrupts identity. No doubt direct observation of the organism has something to do with this conviction. We constantly see beings grow while still retaining their individuality, and we remember quite well having grown under the same conditions ourselves. However this experience, as such, is certainly not sufficient. It is in fact quite incomplete, for although we see organisms grow larger, we do not see them become smaller – the process by which they finally degenerate bears no resemblance to the process that made them grow – and, on the other hand, the resemblance between two successive stages of the same organism is not complete: a man is not simply a child whose proportions have all been enlarged. The truth seems to be that the conviction in question, while it can be suggested by the organism, nevertheless rests in the final analysis on the properties of space itself, mathematical space or, to be more precise, the space of Euclidean geometry, the theory of which is built upon the well-known postulate that guarantees just this possibility of increasing or reducing the dimensions of a figure ad infinitum without in any way modifying the relations of the elements that make it up. In this space, then, everything seems to be proportion and nothing but proportion, without our having any way of imagining any limit or absolute size whatever. Furthermore, at least for those who are perceptive, the habit of seeing our visual image of the object constantly change size may play a considerable role. When I watch from a seated position as a friend walks around the room, his image changes from one to tenfold in size without my noticing it, and I certainly do not experience the shadow of a doubt as to his identity throughout these changes.[4]

The conviction of this fundamental, essential analogy – of this quasi-identity between objects of very different sizes – is obviously what allows the preformationist conceptions. The plant, although enclosed in the narrow confines of the seed, can nevertheless already be the complete plant. It will increase in size, in mass, and it will need external matter to do so. But it will assimilate this matter. Granted, the process of assimilation remains obscure for the time being, but there is no doubt that it will eventually be explained (that is, as we have seen, the way all theories proceed). What is clear is that the assimilated matter will turn out to be arranged in exactly the same way as was that which constituted the seed. It will simply enlarge its dimensions, but without altering its essence in any way.[5] All in all, this supposition resembles what we see occurring for crystals, except that in their case the additional matter simply takes its place on the outer surface, while in the organism it permeates the whole interior.

In the seventeenth and eighteenth centuries people certainly saw no difficulty in using space in this manner; they assumed that physical nature was indefinitely proportioned and identical to itself, like space. That was evidently a necessary condition for the encasement theory of the germ, and once this condition was accepted, they were not loath to admit that the germs, as small as they might seem to us, contained others "in comparison with which they can be reckoned large; for everything in nature proceeds *ad infinitum*," as Leibniz says.[6] We cannot help thinking of infinitely small entities of various orders that can be infinitely large with respect to one another; indeed, it is quite possible that Leibniz himself had this in mind, although such a supposition is obviously much more plausible in physics than in mathematics, for the germ, no matter how small one imagines it, will never be *infinitely* small. Moreover, the conception of the unlimited reduction in size of organic beings antedates Leibniz and thus antedates infinitesimal calculus. It can in fact be found expressed in Pascal's famous passage on the *mite*.

> Let a mite show him in its minute body incomparably more minute parts, legs with joints, veins in its legs, blood in the veins, humours in the blood, drops in the humours, vapours in the drops: let him divide these things still further until he has exhausted his powers of imagination, and let the last thing he comes down to now be the subject of our discourse. He will perhaps think that this is the ultimate of minuteness in nature.
>
> I want to show him a new abyss. I want to depict to him not only the visible universe, but all the conceivable immensity of nature enclosed in this miniature atom. Let him see there an infinity of universes, each with its firmament, its planets, its earth, in the same proportions as in the visible world, and on that earth, animals, and finally mites, in which he will find again the same results as in the first .... (*Pensées* 349 [Krailsheimer 89–90])

An amusing confirmation of the unshakable faith in the proportionality of nature characteristic of that epoch is furnished by Swift's admirable *Gulliver*. His Lilliputians and Brobdingnagians differ from us only by size, but in all other respects entirely resemble men as we know them; likewise everything around them, plants, animals, is on their scale, without otherwise differing from the types familiar to us. One can easily observe, moreover, that even today fantasies of this sort do not seem overly shocking to the imagination of readers in general, even well-educated ones. Thus Wells, in his novel entitled *The Food of the Gods* (which is, incidentally, extremely well-written and psychologically profound), was able to present a conception altogether analogous to that of Swift; his giants, including the animals, also behave – and, in particular, move – except for size, absolutely like the beings with which we are familiar.

Now not only is that inadmissible, but it would have been possible to come to this realization even before the modern discoveries (which we are going to discuss below and which will show the profound reasons for this impossibility). As a matter of fact, it is precisely in the structure of organisms that the lack of proportionality in nature is clearly revealed. A large dog is not simply the enlarged copy of a small dog: the proportions of the different parts is modified and the head, for example, is relatively smaller. If an expert is given a photograph of a dog, he can certainly give a good approximation of the size of the animal. Moreover, the most superficial observation teaches us that large animals move quite differently from small ones. The jumps of a flea appear prodigious to us in relation to its size and we sense instinctively that a larger animal could not perform the same feat. Indeed, as their size increases, animals are built more and more massively and their movements become relatively slower and slower: the elephant usually does no more than walk, trotting only on rare occasions, and then with difficulty.

But to a modern physicist or biologist the indefinite reduction in size of an organism seems much more palpably impossible than could have been imagined in ages past. He is convinced that nature is not continuous, that it is, on the contrary, formed of discrete particles having a definite size. Leibniz's germs, for example, as a result of their successive reductions, would be rapidly reduced to molecular and submolecular dimensions and consequently placed in a world which, although it *constitutes* ours, bears no resemblance to it at all.[7] Moreover, supposing that this germ were no larger than a molecule – since this molecule is an individual, indestructible or (if one supposes it to be composed of electrons, according to more recent ideas) decomposable only by a radical destruction which gives rise to things that are essentially different – it will be necessary for the germ to be composed of a single molecule, which obviously contradicts our idea of it. In addition, although we have only very vague ideas on the constitution of cells, it now seems infinitely probable that they can undergo reduction in size only in strictly limited proportions if they are still to accomplish their specific function. The germ cell in particular certainly has a very complex structure; it contains various groupings made up of subgroupings and finally of large molecules of albuminoid bodies, etc. If one wishes to imagine all that reduced in size but everywhere preserving the same structures, one is forced to admit that the dimensions of the molecules have also been reduced. Now these are by nature irreducible; furthermore, their size is an element determining the properties of the bodies they make up, that is to say, of what appears to us as qualitative. Albumin whose molecules did not have the requisite size would no longer be albumin and could not assume the role that albumin plays in the organism, just as water would no longer be water. The implications of this can immediately be made obvious by appealing to Swift's amusing fancies. What was the water surrounding the coasts of Lilliput? If it was the water of our seas, then on the scale he gives the inhabitants the phenomena of capillary action must already have been very significant, and the form of their boats must have suffered the effects of it.

The more modern science descends toward the infinitely small, the less the world it reveals resembles our familiar world. In it the very principles whose authority seemed most secure, which seemed to us most characteristic of our universe, lose their significance; they are revealed to be relative to the *scale* on which we act and observe. On the scale of Brownian motion, we see particles that have apparently been moving indefatigably for numberless centuries, without any external energy source and without this energy being exhausted – which seems to be a direct contradiction of the denial of perpetual motion, which denial is the point of departure for Carnot's demonstration. We also see thermal agitation of a liquid giving rise to mechanical motion, and Perrin's particles, although they are heavier than the liquid in which they are suspended, can be pushed by this motion in a direction contrary to that required by gravity – which, as we have seen, is a motion we all feel to be absolutely impossible on the human scale in the world as we know it. The authority of Carnot's principle is not affected by the motion of Perrin's particles – it continues to govern our familiar world, where, through the continual increase in entropy which is its immediate corollary, it constitutes the profound source of all change, all becoming. But, as became apparent to those who sought a mechanical explanation for it, and in particular Maxwell and Boltzmann, it is only a statistical principle and consequently valid only where we observe molecules in sufficient numbers to make statistics applicable. On the other hand, where we can observe sufficiently small groups of molecules, as in Brownian motion, the increase in entropy loses its significance.

But underneath the world of Brownian motion and molecules is another world, stranger yet because it is infinitely more different from the one on our scale. It is especially to this world of subatoms and electrons that the 1912 *Council of Physics* turned its attention, and the reader will have been able to see from the few characteristics we cited above how out of date all the norms we instinctively apply to our familiar reality are in that world. It goes without saying, and we have explained why (p. 126), that there can be no question of looking for matter there, but even motion appears to be something essentially different from what we know by that name. Indeed, motion is no longer continuous, since an elementary particle, according to the hypothesis of *quanta* (at least in its original form), can no longer take on just any speed, but only a speed that is an integral multiple of an initial velocity *v*. Likewise, it is being questioned whether the principle of the conservation of energy is applicable there, or whether it will have to be given up. Furthermore, as we have pointed out, there is no unanimity among scientists, nor apparently even any agreement on principle, as to how one must envisage these strange phenomena, which is obviously due to the fact that it has thus far been impossible to set up any kind of consistent spatial image. Even so, everyone seems to recognize that these are phenomena whose consideration will necessarily transform science profoundly. Thus what can be called *our world*, the world of familiar phenomena and even, up to a point, of the laws that govern them, now appears to us to be limited, from the standpoint of size, by a lower limit beyond which there is *something else*.

On the other hand, the supposed upper limit has disappeared. As a matter of fact, curiously enough, neither one of these two determinations was anticipated by the science of the past. Books on the history and philosophy of science often stress how much the science of earlier epochs resembles our own in certain respects, and rightly so, for there are traits that remain immutable, springing directly from the inner structure of human reason, and human reason never changes. Mechanism is certainly as old as science itself and there were many *presentiments* of conservation principles before they were formulated with precision and supported with proofs; on this subject humanity developed a sort of genuine prescience. But this is also because these cases involve questions where there is an agreement between our mind and nature, where nature appears rational. Just the opposite is true in the case of proportionality, and therefore we should not be too surprised that such prescience was entirely lacking there. The statements of the seventeenth and eighteenth century preformationists, as well as those of Pascal and Leibniz, are quite convincing on this point, but this observation can be considerably broadened. Apparently no earlier philosopher or physicist ever doubted that the minute particles of bodies had to behave exactly like the tangible bodies to which we are accustomed. Aristotle and Democritus were entirely in agreement on this point, and Lucretius certainly imagined atomic impact (which was the fundamental phenomenon of nature for him, as it was for all mechanists) to be entirely analogous to impact between tangible bodies, between billiard balls, to use a modern comparison, but one that would not have disturbed Lucretius had he been familiar with the second term of the comparison. But what is perhaps even more significant is that when they wished to introduce a distinction based on magnitude or distance, they placed it not where we now find it, namely in the small, but in the large. For Aristotle, indeed, the heavens appear to be fundamentally different in nature from the sublunar world. They are unalterable, incorruptible, imperishable, and even their motion is distinguished from that of terrestrial objects, being perfect, circular, while terrestrial motion hurries toward a goal. It is well-known what power this conception had over mankind and how hard it was for the Copernicans to break its hold. But finally, particularly following Newton's demonstration that the celestial bodies are governed by gravitation, that is, by the very force that controls all terrestrial matter, this triumph seems complete: in some of the most venturesome speculations of modern science – such as Haeckel's claim that Carnot's principle may not be valid in the stellar spaces, since things happen differently in the cosmos as a whole than they do on earth,[8] or in the devices referred to in Chapter 6 by means of which Arrhenius means to bring about a reconcentration of energy in the stars[9] – we barely recognize a remote and weakened echo of the Peripatetic theory of the radical distinction between the supralunar and sublunar worlds.

Because of this double attitude of modern science, the explanation of phenomena occurring in the immensity of space has become a great deal easier. We measure the speed of light by terrestrial means and unhesitatingly apply the data thus obtained to astronomical observations. The spectral analysis of bodies we handle in our laboratories furnishes us, by comparison, data not only on the composition of the stars but also on their velocities. And it is by means of data drawn from the most delicate and at least apparently most paradoxical experiments of optics and electricity that H. A. Lorentz and Einstein sought to resolve the mystery of the anomaly of the planet Mercury. On the other hand, since everything in nature forms a whole, it is not at all impossible (as we pointed out on p. 183, and contrary to the dogma Auguste Comte was so anxious to impose) that discoveries made in the stellar world have a significant effect on our theories of terrestrial phenomena and facilitate, or at least modify, the way they are explained: the case of helium is a good example. And if Einstein's new *general theory of relativity* really triumphs thanks to agreement between the predictions it allows us to make and astronomical observations, this fact will profoundly influence our whole conception of electrical and optical phenomena.

However, from our present point of view, namely insofar as the means of spatial explanation of terrestrial phenomena are concerned, the double revolution we have just mentioned is certainly a disadvantage. In fact, it is difficult to see how the immensely large could be used in this way, while recourse to the *infinitesimal* is easy and has actually been tried, as we have seen. Now, once again, this recourse is no longer possible for contemporary science, or at least it can no longer be carried out as clearly and openly as in the science of the past. If there are still preformationist conceptions – and we have acknowledged not only that they exist but that they are in some sense inevitable – they must make use of more complex concepts. We shall attempt a little later to discern their texture.

D. More profound, more penetrating as it were, than the three procedures we have just treated, is a fourth, which consists in exploiting the essential properties of geometric figures. Plato's theory was largely constructed on this base and it has thus been justifiably called *metamathematical*. If fire is represented by tetrahedra composed of triangles, and earth by cubes formed of squares, it is because the pointed figure of the tetrahedron seems to facilitate penetration, while cubes placed side by side and then stacked in rows, thus filling space without leaving any gap and even making any slipping difficult, effectively offer an image of the immobility of the element earth. The heat of fire is similarly explained by the acute angles of its particles (*Timaeus* 61E). But the atomists call upon the same resource. "They are distinguished, we are told, from one another by their figures," says Aristotle of the atoms of Leucippus and Democritus, "but their nature is one, like many pieces of gold separated from one another" (*On the Heavens* 275^{b}30–276^{a}2 [J. L. Stocks trans.]). Obviously, as soon as one supposes the unity of matter, matter can no longer be differentiated except by something spatial. As Duhem correctly states, one is necessarily led to imagine "that apparently continuous masses are assemblages of differently shaped small bodies" and "that the different arrangements of these various bodies must explain the properties of the different mixtures studied by the chemist."[10] As we know, the ancient atomists carried this principle to great lengths and attempted to reduce the most disparate properties to the shape of the elementary particles alone. According to Lucretius, hard bodies, such as diamonds, contain intertwined atoms, those of liquids are round, while smoke and flame are composed of pointed but not bent atoms. Sea water is bitter because among its smooth round particles (which are those of fresh water) are others, also round – which makes them behave like liquid particles – but which have rough spots enabling them to hurt the tongue. This is why sea water becomes sweet as it filters through the soil: the rough particles are retained, while those of water pass unhindered. Glass has rectilinear channels running through it, since all images pass through its substance. Milk and honey have round smooth atoms, whereas those of absinthe, on the contrary, are hooked; likewise, pleasant images are transmitted by smooth atoms and hurtful ones by atoms endowed with roughness. If certain impressions affect only one particular sense, it is because they are transmitted by figures whose form corresponds to the shape of the channels of the senses in question (*De rerum nat.* II, 388 ff., 985–988; IV, 603–604; cf. Ch. 6, p. 144 above).

These arguments strike us as bizarre. Depending on the particular mental stance we have adopted, we admire their boldness or smile at their naiveté. But the important thing is to realize that they are entirely within the logic of the system, that their explanatory orthodoxy, if we may put it that way, is irreproachable. Of course everything in these arguments having to do directly with sensation, like the explanation of agreeable images or that of impressions affecting only one particular sense, now appears totally inadmissible to us. That is because we no longer believe in the possibility of a mechanical explanation of sensation; we are much too convinced that a genuine irrational is involved. But for all the rest, the problem for which Lucretius seeks a solution is obviously the "grounds for astonishment" found in the existence of several sorts of matter. Now the same astonishment is also the point of departure for modern chemistry, as Job aptly remarked (cited above, Ch. 5, p. 133).

For this reason, we must above all look to chemistry if we want to know what later became of the spatial explanations of Lucretius, and it is infinitely interesting and instructive to follow their historical development with this in mind.

As soon as science frees itself from the Peripatetic formulas, we see explanations based on the shapes of the elementary corpuscles, barely stripped of their most extreme peculiarities, reappear in Descartes.

Descartes did not, strictly speaking, concern himself with chemistry, but since his theory, as we have seen, explicitly aimed at identifying space with matter, diversity in space commanded his attention still more than it had that of the ancient atomists, if that is possible. "I assume, first," says Descartes,

> that water, earth, air, and all other such bodies that surround us are composed of many small particles of various shapes and sizes .... Then, in particular, I assume that the small particles of which water is composed are long, smooth, and slippery, like little eels, which are such that however they join and interlace, they are never thereby so knotted or hooked together that they cannot easily be separated; and on the other hand, I assume that nearly all particles of earth, as well as of air and most other bodies, have very irregular and rough shapes, so that they need be only slightly intertwined in order to become hooked and bound to each other, as are the various branches of bushes that grow together in a hedgerow. And when they are bound together in this way, they compose hard bodies like earth, wood, or other such things; whereas if they are simply laid on one another without being interlaced at all (or only very slightly), and if in addition they are so small that they can be moved and separated by the agitation of the very fine material that surrounds them, they must occupy a great deal of space, and compose very rarified and light liquid bodies such as oil and air.[11]

In this way Descartes tries to explain both solid bodies and those we now call gases by means of one and the same very ingenious device, founded solely on the shape of the ultimate particles, as also is his explanation of the fluidity of water. The kinship of this whole passage with Lucretius's arguments is obvious. At times Descartes's arguments even turn out to be absolutely identical to those of *De rerum natura*:

> And even though the sea is salty, most springs are not: the reason for this is that the parts of sea water which are sweet, being soft and pliable, change easily into vapors and pass through the by-roads between the little grains of sand and other such parts of the Earth's surface, while those making up salt, being hard and unyielding, are thus less easily vaporized by heat and cannot pass through the pores of the Earth unless these pores are wider than usual.[12]

The part of Cartesian theory concerning the constitution of gaseous bodies is developed and made more explicit by Robert Boyle. The famous English physicist and chemist states that

> The corpuscles of the air must be as well sometimes considered under the notion of little springs, which remaining bent, are in their entire bulk transported from place to place; as under the notion of springs displaying themselves, whose parts fly abroad, whilst, as to their entire bulk, they scarce change place: as the two ends of a bow, shot off, fly from one another; whereas the bow it self may be held fast in the archer's hand.[13]

Lémery, whose *Cours de chymie* (published in 1675) remained the classic chemistry text for more than half a century, likewise appeals to the shapes of the particles.

Since there is no better way of explaining the nature of a thing as hidden as that of a salt than by attributing to its component parts shapes that correspond to all the effects it produces, I shall say that the acidity of a liquid consists in pointed parts of salt, which parts are in agitation; and I do not believe anyone will dispute that acid has points, since all experiences show it; one need only taste it to come to this opinion, for it makes prickings on the tongue like or almost like those one might receive from some material carved into very fine points; but a demonstrative and convincing proof that acid is composed of pointed parts is that not only do all acid salts[14] crystallize in points, but all dissolutions of different materials made by acid liquids take this shape in their crystallization. These crystals are composed of points of different length and thickness, and this diversity must be attributed to the more or less sharp points of the different sorts of acids. It is also this difference in the subtlety of the points that makes one acid penetrate and dissolve a compound that another cannot rarefy: vinegar, for example, makes an impression on lead, which aqua fortis cannot penetrate; aqua fortis dissolves mercury, and vinegar cannot penetrate it; aqua regia is a solvent for gold, and aqua fortis makes no impression at all; aqua fortis, on the contrary, dissolves silver yet does not touch gold, and so on.

As for alkalis, they can be recognized by pouring acid on them, for immediately or shortly thereafter there occurs a violent effervescence which lasts until the acid finds no more bodies to rarefy. This effect can reasonably lead to the conjecture that the alkali is a material composed of unyielding and brittle parts, the pores of which are shaped in such a way that when the acid points have entered into them, they break and separate anything that opposes their movement, and depending on whether the parts that form this material are more or less solid, the acids find more or less resistance and effervesce more strongly or more weakly. Thus we see that the effervescence produced by the dissolution of coral is much less violent than that made by the dissolution of silver.

There are as many different alkaline salts as there are these materials that have different pores, which is why an acid will make one material ferment yet cannot do the same for another; for there must be proportion between the acid points and the alkaline pores.[15]

In Stahl too we find altogether analogous arguments:

Dissolution is nothing more than the division of the body into very fine and very smooth parts, which slip into the pores of the solvent so as to form a single fluid. But this division of the parts making up the whole could not occur if the liquid charged with dissolving or dividing it did not penetrate the pores of the body to be dissolved. It obviously follows that each solvent must be formed of parts that correspond in shape and size, that is, in diameter, to the pores of the body to be dissolved: a given liquid will thus not be able to dissolve all bodies, but only certain ones. Furthermore, a given body is constructed and woven of particles which are not all alike, but on the contrary quite dissimilar; these particles have very different shapes and dimensions. The variations in texture, position and arrangement of these particles give diverse pores to a single body: thus one easily concludes that there must be various solvents whose smallest parts can penetrate the pores of this body. Assuming this to be true, it is easy to understand why aqua fortis dissolves metals but not wax or sulfur, and likewise why it dissolves silver and gold; and why aqua regia dissolves gold and not silver. (*Fundamenta chymiae*, Pt. 1, Sect. 1, Ch. 2, p. 8)

As a matter of fact, we might be surprised at the persistence of these mechanistic arguments in a branch of knowledge as strongly given to qualitative arguments as the chemistry of that era appears to have been. We must note first of all that Descartes holds himself completely aloof from the chemistry of his time, and even Boyle fits only very partially into its framework. The case is different for Lémery and Stahl. But the passages just cited show explicitly that the domination of qualitative conceptions was less absolute than one might at first think; a sort of mechanistic undercurrent persisted, undoubtedly deriving from the ancients but reinforced by the great influence of Cartesian theory in physics. We must also add that in a sense this recourse to spatial explanations was entirely to be expected in a chemist of that time. Indeed, the number of elements at his disposal for the purpose of explanation was quite limited and it was understood – since in any case these elements were above all creations of reason – that one could attribute only a limited number of qualities to them. As a result, in order to explain differences in action between otherwise analogous bodies (like the various acids or metals, for example) that were therefore presumed to be similarly constituted, there was a great temptation to resort to the eternal expedient of shapes.

After Stahl, however, deductions of this type become more and more rare. At the moment when their theories triumph uncontested, the phlogiston theorists seem to have almost forgotten this aspect of the teaching of their master, and neither Lavoisier nor his successors seem disposed to resuscitate a type of explanation that had appeared particularly effective from Lucretius to Lémery. Even after the atomistic conceptions have gained an important place in chemistry – through the work of Dalton, Avogadro and Ampère – and gradually, despite considerable obstacles and resistance, become dominant, science does not return to these erring ways. Most certainly no chemist in the second half of the nineteenth century (to speak only of that period of the past close enough that we can easily judge its tendencies) would have dreamed of explaining why different acids act differently on this or that particular body by attributing a particular shape to the acid particles or to the pores of the body being attacked.

This is because in the meantime chemical theories underwent an upheaval which is no doubt eventually confirmed by the "revolution" of Lavoisier, but which does not coincide with that revolution; it is already largely over before the revolution breaks out and is moreover distinguished from the latter by the fact that it takes place slowly and almost insensibly. This upheaval concerns the concept of the chemical "element." As a matter of fact, its beginnings, or at least its foreshadowings, date back a long way. In Boyle's *Sceptical Chymist* we find a whole series of extracts, beginning with Roger Bacon, in which the indestructibility of gold by chemical operations is clearly affirmed.[16] Thus, because it is recognized that the alleged transmutations are futile, little by little the conviction is established that there are bodies which cannot be decomposed by any means at our disposal. These bodies are not at first considered elements on that account: the conception of a purely qualitative element, one that confers a clearly defined property on the compounds it forms, still has too strong a hold. Even Boyle himself – whose *Sceptical Chymist* is devoted precisely to combatting the conceptions of the Peripatetics and the supporters of Paracelsus and at the same time to establishing the notion of the indestructibility in question – considers that these inalterable bodies (among which he seems to count the metals or at least some of them, such as gold, silver and mercury, but also glass) are nevertheless mixtures, although the parts are so closely bound together that they cannot be separated by laboratory methods. But the common opinion of chemists soon goes beyond this point of view, and while conserving, *pro forma*, Aristotle's or Paracelsus's notion of the elements throughout the eighteenth century, in practice they gradually come to conceive the existence of a considerable number of diverse elementary substances constituting a limit to any attempt at chemical decomposition (or, as Boyle says, "anatomy"). From this point of view, Stahl's scientific career is located astride the two epochs, as it were, and it is interesting to follow the transformation that takes place over the course of his work without Stahl himself, considered the great master and legislator of the chemical knowledge of his time, taking a really active part in it, nor even, it would seem, arriving at a clear awareness of this phenomenon, so pregnant with consequences. At the beginning of his career, Stahl loudly proclaims his faith in the existence of a substance called *Elixir* or *Tincture* whose principal property consisted in transmuting metals; he also explicitly affirms the possibility of a transmutation of lead. Later, on the contrary, he energetically combats the alchemists' claims, although he still does not say that he considers metallic calces to be true elements.[17] It is this latter opinion that becomes current among phlogiston theorists in the second half of the eighteenth century, and Lavoisier, in proclaiming that "the final term at which analysis arrives, all substances that we have not yet been able to decompose by any means, are for us elements," agrees with the current opinion among the contemporary chemists.[18] Of course, in terms of his theory, the nature of these elements has been profoundly modified: it is now the metals themselves we consider to be elements, the calces being only combinations with other substances, the "metalloids," which also appear elementary to us, and in particular with oxygen. Thus chemistry today posits the existence of a considerable multiplicity of substances (more than seventy at the present time) which differ essentially from one another, and a few decades ago, qualified scientists (Helmholtz, Armand Gautier, Étard) were able to write that the "constancy of the elements" constituted a fundamental principle of modern chemistry.[19] We shall see below that the situation was actually never as clear as it seems at first glance and that it has quite recently shifted appreciably in the opposite direction. Nevertheless these conceptions had a decisive influence on the form of explanations in chemistry.

In fact, as soon as sulfur must be accepted as a material differing essentially from nitrogen, as copper does from silver, it becomes useless, to say the least, to seek out whether the different ways nitric and sulfuric acids behave toward these two metals might not be explained by the different shape of the acid particles or by that of the metal's pores, as Lémery had assumed. The divergence must, on the contrary, be founded on the intrinsic properties of these various elementary substances, on their essential properties. On this subject, therefore, we are obliged to consider modern chemistry to be less mechanistic, less inclined to have recourse to explanations by shapes, than was the chemistry of more than two centuries ago. This is due, as we pointed out earlier, to the fact that the chemical elements of today are much more numerous than those of that time and, since they have an infinitely broader experimental basis, enjoy in some sense an indeterminate number of properties. But it is curious to note that while chemistry in general seems to have evolved in a direction unfavorable to the predominance of quality, from the point of view which interests us at the moment it is the opposite that has occurred. That shows, incidentally, that epistemology runs the risk of overschematizing the different scientific trends, and that the details of the history of science must be taken into account by anyone who wants to follow the movement of ideas in this domain and draw conclusions from it. But once again the change in attitude at issue here occurred at the height of phlogiston theory. This explains why Fourcroy, in his admirable exposition of the history of chemistry, opposes Lémery's mechanistic explanation to the conceptions of Stahl (*Enc. méth.* 3:332), even though Stahl himself uses explanations altogether similar to those of Lémery, as we have just seen: he was considering phlogiston theory primarily in its final form, as was only natural.

The most important of the properties that must be attributed to the new elements is the one claiming to summarize how they behave in chemical reactions, in other words, what is eventually defined as their affinity. From that moment on, considerations concerning affinity come to play a considerable role in chemistry. Geoffroy, who publishes the first table of these "affinities" in 1718, immediately has many imitators. In the period before Lavoisier, the best-known are Sénac and Macquer (Cuvier, *Histoire* 1:21). This is not the place to trace the evolution the concept of affinity (originally assumed to be immutable) underwent in its turn. Let us merely note that, especially during a certain period, it appeared to dominate all of chemistry: affinity was considered to be the principal cause of chemical phenomena and was the first thing the theorist thought of when he sought to explain these phenomena.

Fourcroy, speaking of the movement that occurred in chemistry in the first half of the eighteenth century as a result of the work of Stahl, declares that "no discovery is more brilliant in this epoch of great works and uninterrupted research, none has done more honor to this century of renewed and perfected chemistry, none, finally, has led to more important results than that which concerns the determination of affinities between bodies and the specification of the degrees of this force between different natural substances" (*Enc. méth.* 3:333).[20]

At about the same time that the last traces of the old mechanistic explanations tend to fade away in chemistry, qualitative concepts begin to prevail in different fields of physics. This is the consequence of the work of Newton, and Leibniz had understood what was at stake when he fulminated against "*attractions*, properly so called, and other operations inexplicable by the natural powers of creatures; which kinds of operations, the assertors of them must suppose to be effected by miracles or else have recourse to absurdities, that is, to the occult qualities of the schools; which some men begin to derive under the specious name of *forces*; but they bring us back again into the kingdom of darkness."[21] Newton himself had retained the notion of heat as motion, but, as Duhem has correctly pointed out, it is a holdover from Cartesian physics that did not fit into the system very well.[22] Therefore this concept (helped along by the works of Black) yields to that of heat as fluid, and soon there is nothing left except forces and fluids of all sorts in all areas of physics. These fluids, moreover, are imagined to be semimaterial, and in Lavoisier's tables we still see caloric, electric fluid, etc., listed alongside oxygen and nitrogen, as entering into combination with other elements. Of course explanations under the jurisdiction of these qualitative theories also use the concept of displacement: we have seen this with reference to phlogiston, and it is clear that Black's theory likewise draws all its force from the fact that caloric can be *followed* when it passes from one body to another. In the same way, finally, the entire concept of the chemical element rests on its indestructibility; and though one would be hard put to say *why* it confers this or that property on this or that compound into which it enters, one at least knows that these properties will always be the same and that if one succeeds in isolating the element itself, it will always have identical properties. But what these fluids and elements transport are no longer particles with characteristic shapes but essential properties for which no explanation is sought, or at least not for the time being.

However, this triumph of qualitative ideas certainly in no way precludes an appeal to directly spatial explanations under another form. Already Lavoisier and his immediate collaborators sense that it is not sufficient to know what elements form a compound, but that it is also indispensable to describe how they are grouped together; no matter how irreducible to one another "elements" are thought to be (we shall see below that this was not even Lavoisier's position, at least not his whole position), one must recognize that between their compounds there are striking analogies which can obviously derive in large part only from the way they are grouped. The new nomenclature, which Guyton de Morveau works out in collaboration with the master, immediately translates the ideas the antiphlogiston theorists are developing on this subject. These ideas are picked up faithfully by the next generation which, although it adds modifications and refinements suggested by important discoveries (such as the replacement of murium by chlorine, the determination of the true nature of alkalis and especially the laws of definite proportions and multiple proportions), nevertheless retains their major principles. Finally Berzelius codifies the new nomenclature in a body of doctrine, which he links to electrical phenomena. But Berzelius also means to apply his electrochemical theory to the bodies of organic chemistry and there it becomes clear from the outset that we must also pay much more attention to the way things are grouped in inorganic chemistry. Indeed, it is seen immediately that a great many substances, though containing the same elements in identical proportions, nevertheless exhibit quite different properties. Obviously, whatever qualities one may attribute to the elements, they will be inoperative under the circumstances, and grouping will be the only possible explanation. Therefore, the vague attempts of Berzelius's theory are soon replaced by the conceptions of Dumas, Laurent and Gerhardt, which are much more precise and better adapted to the facts. In considering this evolution, it is interesting to note the confrontation between the concept of the influence of grouping on the one hand and that of the qualities inherent in the element on the other. When J. B. Dumas, following his memorable discovery of trichloroacetic acid, develops his ideas on *substitutions*, the assumption that apparently dissimilar elements can replace one another in a molecule and thereafter play a role analogous to that of the original element so shocks the prevailing opinion among chemists that Liebig, who, although he was far from sharing all the ideas of Berzelius, had a short time earlier concluded a sort of alliance with Dumas and even appeared disposed to admit, up to a point, his way of interpreting the genesis of trichloroacetic acid, does not hesitate to publish, in his *Annals*, a truly crude attack against his former associate. It is the famous letter signed S.C.H. Windler (*Schwindler*, meaning joker or swindler) in which the author announces that he has replaced not only the hydrogen in manganese acetate, but also the metal, the oxygen and even the carbon by chlorine, and that the product, made up solely of chlorine, still retains all the properties of manganese acetate. He adds: "I have just learned that the London stores are selling fabrics made of spun chlorine, much in demand in hospitals and preferred over all others for nightcaps, drawers, etc."[23]

Furthermore, we know how vain such resistance was. A few years later Liebig himself retracted his error and praised the scientific merits of J. B. Dumas in dedicating one of his works to him (Ladenburg 17, *Supplément* by A. Colson).

Since that time, hypotheses as to how the elementary atoms are grouped, or as we now say, the *structure* of the molecule, have assumed a larger and larger role in chemistry. Genuine geometric figures are now being utilized and one need only glance at an organic chemistry text to see that chemists really are appealing to the properties of these figures to explain, first of all, the existence of isomerism and then, at least insofar as that appears possible, the particular properties that characterize them.[24] Thus Kekulé's great discovery, the one that assured the definitive triumph of "structural" conceptions, was the observation that the number of isomers of the benzene group could be "explained" by the properties of a hexagonal figure; and chemists ever since have found it a serious drawback that this conformity is not altogether complete, since the alternation of single and double bonds in the hexagon does not produce isomers. In the same way, the ease of condensations in two lateral chains having the *ortho* position with respect to one another is justified by "proximity," and the relative ease with which this or that atom lends itself to substitution is explained by the influence that this or that "neighboring" group is able to exert upon it.

But perhaps the close relationship of these explanations by figures with those of the past will become even more obvious if we address ourselves to the chemists' most recent phase of "atomic theory," to the conceptions known as *stereochemistry*. As early as the admirable work at the very beginning of his scientific career by which he had worked out the isomerism of the two tartaric acids, Pasteur had noted that the asymmetry observed in the hemihedral crystals of the two varieties persisted in the liquid state, deducing from this that it must be of molecular origin. "There can be no reason for doubting," he said, "that the grouping of the atoms has an unsymmetrical arrangement with a nonsuperposable image. It is not less certain that the atoms of the laevoacid realize precisely an unsymmetrical arrangement the inverse of the above [that is, of the dextroacid]."[25] Pasteur was thinking rather of figures on the order of two spirals, right-handed and left-handed.[26] On the other hand, Kekulé's theory of atomic bonding was erected on the fundamental assumption that carbon is quadrivalent and that the four valences are of the same nature in all respects. Thus four atoms bonded to a carbon atom would be placed around it symmetrically, that is, at the four corners of a tetrahedron. In order to fit these two currents of thought together, it was enough to notice that if the four atoms or groupings bonded to the carbon in this tetrahedron were all different from one another, the result would be a figure satisfying Pasteur's requirements of asymmetry. Therefore it is not too surprising that two men of science, Le Bel and Van't Hoff, simultaneously came up with this idea.[27] We know, moreover, that the conception proved to be particularly fruitful later on. Of course, like any theory, it encountered obstacles, and, like at least the immense majority of them, it did not always succeed in overcoming them. Anomalies have been pointed out, and "physical" isomerism neither appears nor disappears exactly where the formula would demand. Nevertheless it has been and remains an acquisition of inestimable value for the chemist and the physicist; it has made possible whole series of highly important discoveries, particularly those of E. Fischer on the synthesis of sugars; and it promises still more discoveries by its extension to compounds of elements other than carbon, such as pentavalent nitrogen and quadrivalent tin and sulfur. Most certainly the modern scientist has the impression that this theory offers a profound view into the inner structure of matter.

A recent historian of science, in speaking of the tetrahedron that characterized the element fire according to the Pythagoreans and Plato, calls it the basic idea for Le Bel's and Van't Hoff's "stereochemistry."[28] This is obviously something of an exaggeration. Yet we realize upon closer examination that the comparison is not entirely spurious. As soon as one sets out to explain the diversity of substances by arrangements in space, one is necessarily led to use stereometric figures; and since the tetrahedron is the simplest of these figures, it will always play a considerable role in a system of this kind. Of course one does not employ the same properties in the two cases. Plato is struck by the fact that the tetrahedron is pointed, and thus finds it suitable for an element to which he attributes a great force of penetration, while the stereochemists utilize its property of furnishing asymmetrical figures under certain conditions. But in both cases, there is a tendency to explain phenomena by reducing them to properties of a geometric figure in space. Furthermore, it is too early to say whether the property used by Le Bel and Van't Hoff will remain the only one stereochemistry will utilize: on the contrary, we can see that in the modern conceptions dealing with the architecture of atoms, like the one put forward by Bäyer, for example, theorists calculate the direction that two carbon valences must take (Bäyer imagines them sticking out from the atom like rigid stems) and the curvature these stems will have to undergo when there are double bonds, in which case they will bend like springs.[29] Similarly, researchers who cultivate the field of what is called the "new crystallography" and hope to use their methods to find out not only the actual arrangement of the particles in a crystal, but also the molecular and atomic structure of the bodies, after having arrived at observations confirming the chemical theory of the carbon tetrahedron, have constructed models by means of which, according to one of the experts in this field, "it is interesting ... to observe ... how readily the carbon atoms can be seen linking themselves together in chains of six,"[30] a peculiarity obviously calculated to explain the frequency and particular solidity exhibited by the benzene nucleus.

Furthermore, we know that Le Bel's and Van't Hoff's conception of the tetrahedron has not remained isolated in science. Werner constructed his theory of perfect inorganic complexes, to which we referred in Chapter 3 (p. 56), on the fundamental hypothesis that the six atoms or groups called for by the principle of the "hexacoordination" of cobalt complexes are placed at the corners of an octahedron. And it can be seen that the reason that dictated this hypothesis to him is strictly analogous to the one put forward in favor of the tetrahedron by Le Bel and Van't Hoff, namely the fact that the asymmetry produced in an octahedron by the presence of dissimilar substituents at the corners would be precisely of a nature to account for the observed isomerism.[31] It should be pointed out that the analogy in this case is not nearly as strict as it is in the case of the tetrahedral carbon atom. Experiments are a long way from revealing isomers everywhere the model suggests that they exist; supporters of the theory therefore suppose that they are encountering undecomposable *racemisms* (that is, mixtures in equal proportions of two compounds having an equal optical activity but in the opposite direction).[32] Corroborations such as the one considered decisive in establishing the theory, namely the explanation of the isomerism of the violeotetrammonia and praseotetrammonia salts, which did not fit into the old theory formulated by Jörgensen,[33] are rather few in number. And already we have discovered quite a number of facts that Werner's theory, in its turn, does not explain or explains only with great difficulty and with heavy use of auxiliary hypotheses, such as that of hydrolysis; this hypothesis, as one can easily see, actually breaks through the framework of the theory, which, moreover, its supporters themselves are forced to characterize as "too rigid."[34] Nevertheless, Werner's theory, as we explained above in referring to the notion of valence, has come to be almost universally accepted by chemists, who say that "the theory could be shaken only if a derivative, allowing an octahedral formula, superimposable on its mirror image due to a plane of symmetry, were endowed with rotatory power. But such a case seems improbable" (Urbain and Sénéchal 169). The success of Werner's conceptions is undoubtedly due above all to the fact that they brought order into a particularly complicated field of science where it had been lacking; but most certainly the great explanatory power which is the legitimate prerogative of all direct spatial explanation is not unconnected with their triumph. It is impossible to look through a conscientious study of this question (like the book by Urbain and Sénéchal, on which the present observations are based) without being struck by the enormous role of the *figure* and by the immense prestige that the introduction of *Werner's octahedron* seems to have bestowed on the theory.[35] Let us add that since that time this theory, in spite of a few weaknesses, has largely justified the confidence of its supporters by allowing them to make and classify a considerable number of highly interesting discoveries, and that its potential for doing so does not at all seem to have been exhausted at the present time.

Nevertheless, a fundamental divergence seems to persist between the tetrahedron or octahedron of modern stereochemists on the one hand and the geometric figures of the thinkers of antiquity on the other: it is that resulting from the evolution referred to above which led to the acceptance of the concept of multiple elementary substances by science. Thus in the case of Plato's tetrahedron of fire, there could be nothing specific except this tetrahedral form, whereas at the center of the modern tetrahedron there is an atom of carbon, that is, of a substance essentially different from any other. But here again, upon closer examination the difference tends to diminish.

In the preceding pages we have touched several times upon this question of the concept of the element, which is obviously of prime importance from the standpoint of chemical theory. We saw in particular in Chapter 5 (p. 119 ff.), with respect to the explanation of *diversity* in space, that the specificity of the chemical elements is an obstacle standing in the way of this explanation, an irrational, or rather an indication that science, as it advances in this area, will probably someday succeed in determining a clearly demarcated irrational there. Then, in the present chapter, we chronicled the birth of the concept of the chemical element as modern science knows it, following its evolution up to a quite recent period when the specificity, the "constancy" of the elements seemed to be universally recognized as one of the fundamental bases of chemistry (p. 222). But at the same time we pointed out that the situation was actually more complex. In fact, underneath the quite obvious current inspired directly by experimental data (as we have seen), we discover, if we are willing to take heed of it, a countercurrent tending to deny or at least to reduce the specificity of the elements, a countercurrent moving exactly in the direction of that aprioristic tendency we saw at work in the case of diversity in space (Ch. 5, p. 134). Despite the fact that practicing chemists based everything on this theory, despite the fact that they seemed to consider the properties of elements to be ultimate and to regard any explanation succeeding in going back to these properties as perfect, deep down inside, and no doubt sometimes unbeknownst to themselves, there was a lingering hope of someday returning to the concept of the unity of matter. It is a tendency whose numerous manifestations are quite clear and can easily be followed throughout this whole period. Already for Lavoisier, not all elements appear to have the same rank; some of them – oxygen, nitrogen, hydrogen – seem simpler than the other bodies,[36] which would consequently have been only compounds more strongly constituted than those we form and analyze in our laboratories (this was, as we know, more or less Boyle's position). The second decade of the nineteenth century gives rise to Prout's theory, according to which hydrogen is the unique primordial element. After slight modifications by J. B. Dumas, it immediately gained many supporters, and although later research was far from confirming its experimental basis,[37] it did not disappear from science (cf. IR 266, n. 3 [Loewenberg 255, n. 28]). Along the same lines are attempts to establish relationships between the atomic weights of the chemical elements and their various properties. Indeed, all such efforts tended to reduce the specificity of each particular chemical element, to make it appear as if it depended on the variation of a factor that could only be characteristic of a sort of ultimate element.[38] It is well-known that these efforts culminated in the establishment of Mendeleev's famous *periodic system* of the elements, and that this grandiose conception, after initially meeting great resistance from chemists, finally compelled recognition, to the point that it is today an integral part of the theoretical edifice of science. Finally, through Moseley's discovery (cf. Ch. 6, p. 176, n. 36) these two currents in some sense merge, mutually defining and confirming each other.

These tendencies toward the unity of matter have gained still more strength following recent discoveries in the field of electricity and especially in that of radioactive bodies, discoveries tending to demonstrate that the helium atom enters into the composition of substances which in all other respects behave like true chemical elements and that, on the other hand, the elementary electrical particle, the electron, is a component of all bodies without exception. To these observations is linked a whole series of modern speculations on atoms, which tend to explain their affinity and their valence.[39] Perhaps the most thorough conception in this area of the structure of the chemical atom, the one in which the true nature of these hypotheses is seen most clearly, is that of Sir J. J. Thomson. The illustrious physicist starts from a curious experiment by Mayer in which a certain number of minuscule magnets placed in corks are floated in a container of water. If we stir the liquid, we soon see these magnets arrange themselves in regular figures, which vary depending on the number of magnets. Thus, by adding one or several magnets, we generally see the shape of the arrangement change completely. But, remarkably enough, if we apply ourselves to observing the entire series of these figures, adding only a single small magnet at a time, we see similar figures which differ only by their dimensions, so to speak, that is by the number of component magnets, recurring from time to time, separated in the series by entirely dissimilar figures. This would explain why, in Mendeleev's periodic system, elements presenting highly analogous properties appear at intervals separated by others quite different in nature. It is obvious that this is a case of the purest type of explanation by geometric figures, since the properties of the chemical elements are reduced to those that can be offered by arrangements in space, that is, having to be explained by the properties of space. Leaving aside, of course, the properties of the basic component – the small magnet in the present case – we dealt in Chapter 3, p. 55 ff., with the true nature of this point of departure for any scientific theory. Mayer's figures are arranged in a plane, but it goes without saying that this is only a first approximation. The actual atomic figures will necessarily have three dimensions (as, moreover, the author of the theory anticipates) and it is perhaps not going too far to predict that if the conception is some day pursued far enough in this direction, we shall necessarily see the tetrahedron reappear, in this domain as well, as the simplest spatial figure of all, perhaps accompanied by the octahedron and the other stereometric figures of Platonic "metamathematics."

E. The various ways just discussed in which spatial functions are used for causal explanation have in common that they seem to come to mind spontaneously. Therefore we see procedures derived from them being used in almost all epochs, and no doubt almost from the dawn of human intelligence, sometimes less, sometimes more, depending on conditions resulting from the progress of human knowledge. But there is one way that belongs almost exclusively to modern science, in the sense that ancient physics could glimpse it only fleetingly and indistinctly, if at all, whereas in our science it plays a quite considerable role that, in all likelihood, can only increase. We mean to speak of causal explanations founded on the equivalence of motions. We know, as a matter of fact, that the ancients were entirely ignorant of the principle of inertia, and were therefore *a fortiori* unable to conceive anything analogous to our notion of the conservation of energy. Of course, in spite of Aristotle's theory, there may have been a vague notion of the continuity of motion, particularly for a few of the atomists; moreover, Aristotle himself feels that circular motion persists and draws from it his theory of the "natural" circular motion of the celestial spheres (cf. IR 118 [Loewenberg 116]). But, once again, everything actually involving the transformation of spatial movement and the use of this notion for causal explanations can only be of modern origin.

In modern physics, on the other hand, there is no lack of examples. Take the piston of a steam engine that moves in a cylinder thereby setting in motion a whole series of devices. What is the source of the force that pushes it? It is the force of the expansion of the water vapor. Now the kinetic theory of gases teaches us that this expansion is the result of small impacts produced by the gas molecules on the walls of the chamber that encloses them. Thus the powerful visible motion of the piston is only the transformation of a great many motions that animate very small and consequently invisible corpuscles. But there is no doubt that there must be, at least in some sense, an equivalence, or even an identity, between the two motions, *molecular* and *molar*, since energy, in virtue of the well-known principle, has to "be conserved." Or take a rotating machine part to which a brake is applied. Part and brake heat up, and the theory tells us that the increase in temperature is only an acceleration of the motion of the particles that constitute the visible bodies, this molecular motion being a transformation of the visible molar motion of the part. In the first case we have "explained" the birth and in the second the disappearance of the visible motion by connecting it to an invisible motion and assuming that what was happening on our scale, in spaces that we can discern with the naked eye, had an equivalent on the molecular scale, in minute spaces barely accessible to our microscopes. Insofar as this last characteristic is concerned, the explanation, as the reader has no doubt already noticed, bears some resemblance to the explanations by reduction in scale discussed earlier, as exemplified by the preformationist hypotheses, whereas the transmission of movement has characteristics in common with the *passing* of an immaterial principle, as we pointed out earlier. These resemblances are quite understandable, for in all these processes it is basically a question of one and the same thing: the use of spatial functions in order to explain phenomena. The class of explanations we have just introduced, however, has the distinctive and characteristic feature of involving bodies *in motion*, whose properties are explained by assuming that their particles are in a *state of motion*. The theory originally proposed for radioactive bodies – that of the capturing of ethereal waves – as well as the one which subsequently prevailed, both depend on this point of view, since they both transform invisible motion into visible motion. But the second of the two theories is more like the one that explains the transformation of heat into motion, while at the same time pushing what was characteristic of this theory to its limit. Indeed, in the case of radioactive bodies, the occult movement, the source of the prodigious energy they unceasingly show, is presumed to take place within the atom, that is, within the limits of a space bearing approximately the same relation to the molecular distances treated by the kinetic theory of gases as the diameter of our earth bears to the dimensions of stellar space.

Obviously today we find ourselves in the heyday of this type of explanation, which is so characteristic of modern physics and unquestionably destined to become more and more frequent in the very near future. This will be true especially if, as is not impossible, physics returns to the tradition of pure kinetics (if we dare use this term) which long seemed to dominate it and which strove to rid it of all recourse to the concept of force, which concept posits the existence, as of a real being, of a *tendency* to movement or (as it is more often expressed) of a *potential* or *in posse* motion. We shall try to clarify the true nature of this concept a little later. But it is clear that any theory seeking to replace a potential state, a *tension*, by a *vis a tergo* of which this state would be the apparent consequence, will have to appeal to a preexistent motion of minute particles. For example, all possible *mechanical* explanations of Newtonian gravitation will always necessarily be of this type; we shall be led to attribute the attraction to small molecular impacts (as we saw that Le Sage did). In a recent writing, a most interesting note appended to Arthur Balfour's immediately and justifiably famous work entitled *Theism and Humanism*, Sir Oliver Lodge presents this purely kinetic point of view with singular force and clarity,[40] and most certainly, given the authority of the illustrious English physicist, his position, so true to the principles of Descartes and Leibniz, is destined to win much public support.

It certainly seems, however, that even supposing the most favorable combination of circumstances, the most recent developments in physics tend to limit the use of this means of explanation, at least from one point of view. We need only recall in this regard the comparison we just made between explanations by the motion of the infinitely small and explanations by simple reduction in scale. If, as the results of the *Council* of Brussels seem to indicate, we are forced to accept the idea of an atomic world fundamentally different from our familiar one, a world in which motion itself would be discontinuous, we can foresee that, for phenomena occurring on our scale, the motion of these atoms will no longer have the same explanatory power as the ordinary movement to which we were accustomed, that is to say that phenomena will seem to us to be less explained as soon as we are forced to use this atomic motion in our explanations. But that is perhaps trying to look a bit too far ahead.

Need we point out that our enumeration of the different forms of spatial explanation in no way pretends to be exhaustive, nor even to have established a durable classification in a subject that perhaps does not require one, since at bottom it is a question of a single identical concept, that of space? We simply wished to give the reader an idea of the way this process of explanation is carried out and of the resources it puts at the disposal of science.

NOTES

1. See John Burnet, *L'Aurore de la philosophie grecque*, trans. Reymond (Paris: Payot, 1919), pp. 251–252, 260, 280 [*Early Greek Philosophy*, 2nd ed. (London: Adam and Charles Black, 1908), pp. 253–254, 263, 284].
2. Maxwell, 'Atom,' *Encyclopedia Britannica*, 9th ed. (London, 1909), 3:47.
3. Schelling likewise wondered whether "suns are only the light magnets of the universe which gather around them all the light produced by nature, by drawing it from all the spaces" (*Weltseele*, I, 2:391).
4. Trendelenburg (*Log. Untersuch.* 1:267), after having pointed out that, according to Plato, man's superiority over animals is based on the fact that he alone can count, expresses the opinion that the geometric similarity of different-sized figures plays an analogous role, being conceived only by man. Nevertheless there is no doubt that a carnivore, whose eye resembles the human eye in all respects and whose visual impressions must thus be altogether analogous to our own, likewise experiences no hesitation as to the identity of the prey whose size he sees continually change as he pursues it.

5. Hegel was also well aware of the fact that causal explanations frequently make use of a modification in size. "In thinking about the gradualness of the coming-to-be of something," he says, "it is ordinarily assumed that what comes to be is already sensibly or actually in existence; it is not yet perceptible only because of its smallness. Similarly with the gradual disappearance of something, the non-being or the other which takes its place is likewise assumed to be really there, but not yet observable .... In this way, coming-to-be and ceasing-to-be lose all meaning ....[or are] transformed into a smallness of an outer existence" (*Wiss. der Logik*, 3:450–451 [Miller 370]).
6. [Leibniz, *Principles of Nature and Grace*, § 6, Parkinson 198–199, as quoted on p. 122 above.]
7. Maxwell, in a passage already cited above (Ch. 5, n. 30, p. 140) with regard to the distinction between the properties of palpable matter and those of molecules, has stressed this difference between the molecular world and our world.
8. Ernst Haeckel, *Les Enigmes de l'univers*, trans. Camille Bos (Paris: Schleicher Frères, 1902), pp. 283–284 [*The Riddle of the Universe*, trans. Joseph McCabe (New York: Harper & Brothers, 1902), pp. 246–248].
9. See above, pp. 155–159. Hegel, whose basic hostility toward the science of his time sometimes results in an actual regression toward the outdated science of the past, did not hesitate to state that the celestial bodies had to be considered as obeying laws other than those governing terrestrial bodies. "Thrust, pressure, resistance, friction, pulling and the like, apply to an existence of matter other than celestial corporeality," and one could not, because a stone is inert and because the earth and other celestial bodies are composed of stones, set "the qualities of the whole equal to those of the parts" (*Naturphilosophie*, 7_1:97 [Miller 65]). Auguste Comte, whose profound admiration for Newton is well-known, nevertheless declared that it was "rash" to extend the concept of gravitation to celestial bodies situated outside the solar system (*Cours* 2:174 [cf. Martineau 168]; see also 2:244 [Martineau 187–188]). But that point of view was bound up with his general idea of a limit to be imposed upon science; in astronomy, research was to be restricted to what was relevant to the solar system (*Cours* 2:12–13 [Martineau 133–134]; cf. *Système de Politique positive*, Paris: L. Mathias, 1851, 1:510 [*System of Positive Polity*, trans. John Henry Bridges, New York: Burt Franklin, 1968, 1:412–413]).
10. Pierre Duhem, *Le Mixte et la combinaison chimique* (Paris: C. Naud, 1902), p. 7. Moreover, this deduction is quite clearly laid out by Lucretius; given that there exist different materials and our sensory organs receive varied impressions, the only possible explanation is that this arises from the diversity in shape of the corpuscle: *Quapropter longe formas distare necessest / principiis, varios quae possint edere sensus* (*De rerum nat.* II, 442–443; cf. also II, 478–599 and IV, 654–655). Schelling saw quite well that all scientific theories intending to explain the quality of substances, with the exception of those drawn from simple analytic formulas of mathematics, basically amount to attempts "to express qualities by figures, that is, to substitute a specific figure for each primordial quality of nature" (*Einleitung zu dem Entwurf*, I, 3:295).
11. Descartes, *Les Météores*, Discourse 1, ¶ 3, *Oeuvres* 6:233–234 [*Discourse on*

Method, Optics, Geometry, and Meteorology, trans. Paul J. Olscamp (Indianapolis: Bobbs-Merrill, 1965), pp. 264–265].

12. Descartes, *Principes* IV, 66. For Descartes (as for Lucretius, moreover) sea water made its way into springs not (as for us) through clouds and rain, but by infiltration through the earth. We know that Descartes consistently denied following the principles of Democritus and Epicurus. "I admire those who say that what I have written are only *Centones Democriti*," he writes to Mersenne in 1640 (*Oeuvres* 3:166; cf. the earlier letter, dated 1638, 2:396), and in one of the last paragraphs of the *Principles* (IV, 202) he declares: "It is evident that this way of philosophizing has no more affinity with that of Democritus than with that of all the other sects." Nevertheless his contemporaries were not completely mistaken in being struck by this similarity.
13. Robert Boyle, *New Experiments Physico-Mechanical Touching the Spring of the Air, and its Effects; Made, for the most part, in a New Pneumatical Engine*, Exp. 1, *The Works of the Honourable Robert Boyle in Five Volumes* (London: A. Millar, 1744), 1:10.
14. "Acid salts" in the nomenclature of that time are what we call acids, while "alkaline salts" are our alkalis. Stahl still uses the same terminology: "*Menstrua salina sunt vel acida vel alcalia*" (D. D. Georgii Ernesti Stahlii, *Fundamenta chymiae dogmatico-rationalis et experimentalis*, Norimbergae [Nuremberg]: impensis B.G.M. Endteri filiarum, & Vid. B.J.A. Engelbrechti, 1732, p. 11).
15. Nicolas Lémery, *Cours de chymie* (Paris: Jean-Thomas Herrisant, 1756), pp. 21–22.
16. Robert Boyle, *The Sceptical Chymist* (London: Dent, 1911), pp. 100–101.
17. See Hermann Kopp, *Die Alchemie in älterer und neuerer Zeit* (Heidelberg: Friedrich Vieweg und Sohn, 1886), 1:69–72.
18. Werner gives this definition: "We designate as elements homogeneous substances which cannot be decomposed by the majority of our analytic methods" (*Neuere Anschauungen auf dem Gebiete der anorganischen Chemie*, Brunswick: Friedrich Vieweg und Sohn, 1913, p. 1; hereafter *Neuere Anschauungen*). We can see here an attempt to adapt this traditional definition to conceptions concerning the variability of chemical elements.
19. Cf. IR 263 [Loewenberg 236]. Auguste Comte certainly interpreted the chemistry of his time quite faithfully in stating that "chemical properties are specific," but that "all the data must finally be reducible to the knowledge of the essential properties of simple substances," so that chemistry "has for its object, – the properties of all simple bodies being given, to find those of all the compound bodies which may be formed from them" (*Cours* 3:12, 15, 18 [Martineau 251, 252, 253]).
20. An amusing confirmation of the high esteem in which the concept of chemical affinity was then held by the educated public is furnished by the title of Goethe's famous novel *Elective Affinities* and by the way in which the title is explained in the text of the work itself. Without falling into the excesses that sometimes characterize the judgment of the Germans where their great poet is concerned, and without wishing to attribute to him too large a role in the field of science, one cannot help admitting that though (unlike Kant) he remained closed to the

understanding of all that touched the most advanced part of science, namely the physicomathematical sciences (which explains his inability to grasp the work of Newton and, later on, the formidable aberration of the *Farbenlehre*), Goethe otherwise had an excellent and sometimes surprising knowledge and generally proves to be abreast of the science of his time.

21. Leibniz, *Opera* 777 [*Leibniz-Clarke Correspondence*, Leibniz's Fifth Paper, § 113, Alexander 92].
22. Pierre Duhem, *Le Mixte et la combinaison chimique* (Paris: C. Naud, 1902), p. 61.
23. See Albert Ladenburg, *Histoire du développement de la chimie*, trans. A. Corvisy, 2nd ed. (Paris: A. Hermann et Fils, 1911), pp. 127, 152, 169 [*Lectures in the History of the Development of Chemistry since the time of Lavoisier*, trans. Leonard Dobbin (Edinburgh: Alembic Club, 1900), pp. 135, 163, 179].
24. Alfred Werner, whose authority in this field is well-established, states that "an examination of the development of structural formulas teaches us that the gradual transformations they have undergone has resulted in a continual perfecting of their spatial representation" (*Neuere Anschauungen* 14).
25. J. H. Van't Hoff, *Dix Années dans l'histoire d'une théorie* (Rotterdam: P. M. Bazendijk, 1887), p. 29 [*Chemistry in Space*, trans. J. E. Marsh (Oxford: Clarendon Press, 1891), p. 27; Meyerson's brackets].
26. Curiously enough, in this passage, concurrently with the conception of two spirals, Pasteur advances the idea that the atoms could turn out to be placed "at the vertices of an irregular tetrahedron." Although it is at least doubtful that he thought of the very representation that Le Bel and Van't Hoff later proposed, we can see, at any rate, how close he came to it.
27. Need we point out that we have no intention of belittling the merit of these two scientists by this brief historical account? The most admirable thing in the evolution of science is its continuity. Anyone who examines at all closely the history of human thought and, in particular, of scientific thought, cannot fail to be struck with a sort of religious awe before the fundamental unity of the intelligence in its persistent effort toward the penetration of the unknown that surrounds us. But the individual minds by which this universal intelligence is manifested, although they are only particular links of a strongly riveted chain, are nonetheless worthy of the highest admiration. As for Van't Hoff in particular, it is interesting to note that he started out as a disciple of Kekulé at Bonn and later worked in Wurtz's laboratory in Paris, where he became strongly impregnated with the ideas of Pasteur (G. Bruni, 'L'Oeuvre de J. H. Van't Hoff,' *Scientia* **10** (1911) 32).
28. Max Simon, *Geschichte der Mathematik im Altertum* (Berlin: B. Cassirer, 1909), p. 131.
29. See André Job, 'Les Progrès des théories chimiques,' *Bull. Soc. fr. phil.* **13** (1913) 54, and Alfred Werner, *Neuere Anschauungen* 78 ff. Moreover, Werner also speaks of "the direction of the action of the force of attraction" and of "privileged directions from the standpoint of chemical affinity, which will determine the place" where the transformation will occur (p. 313).
30. W. H. Bragg, 'The New Crystallography,' *Scientia* **18** (1915) 381–382.

31. Urbain and Sénéchal, *Introduction à la chimie des complexes* (Paris: A. Hermann et fils, 1913), pp. 155, 162, 165.
32. Urbain and Sénéchal 164, 169. See Werner, *Neuere Anschauungen* 76, on the difficulties involved in this assumption.
33. Urbain and Sénéchal 158, 159, 164, 295, 297.
34. Urbain and Sénéchal 176, 179, 188, 229, 240–241, 263, 283, 321. Furthermore, one can see in Werner himself how certain correlations, such as that of accessory "ionogenic valences," are hard to fit into his theory (*Neuere Anschauungen* 62–63). Cf. also *Neuere Anschauungen* 205, 208 ff., on the difficulties presented by hydrates.
35. The feeling that construction – spatial deduction – constitutes the truly essential part of the theory is no doubt what explains statements like the one we quoted above (p. 70, note 17) which claims that Werner's conceptions are modeled on those of organic chemistry, when this opinion can only refer to the explanation of isomerism by means of a geometric figure. Werner, moreover, seems perfectly aware of the explanatory value of the spatial image and uses such images wherever possible. Thus he judges that the two observed limiting radicals MO_4 and MO_6 must "correspond to the coordination numbers in the plane and in space" of the elements (*Neuere Anschauungen* 121). Similarly, after having noted that we have never observed more than three "bridging bonds," he states that this fact "can be simply explained by the octahedral grouping of the different groups around metallic atoms as centers" (*Neuere Anschauungen* 287). As for the octahedron in particular, he declares: "This consequence, which is basic from the standpoint of the spatial conception of the MR_6 complexes, has been experimentally confirmed to the point that we can no longer doubt that the spatial formulas which have been established are justified" (*Neuere Anschauungen* 342).
36. See Henri Bouasse, *Introduction à l'étude des théories de la mécanique* (Paris: Georges Carré, 1895), p. 166.
37. It would have required all atomic weights to be integral multiples of that of hydrogen, or at least, according to Dumas, multiples of half that atomic weight.
38. Alfred Werner (*Neuere Anschauungen* 1), with that characteristic fear frequently manifested by contemporary scientists as soon as they turn their attention to theoretical conceptions, the fear of being called "philosophers" or "metaphysicians" (apparently the greatest insults of all), states that "it is not in basing itself on representations of the unity of matter – although these are certainly remarkable from the hypothetical and philosophical point of view – that modern chemistry comes to suppose that the elements are only different forms of one and the same material, distinguished perhaps solely by the conditions in which they arise, but in basing itself on certain relations between the properties of the elements, which would remain entirely incomprehensible without the supposition of a common origin." But it is clear that what we have here is not, as Werner seems to suppose, two distinct, or even opposing, currents of thought, but one and the same current: it would have been absurd to seek relations between the properties of the elements if one had not at bottom entertained the idea of the unity of matter. We have seen, moreover, that the history of this conception confirms its perenniality.

39. André Job, 'Les Progrès des théories chimiques,' *Bull. Soc. fr. phil.* **13** (1913) 52 ff.
40. Arthur James Balfour, *L'Idée de Dieu et l'esprit humain*, trans. Bertrand (Paris: Bossard, 1916), pp. 329–331 [*Theism and Humanism* (New York: Hodder & Stoughton / George H. Doran Co., 1915), pp. 240–243].

CHAPTER 9

THE POSSIBILITIES OF SCIENTIFIC EXPLANATION

Will these resources be considered quite varied or somewhat limited? That will no doubt depend largely on the observer's point of view and individual predispositions. But is it possible, in general, to arrive at an idea, even if only a completely general and approximate one, of the possible relation between the power of these means of resolution and the extent of the problem or problems to be resolved? We hope that the reader will understand that we venture onto this terrain only with great hesitation and will consider the following remarks as mere suggestions.

Lucretius, in examining the number of possible combinations of the primitive elements in space, judges that it must be very limited; moreover, he considers this an advantage of his theory from the standpoint of the endurance of the species and the maintenance of the laws of nature.[1] It is obvious that in all respects we have changed our minds on this subject. We have greater confidence in the stability of the order of things in general, and we feel no need to ensure it by means of this sort. On the other hand, the immense diversity of things probably strikes us much more strongly than could possibly have been the case for the ancients; that is why we would like the diversity of spatial combinations to be as great as possible in order to explain the diversity of things. Furthermore this desideratum does not seem impossible to satisfy because, first of all, these elementary parts obviously appear to us to be a great deal smaller and thus infinitely more numerous than Lucretius imagined, and then we have a more accurate idea of the number of permutations likely to be provided by a given number of elements, that is, of the diversifying power of these permutations. But it is perhaps also because we are less preoccupied with the fundamental idea of the unity of matter, which the ancient atomists never lost sight of, whereas we, although we believe in it wholeheartedly, consistently relegate it to the background, while operating *provisionally* with the atoms of chemical elements, that is, with qualitatively diverse particles. However that may be, contemporary chemists clearly seem to agree that the spatial combinations at our disposal are more than sufficient to represent everything diverse, specific, and even individual, to be found in nature. For example, we know that E.

Fischer and the scientists of his school have successfully decomposed albuminoids into various amino acids and recombined the latter to obtain polypeptides, and that the synthesis of these substances is generally considered an important step towards the synthesis of albuminoids. Supposing that these albuminoids are composed according to the schema in question, we can easily see, since it is a question of combining twenty different amino acids, that the number of possible permutations able to furnish isomeric forms is on the order of two trillion – not counting the equally immense number of stereochemical isomers resulting from the considerable number of asymmetrical carbon atoms included in the molecule of each amino acid. Thus, with a certain amount of good will, we are able to imagine with Hollemann that there is a probability that each living being possesses his own individual albumin and that the variety of forms offered by nature, a variety that at first sight seems almost breath-taking, is in the end due only to the isomerism of the albumin molecules.[2]

By thus linking an arrangement, a specific spatial figure, to each example of diversity observed in nature, we shall certainly have taken a great step forward. However, it must be realized that this is only the smallest of the steps that will have to be taken. In fact, it will then be necessary to show that all the properties of the entity represented follow from this figure necessarily and intelligibly, that is, that they can be *deduced* from it by strictly rational means. The examples of reasonings of this type that we have cited, from Lucretius to Lémery to Stahl, allow no doubt: it is precisely because the problem appeared infinitely less complicated to them than it does to us that our predecessors, believing they had solved it, have clearly shown the goal toward which we are aiming. The figures, says Lémery, must "correspond to the effects" produced by the bodies; this is why the particles of salt which sting our mucous membranes must be pointed and the particles of acid more or less "subtle" depending on whether they penetrate more or less easily into the interstices of the bodies to be dissolved (Ch. 8, pp. 219 ff.). Now of course the attributions of formulas "of composition," supposing them to have been determined, will not have been made without reason; it will have been found that, on a specific occasion, the substance lends itself to this or that reaction, which fits this formula exactly. But it will then be necessary to show that *all* the reactions of the aforesaid body, all its properties, all its peculiarities, are also deducible from the formula.

In order to make clear how different this problem is from the first (which consists in merely *determining* the formula) and how much more difficult it is, we need only consider what happens in the cases – infinitely simple, it goes without saying, compared to those that will be at issue in future explanation of the organism – to which contemporary chemistry applies itself. We know that chemists have succeeded in synthesizing a significant number of products which, from the standpoint of their useful properties, are considerably more efficient than the substances man found ready-made in nature or that he extracted from nature by very simple procedures involving rudimentary technique. These products – dyes, pharmaceuticals, perfumes – were not discovered completely at random in experiments probing the almost unlimited field of organic chemistry. On the contrary, and more and more so as investigations proceeded, researchers let themselves be guided by reasons based on considerations of formula: the presence of this or that group in a given position, the possibility of this or that "condensation" indicated to them the probability, and sometimes the quasi-certitude, that the body they wanted to produce would have this or that property as a colorant, an antipyretic, a fragrance. Thus their research could unquestionably be called in large part rational, since it was actually based on formulas designated as such. But the relation between these formulas and the most important properties of the bodies they were supposed to represent remained itself entirely empirical.

For instance, to speak only of optical properties, which are the easiest to grasp, the theory of *chromophores* by which Witt had succeeded in reducing the observations quite satisfactorily to a system as early as 1876, which system, subsequently more or less completed, long served as a guide for research in this field,[3] did not try to understand *why*, for example, the entry of the two hydroxyl radicals transforms almost colorless anthraquinone into a red material with incomparable coloring power (alizarin or synthetic madder dye), nor what connection there is between the presence of amide groups and the color of rosaniline (or fuchsin). Yet – and this is one of countless examples encountered at every step in the evolution of science which prove how little lawlike generalization alone, no matter how far it is pushed, is able to satisfy the scientific mind – scientists immediately strained their ingenuity in every possible way to transform the purely empirical observations into explanations, that is, to connect optical phenomena to chemical composition by genuinely rational theories. It is a formidable task, and it will surely take generations of physicochemists to arrive at anything like precise and satisfying solutions in this area. But we can be assured by Victor Henri's admirable résumé of the work of his predecessors, his own work, and that of his students, that the task is well on its way to being accomplished and that the research projects indeed pursue the goal we have just indicated: the different "vibrators" whose existence is assumed, through hypotheses connected directly to the formula of composition, must account for the appearance of this or that band in the spectrum.[4]

Clearly, once this task has been accomplished, the optical properties of chemical substances will really have been *rationalized*. But for the time being we must not be taken in by certain expressions in current use in science. It is certainly not without reason that the formulas of these dyestuffs are called rational: they in fact account for certain properties of these bodies, for their kinship with certain other bodies, for their synthesis, for the way they behave in specific reactions, etc., and they reduce these properties, at least gradually and partially, to spatial arrangements. But the properties explained in this way (even insufficiently) are few in number, while all the others remain purely empirical and at times their rationalization even appears very remote.

However, at the same time it is important to note that this rationalization seems to us to be possible, that is, as far as we can judge at the present time, science does not appear to have to confront any new irrational on the way to this goal. Even without attempting to take into account the progress already achieved that we have just mentioned, purely theoretical deliberation suffices to show this. Leaving aside sensation (which is an irrational of the highest order), the light that strikes the dyes and is reflected by them is a motion. The change produced by reflection must therefore result from movements within the molecule, and the same must be true for properties that substances exhibit from the point of view of taste or smell, for supposing that our sensations are due first and foremost to chemical reactions occurring in certain cells of our mucous membranes (which is possible), these reactions in their turn must be reducible to motion. Let us observe, however, that this is only a quite general schema resulting from the universal postulate that phenomena consist only of figures and motion – that is, spatial functions – a postulate that is, as we know, only a form of the postulate of the general intelligibility of nature. Consequently, it is clear that we can draw from this schema no guarantee that it is really so, that is, no guarantee that we shall someday succeed, in the precise case of the color red of the synthetic substances to which we have just referred, in establishing an unbroken chain of deductions. In other words, in seeking to establish this chain we may very well come up against irrationals. Thus, since it is a question of how the molecule and its atoms react to light, absorb and reflect it, the irrational or irrationals whose existence is anticipated by the debates at the Council of Brussels can play a role in the phenomenon. Or, again, since light rays are involved, there may be anomalies of a different order, as Victor Henri seems to suppose (Ch. 6, p. 175, n. 34). In general, we are obviously not in a position to make any claims on this subject until the complete reduction of the phenomenon has been carried out, since the irrationals in question here are, as we have seen, essentially unforeseeable.

If we now set aside or, as it were, forget these present or future irrationals and try to take in the whole body of *possible* explanations with one glance, we shall arrive at a somewhat surprising discovery: in general, we find the explanation of being less remote, less inaccessible than the explanation of becoming. No doubt in the absolute sense one is as unrealizable as the other. It is as absurd to try to demonstrate that nothing new has been produced when the two gases we call oxygen and hydrogen have combined to form water as it is to try to reduce the properties of each of these two gases to those of a unique substance itself having only geometric properties, that is to say constituting a hypostasis of space. But we know only too well that science more or less consciously pushes the latter ambitions outside its customary field of vision, so to speak, contenting itself with more immediate satisfactions by way of explanations. Now in this case it certainly appears less arduous, less paradoxical to seek the rational explanation of the red color of fuchsin in a peculiarity of the chemical structure than to attempt to demonstrate (to consider only the simplest case) that each time we see a mechanical motion begin, it must have preexisted and that when we see it stop, it must nevertheless, in reality, continue.

Thus, to consider only a phenomenon we continually have before our eyes, we constantly see motion transformed into *tension* and vice versa; we need only consider a pendulum to observe how easily this metamorphosis takes place in both directions. Now it is clear that if we want to understand anything about this phenomenon, we must suppose that at the moment when the pendulum has arrived at the highest point of its path and when as a result its visible velocity is strictly zero, the motion has nevertheless persisted. This statement is not absurd in itself; indeed, we saw in the preceding chapter (p. 232) how kinetic theory explains the force of expansion of a gas by molecular impact. Consequently, in the case of the motion of a piston which compresses a gas (or which, inversely, is pushed by an expanding gas), the transformation of motion into tension or tension into motion amounts to a real persistence of motion, which alternately assumes molar and molecular form.[5] But one need only consider what this model explanation is to realize what difficulties it will encounter wherever gravitation comes into play (as is the case for the pendulum). As a matter of fact, the only consistent kinetic theory of gravitation ever formulated is, according to Maxwell's expert opinion, that of Le Sage,[6] and it entails consequences such that surely very few contemporary physicists would dare consider it capable of ever being taken up again (cf. IR 80 [Loewenberg 80]). Thus the case of the simplest motion of a heavy body appears totally hopeless to us from the standpoint of a truly causal explanation.

However there is no doubt that, from the logical point of view, the explanation of being is a less pressing, less immediate need for our reason than is explanation of becoming. As Riemann pointed out, the concept of cause, the need for explanation originally arises in connection with change. Moreover, the history of science confirms this view, for it presents us with a whole series of qualitative physical theories: the whole of Hindu atomistics seems to belong to this class, and the systems that grew out of Peripatetic philosophy in the Middle Ages, insofar as they were not satisfied with the purely logical apparatus of Aristotle, but attempted to develop a genuine physics, were clearly qualitative. The Renaissance returns to the traditions of atomism, but there again one can point to Bérigard's curious system, which is atomistic and qualitative at the same time (cf. IR 370 ff. [Loewenberg 326 ff.). Furthermore, even within the apparently mechanistic science of the seventeenth and eighteenth centuries, qualitative conceptions continue to exist, as we have seen, in the theory of quality-bearing fluids. Now obviously a qualitative theory, by the fact that it posits the quality that passes as something persistent and fundamental, refuses to explain it, that is, renounces this aspect of the explanation of being, seeking solely to reduce phenomena to the displacement of this same quality, or, in other words, making do with the explanation of becoming.

NOTES

1. Lucretius, *De rerum nat.* II, 478–507. H.A.J. Munro (*De rerum natura*, 3rd ed., Cambridge: Deighton Bell and Co., 1873, 1:373 [4th ed., 1886, 2:80]), while

noting that Lucretius nowhere makes clear how he conceives the dimensions of the atoms, nevertheless thinks he would have had no difficulty accepting the modern opinion that if a drop of water were enlarged to the volume of the terrestrial globe, the size of the molecules would vary between that of billiard balls and that of bird shot. We believe, on the contrary, that the English commentator, who is almost always able to fathom his author's ideas, was in this case led astray by a false analogy with modern science and that passages such as the one we cite here prove that Lucretius imagined his particles to be much larger than Perrin's.

2. See José R. Carracido, 'Les Fondements de la biochimie,' *Scientia* **21** (1917) 132–133.
3. H. Ley, *Die Beziehungen zwischen Farbe und Konstitution bei organischen Verbindungen* (Leipzig: S. Hirzel, 1911), passim.
4. Victor Henri, *Etudes de photochimie* (Paris: Gauthier-Villars, 1919), passim, esp. p. 151.
5. Lodge, in the note we cited on p. 233, felicitously invokes this typical example of the kinetic explanation of potential energy.
6. Maxwell, 'Atom,' *Encyclopedia Britannica*, 9th ed. (London, 1909), 3:47.

CHAPTER 10

STATE OF POTENTIALITY

The proof of the primacy of this explanation of becoming, of the unequaled ardor with which our reason seeks it, is also furnished to us by the quite remarkable fact that wherever such explanation is too obviously lacking, wherever the task of equating the antecedent and the consequent appears too arduous, the human intellect has forged a special concept to compensate, or at least to seem to compensate for what we necessarily see as an anomaly in the order of things. This concept is that of the *state of potentiality*, and one need only examine the circumstances under which it emerges in the history of human knowledge to recognize that it was indeed created for the purpose we have just indicated. For Aristotle it is "especially from something potential that the actual and real body comes"[1] (cf. IR 410 [Loewenberg 360]). It comes from this potential state, or in other words, according to the logical system of the Stagirite, must be deduced from it, which in its turn, if we abandon metaphysical formalism and try to see how that can be translated into properly physical theories, can, as we have seen, mean only one thing, namely that there is a fundamental identity between the two states. We all know how medieval science used and abused this stratagem, and it is easy to see that each time it was brought into play, it was to explain the appearance or disappearance of something existing, or its nonexistence when it *could* appear. Furthermore modern science has made just as copious use of similar notions. In fact, Black's latent heat clearly follows from the same way of thinking, as does Rankine's *potential* energy. The latter concept is applied wherever we see a motion transformed into tension or vice versa. We noted above that physicists had not abandoned the hope of arriving at a kinetic explanation in this case, that is, of showing that tension itself is the effect of a motion. But we have also pointed out the formidable difficulties of such an explanation. Certainly in the meantime mechanics is obliged to treat motion and tension as two different things. But, as a matter of fact, it tries to assimilate them by declaring that they are only two forms of one and the same thing: *energy*. Now only the energy of motion is visible; it is measured by the square of the acquired velocity. Potential energy, on the other hand, eludes our direct perception and can be measured only by the energy of the motion it is capable of engendering; it is thus only a *tendency*, something whose existence we suppose in order to explain the appearance of what can result from it.

It is obvious, moreover, that this artifice is the only thing that allows us to speak of the conservation of energy. Indeed, even in cases where this conservation is generally considered to be directly demonstrated, what is actually being demonstrated is only the fact that the energy is capable of reappearing. But that it has continued to exist without manifesting itself in any way during the time that elapsed between its disappearance and its reappearance is a pure fiction that only our causal propensity transforms into reality.

The same is true for a concept related to that of energy but formulated much earlier, namely the concept of force, such as is required by the supposition of action at a distance. The force of attraction between celestial bodies undoubtedly manifests itself through motion, but clearly is not itself a motion, nor anything directly perceptible; it is only the cause of motion, potential motion. Furthermore we know it only by this motion; the motion is what makes us think of it in the first place, and the concept is surely created only to explain it.

The case is analogous for the conservation of matter. Of course we are able (and have been ever since we learned to weigh gas) to demonstrate directly that the weight remains constant. But as we saw on page 119, this proposition does not really exhaust the content of the principle: we also claim that the qualitative element is conserved. Now to all appearances the element certainly disappears in the compound; sulfurous acid in no way exhibits the properties of either sulfur or oxygen. Yet they must be there, since they can reappear; therefore they are there potentially. It is this conviction we express when we write the formula of the compound SO_2.

Here is another example, just as convincing, that we shall borrow from the modern philosopher whose position is generally considered the farthest from that of science: Hegel. Hegel, as we saw at the very beginning of this work (p. 9 above), deals in his *Philosophy of History* with the "world-spirit" which "unfolds this its one nature" throughout the course of history though at the same time its nature "is always one and the same." A few pages later Hegel clarifies this conception.

According to this abstract definition it may be said of Universal History, that it is the exhibition of Spirit in the process of working out the knowledge of that which it is in itself. And as the germ bears in itself the whole nature of the tree, and the taste and form of its fruits, so do the first traces of Spirit virtually contain the whole of that History. (*Phil. der Geschichte*, 9:14, 23 [Sibree 11, 18])[2]

Thus history is only developing what was already found in the spirit, but was found there only *in itself, virtually*. By translating this as *potentially* we would surely not be betraying the author's thought. Here, indeed, is how he explains this existence *in itself*: "That which exists in itself only, is a possibility, a power; but has not yet emerged into Existence" (*Phil. der Geschichte*, 9:28 [Sibree 23]).[3] Here the use of the term *power* (*Vermögen* could also be translated *potentiality*) is just as significant as the image of the tree and its fruits, which is the same one used by Bossuet, one in which, in fact, our innate conception of causal development is revealed with particular clarity. Moreover, Hegel came back to this image in another part of his work and there treated it more extensively. In speaking of the notion he says, in the *Logic* of the *Encyclopedia*:

The movement of the *Notion* is *development*: by which that only is explicit which is already present in itself. In the world of nature it is organic life that corresponds to the grade of the notion. Thus *e. g.* the plant is developed from its germ. The germ virtually involves the whole plant, but does so only ideally or in thought: and it would therefore be a mistake to regard the development of the root, stem, leaves, and other different parts of the plant, as meaning that they were *realiter* present, but in a very minute form, in the germ. That is the so-called 'box-within-box' hypothesis; a theory which commits the mistake of supposing an actual existence of what is at first found only as a postulate of the completed thought. The truth of the hypothesis on the other hand lies in its perceiving that in the process of development the notion keeps to itself [*bei sich selbst*] and only gives rise to alteration of form, without making any addition in point of content. (*Enc.*, *Logik*, 6:317, § 161 [Wallace 289])[4]

If we now descend from the heights of Hegelian metaphysics to the simplest notions of common sense, it is easy to see that this is the same process we use continually, that in fact the entire conception of the reality of the world of our direct perception, as it is grasped by common sense, is nothing but the result of its unconscious application. Indeed, the world can be nothing more than our sensation. Now common sense supposes that the world exists, in itself, outside this sensation, that it continues to exist when it is not present in our sensation.

Contrary to what is sometimes claimed, then, objects are not simple possibilities of sensation; just like the objects created by science, they are (as we stated in Ch. 1, pp. 23 ff.) something more. But the proposition is true in the sense that this ontology is nevertheless *constituted* with the aid of possibilities of sensation: it is because we know that sensations which we are not experiencing at the moment can appear or reappear under given conditions that we hypostasize them into objects. From this point of view, the object is indeed (as the term *possibility* suggests) a potential sensation or group of sensations.

Common sense, we are well aware, does not hesitate to endow objects with properties that have been shown by the most direct scientific experience (in agreement on this point with aprioristic analysis such as that employed by the founders of Greek mechanism) to belong exclusively to our sensation, like color, for example. These properties or qualities are thus also sensations and insofar as we do not experience them directly, potential sensations. To claim, as common sense does, that our sensations exist outside us, independently of our sensation, is obviously absurd, but quite useful from the standpoint of forming a coherent representation of reality, in the sense that we know that these sensations which we are not experiencing for the moment *can* appear or reappear under given conditions.

But the concept of potential being, as Hegel's example reminds us, is not solely applicable to the domain of common sense and to that of the physical sciences. It arises spontaneously, as it were, wherever the task of equating the antecedent and the consequent seems too arduous. Did the seed contain the preformed tree? We no longer dare make this claim. But did it contain it *potentially*? Of course. And in the same way, the theory of the descent of organisms implies that the mammal was found *potentially* in the amoeba since it grew out of it by the simple action of external circumstances (as for Darwin) or of faculties inherent in the primitive organism (as for Lamarck).

It is likewise easy to see that the historical sciences (history strictly speaking, biography, etc.) more or less openly make almost constant use of analogous concepts. It seems very natural to suppose that barbarian humanity harbored civilized humanity within itself, and that each people as we imagine it in the deepest past contained as potentiality this same people as we know it today. Ever since nationalistic ideas have come to assume such importance in political life, this last conception in particular has become extremely popular and serves as a more or less acknowledged basis for countless harangues. Of course a historian worthy of the name is perfectly aware of the almost unimaginable multiplicity of factors that have contributed to creating modern *nations*, but the orator or the journalist, having a more simplistic soul, is content to project back into the past the image that is familiar to him. Especially the Germans – for whom the nationalistic doctrine has assumed its most excessive forms – have indulged themselves in this favorite pastime. It is not only in the childish lucubrations of the *Deutschthümler* that Arminius appeared as the prototype of the *Turner* and the *Freiwilliger* of 1813;[5] so-called serious historians invested the barbarians invading the Roman Empire with all the virtues they attributed to the modern German, including, and even especially, the famous "idealism" (the term taken in its ethical sense) thus considered as a prerogative of the "superior" race. But even elsewhere this state of mind is not totally unknown, and such oratory on the ancestral virtues displays it obligingly enough. Here, as always, the concept of potentiality is only a substitute, a stopgap for the concept of identity, to which it aspires to return, to be assimilated.

Moreover, it is easy to see that the qualities with which we endow these primitive peoples, just like the qualities of common sense objects, are quite often hypostases, only what is being hypostasized is no longer simple sensations but historical events in which the people was the hero. Taine has ridiculed statements such as "Rome's destiny was to conquer the universe." He quite rightly adds that the sentence means simply that "the Roman people conquered the Mediterranean basin along with a few countries to the northwest and that this was necessary" (*Les Philosophes classiques* 329). He then seeks to explain the conquest by the military and political superiority of the Romans and concludes that "the destiny of a people is nothing more than the combined effect of the circumstances, its faculties and its inclinations."

That is evident, and it is just as certain that every time we can get rid of a quality, such as this alleged *destiny* of the Roman people, it will constitute progress. But it is no less clear that if the expression was coined, it was in order to use it to explain the history of Rome. Its history is prodigious: how could this small community, made up of a population of uncertain origin, almost constantly torn by civil discord, located on an indifferent site in the middle of a not so fertile countryside on the banks of an insignificant river, win a series of brilliant and almost uninterrupted successes century after century and finally subjugate and absorb the oldest empires of the then known world? And then, in order to diminish this astonishment, in order to conceive that this was *necessary* (as Taine says),[6] one imagines that this entire sequence of events was preformed *ideally* (as Hegel would have said): the original small group of Romans contained it potentially; it was an attribute, a quality of this group, its *destiny*. And similarly, when one speaks of a man's genius, one uses the term to sum up the whole succession of works and acts by which he manifested his greatness. But at the same time one means to suggest to the reader that although none of these manifestations had yet occurred, they already *subexisted* [see p. 82 above], they existed potentially, in their author; since they issued from him, they must already have been there, we shall say, in recalling once again Maeterlinck's argument (cf. Ch. 5, p. 124). Napoleon at Brienne was potentially Marengo, the 18th of Brumaire and Austerlitz, just as Jean Jacques Rousseau, when he arrived at the home of Mme. de Warens, was carrying *Emile* and the *Social Contract* ideally in his meager luggage.

If we could entertain the slightest doubt as to the true nature of these conceptions, we would need only consider the way in which Taine, in the above-mentioned work, speaks of the *destiny* of the Roman people. Indeed, he compares it to expressions current in physics such as "heavy air is a force," "heat has a force of expansion" or "iron and oxygen have a reciprocal force of affinity" (*Les Philosophes classiques*, p. 327). Now it is clear (as we pointed out above in the case of the force of gravitation) that this force, insofar as it represents anything other than a simple function of the actual movement, can only be future change, potential change.

All these conceptions evidently contain a goodly share of fiction. Does it follow that it is a mistake to use them? We could respond by challenging advocates of this point of view to try to reason without any use whatever of fictions of this sort. We saw above what the situation is in allegedly positivistic physics, and without even digging very deeply into the historical sciences, one can see that by forgoing any attempt at truly causal explanation (which would be inevitable if one had no recourse to the concept of potentiality) one would do exceedingly dull work. But we can focus a bit more sharply on the question by simply analyzing the processes peculiar to common sense. Let us set aside the actual constitution of the world of objects, since common sense and science are in agreement on the subject, as we have seen. Let us consider common sense where it is, on the contrary, in disagreement with science. Is the object yellow? On the contrary, there is no doubt that in this case the *quid proprium* of the quality belongs entirely to my sensation. What there is in the object is a surface constituted in such a way that certain radiations which strike it are reflected so as to produce on the human eye a sensation resembling the one it receives by contemplating a piece of sulfur under analogous conditions. But common sense is not constituted for the sake of science, it is constituted for the sake of everyday life. Would it serve any purpose in everyday life to substitute the phrase we have just written for the simple term *yellow*? It should be noted that the circumlocution contained in that phrase is not only extremely awkward, but also suffers from a great indetermination, and that this defect is inevitable. Since the color yellow is a luminous phenomenon, in order to say clearly what conditions it in the object, one would have to know what light is. Now that we do not know, and the vague ideas that science forms on the subject change constantly. At one time it was an undulation of the ether, which was thought to be purely mechanical; at the moment it is an electrical vibration. Moreover, if the most recent scientific ideas are accurate on this point, we shall never be able to say what light is, since to do so would require a mechanical theory of electricity, and because electricity, on the contrary, does not appear capable of being reduced in this way, it thus remains inexplicable (cf. Ch. 6, p. 145).

Therefore, even though we know that color is not really a quality of the object, it is unquestionably advantageous in everyday life to use a locution claiming that it is, because this locution is much shorter and more precise than the one that would express the scientific truth and because the odds are overwhelming that it will nonetheless turn out to be verified in the immense majority of cases we usually encounter. In order to avoid the risk of error, all we need do is remember that the statement expresses only a relative truth and that in extraordinary cases, for example if the object happened to be lighted by monochromatic light, we would see this supposedly inherent quality of the object modified.

In the same way practical advantage will decide in other cases. Is it useful to assume the existence of potential energy? Contemporary science undoubtedly answers in the affirmative: the principle of the conservation of energy is one of its most essential foundations and it would disappear if the conception of a hidden yet real energy were abandoned. But insofar as force is concerned, there are certainly physicists who believe that science ought to dispense with this hypostasis. And if current chemistry indeed seems to proclaim the conservation of the qualitative element, there have nevertheless been chemists like Henri Sainte-Claire Deville who claimed, on the contrary, that the element disappears in compounds.

Similar observations can be made with respect to the historical concepts we have mentioned. By taking Napoleon's genius into account, we shall perhaps better understand his conduct in the affairs of Corsica prior to the siege of Toulon, which marks the beginning of his career in continental France, and the particular genius of Rousseau will be able to help us account, at least to some extent, for a few strange character traits that appear from his early youth onward. To speak of the destiny of the Roman people will be dangerous if one means to suggest that this is a true and complete explanation of the wonders of Roman history. But it can be useful if the author intends, on the contrary, to use this expression as a simple metaphor, for that can allow him to explain events in antiquity more conveniently, to show forces at work in primitive Rome that the organism actually revealed only much later. Of course he can also, in speaking of earlier events, directly invoke those that followed much later. But the use of the term *destiny* to sum up the whole subsequent development can certainly be of service in this way. This use can also be of service if the author, far from supposing that the *destiny* of Rome fully explains its history (which is the intention Taine attributes to him), intends to imply, on the contrary, that there is some sort of mystery in the history of the Eternal City, something for which the elements we know cannot furnish a truly satisfying explanation. Taine, in a few extremely well-written pages, believes he can summarize all that must have characterized early Rome and that, according to him, explains her history (*Les Philosophes classiques* 364 ff.). It does him no injustice to say that, without mentioning the fact that some of the essential strokes of the picture he paints are unconfirmed (for example, we are no longer as sure today that Rome was originally an asylum), the explanation remains quite incomplete. Taine laughs at Virgil who, he says, makes prophecies after the fact in his capacity as a "poet and official poet of Rome." But if Virgil by his *tu regere imperio populos, Romane, memento*[7] simply meant to express the astonishment, the feeling of mystery that any observer cannot fail to experience before this piece of history, one must recognize that the impression of modern man, though certainly less immediate and less intense than it was for someone who actually beheld the splendors of Augustan Rome, still remains of the same order.

What we have just set forth – the reader will probably have already noticed – is only a development of what we suggested earlier (Ch. 3, pp. 61 ff.; Ch. 5, p. 120) concerning hypostases. Indeed, it is evident that no hypostasis is conceivable if we do not allow the possibility of a potential existence. Thus, in examining the question of knowing where we can and where we cannot assume a state of potentiality, have we not in fact merely clarified the use of "Ockham's razor"?

What is also quite remarkable in all cases where the concept of potentiality consciously or unconsciously comes into play is the unavoidable confusion it creates between the temporal, historical point of view and the rational, logical point of view. It is the same confusion we noted earlier within the boundaries of logic itself in the use of terms such as *prior*, *posterior*, and *consequence* (pp. 55, 90 ff.). It reappears here, in an even more complete form if that is possible. So complete is the confusion that we would sometimes be hard put to say which of the two conceptions, as fundamentally distinct as they are, we had in mind. One thing "existed in germ" in another. What does this mean? That it followed it in time? It should mean that, for all we know of a germ is that, given time and favorable circumstances, it produces a plant. We know of no *logical* connection between the germ and the plant; we certainly cannot say, even in the most general way, *why* a germ must necessarily become a plant. However it is to a conception of this sort that the word "germ" ordinarily refers. What one means to say, or at least suggest, is that there was something *potential* there from which what followed had to issue necessarily, logically: in other words one means to *explain* what happened.

The same is true (although in an opposite sense) for the terms *evolution* or *development*. We saw in Chapter 5 (p. 125) that the image underlying these locutions is preformationistic. But, in the absence of complete identity, it is the concept of potentiality that intervenes, and given its lack of precision, the confusion becomes inextricable, just as it was in the case of the "germ." Here, etymology suggests, it should be a question only of *necessary* relations. But the single fact that Bergson could, without shocking us, speak of *creative* evolution – in a sense that is therefore directly counter to the etymological, preformationistic sense – proves we also use this term to characterize the phenomenon from the purely external point of view of change occurring in time. Of course the etymological sense, that of necessary connection, is never totally absent, but it becomes, as it were, a hidden, secondary sense, which we are ready to recall or forget at our convenience, and this duality is precisely the source of the more or less permanent confusion of which we just spoke.

A question that naturally arises as soon as we recognize the considerable role the state of potentiality plays in our explanations is the following: How can this state be understood as both identical to the state of actuality and different from it?

Let us note first of all that the degree of identity (if we may be allowed this expression) is far from being the same in the various circumstances where we have pointed out the use of the concept. At one end of the chain, so to speak, we have the *objects* of the external world, and there it is obvious that common sense affirms a complete identity between what is or for the moment is not part of our sensation. Whether I look at it and touch it or turn my back on it, the table, in my immediate consciousness, remains just what it is; my sensation adds to it no reality of any sort. As Spinoza pointed out, the very term *perception* seems "to imply that the mind is passive in respect to the object" (*Ethics*, Pt. 2, Def. 3), and Schelling is absolutely right when he states that it is from this identity between the object and the perceived and from the inability to distinguish the perceived from its object during the act of perception that common sense draws the conviction of the reality of external things (Schelling, *Ideen*, I, 2:15 [Harris 12]). This conviction, as is easy to see, resists all subsequent conscious arguments, or at least yields to them only in appearance, ready to reassert its rights as soon as the pressure exerted by these arguments weakens the least bit. This is why for the most rigorously idealistic philosophers the concepts claimed to be carefully separated out from the realistic gangue still have an unfortunate tendency to be transformed into notions only verbally distinguishable from those of the grossest common sense, to the point that the entire system becomes simply an "inverted [*umgekrempelt*] naive realism," as Eduard von Hartmann maliciously remarked.[8] This return to the "inevitable" notions of common sense (to use Balfour's excellent expression)[9] operates, of course, with particular facility in thinkers whose feel for science is very strong. The conviction of the reality of the external world dominates them to the point that they cannot free themselves of it. This is why in John Stuart Mill the "possibilities of sensation"[10] are transformed, one might say automatically and instantaneously, into veritable objects, while Bertrand Russell, for whom these possibilities are called *sensibilia*, goes so far as to state that he personally does not feel that it is "monstrous to maintain that a thing can present any appearance at all in a place where no sense organs and nervous structure exist through which it could appear."[11] It is quite obvious, on the contrary, to speak with Balfour (who applies these expressions to the "implicit thought" which certain metaphysical deductions assume to exist – but it is even more true, if possible, in the domain of sensation), that the very definition of the possibility of sensation implies that it has no existence qua sensation; the affirmation of this existence is basically only a simple rhetorical figure, it is in no way a sensation, it is "a creation of language, which can constitute nothing because it is nothing."[12]

At the other end of the chain we have scientific notions and, given their precision, we shall not be surprised to discover that here the nonidentity is avowed, patent. No supporter of Black ever claimed that latent heat was really what we call heat: it was always understood that it was detectable neither by our direct thermal sensation nor by any measuring instrument, that it was hidden, secret, as its name indicates. Likewise, the most intransigent partisan of kinetics could not treat potential energy as something really identical to the energy of motion; the formulas suggest otherwise, given that the two appear there in different forms. What physicists claimed and continue to claim in the two cases is that there can be transformation, reappearance: latent heat is indeed a possibility of heat and potential energy is a possibility of motion, just as the object absent from our immediate perception is a possibility of sensation, but with the important nuance that in science there is no longer any fear of confusing the possible and the real, the potential and the actual.

Between these two extremes can be found the various conceptions of the state of potentiality used by philosophers. It would be entirely beyond the scope of the present work to attempt to follow them and examine them in detail. Perhaps such a study would not even be very useful, except for revealing the power of auto-suggestion and auto-deception (if we may use these disparaging terms) to which the most exalted and vigorous minds of which mankind is justifiably proud have sometimes succumbed.

This is because the thinker, as soon as he has recourse to the concept of the state of potentiality, finds himself the victim of two opposing tendencies: the potential *must* be distinguished from the actual and yet in spite of this distinction *must* be able to give rise to the actual, which is possible only if the two are identical, if they can be confounded with one another. It is necessary then that thought simultaneously posit these states as alike and different and that it reconcile or seem to reconcile this contradiction, to resolve it. That is the task to which the philosophers' ingenuity has untiringly applied itself. How does Hegel intend to explain the development of the plant? By the supposition that all its characteristic traits were already found in its seed. But they were there only *ideally*. Let us not attempt to fathom the miracles of dialectical artfulness by which the philosopher strives to define this ideal state. Suffice it to note that this ideal inclusion, while it is fundamentally different from the hypothesis of encasement (which Hegel scornfully rejects as infinitely too crude) must nevertheless lead, in the final analysis, to the same result, namely to make us conceive as preexistent what we initially judged to be entirely new.[13] Moreover, all we need in order to convince ourselves that it is really a question of a fundamental identity, assumed and affirmed in spite of everything, is to note what pains Hegel takes to forewarn us against this interpretation of his position. He comes back to it on several occasions, renewing his warnings and protests. Thus in his *Philosophy of History*, after first setting forth, in the passage we cited at the very beginning of our work, how the world-spirit *explicates* its nature in history, he recalls a few pages later:

> I have already directed attention to the distinction here involved, between a principle [of freedom] as such, and its application; *i.e.* its introduction and carrying out [*Durchführung*] in the actual phenomena of Spirit and Life. This is a point of fundamental importance in our science, and one which must be constantly respected as essential.

On the next page, again the same injunction in almost identical terms: "Attention was also directed to the importance of the infinite difference between a principle in the abstract [*an sich*], and its realization in the concrete." And three pages later, again this explanation:

> Principle – Plan of Existence – Law – is a hidden, undeveloped essence, which as such – however true in itself [*in ihm*] – is not completely real .... That which exists for itself only, is a possibility, a potentiality; but has not yet emerged into Existence. (*Phil. der Geschichte*, 9:24, 25, 28 [Sibree 19, 20, 23; Meyerson's brackets])

Obviously if it is necessary to bear the distinction between the two states constantly in mind in order to keep from confusing them, it is because the reasoning process itself, just as constantly, leads to this confusion.

Is this turn of thought peculiar to the great dialectician? Not at all. Here is how a historian of philosophy summarizes the thought of the Stoic Mnesarchus:

> What truly exists is the primitive *pneuma*. This is imperishable and exempt from both increase and diminution. These changes actually affect only the particular beings that form or are formed from this *pneuma*; they do not differ from the latter by their proper substance, yet, insofar as they constitute species and modalities of that which exists, they are not the same thing as it is either.[14]

And Spinoza says, altogether in the same spirit, in attempting to reconcile the Cartesian oneness of matter with the apparent diversity of things in space, "that matter is everywhere the same, that its parts are not distinguishable, except in so far as we conceive matter as diversely modified, whence its parts are distinguished, not really, but modally" *(Ethics*, Pt. 1, Prop. 15, Scholium [Elwes 2:59]). Here then, even more openly than for Hegel, are modalities and species distinct from the primitive substance and at the same time identical to it.

Should we be surprised at these ever recurring efforts to reconcile the irreconcilable, to bring reason to conceive simultaneously notions that, deprived of metaphysical prestige, certainly appear contradictory, mutually exclusive? Should these efforts even be seen as lapses on the part of the great minds that indulged in them? Does not the whole of our exposition tend to show us, on the contrary, that this is an inevitable condition of the very functioning of our reason, a functioning whose processes we can follow from the genesis of the most rudimentary conceptions of common sense? What indeed is common sense if not (as we have seen) the supposition that objects, which are only groupings of sensations, exist independently of sensation, that is, according to Hume's famous formula, "that the senses continue to operate, even after they have ceas'd all manner of operation," or again, in other words, obviously "a contradiction in terms."[15] And does science proceed any differently when, seeking to form its atoms out of "singular points" of the ether, it conceives this ether to be both different from and at the same time identical to the ether that surrounds it? Or when it treats kinetic energy and potential energy and even fundamentally different energies, such as heat, electricity, mechanical energy, as "forms" of one and the same basic essence (or as modalities, to use Mnesarchus's terminology), capable of being transformed into one another without ceasing to be *conserved*? How then could philosophy, whose proper task it is to try to form a coherent image of the great Whole, avoid this ineluctable necessity?

NOTES

1. [We have translated Meyerson's French. The same French passage is identified in IR as *De Caelo*, Bk. 3, Ch. III, Sect. 1., and Loewenberg gives the J. L. Stocks translation for that citation: "An element, we take it, is a body into which other bodies may be analysed, present in them potentially or in actuality."] Höffding also notes that the passage from possibility to reality, which Aristotle calls κίνησις (motion), "forms an analogy with what we today call the passage from potential energy to actual energy" (*La Pensée humaine*, trans. Jacques de Coussanges, Paris: Félix Alcan, 1911, p. 231).
2. [To conform to Meyerson's translation, we have substituted "in itself" for Sibree's "potentially" at the end of the first sentence (the German is *an sich*) – Sibree's actual translation certainly makes Meyerson's point for him!]
3. [To conform to Meyerson's translation we have changed Sibree's "for itself" (for *an sich*) to "in itself," and "potentiality" (for *Vermögen*) to "power" – again

Sibree's translation makes Meyerson's point directly.]

4. [To conform to Meyerson's translation, we have substituted "in itself" for Wallace's "implicitly" at the end of the first sentence (the German is *an sich*); Meyerson's brackets.]
5. Even Victor Scheffel (as nationalistic as he was on occasion) felt compelled to make fun of this grotesque manner of parodying history. In the song of the battle of the Teutoburg forest, so popular among German students (it is included in the *Kommersbuch*, the official "key to the wine cellar" for drinking parties), the Cherusci fight "with God, for king and country," which was of course the motto of the Prussian volunteers in 1813.
6. However, this term gives rise to ambiguity in Taine's reasoning. For example he states that "when air presses against the barometer, the mercury will necessarily rise"; likewise "the bar, when heated, will necessarily expand" and "iron exposed to humid air will necessarily combine with the oxygen." Necessity here would thus be merely drawn from experience. But it must be noted that these statements are placed in the intentionally positivistic presentation of Mr. *Peter*. Mr. *Paul*, on the other hand, taking his inspiration chiefly from Spinoza's *Ethics* or Hegel's *Logic* (*Les Philosophes classiques* 348), strongly insists on the importance of deduction, which is obviously understood as logical (*Les Philosophes classiques* 351 – Taine appeals to passages from Cuvier that we ourself have cited and whose implications are unmistakable – cf. also Appendix 3, p. 564 below, on the role of deduction in Taine's positivism). There we catch him in the act of trying to superpose an idealism approaching Hegel's on Comte's positivism, and the ambiguity to which we have just referred is clearly a consequence of this.
7. [*Aeneid* VI, 851: Roman, be this thy care, to bear dominion over the nations.]
8. Eduard von Hartmann, *Das Grundproblem der Erkenntnistheorie* (Leipzig: Wilhelm Friedrich, 1889), pp. 14 ff. [Meyerson's brackets; this precise quotation is not to be found].
9. Arthur James Balfour, *L'Idée de Dieu et l'esprit humain*, trans. Bertrand (Paris: Bossard, 1916), p. 15 [*Theism and Humanism* (New York: George H. Doran, 1915), p. 29].
10. [E.g., John Stuart Mill, *An Examination of Sir William Hamilton's Philosophy* (Boston: William V. Spencer, 1865), 1:238.]
11. Bertrand Russell, 'The Relation of Sense-Data to Physics,' *Scientia* **16** (1914) 4, 11.
12. Arthur James Balfour, *A Defense of Philosophic Doubt* (London: Macmillan, 1879), pp. 95, 99, 101.
13. Schelling, moreover, had set an example for him, fulminating against the theory of encasement in the sense in which it is ordinarily understood, which did not prevent him from declaring that there is preformation; however, this preformation is not *individual*, it is only *dynamic* (*Erster Entwurf*, I, 3:46 ff., 60).
14. August Schmekal, *Die Philosophie der mittleren Stoa* (Berlin: Weidmann, 1892), p. 297.
15. Hume, *A Treatise on Human Nature* (London: Longmans Green, 1878), 1:479 [*A Treatise of Human Nature*, ed. L. A. Selby-Bigge (Oxford: Clarendon Press, 1967), Pt. 4, Sect. 2, p. 188].

BOOK THREE

GLOBAL EXPLANATION

CHAPTER 11

HEGEL'S ATTEMPT

In order to get a better idea of the mechanism reason uses when it seeks to understand phenomena, we are going to devote this chapter to a study of Hegel's system of the global explanation of nature from that particular perspective.

This choice may be surprising in several respects. But to speak first of the considerations most foreign to philosophy, no matter how much disapproval one may wish to register toward Hegel's political philosophy (see Appendix 5), it seems difficult to extend it *a priori* to his logic and epistemology.

Some, on the other hand, have felt obliged to reject *a limine*, as it were, anything in any way derived from Hegelianism, in virtue of the deliberately paradoxical appearance of the doctrine. The late Duhem formulated this protest in vehement terms. "Is it possible," he exclaims, "to ride roughshod over the most important evidence of common sense more harshly or insolently than Hegelian metaphysics does?" He continues:

> And what deserves to be noted is not that a Hegel happened to be found among the German people; at all times and in all peoples are encountered unfortunate maniacs who reason interminably on absurd principles. The disturbing thing is that the German universities, instead of taking Hegelianism for the dream of a madman, welcomed it enthusiastically as a doctrine whose splendor eclipsed all the philosophies of Plato or Aristotle, of Descartes or Leibniz.[1]

There is little need to point out that summary judgments of this sort (that of Duhem was certainly influenced by his legitimate indignation at German aggression) can in no way lay claim to serious consideration in philosophy. Whatever may have been said of it, metaphysics is no more under the jurisdiction of the conceptions of common sense than is mathematics. Now here we have a system which for a whole generation exerted a quasi-absolute dominion over a significant part of the educated world (see Appendix 6), and which for this reason, if for no other, must appear eminently worthy of commanding the attention of those who seek to deduce laws of thought from the manifestations of this thought in the various fields of human knowledge. Furthermore, we hope that by way of conclusion to the study we are about to undertake, the reader will come to recognize with us that important lessons can be drawn from the examination of this doctrine, which at first sight seems so profoundly dissimilar to the one that inspires contemporary science.

We can begin by observing that as regards the notion of the state of potentiality that we have just treated in the preceding chapter, although Hegel widely used and sometimes unquestionably abused it, at the same time he clearly discerned the thought process that engenders it. Indeed, he recognized, as we showed in Chapter 5 (pp. 104 ff.), that at the foundation of our reasoning process there is a constant contradiction which the very necessity of the progress of this thought makes inescapable. By breaking down the mechanism of our thought, the Hegelian *dialectic* shows us, from the inside as it were, the profound reasons for the seemingly paradoxical phenomena we have just observed. If science to some extent tends to confuse potential energy, which is a simple possibility without any external manifestation, since it is the attribute of a body at apparent rest, with kinetic energy, which is, on the contrary, a function of motion, and if common sense posits this enormous contradiction of the object that exists and yet is unperceived even though composed of perceptions, it is because our thought is accustomed to functioning in this manner, because that is its normal way of proceeding. Hegel felt very intensely all the generality of this process, the strength of the current which in all cases of this kind sweeps the mind along beyond what is strictly rational, logical. So intensely did he feel it that he believed that logic, as it had been known up to his time and as it had been codified by Aristotle in particular, was, on this account, if not entirely abolished, at least subject to amendment,[2] in the name of a sort of higher logic, whose propositions he tried to formulate in his turn. In order to provide grounds for the coexistence of these two logics Hegel has them proceed from two entirely distinct faculties of our mind. For him, the faculty invoked to direct our thoughts, or in other words *reason*, actually turns out to be twofold: on the one hand, there is *abstract reason*, which, in its search for perfect identity, is led to deny diversity and, on the other hand, there is *concrete reason*, which, by going beyond this diversity, brings about reconciliation (see Appendix 7). From the standpoint of form, therefore, logic presents a triple aspect: the abstract or rational, the dialectical or negatively rational, the speculative or positively rational (*Enc.*, *Logik*, 6:146 [Wallace 143]). Obviously the first of these aspects conforms to abstract reason, the second is the one in which the inner contradiction of thought conceived according to that reason is revealed, while the third is destined to resolve the contradiction.

The intervention of concrete reason certainly constitutes the most striking feature of the Hegelian system, which could rightly be called the philosophy of *concrete reason* (*Vernunftphilosophie*).[3]

There is no question but that Hegel is extremely original in this area. Of course the germs of these logical conceptions can be found in earlier thinkers. Herbart stressed that the concept of thing is contradictory in itself, a contradiction that thought must remove,[4] and already in Kant, as Boutroux has admirably shown, the notion of a transcendental logic or logic of concrete being, with its trichotomic division, paves the way for the Hegelian logic on this point. Boutroux has also brought to light more remote origins of Hegel's logic in Leibniz, Pascal, Descartes, and the thinkers of the Renaissance and antiquity.[5] Nevertheless the distance to be covered between these somewhat sketchy arguments and the Hegelian system is considerable.

But what must be especially emphasized is how independent Hegel is from Schelling here – at least insofar as the truly logical side of his position is concerned. That is not an empty remark, for Schelling himself, when he set himself up as the declared adversary of his rival after the latter's death, brought the weight of his claims of priority to bear primarily on the Hegelian dialectic. After providing an admirably clear and precise résumé of this dialectic, he says it is only an "imitation" whose "original" is to be found in his, Schelling's, own work (*Zur Geschichte*, I, 10:137), and moreover that Hegel had always intended to establish a system identical in its general outline to his (*Zur Geschichte*, I, 10:127–129). For this reason Schelling cannot admit that "someone else should boast of having invented it."[6]

Is that claim entirely unfounded? Not at all. As a matter of fact, it is impossible to read through Schelling's description of the process by which the primitive identity of nature splits into opposition and finally resolves this same opposition – although without ever succeeding in making it disappear altogether (this opposition being infinite in nature), so that each successive attempt at reconciliation by means of a third term gives rise to a new opposition[7] – without discovering the inmost rhythm of the Hegelian dialectic. But there is the fundamental difference that in Schelling all this is understood to occur in the relationship of the mind to nature, and thus that Hegel, in borrowing the process from his predecessor (as seems extremely likely), transposed it by placing it on the terrain of logic proper, that is, within the mind. Furthermore, Schelling himself acknowledges this in the passages we have just cited, affecting to consider that what is involved is, on the whole, a minor modification. At other times he is more fair; he declares that Hegel deserves credit for developing the logical side of his (Schelling's) philosophy, but that he ought to have confined himself to that and refrained from any attempt to go on to a positive philosophy.[8] But above all he draws from this observation a whole series of objections to the Hegelian philosophy, objections we shall have occasion to consider below. As a matter of fact, these objections seem to us to be likely to cast a strong light on the subject that interests us, which is why we have felt obliged to dwell at somewhat greater length on the question of the relations between the ideas of Hegel and those of his predecessor. We need not go any further into Schelling's ideas themselves, since we have just seen by Schelling's own admission that they do not touch on logic at all. It must be noted, however, that Hegel's independence from his rival does not extend to the triadic form, the idea for which comes from Schelling, as orthodox Hegelians themselves acknowledge.[9]

As an example of Hegel's dialectical deduction, let us consider the deduction of the concept Hegel himself considers (and rightly so) fundamental for his system, namely, the concept of *becoming*. Becoming (which is the speculative concept) is the unity of being (the abstract concept) with nonbeing (the dialectical concept).[10] No doubt it is generally claimed that we are unable to arrive at a representation of the unity of being and nonbeing, but this is not correct; on the contrary, each of us has an infinite number of these representations:

> The readiest example of it is *Becoming*. Every one has a mental idea of Becoming, and will even allow that it is *one* idea: he will further allow that, when it is analyzed, it involves the attribute of Being, and also what is the very reverse of Being, viz. *Nothing*: and that these two attributes lie undivided in the one idea. (*Enc.*, *Logik*, 6:174 [Wallace 165–166])

Furthermore, "pure being and pure nothing are … the same …. Their truth is … this movement of the immediate vanishing of the one in the other: becoming."[11] Thus Heraclitus's doctrine surpasses that of Parmenides while preserving[12] it (this is the Hegelian concept of *aufheben* mentioned in Chapter 5, p. 107), and it cannot be denied that in some respects the example is well taken, for Parmenides's absolute being does indeed bear an amazing resemblance to nonbeing, given that, as

Renouvier aptly remarked, "this homogeneous and immobile uniform matter is neither cause nor reason for anything whatsoever in the world"[13] and that, on the other hand, as Hegel himself emphatically pointed out, *ex nihilo nihil fit* completely excludes all becoming.

It is on this base that Hegel constructed a logic, which he means to set over against Aristotle's. But he immediately goes further. Indeed, he is so convinced that the idea is the absolute, that spirit is the cause of the world,[14] that this new logic, in conformity with concrete reason, seems to him to reveal the very essence of things, in other words to be necessarily inseparable from metaphysics, which, according to his definition, is "the science of things set and held in thoughts, – thoughts accredited able to express the essential reality of things" (*Enc.*, *Logik*, 6:45 [Wallace 45]). Therefore, for him "the Philosophy of Nature and the Philosophy of Mind take the place, as it were, of an Applied Logic."[15]

As Seth rightly makes clear, the language used by Hegel "can only be interpreted to mean that thought out of its own abstract nature gives birth to the reality of things," since Hegel "sought to demonstrate the existence of the worlds of nature and of spirit by a pure synthesis starting from the world of logic." He has also "once more repeated ... the extraordinary but apparently fascinating attempt to construct the world out of abstract thought or mere universals."[16]

Some of Hegel's commentators, in a commendable attempt to defend the memory of the master against the accusations of extravagance to which it is only too vulnerable, and to bring his way of thinking closer to that of our contemporaries insofar as possible, have sometimes tried to tone down the more shocking aspects of these declarations, although they are explicit enough. Thus McTaggart, while stating that for Hegel reason "is the key to the interpretation of the whole universe; it finds nothing alien to itself wherever it goes," and while recognizing that it "is beyond doubt that Hegel regarded his Logic as possessing, in some manner, ontological significance," nevertheless seeks to demonstrate that "he endeavors to find the idea in everything, but not to reduce everything to a manifestation of the idea" (*Studies* 31, 26, 29; cf. also pp. 65 ff., 204, 234). There is some truth in these words in the sense that, as we shall see below (pp. 275 ff.), Hegel did not attempt to deduce *everything* in phenomena, but, on the contrary, left a sort of margin devolving to the activity of experimental science. Similarly, he certainly tried to separate out the essential (and therefore, for him, necessarily *a priori*) traits of reality by an *a posteriori* examination of phenomena and of science, and in so doing, whatever may have been said, he committed no illogicality (cf. Ch. 14, pp. 368 ff. below). But once these reservations have been made, and giving McTaggart's claim its full meaning, one is forced to recognize that it tends to establish a purely verbal distinction. For finally, setting aside nonessential phenomena, or rather the nonessential side of phenomena, and once the thread of the idea in reality was discovered, it is certain that this idea had to be taken as *pure*, that is, free of all empirical elements, and Hegel's declarations, as well as his procedure itself, leave no doubt that he in fact understood the whole of this essential thread to be deducible in its entirety by the work of thought alone, proceeding according to the rules of the dialectic, starting from foundations themselves necessary. Edward Caird, after first suggesting the same sort of toning down as McTaggart's, nevertheless concludes that "if to find thought in things be more than an empty word, then the movement or process, which thought *is*, must explain at once the transition from thought to what in opposition we call '*things*,' and must give us the means of reconciling that opposition."[17] Disregarding the general demonstration contained in this passage, it is certain that it accurately identifies Hegel's position on the subject.

Thus Hegel ends up with a *panlogism*. René Berthelot, certainly not without justification, has challenged the application of this term to the Hegelian corpus.[18] As a matter of fact, the designation, which has become almost classic in the history of philosophy,[19] might tend to suggest the ordinary logic, which is that of Aristotle and against which Hegel wages a constant battle, declaring it inadequate. It remains no less true that what Hegel set out to do was in fact to resolve reality into a body of purely mental and *necessary* concepts, that he gave the title *Logic* to the work in which he sets forth the way he believed these concepts had to be connected, according to the precepts of *concrete reason*, and that finally he himself quite explicitly compared his work to that of the Stagirite, claiming that "Aristotelian physics ... is far more a *Philosophy of Nature* than it is physics."[20]

Of course, since it is the idea that creates nature,[21] Hegel has no doubt that nature must appear perfectly intelligible. "The hidden essence of the Universe is not vigorous enough to resist the courage of knowing; it will necessarily have to open itself to this courage, reveal its richness and depth and make itself enjoyed," he declares in outlining the plan for his *Naturphilosophie*, and in the work itself he states that "the inscription on the veil of Isis[22] ... melts away before thought." Finally, in the conclusion of this work he repeats that "Concrete Reason must have confidence in itself, confidence that in Nature the Notion speaks to the Notion and that the veritable form of the Notion, which lies concealed beneath Nature's scattered and infinitely many shapes, will reveal itself to Reason."[23] Furthermore, like nature, history is the expression of concrete reason. "The only Thought which Philosophy brings with it to the contemplation of History, is the simple conception of Concrete Reason; that Reason is the Sovereign of the World; that the history of the world, therefore, presents us with a rational process" (*Phil. der Geschichte*, 9:12 [Sibree 9]).

This theory, it is hardly necessary to point out, is only a precise expression of the eternal postulate of the rationality of the external world, which Spinoza had summed up as follows: *ordo et connexio idearum idem est ac ordo et connexio rerum.*[24] The conception came down to Hegel by way of Schelling, whose disciple Hegel was early in his career and who had stripped it of all that was connected in Spinoza to the Cartesian dualism of thought and extension.[25]

Of the amazing edifice constructed by Hegel, whose outlines we of course have no intention of retracing, even in the most general terms, the only thing that interests us is its logical and scientific part. For him the two are closely connected; in fact, as we know, in the *Encyclopedia* the *Philosophy of Nature* immediately follows the presentation of the *Logic* and is rigorously connected to it. We believe we can say without exaggeration that this aspect of the Hegelian philosophy has been somewhat neglected by the commentators and even by the continuators of the philosopher. In the first place, the *Logic* itself, as a contemporary Hegelian observes, generally occupies only an altogether insignificant place in the commentaries and criticisms, as compared to its real importance in Hegel's work, where everything actually depends on it.[26] But as for its application to nature – which is a direct consequence of the *Logic* and to which Hegel attributed such an eminent place, consecrating to it about one-third of the abridged version of his system, which is what the *Encyclopedia* is – it is often barely mentioned. This state of affairs is to some extent understandable (at least insofar as the *Naturphilosophie* is concerned), for, unlike what occurred in many other areas, such as law, history, sociology or politics for example (see Appendix 9), Hegel's thought never wielded any influence in the domain of science. That is why the very people who claimed kinship with the philosopher and professed to follow in his footsteps had a tendency to leave in the shadow everything having to do with his conception of science, considering that part of his work its "*partie honteuse*" [shameful part], to use Höffding's apt and colorful expression (*Histoire* 2:185 [Meyer 2:183]). From our present point of view, on the contrary, Hegel's *Naturphilosophie* seems to afford considerable interest. Indeed, among other remarkable characteristics, it is the most recent of the great attempts at a global explanation of nature. Moreover, as we hope to be able to show, the deep foundations of Hegel's theory have points in common with the ideas we have expressed above. Thus our study will have a specific and precise goal, and consequently it could not result in a truly complete picture of Hegel's effort, even in this limited field. We shall attempt to set forth only one aspect of this effort, and some Hegelians may judge that we are missing what constitutes the true essence of the master's thought. But perhaps also some of them will be willing to concede that this quite specific aspect is not without importance for the general comprehension of Hegelian philosophy.

We have observed – and it is common knowledge – that Hegel's attempt in the area of science was a failure. But it may be of some use to make clear how complete this failure was. For, all things considered, Descartes's system, after prevailing for a time, could not sustain itself in science either. But in fact it is impossible not to see the immense distance separating the two failures. Certainly no contemporary physicist believes in the Cartesian vortices, in the three elementary substances, nor in the channeled parts, all things for which Descartes claimed a "more than moral certainty" (*Principes* IV, 205, 206), and the sumptuous palace erected by the author of the *Principles* is now only a ruin beyond repair. But it is a truly grandiose ruin and one that inspires profound respect. It is not only that certain stones of the collapsed edifice – the principle of inertia, that of the conservation of motion (which is indeed, despite what Leibniz called "Descartes's memorable error," the true starting point for the developments culminating in our principle of the conservation of energy), a considerable part of optics, etc. – have served as foundation stones for the establishment of some of the highest achievements of our present science; it is also that the general conception of the work and the spirit that animates it could not be truer to the thought which inspires the scientists of today, and, through their influence, our educated contemporaries in general. The rapidity and simplicity of certain deductions may shock us, but still, almost in spite of ourselves, we are struck with admiration, and sometimes we catch ourselves regretting that Descartes was not closer to the mark and that nature (as the subsequent evolution of physics showed) proved to be so much more complicated than he imagined it. To enumerate all the inspiration science has drawn from his work would be to reproduce the history of the scientific development of the ensuing centuries. And perhaps this spring is not entirely dry and a modern physicist could still find many subjects for fruitful meditation by reading through the *Principles*.

The impression made upon a present day reader by the work of Hegel is totally different. It is – one is forced to admit, in spite of the great respect one may feel for an otherwise so powerful mind – one of profound bewilderment. There is nothing that recalls the science of today, nor the science of the author's contemporaries (which was, moreover, moved by almost the same spirit), nor even genuine science of any period of human endeavor, for instance Peripatetic physics or the chemistry of the alchemists. It is as if someone presented us with a series of absurdly grimacing monsters where we expected to see human figures. Sometimes one begins to doubt and rereads several times to convince oneself that the phenomenon treated by the author is indeed the one known by science, so fundamentally different is his interpretation from anything that science imagines or has imagined. There are so many examples of this that it is difficult to choose. Let us cite the one that Hegel himself sees as fundamental and that he mentions as such in the exposition of the *Logic* in the first part of the *Encyclopedia*. What is a magnet? It is an "exemplification of the syllogism [*Schluss*]." For "the syllogism (not as it was understood in the old formal logic, but at its real value) ... gives expression to the law that the particular is the middle term which fuses together the extremes of the universal and the singular." In the same way, "in the middle or point of indifference of a magnet, its two poles, however they may be distinguished, are brought into one."[27] Does the reader find this parallelism inadequate? It is solely because of the impotence of nature, replies Hegel. Indeed, this impotence entails the consequence that nature "fails to exhibit the logical forms in their purity." Nevertheless this syllogistic form is "a universal form of all things."[28] In spite of the clarity of Hegel's declarations one could, given the bizarre nature of his theory (from the standpoint of current scientific conceptions, obviously), believe that it is perhaps only a flight of fancy. But that would be to misjudge the serious and sustained effort he brings to the subject. As a matter of fact, in the *Naturphilosophie* Hegel returns to this fundamental theme several times and declares in particular that the representation of the syllogism in the magnet comes about "in simple, naïve fashion" and that therefore magnetism necessarily had to occur to one as soon as one had conceived the idea of a philosophy of nature, that is to say as soon as one suspected the existence of the concept in nature. Moreover, the chemical process is likewise a syllogism, "and not merely the beginning of the process but its entire course is syllogistic. For the process requires three terms, namely, two self-subsistent extremes, and one middle term in which their determinatenesses come into contact." And here is the proof of this assertion, which at the same time shows us how Hegel understood these three terms:

Fully concentrated acid, which as such contains no water [Hegel considered what we now call the anhydride to be the true acid, as did all chemistry in his time], when poured on metal either fails to dissolve it or else has only a weak action on it; if, on the other hand, it is diluted with water, it attacks the metal vigorously, simply because the process calls for three terms.[29]

After that, need one wonder that Hegel's work has remained without the slightest influence on the course of the physical and biological sciences? It seems clear that no progress in the science either of his time or later is connected with it in any way,[30] and this fact is all the more significant because it can certainly not be attributed to external circumstances, for example to lack of encouragement. The Hegelian philosophy, as is well-known, exercised incomparable attraction in its own time; as a matter of fact, in Germany it was the true scientists who had to suffer cruelly from preferential treatment granted to supporters of the doctrine in vogue,[31] as we see in the correspondence between Liebig and Wöhler for example.[32] Furthermore, the absolute sterility of Hegelian thought in the scientific domain stands out even if one compares it to other more or less analogous positions from approximately the same period. Hegel's *Naturphilosophie* is of course not a unique phenomenon in German romantic philosophy. At the same time or earlier, other philosophers also tried to fathom nature. Hegel has nothing but scorn for these attempts by his rivals:

What has in modern times been called the Philosophy of Nature consists principally in a frivolous [*nichtig*] play with empty and external analogies, which, however claim to be considered profound results. The natural consequence has been to discredit the philosophic study of nature.[33]

It is "sleight-of-hand," and "the trick of wisdom of that sort is as quickly acquired as it is easy to practice."[34]

In reading these lines, the modern reader would be tempted to wonder at the blindness of the author, crying *Quis tulerit Gracchos ...*,[35] particularly since Hegel's explanations and "constructions" often turn out to be borrowed directly from his predecessors (we shall see more than one example of this in the present chapter).

But Hegel quite sincerely believed that there was no possible connection between these attempts, which he considered more or less haphazard, and his own, which began from a unique and well-defined principle.[36] Now, curiously enough, these vague attempts remained less vain than Hegel's coordinated system. In fact, whatever may have been said about it, the *philosophy of nature* was not entirely fruitless. Oersted, the Dane who discovered electromagnetism, was surely a perfectly authentic *philosopher of nature*. And although, at the mere announcement of the new phenomenon, its true nature and principal laws were immediately established by the superior genius of Ampère, Oersted nonetheless deserves considerable credit. Similarly, Schönbein, in his research culminating in the discovery of ozone, seems to have followed a line of thought connected to the *philosophy of nature*. Finally, it would seem that the philosophy of nature can legitimately claim a notable share in a much more important scientific enterprise than those we have just mentioned, namely in J. R. Mayer's work on the principle of the conservation of energy. For anyone at all familiar with the scientific writings of the school who reads through Mayer's two short treatises and observes the way he stresses the universal relationship between forces and the close link between the inorganic and organic worlds, as well as his completely and boldly aprioristic way of deducing his statement of the principle from the equivalence of cause and effect, there could be no doubt: this is *philosophy of nature* – applied, obviously, with an infallible scientific instinct, where the argument was in its proper place, namely, to the deduction of a principle of conservation – but unadulterated *philosophy of nature*. That Mayer is not generally considered to be associated with these doctrines stems from the fact that he came to them rather late; in 1842 the vogue of philosophic deductions in science had quite clearly passed, as the poor author soon found out. His discovery was misunderstood and, characteristically, in the attacks to which Mayer was subjected, even in an epoch when the principle of the conservation of energy was already universally recognized, the accusation of being a speculative philosopher or metaphysician always played a large part.

If we now ask ourself to what current within the *philosophy of nature* these discoveries belong, there does not appear to be any doubt as to the reply. Indeed, it seems to be quite difficult to establish any link whatever between the work of these three scientists and Hegel's scientific theories. Their efforts appear to have been directed rather by considerations founded on simple analogies, a subject on which Hegel expresses himself with such severity (see Appendix 10). However it may seem at first glance, this fact is not at all paradoxical, for the analogies between nature and our mind are manifold, and to a man whose scientific instinct is sufficiently vigorous, insights of this kind can suggest the existence of highly important relationships. But once again these results only further point up the basic sterility of the Hegelian theory.

Perhaps it will not be superfluous to illustrate this surprisingly troublesome characteristic by a particular example. We choose it, not in the physical sciences where, as we saw by the example of the magnet, Hegel was obviously at odds with the convictions of his time and where the two directions were therefore too divergent for any mutual influence to have been possible, but in the field of biology, where theories and methods were still not at all firmly established. Here is how Hegel explained what we today call the infection of wounds: "Through its breathing, the organism is ... in conflict with air as it is with all the Elements generally. A wound, for example, only becomes dangerous through exposure to air."[37] It is perfectly clear that Pasteur could never have arrived at the theories with which we are familiar by starting from such a conception.

However, let us try to understand from a more general perspective why Hegel's attempt could not succeed, any more than could Descartes's for that matter. To this end, let us resolutely strip the German philosopher of all the prestigious metaphysical apparatus he is able to use to such advantage and consider his effort in the light of notions provided by the examination of science past and present. At first glance it might seem that Hegel undertook his venture under rather favorable conditions. Descartes – as attested by all his arguments from one end of his scientific work to the other – admits only an entirely rational physical world. Everything must be rigorously deducible from the premises he posited, which themselves appeared to him as necessary. It is from these "principles alone" – or, as he says elsewhere, "from first causes," from "germs of truth which are naturally existent in our souls,"[38] that Descartes claims to "deduce the explanation of all phenomena, that is of the effects that are in nature and that we perceive by the mediation of our senses" (*Principes* III, 1), and he announces that he has actually succeeded in drawing from them "the heavens, the stars, an earth, and even on the earth, water, air, fire, the minerals and some other such things, which are the most common and simple of any that exist, and consequently the easiest to know."[38] Of course, guided by his powerful scientific instinct, he does leave a place for experiment, but it is a really negligible one.[39] Hegel, at least in theory, is somewhat less trenchant in this regard than Descartes. No doubt he considers nature to be accessible to aprioristic reason, just as the latter did. But he thinks that reason must aspire to know only the broad outlines of nature in this way. He admits that "the Idea of Nature, when parcelled out in detail, is dissipated into contingencies" which "are not determined by concrete reason but by sport and adventitious incidents" (*Enc.*, *Logik*, 6:24 [Wallace 26]). There is here, on the part of nature, an "irrational way of playing up and down the scale of contingent quantity between the moments of the notion" (*Phänomenologie*, 2:207 [Baillie 308]). Similarly history contains much that is accidental, many historical facts impossible to justify (cf. Berthelot, 'Sur la nécessité' 132). Philosophy must abstain from dealing with these contingencies, on pain of losing all credibility.[40] Thus for him it is "a matter of complete indifference which bodies manifest magnetism" and, likewise, one must not wish to understand everything in geology; the historical element must be accepted there as a fact, "it does not pertain to philosophy." In general, it is necessary to forgo the desire to explain everything: "we must be content with what we can, in fact, comprehend at present." The *philosophy of nature* was bent on confronting all phenomena, but that is a mistake. It is the "finite sciences" that proceed in that way; since for them experience is the only thing that constitutes the guarantee for the hypothesis, everything must be explained. "But what is known through the Notion is clear by itself and stands firm; and philosophy need not feel any embarrassment about this, even if all phenomena are not yet explained" (*Naturphilosophie*, 7_1:252, 437, 93, 124 [Miller 166, 282, 62, 82]).

Thus Hegel means to leave a good deal of leeway for experimental science. He energetically rejects the idea that he wished to set himself in conscious opposition to it. That is a prejudice, he says (*Enc.*, *Logik*, Preface, 6:xiii). Speculative science, far from ignoring the empirical content of the other sciences, recognizes it and uses it; the empirical sciences prepare the content of the particular so that it can be included in philosophy (*Enc.*, *Logik*, 6:15, 20 [Wallace 16, 21–22]). Not only must philosophy agree with the experience of nature, but the birth and formation of philosophic science have empirical physics as their presupposition and prior condition (*Naturphilosophie*, 7_1:11 [Miller 6]). Of course he frequently takes to task the science of the scientists, sometimes not without reason – as when he points out that the concepts it uses, such as matter, force, atom, etc., are metaphysical concepts and that it is futile to try to exclude this metaphysics from science, as Newton had advised, though without following his own advice, for which he is to be highly commended (*Enc.*, *Logik*, 6:80, 83, 194 [Wallace 78, 80–81, 183]) – most often completely erroneously, not to mention interminably. However, in so doing, he means to *act* on this science, to bend it to his views. For instance he strongly accuses Kantian philosophy of having exercised no influence on the course of science. When the principles of this philosophy were set forth in scientific works, it was obviously only extraneous material which had nothing to do with the true content of the work (*Enc.*, *Logik*, 6:121 [Wallace 117]).[41] In 1848 Karl Michelet, in his preface to the *Naturphilosophie*, while noting that scientists had greeted Hegel's work with "a commiserating smile" (7_1:vii [Petry 180]), was still pleading for an understanding between scientists and philosophers, an understanding toward which he felt this book was making a step in the right direction (see Appendix 12).

Moreover, it is easy to see how seriously Hegel took his program for science by reading through the *Naturphilosophie*. This book contains a considerable mass of empirical facts from almost all fields, and the author must certainly have taken great pains to become so well acquainted with the principal writings of the leading scientists of his time. Newton (whose ideas dominated this science to a much greater extent than it does our own) is cited many times, and sometimes Hegel summarizes his mathematical deductions at great length (see for example *Naturphilosophie*, 7_1:111 ff. [Miller 56 ff.]). Similarly, at each step we find the names of contemporary physicists and chemists, especially French scientists, for whom Hegel appears to show some predilection, which is understandable, since the German science of the time was of rather inferior quality and Hegel apparently did not know English.[42] There are innumerable citations of Biot, from whom Hegel sometimes furnishes long extracts. But Lavoisier, Berthollet (especially the *Statique chimique*), Guyton de Morveau, Haüy, Francoeur, Pictet, Lagrange and Laplace are also frequently invoked. In addition, Hegel knows, at least secondhand, the works of Galvani, Volta, Dalton, Davy, Berzelius, Wollaston, Rumford, Chladni, Tortini, the abbé Conti, Richter, etc. In the biological sciences

French names are less predominant and the works on which Hegel relies for his documentation are chiefly German: Treviranus, Willdenow, Link, Schulz and others. Yet there too Cuvier, Bichat, Jussieu and Lamarck, all generally mentioned with high praise,[43] occupy a considerable place.

Thus Hegel's lack of success cannot be attributed to hostility toward science on principle, nor even to ignorance of the results achieved by the science of his time.[44]

Furthermore, if we examine more closely the position from which Hegel begins the whole of his deduction, we are forced to admit that, from the theoretical point of view, it has the advantage over that of his predecessors and of Descartes in particular. Indeed, Hegel, unlike the latter, admits the existence of an irrational element in the physical world. For, from the standpoint of scientific explanation, this admission is obviously what the theory set forth in the *Logic* (whose main features we tried to sum up in Chapter 5, pp. 104 ff.) amounts to. According to Boutroux's excellent formulation, "Hegelian logic wants the irrational, with the antinomies it engenders, to be the condition, the father of the concept" (Berthelot, 'Sur la nécessité' 146).

The element of contradiction whose presence Hegel establishes in each application of the principle of identity, that is to say in all reasoning and particularly in all scientific reasoning, is certainly only an initial manifestation of the irrational. It is the discovery that *diversity* exists, while our reason would prefer that there be none. Of course, Hegel asserts that this reason is not the right one, that it is only abstract reason and that above it is another reason, concrete reason, which understands this diversity perfectly and makes use of it in order to explain things. We shall see shortly the true implications of this last claim. But what can be seen immediately is that by enlarging the scope of his explanation in this way, Hegel includes at least something of the irrational element constituted by the existence of diversity.

Thus what we conceive as irrational is divided into two parts for Hegel. One part is declared to conform to reason in a broader sense, to concrete reason; that part is *reasonable* (*vernünftig*) and can therefore be deduced from the essence of this reason itself. On the other hand, the rest is dismissed as outside the limits of this deduction; it is undoubtedly knowable, but it is knowledge of an inferior sort, left for experience, *raw experience*, and excluded from what Hegel takes to belong to the domain of true science (*Wissenschaft*), which for him (just as for Aristotle) includes only what is completely dependent upon deduction.

Let us note in this connection what is at first glance a somewhat unexpected parallel between Hegel's ideas and those of Newton. Newton had expressed the profound idea that diversity, whether in time or in space, could not be deduced from "blind metaphysical necessity" (Ch. 6, p. 152). Hegel, as we see throughout the *Naturphilosophie*, has read Newton's *Principia* extensively, and it is not impossible that he was inspired by the passage in question. But he may also have arrived at his observations independently, for in Newton as well we are dealing with an entirely a priori deduction. What is certain is that the diversity Hegel includes in the domain of his "concrete reason" is just like this Newtonian diversity.

Thus Hegel completely admitted the irrational, placing it within his principle of explanation itself. In spite of this fact, which one would think ought to give him a considerable advantage over his predecessors, he failed completely, fundamentally, as completely and fundamentally as it is possible to fail in this domain, and most certainly failed infinitely more completely than Descartes. How can we account for the enormity of this failure?

A particular circumstance that strikes us as soon as we embark upon an examination of the *Naturphilosophie* is that Hegel does not hesitate to appeal to direct sensation in his deductions. Thus in his electrical theory, the *odor* and *taste* of electricity are an important element (*Naturphilosophie*, 7_1:346–347 [Miller 224]). His theory of light is actually founded entirely on sensation. In fact, it is nothing other than the one presented by Goethe in his *Farbenlehre*.[45] Hegel subscribes completely to all the conclusions contained in that work, lavishing the most dithyrambic praise on its author and even improving on Goethe's invectives against Newtonian optics. It is impossible to pass too harsh a judgment on the "conceptual barbarism" of Newton. Newton observed and experimented with "ineptitude and incorrectness" and his exposition is marked with "dishonesty." Newton's theory is only "metaphysical hotch-potch." In approving of him science shows "blind prejudice," "thoughtless inconsistency" and "stupidity" (*Naturphilosophie*, 7_1:303 ff. [Miller 198 ff.]).[46] This attitude of Hegel's (apart from the exaggerations, of course) is quite understandable. We saw, in fact, that if scientific theories exclude sensation from their domain, it is not in virtue of an explicit decree, but as the result of a tacit decision implied by the foundations of mechanism. As soon as one denies the supremacy of mechanism, as Hegel did, the exclusion of sensation is no longer justified in any way. Indeed, Hegelian idealism understands nature exclusively as what appears in consciousness and for consciousness; the nature that actually exists will have to possess all the properties sensation attributes to it and, in particular, the specific qualities science attributes to the action of the sense organs.[47] The opposite procedure, in Hegel's opinion, constitutes a serious shortcoming of empirical science, and here, of course, he is not entirely wrong, at least in a certain sense, since that involves the admission of an irrational, which is to say, in Burnet's language (Ch. 6, p. 150 above), a sort of "suicide" of reason. For Hegel, who fails to recognize the need for such a sacrifice, philosophy (of which true science, the *philosophy of nature*, is only a branch) agrees with the common understanding on this point and is entirely opposed to the theory of the physicists. "What is concretely actual is not something spatial, such as is treated of in mathematics. With unrealities like the things mathematics takes account of, neither concrete sensuous perception nor philosophy has anything to do" (*Phänomenologie*, 2:34 [Baillie 103]). Hegel, unlike Kant, considers space and time to be attributes of things themselves. However, anything relating to these considerations he finds of only minor importance.

> To those narrow-minded enough to attribute a quite singular importance to the question of the reality of space and time, it is fitting to reply that space and time are extremely poor and superficial determinations and that objects, in possessing these forms, thus possess very little .... Thought which seeks knowledge does not stop at these forms.[48]

Obviously, in virtue of these principles, Hegel found himself a thousand leagues away from science, for which nothing is more essential than temporal and spatial determinations. And one sees at the same time that if he theoretically has the advantage over Descartes by his (at least partial) admission of the irrational element of *diversity*, he is, on the contrary, greatly inferior insofar as he fails to admit the irrational element of sensation.

But this shortcoming of the Hegelian theory is not the only one, nor perhaps the most important. And first of all it should be noted that Hegel himself practically nullified any possible advantages his irrational might provide by the fact that, having included it in his explanation, he therefore had to consider it explicable or *reasonable*. He *reasoned* about this irrational, reasoned exactly as if it had been rational (for when all is said and done there is only one way of reasoning). Thus he rationalized it, disciplined it so to speak, attempting through pure reasoning to assign it its share and its limits.

The point we have just made can be recast in a slightly different form that will perhaps show more clearly the relation of Hegel's attempt (a relation of opposition to be sure) to tendencies in contemporary science. In Chapter 6 we saw that science has encountered or is in the process of encountering a whole series of these *irrationals*. Is it possible to *locate* Hegel's irrational with respect to them? At first sight the task does not seem to be an easy one. Above we found an analogy between Hegel's thought and that of Newton, which suggests that what the former would have allowed might be the diversity that found its expression, on the one hand, in the "improbable state" we must posit as the initial state of the world if we admit the mechanistic interpretation of Carnot's principle according to Boltzmann and Maxwell, and, on the other hand, in Perrin's absolute magnitudes of atoms. But that would certainly be to take the wrong tack and to attribute to Hegel's scientific thought a precision it is far from offering. Yet his expositions are not without occasional quick insights showing how deeply he reflected on the obstacle faced here by scientific reason (*abstract* reason, of course, in the Hegelian nomenclature). For example, speaking of the Newtonian demonstration of Kepler's laws, he states that

> mathematics is altogether incapable of proving quantitative determinations of the physical world in so far as they are laws based on the qualitative nature of the moments [of the subject matter]; and for this reason, that this science is not philosophy, does not start from the Notion, and therefore the qualitative element, in so far as it is not taken lemmatically from experience, lies outside its sphere.[49]

Leaving aside the reference to philosophic deduction through the notion, it is clear that some of his views in this passage are quite sound, especially concerning the essence of physical magnitudes (we shall say coefficients) which translate into quantity what in nature is qualitative, as well as concerning the impossibility of deducing these coefficients completely (since it would be necessary to deduce them from the undifferentiated), which means that scientific explanation must definitively forgo explanation of the qualitative (at least insofar as it cannot be deduced from pure arrangements in space – but Hegel does not believe in such spatial deduction). However, once again, all that is only a sort of ephemeral insight. Hegel does not follow it through, for the simple reason that he would certainly find it quite unnecessary to study in depth the methods applied by explanatory science, a subject he finds entirely unworthy of the philosopher's serious attention.

Therefore, for the most part, he envisages diversity in the most general fashion, being content simply to posit that it is *other* (*Anderssein*). That being the case, then, it is this irrational that characterizes all our reasoning without exception, as we have seen, or, if one prefers (and whatever Hegel himself may think), the irrational such as it appears in mathematics. Now that is a side of the irrational with which physical science does not even deal. Physical science does not have to deal with it since it concerns mathematics, since it is included in mathematics, so to speak, and since reason, by the very fact that it uses mathematics for the explanation, the rationalization of the physical sciences, shows that it has, as it were, *passed beyond* this first stage.

But in order to get a better grasp of the difference between the standpoint of science and that of Hegel, we must look more closely at the philosopher's attitude toward the mathematical sciences.

The first thing that strikes us is the psychological fact that Hegel's mind seems astonishingly closed to mathematical explanation of physical phenomena. In the case of Kepler's and Newton's discoveries, he claims to have applied himself to these matters for twenty-five years (*Naturphilosophie*, 7_1:100 [Miller 67]), and it is certain that he must have sometimes taken great pains to understand, even if only superficially, certain deductions of mathematical physics in order to summarize them as he did. But it is just as certain that he never truly grasped their spirit or (what amounts to the same thing) that these deductions never really spoke to his own spirit. Not only was he often puzzled by their finer points, as he himself confesses with regard to the decomposition of the motion of the planets (*Naturphilosophie*, 7_1:112 [Miller 75]), but these deductions always appeared to him to be something artificial, something entirely external to the true essence of things: "the mathematical exposition" is "a tortuous one" (*Naturphilosophie*, 7_1:90 [Miller 60]).

We saw in Chapter 5 (pp. 111 ff.) his observations on mathematical demonstrations and recognized that his objections contained something sound and moreover inevitable, inherent in the very character of mathematics and in the nature of the services it must render (of which Hegel himself was perfectly aware). But Hegel concludes that "The process of mathematical proof does not belong to the object; it is a function that takes place outside the matter in hand." In general, for him, "in mathematical knowledge the insight [*die Einsicht*] required is an external function so far as the subject-matter dealt with is concerned. It follows that the actual fact is thereby altered." For example, in a proof a triangle is "taken to pieces." In mathematics "the process of knowledge goes on, therefore, on the surface, does not affect the concrete fact itself, does not touch its inner nature or notion, and is hence not a conceptual way of comprehending" (*Phänomenologie*, 2:32–34 [Baillie 101–103; Meyerson's brackets]).

The following example will make clear how far this intellectual bias extends. After summing up, on the whole correctly, the deduction of the laws of falling bodies (citing Lagrange's exposition in the *Théorie des fonctions*), he declares that the separation of the two motions "has no reality and is merely an empty fiction" conceived solely for the mathematical representation. Now the law of falling bodies is a *free* law of nature and therefore conceals an aspect that must be determined by the *notion* of body. Galileo's law according to which the spaces traversed are proportional to the squares of the times elapsed must in particular be deducible from this determination of the concept. It would be completely impossible for us to exhibit the process of this deduction without penetrating into the very depths of Hegelian metaphysics; for our present purposes suffice it to say that Hegel justifies the introduction of the square of the time by the fact that this is

> magnitude as coming outside itself, raising itself into a second dimension and thus expanding itself, but solely in accordance with its own determinateness; in this expansion it sets its own self as limit, and thus in becoming an Other, it is related solely to itself.[50]

Deductions of this sort are what Hegel means to substitute for the one adopted by science since Galileo. After that it will not surprise us to discover that he finds the Newtonian deduction of planetary motion *imbecilic* [*läppisch*] and that he declares it inconceivable that the celestial bodies are pulled "this way and that," given that they must, on the contrary, move like free gods (*Naturphilosophie*, 7_1:96–98 [Miller 64–66; Meyerson's brackets]).

Thus mathematical deduction is inapplicable when it comes to explaining physics. "The distinctions and determinations brought forward by mathematical analysis, and the course it has to follow in accordance with its method, are wholly distinct from what is supposed to have a physical reality" (*Naturphilosophie*, 7_1:100 [Miller 67]). This is an altogether fundamental conviction for Hegel; it is firmly established as early as *De orbitis*,[51] and he never seems to have varied in that respect.

The first explanation that might occur to one for this curious "psychological case" is that of a total inaptitude for mathematical reasoning in general. And it is highly likely that there was indeed something of the kind in Hegel, a sort of innate lack of mathematical sense, perhaps aggravated by an education short on the mathematical sciences,[52] leaving gaps that could never be entirely filled by subsequent efforts. There certainly must have been some such anomaly in the mind of a man for whom the concept of the power of a number was not justified in itself but who felt the need to deduce it in the following way:

> The quantum as an indifferent determinateness undergoes alteration; but in so far as this alteration is a raising to a power, this its otherness [*dies sein Anderssein*] is limited purely by itself. Thus in the power, quantum is posited as returned into itself; it is at once its own self and also its otherness. (*Wiss. der Logik*, 3:390 [Miller 322; Meyerson's brackets])

Obviously this is no longer a question of mathematics applied to physics, as in the analogous case of the square of the times for the law of falling bodies, but of pure mathematics, and thus we find the argument even more shocking.

However, it would certainly be misleading to overestimate the role of this turn of mind, be it innate or acquired. For Hegel was not at all a mathematical ignoramus. During his professorship at the University of Jena he offered a course in pure mathematics (arithmetic and geometry) three times and seems to have actually taught it twice. In his teaching he followed standard texts;[53] therefore one must assume that he was really teaching arithmetic and geometry in the usual sense of the terms and not philosophy of mathematics. As far as one can surmise, it was probably only elementary mathematics, but that still attests to a thorough knowledge in this field. Moreover, at many points in his work Hegel avails himself of the proofs of pure mathematics, often understanding them quite well. He has – especially in the two *Logics* – many pages filled with mathematical formulas and proofs borrowed from the best authors and accurately summarized; there are also a great many extremely interesting remarks, particularly on the infinitesimal calculus (he is even familiar with its history, citing Descartes, Fermat, Cavalieri, Barrow, Euler, Newton, Lazare Carnot, Lagrange).[54] Of course not everything he includes is correct – at the present time the theory of the infinitesimal calculus is still not completely established, even among the most authoritative mathematicians – but the general impression is clearly one of a very sustained and largely successful effort at comprehension. The contrast between these expositions and those devoted to applied mathematics, to mathematical physics, could not be more striking: there is unquestionably some kind of absolutely clear dividing line in Hegel's mind on this issue.

We have already, in discussing the Hegelian conception of the principle of identity (Ch. 5, pp. 106 ff.), alluded to the particular place mathematical reasoning holds for Hegel. It appears to him to be fundamentally different from the reasoning followed in all other branches of human knowledge, and particularly in physics. Everything we have said on the subject of the path followed by *concrete reason* actually applies (within the scientific domain, obviously) only to the physical sciences. As for mathematics, it is governed solely by abstract reason, whose principle is pure and simple identity, and in mathematical knowledge it is this identity that conditions the passage from one determination to another. One confines oneself above all to determinations of magnitude, putting aside all others, and so it is that geometry compares figures by showing what is identical in them, that it declares, for example, that a triangle and a quadrilateral are equal in size.[55]

Generally, in mathematics "knowledge advances along the lines of bare equality," and "herein consists the formal character of mathematical evidence." But of course herein also lies its chief defect.

> For what is lifeless, not being self-moved, does not bring about distinction within its essential nature; does not attain to essential opposition or unlikeness; and hence involves no transition of one opposite element into its other, no qualitative, immanent movement, no self-movement. (*Phänomenologie*, 2:35 [Baillie 103])

It must also be added that although Hegel sometimes deals at length with certain mathematical concepts in the two *Logics*, anything having to do with the internal mechanism of mathematical demonstration is treated more or less in passing and without great rigor. This is apparently because the way thought proceeds there seems almost insignificant to him, at least in comparison to the *dialectical* process, which is the one he believes conscious thought must use in order to arrive at results having a real value from the standpoint of comprehension.

Mathematical reasoning must not be imitated in any other science; in particular "in the concrete philosophical sciences philosophy must take the logical element from logic, not from mathematics" (*Wiss. der Logik*, 3:250 [Miller 216]). "The evidence peculiar to [mathematics] ... rests solely on the poverty of its purpose and the defectiveness [*Mangel-*

haftigkeit] of its material, and is on that account of a kind that philosophy must scorn to have anything to do with" (*Phänomenologie*, 2:34 [Baillie 102; the German was inserted by Meyerson]).

Certainly Hegel's bias does not go so far as complete failure to recognize the existence of mathematical deductions in physics. But the fact that demonstrations such as the ones provided for the law of the lever, the laws of falling bodies, etc., are accepted by the mind "is itself merely a proof of how great the need is for knowledge to have a process of proof, seeing that, even where proof is not to be had, knowledge yet puts a value on the mere semblance of it, and gets thereby a certain sense of satisfaction" (*Phänomenologie*, 2:35 [Baillie 104]). But it is important to purge mathematics of these erring ways. Of course he heartily disapproves of Spinoza's and Wolff's attempts to give philosophy a mathematical form.[56] The thought process operating according to traditional logic could quite properly be assimilated to arithmetic calculation, but that is precisely what proves it to be a "spiritless" work (*Wiss. der Logik*, 3:39 [Miller 52–53]).

We need hardly point out the flagrant prejudice in this distinction between mathematical reasoning and all others. How does Hegel, after speaking of *dialectic* with reference to all concepts, propose to exempt mathematical conceptions? In any case, one can to some extent remove the external contradiction this attitude seems to entail. Indeed, Hegel admitted the existence of an inner contradiction in the mathematical proposition. He may then have meant to refer only to this contradiction (following the definition we suggested in Chapter 5, p. 107) in speaking of *dialectic* with respect to reasoning in general, so that the *dialectical method* properly speaking would indeed remain limited to extramathematical reason. Let us add, however, that Hegel's text does not always permit this interpretation; we shall see shortly, through the revealing testimony of a commentator, that in the *Logic* in particular the dialectical method is unquestionably sometimes applied to mathematical concepts. But even if it were possible to render the philosopher's position internally consistent, its foundation would remain no less paradoxical.

So true is this that one of Hegel's direct disciples, and one might even say his most faithful disciple, the one who became his biographer and apologist, who consecrated the best part of his career to celebrating the glory of the master in work after work, Karl Rosenkranz,[57] could not entirely adopt his viewpoint on this subject. Wishing to explain a point of Hegelian logic more clearly, Rosenkranz chances to use an example borrowed from mathematics. He immediately recalls that Hegel intended to place mathematics under the jurisdiction of abstract reason alone. He continues:

But when, as Hegel affirms, truth can become certain of itself only in the form of dialectic method; when, further, according to him, mathematics forms a necessary member in the system of science; when, finally it is the conception of space with which the idea as nature first found its existence, – it is hard to see why mathematics should be an exception to all other content. That it never has been, is no reason why it never should be treated dialectically. The conception of the one, of quantity, etc., i. e. of arithmetic, Hegel has already presented dialectically in the first part of his *Logic*: why should geometry dispense with the dialectic?[58]

Thus Hegel's position presents a genuine anomaly on this point.

It is understandable, psychologically speaking, how this anomaly must have forced itself upon the mind of the philosopher, once the foundations of his theory of science had been laid. To be sure, neither the arithmetician nor the geometer actually reasons according to the schema of concrete reason and, moreover, thanks to the basic simplicity and pellucidity of mathematical reasoning (at least compared to that used by the physical sciences), it is not possible to entertain any illusions on this subject: we see clearly that geometry (in line with what Hegel himself recognized in the passage we quoted from the *Logic*), while it is aware of the diversity that exists, simply sets it aside, reasoning as if it did not exist and obviously obeying the rules of the old logic, that is, the logic which conforms to the dictates of abstract reason – or, in the words of the *Phenomenology*, "the formal character of mathematical evidence" consists in the fact that it proceeds "along the lines of bare equality" (2:35 [Baillie 103]; cf. p. 284 above). Therefore mathematical reasoning *cannot* be essentially analogous to that of physics; there *must* be a break between them that differentiates them profoundly – this is in fact the point of view Hegel finally adopted.

If we now try to delve still deeper and ask ourselves what Hegel's source could be for this epistemology, an epistemology which would thus be psychologically prior to his logic, we shall have to advance further into the realm of conjecture. First of all, we can see that Hegel recognized full well that explanation, as used by the physical sciences, rests on persistence, identity or, as he says, on "the quest for what is similar and the continued pursuit of it" (*Naturphilosophie*, 7_1:621 [Miller 398]). For example, chemical theory presupposes the substantial and unalterable diversity of the elements. "Where higher transitions occur in these finite processes, where, for instance, water is solidified in a crystal, where light and heat vanish, and so on, Reflection has recourse to the nebulous and meaningless conceptions of solution, of becoming bound or latent, and the like." These explanations are the work of abstract reason and, of course, a reprehensible work. He says in another passage that "it is only in the Abstract Reason's 'system of identity' that matter perdures .... Still, empirical physics has no right to credit the imperceptible with existence; and if it wishes to proceed purely empirically, it must admit that matter passes away." But it is also a work that cannot succeed; the fixity of the elements constitutes a "fundamental defect" of scientific theory and "it is on this immediate transition and transformation" of substances "that all chemical and mechanical explanations founder."[59] The facts Hegel cites in support of his thesis appear utterly fantastic to us today and no doubt must have seemed so to the sound minds of his time formed in the school of Lavoisier and his disciples. For example he does not believe in the indestructibility of the chemical elements. Air is transformed directly into water and vice versa: "the rain comes, so to speak, from dry air."[60] Chemistry is incapable of explaining such a transformation, for it would have to admit that nitrogen is transformed into hydrogen, instead of considering these two substances immutable. Similarly, oxygen and hydrogen "are turned into water by an electric spark; but water is not composed of them." Meteorites are formed of clouds. "Water on which polyps ... feed is directly turned into lymph, jelly." Thrushes fatten in a single foggy morning by "a direct transformation of this moisture into animal matter."[61] This does not prevent Hegel's views on the inmost nature of mechanistic explanation from being as accurate as they are profound. One would almost dare say too accurate. Indeed, since he has unlimited confidence in the power of thought and the infallibility of the logical sequence; since, on the other hand, he fails to recognize the explanatory force of the "power of grouping"; and since, finally, he clearly perceives that scientific explanation proceeds and can only proceed in seeking the cause or the reason, in equating it with the effect and in thus progressing from identity to identity, this whole work seems to him to be an immense tautology.

Because of this identity of the reason and the consequence, ... the reason is *sufficient* ...; there is nothing in the reason that is not in the consequence, and there is nothing in the consequence that is not in the reason. When we ask for a reason, we want to see the same determination that is content, double ... the assignment of a reason remains a mere formalism and empty tautology .... Such an assigning of reasons is therefore accompanied by the same emptiness as the talk which restricts itself to the law of identity. The sciences, especially the physical sciences, are full of tautologies of this kind which constitute as it were a prerogative of science.[62]

By way of example, Hegel cites the explanation of the earth's motion by means of a force of attraction acting between the earth and the sun, as well as the supposition of arrangements of molecules assumed to give rise to different crystalline forms. "In ordinary life, these aetiologies, which are the prerogative of the sciences, count for what they are, tautological empty talk." So empty is it that "in this way, knowledge has not advanced a step; its movement is confined within a difference of form which this same procedure inverts and sublates." For that reason a well constituted mind has difficulty conforming to this way of reasoning.

One of the main difficulties in the study of the sciences in which this method prevails comes from this perverse method of premising as reason what is in fact derived, and then actually placing in the consequents the reason of these supposed reasons .... Therefore he who aims to penetrate such sciences must begin by instilling his mind with these reasons, a distasteful business for concrete reason [*Vernunft*] because it is asked to treat as reason what has no reason.[63]

In this way one creates confusion and comes to express oneself on hypothetical concepts – such as molecules, interstices, centrifugal force, the ether, the light ray, electrical or magnetic matter, etc. – as if they belonged to direct perception; it is a "witches' circle." Nevertheless, one is admitting that one does not know the inner essence of these forces or these substances. "This amounts," concludes Hegel," only to a confession that this assigning of reasons is itself completely inadequate."[64]

It is obvious that in all the preceding the term *reason* (*Grund*) is understood as synonymous with the term *cause* in the sense science generally uses it – which is also, as we have seen, perfectly consistent with the way it is used by Spinoza and Leibniz. But here, in addition, is a passage in the same work where cause is spoken of in entirely analogous terms:

Effect contains nothing whatever that cause does not contain. Conversely, cause contains nothing which is not in its effect .... The self-identity of cause in its effect is ... the unity which is indifferent to the differences of form, that is, *content* .... Through this identity of content, this causality is an analytic proposition .... Since these determinations of form are an external reflection, it is, in point of fact, the tautological consideration of a *subjective* abstract reason [*Verstand*] to determine a phenomenon as effect and from this to ascend to its cause in order to comprehend and explain it; it is merely a repetition of one and the same content; there is nothing else in the cause but what is in the effect.[65]

But perhaps the strict relationship between these thoughts will become still clearer if we use a precise example. Let us consider chemical reaction, a subject on which the reader is sufficiently acquainted with Hegel's views through our many citations. Chemists *explain* reactions by assuming immutable elementary substances. But what can be the use of an "explanation" of this sort? As a matter of fact, it is clear that if it could be a real, a complete explanation, it would succeed in indicating the source of *all* the properties of the products resulting from the reaction. Now it could do this only by stating that nothing was created, nothing was lost, that everything consequently preexisted. Therefore there would have been no reaction at all – which is absurd, for if there had been no change, chemists would not have worried about an explanation. Is it not then *simpler* (a term Hegel uses several times in similar contexts) to assume that air is transformed directly into water and water into nitrogenous organic lymph? For nature surely must be in conformity with our reason, *reasonable*. Would it be reasonable for nature to have maintained elementary substances, thus paving the way for the semblance of an explanation, when this explanation can only lead to absurdity? Is it not infinitely more probable that this is not the case and that all the alleged demonstrations of the chemists are merely apparent?

We need hardly warn the reader that there is a measure of hypothesis in our attempt to reconstitute the filiation of Hegel's ideas on chemical reaction. We cannot provide textual corroboration for each phase of this thought, as is our usual practice. We hope nevertheless that those who have followed our exposition of the philosopher's general ideas will be willing to acknowledge that the hypothetical portion is at least based on strong presumptive evidence.

Is it too far-fetched to see these deductions on the futility of explanatory science as it is practiced by scientists themselves – ideas that were thus largely based on the fact that he had really laid bare, with admirable insight, the deep logical foundations of sciences – as the starting point for Hegel's epistemological conceptions? We do not, of course, exclude properly metaphysical motives that might push him in the same direction, but they are outside the area that interests us here. What we mean to suggest is that, judging the explanatory effort of science to be futile, he may have concluded that its work as a whole could only be a fundamental error, a formal aberration of the human mind, and that it was therefore necessary to seek genuine explanations in a completely different way, indeed even to leave phenomena entirely unexplained.

In support of this supposition, let us note that competent critics have long recognized that Hegelian logic, contrary to the claims it so audaciously flaunts, is far from being a spontaneous development of pure reason, but seems, on the contrary, directly inspired by the use its author meant to make of it. "One can legitimately doubt," says Trendelenburg,

> that these alleged logical categories are purely logical and that they are nothing more than the products of thought relying strictly on itself and drawing only from within itself. Hegel's *Naturphilosophie* has been considered an applied logic given that in it the abstract categories of the *Logic* reach the concrete state. But it is the opposite that takes place. The *Logic* ... is in many places a purified perception, an anticipated abstraction of nature. (*Log. Untersuch.* 1:68)

Let us point out, furthermore, that the chronology of Hegel's work, although it does not directly corroborate our hypothesis, in no way contradicts it. As a matter of fact, our citations from *De orbitis planetarum* (cf. p. 306, notes 50 and 51) show that the very foundation of his epistemological position, especially the conviction that mathematical deduction must be entirely separated from physical explanations, dates from the very beginning of his philosophic career. It is perhaps unlikely that his logical conceptions were entirely formed as early as that, however.[66]

Wherever the truth lies in these conjectures about Hegel's attitude toward science and epistemology, the attitude of science and epistemology toward Hegel's theories appears clear enough. What must be retained from the Hegelian *process* is, first of all, the very important fact of the existence of a contradiction, an antinomy, in every instance of reasoning; and then it is the way in which reason goes beyond this conflict, putting it aside, so to speak, since, on the whole, although we are aware, if we ask ourselves, that there is a contradiction, the rest of the time we do not hesitate to act as if it did not exist, as if everything were perfectly rational. And Hegel was also correct in believing that this process of *going beyond* is indispensable for our scientific knowledge, that it is applied quite generally in the domain of science. But this generality is even greater than he supposed. For there is more uniformity in the procedures of our understanding than follows from his system. Indeed, for him the *going beyond* took place by means of two essentially different procedures: on the one hand – in mathematical reasoning – by simply setting diversity aside and, on the other hand – in all other reasoning, including that of the physical sciences – by the "dialectical" process properly speaking, which enriches the concept by means of negation. Now, such is not the case, and our reasoning – mathematical, physical or otherwise – is entirely analogous in this respect. For the alleged "dialectical" process actually amounts quite simply to the same *setting aside* that Hegel did indeed observe in mathematics but disdained, as it were, declaring it unsuitable for application elsewhere. In reality it is to mathematical reasoning that science conforms, or at least tries to conform, in all its parts.

The Hegelian process, which consists in arriving through negation at a higher affirmation, is basically quite mysterious, as Höffding astutely pointed out. For Hegel, negation, being *determined*, must have a content; it constitutes "a fresh Notion but higher and richer than its predecessor; for it is richer by the negation or opposite of the latter, therefore contains it, but also something more, and is the unity of itself and its opposite" (*Wiss. der Logik*, 3:42 [Miller 54]). But as soon as one conflict has been resolved in a higher synthesis, the dialectic opens another one, which must in its turn be resolved in an analogous way. Thus, as McTaggart aptly remarks (*Studies* 1), thought, according to Hegel, "advances, not directly, but by moving from side to side, like a ship tacking against an unfavorable wind." This method of reasoning (while admitting of improvements) is, moreover, the only correct one, for it is one with its object and its content, it is the content itself and the dialectic which is in it (that is, the negative element it conceals) that make it progress. This is how "cognition rolls onward from content to content." It begins by simple determinations, while the following ones become richer and richer; "by its dialectical advance it not only does not lose anything or leave anything behind, but carries along with it all it has gained, and inwardly enriches and consolidates itself." "It is clear," concludes Hegel in one of the passages where he sums up his views, "that no expositions can be accepted as scientifically valid which do not pursue the course of this method and do not conform to its simple rhythm, for this is the course of the subject matter itself" (*Wiss. der Logik*, 3:42, 44; 5:336, 349 [Miller 54, 56, 830–831, 840]).

Now in reality, as Höffding says,[67] in denying negation by positing *not*-(*not-A*), the "negation of negation,"[68] one returns quite simply to affirmation, that is, to *A*, instead of arriving at a new content *B*. Höffding therefore concludes with overwhelming soundness: "the triad (*position, negation, higher unity*) is only a *schema* into which he [Hegel] presses the empirical content more or less arbitrarily." As a result, the empirical content, which comes to reason from outside, from sensation, from the external world, remains, in spite of everything, entirely foreign to its essence.

In order to be convinced of this one need only observe the cruel quandary of the commentators when it comes to actually justifying the way one concept gives birth to another, through dialectic, in the *Logic* or the *Philosophy of Nature*. The most conscientious of them are content to follow anxiously and step by step the working of Hegel's own reasoning (cf. p. 302, n. 26), so unsure are they of having grasped its true content. Benedetto Croce, whose great admiration for Hegel we have already noted, is more candid. He admits that it "is not easy to hold Hegel's *Logic* in one's mind, unless recourse be had to learning it mechanically: for there is no necessary generation of its successive parts from one another."[69] If Hegel's claim of having found a method of reasoning "that is one with the object and its content" had the slightest basis, surely, once the relationship of the concepts has been pointed out, it ought to engrave itself indelibly on the mind of anyone reading him in good faith. Now we see that this is not at all the case, and we are therefore forced to the conclusion that Höffding was correct and that the alleged deductions are in fact only clever prestidigitation introducing an empirical content more or less arbitrarily into the deduction (Höffding, *Histoire* 2:184 [Meyer 2:181–182]; see Appendix 14).

Hegel imagined he had found a method, his dialectical method, that allowed him to deduce reality, or at least the essential part of it, starting exclusively from the concept. That is why he was able to consider it a logical shortcoming of mathematics that it "does not arise from the nature of the theorem" (Ch. 5, p. 111 and this chapter, p. 284) – a strange reproach for, as we have seen, it is only through precisely this "synthetic" character that mathematical deduction can serve us by creating something new. In the same sense he reproached Kant for declaring that attraction is part of the concept of matter, although it is not contained in that concept. Hegel maintains that "a determination which belongs to the concept of anything must be truly contained in it," and he adds that Kant's assumption to the contrary is only a "vain subterfuge [*leere Ausflucht*]" (*Wiss. der Logik*, 3:203 [Miller 180; Meyerson's brackets]). Clearly the stumbling block for the Hegelian process is that one does not understand how it can lead to something new, something the concept did not specifically contain. Indeed, that seems to have been the obstacle at which Hegel faltered.

Judging that *dialectical* deduction is applicable to the physical sciences, Hegel consequently believed that everywhere in this domain it was the *same* conflict that was being resolved in the same fashion. Now we know that this is not the case either. Science, in its patient and unceasing effort to make the real rational, continually runs into irrationality, but it is not always irrationality of the same sort, or, if one will, the same irrational. The irrational as Hegel conceives it is sufficient, strictly speaking, only to establish mathematics. If we want to proceed from there to mechanics, we encounter the new obstacle of transitive action; since it is a question of *rational* mechanics, that is, of an entirely deductive science, we can safely say that the obstacle, as for mathematics, is unique. But if we then wish to enter into physics, we have to list a whole series of obstacles, a series that we have seen can never be complete, since new anomalies may crop up where least expected. Likewise there will quite likely be some for the transition between physics and chemistry and, almost certainly, one or several that will someday characterize the life sciences when they have developed far beyond their present stage.

It is because he considered the obstacle that had revealed itself to him to be unique, because he was convinced that once this obstacle had been overcome reason would thereafter need only proceed with its work without having to fear any new impediment, that Hegel posited his concept of concrete reason. Now concrete reason, *Vernunft*, insofar as it is a principle higher than abstract reason, one that governs the course of our thoughts in a way essentially different from the latter, does not exist.[70] What Hegel adorned with this name is just plain reason (or abstract reason, *Verstand*, according to his nomenclature), after its first compromise with external reality. Reason does not overcome this compromise, it merely suffers it; it makes the best of it in one way or another because it cannot do otherwise, because it is forced to recognize that "the world is like that." That is to say that it does not in any way make this acceptance a new principle of understanding. On the contrary, its principles remain what they always were. Nowhere does this stand out more clearly than in the example Hegel himself chose as fundamental from the standpoint of his theory: becoming.

Hegel, as we showed above (p. 266), seeks to force our consent in this matter by pointing out that "everyone has a mental idea of Becoming and ... it is one idea" (*Enc.*, *Logik*, 6:174 [Wallace 165–166]). But it goes without saying that the second part of the sentence constitutes mere sleight of hand: Hegel does repeat at the end of the passage that the idea is *unique*, but nowhere does he attempt to demonstrate it. Furthermore, this would have been impossible, and Hegel basically takes advantage of the fact that the language designates this process by a particular term; but one need hardly point out that all languages possess terms for quite complex collections of concepts – terms such as *war*, *course* [*marche*] and countless others. He also obviously intends to take advantage of the fact that this process is *familiar* to us, which is true but irrelevant, for that does not prove that we understand it, that it is *rational*. Yet this pseudodeduction is what seems to have seduced many disciples. For example, Engels declares that

dialectics ... comprehends things and their mental images essentially in their connection, in their concatenation, in their motion, in their coming into being and passing away. For dialectics the processes referred to above [the phenomena of becoming] are just so many corroborations of its own method of procedure. Nature is the proof of dialectics.[71]

And quite recently a Polish Hegelian, in what is in fact an excellent presentation of the master's doctrine, gives a preeminent place to this same argument: "Reality is a continual transition of contrasts, their uninterrupted flowering in the midst of uniformity. *Reality is dialectic*."[72] But since the dialectic claims to *explain* nature, the attempt to prove it rational by invoking what occurs in nature obviously constitutes the elementary logical fallacy called a *hysteron-proteron*. That might lead us to suspect that there is, at the foundation of the Hegelian dialectic, nothing more than concepts deduced from reality and emptied of their content by abstraction, that is, in short, nothing truly rational. This would be a mistake, for, as we have seen, the Hegelian dialectic contains a substratum of rationality. But it is certain that in attempting to justify it as these disciples do, this relationship tends to become obscured. By dwelling on what Hegel had been content to suggest, they make the defects of the argument more obvious.

Trendelenburg had already subjected the Hegelian conception of becoming to rigorous criticism from a strictly philosophic point of view. "Pure being," says this philosopher,

is immobility; nonbeing, which is always equal to itself, is also immobility. How can mobile becoming arise out of the unity of two immobile representations? Nowhere in its preparatory phases is motion found preformed, without which, however, becoming would be only being. Given that both pure being and nonbeing express immobility, the most immediate task of thought, as soon as it must posit the unity of these concepts, can logically only be the following: to find an immobile union. If, on the contrary, thought creates something else in bringing about this union, it must be because it provides this different element itself and surreptitiously introduces the motion in order to put the being and the nonbeing in flux.[73]

The argument is not very original. All it does is develop what Parmenides had already pointed out. When the ancient Eleatic said *being* and *nonbeing*, he certainly meant to affirm that no third term could be slipped between the two and that consequently all becoming is irrational. Need we insist that the antiquity of the objection does not entail its senility? Indeed, we are on the immutable rock upon which eternal reason takes its stand, and it is only natural to see the epochs blend into one another.

So far is the objection from senility that we have only to follow the subsequent evolution of the Hegelian philosophy to see the curious spectacle of the Eleatic position triumphing within this school itself. To many readers, of course, considerations of this type will not appear very convincing as an argument against the Hegelian doctrine, since they do not apply directly to Hegel himself. Perhaps by way of compensation, however, they will particularly appeal to the minds of those who are accustomed to seeking evidence on the implications of a doctrine in its historical evolution; this is, moreover, a current of thought quite closely related to the Hegelian position, as we know. Now it seems quite significant, in this regard, to observe what the Hegelian doctrine became in the hands of one of his disciples, and certainly not the least of them, who belongs to our own era: we mean McTaggart, whose name the reader has already frequently encountered in the course of this work. This philosopher has the very great merit – one which sets him somewhat apart among Hegel's commentators and continuators – of having principally applied himself to what truly constitutes the foundation of the doctrine, namely, Hegelian logic and *dialectic*, and of trying not only to set forth the whole of the system from this point of view – which he does with rare depth, good faith and clarity – but also, as it were, to recreate it, and even to modify it wherever Hegel's thought does not seem to have remained sufficiently faithful to itself. One of the most important modifications he introduces thus has to do with the role of time. Time, for McTaggart, is not part of "ultimate reality"; the temporal process that characterizes apparent reality has its source in a "timeless state" of the universe, which is entirely rational, perfect, and to which judgments implying the existence of a past or a future, of a better or a worse, are therefore inapplicable. McTaggart does not hide the fact that these affirmations disagree with certain developments of Hegel himself, but he maintains that they follow ineluctably from the foundations of the system, since "the theory that the Absolute Idea develops in time" leads to "a hopeless difficulty."[74]

Anyone at all familiar with McTaggart's works will hesitate to declare him entirely unfaithful to the work of his master.[75] Is it not striking, then, to observe that the image of reality at which he arrives, this *perfect and timeless* universe (which is obviously identical to Parmenides' famous sphere), rigorously excludes all becoming? If we really judged things according to "concrete reason," if becoming appeared truly "reasonable" to us, or if Hegelian logic merely succeeded in making it appear so in our eyes, could one of those who best understood its motivating force conceivably have resigned himself to completely excluding it from his "rational universe"? Are we not forced to conclude instead that, according to Hegel's *Logic*, becoming remained for our reason what it was before, namely a stumbling block, an obstacle it seeks to brush aside or to evade in any way it can?

But this analysis can be confirmed and clarified using the procedures we have applied in this work and thanks to the results they have enabled us to attain. It is no doubt correct that we have an altogether clear idea of becoming. It is in fact a perfectly familiar phenomenon. But at the same time it is a phenomenon we know only from the outside and whose essence remains enigmatic to us. All of science, by its history as well as its present state, eloquently proclaims this. Scientific theories seek first and foremost to explain becoming – which would be contradictory, strictly speaking, if our reason could conceive of becoming itself as consistent with its postulates. The appearance and perpetual reappearance of theories of preformation in biology would then be a pure and simply enigma, as would be the prestige enjoyed in physics by mechanism, whose real aim, as Aristotle said, is to maintain the permanence of being. Furthermore, the evolution of physics in the nineteenth century offers a particularly valuable demonstration of this in the history of Carnot's principle. The lack of comprehension encountered by Sadi Carnot, the long neglect the proposition suffered after his death, the difficulty with which science accepted it, even after the work of Clausius (cf. IR 298 ff. [Loewenberg 266 ff.]), clearly prove how inconsistent it is with the rationality our mind would impose on nature. And the same applies to science's unceasing efforts (some of which we have mentioned) to free itself from the grip of the principle and to return to the concepts of conservation, either directly or by way of cyclical change. Finally we come to the theory of Boltzmann and Maxwell (Ch. 6, pp. 156 ff.). If recourse was had to such a laborious rationalization of the phenomena of becoming, and if this device incontestably satisfies our reason, it is because becoming is not in itself rational.

We know, moreover, that – Hegel to the contrary – this initial compromise is followed in science by a whole series of others. Indeed, each time there is an irrational, it can undoubtedly be recognized if we take the trouble to study the corresponding chapter of science; we recognize it especially in the passionate efforts that were made or are still being made to reduce it, in the multiplicity of theories that have sprung up around it, in the difficulty the textbooks have in dealing with the phenomena related to it. Any competent scientist in a specific area of science, if asked, will frankly admit that there is a problem at the point in question; he may not be aware of its permanent nature, he may believe, on the contrary, that it will be possible to clear it up later, but he will certainly not deny its existence. And yet the following chapter of science will nevertheless consider this obstacle, the obstacle of the preceding chapter, as nonexistent, so to speak, and the science of the preceding chapter will be treated, by implication, as entirely rational, since it serves as the rationalization for the following chapter. For example mechanics does not concern itself in any way with possible irrationality, either in reasoning in general (such as the contradiction exposed by Hegel) or in geometry (by the limitation of space to three dimensions). That is the business of mathematics; for mechanics the whole of mathematics is purely rational and it considers its task of rationalization to have been completed if it has been able to deduce its phenomena mathematically, to reduce them to mathematics. Likewise in physicochemistry, at the time when everyone was convinced (as Cuvier said) that all phenomena had to be reducible to mechanical impact, there was no need to worry about whether impact itself was explicable, rational. And also in the same way a biologist would consider a phenomenon explained if he could reduce it to physics and chemistry, without asking himself whether reduction of the physical or chemical phenomena themselves is possible.

It is not even literally correct, we should point out, that reason, as Hegel would have it, in going beyond the obstacle, in *elevating* it, actually *conserves* it, and the pun on the double meaning of the word *aufheben* (Ch. 5, p. 138, n. 10) in which the philosopher indulges here rests totally on a false premise. Of course science, in the most general acceptance of the term, is a work of good faith – despite what Goethe and the "philosophers of nature" may have said in their passionate invectives against Newtonian physics. Science does not seek to dodge the obstacle, nor even to cover it up, and if we return to the corresponding chapter we are sure to find the obstacle there intact. But it is nonetheless true that, in the following chapter, not only does science not seek to keep the obstacle in mind, but one might say that it does everything in its power to obliterate it (in obedience, of course, to the inherent tendency of the human mind toward the progressive rationalization of the real), by the very fact that it seems implicitly, in its entire attitude, to brush aside the obstacle in question, to affirm that the entire content of the preceding chapter was explicable and thus contained no irrational.

There is more than a simple nuance separating this attitude of science and the scientists from the one Hegel attributes to reason. Indeed, if he had not been convinced that science, as it progresses, constantly retains the living memory of the first conflict of reason with reality, he would perhaps have discovered that this conflict is followed by a whole series of analogous conflicts.

Thus one could say that what distinguishes Hegel's attitude from that of contemporary science is that Hegel allowed within his "science" only a single irrational, the "*Anderssein*," and, what is more, called it *reasonable*, whereas our science recognizes a whole series of them. But this would already be to express oneself too precisely and to try to formulate logical categories, as it were, in a subject which by its very essence seems radically refractory to them. Is the irrational one or is it multiple? How can we decide? Certainly science encounters it at very diverse points in its development and as a result one is tempted, if one follows this development, to enumerate it, so to speak. But that does not prove that there is a fundamental multiplicity involved. For example, sensation and transitive action appear in science as things of a quite different order. Physics resolutely dismisses the former from its domain, as if it had called the previous question, whereas transitive action serves as the foundation for mechanics and thus, despite all demonstrations to the contrary, is even today very often considered explicable, not to say explained. And yet we have only to imagine that nonliving matter reacts, as we do, by sensations and volitions, imagine that in exercising an act of will we find ourselves, as Schopenhauer said, "behind the scenes of nature," in order to compare, to make identical, these two concepts that initially seemed so far apart. We can entertain analogous ideas concerning the concept of *diversity*. Applied to space and in its simplest form (we have already pointed out that Hegel often understands it in this way), it is the concept of diversity of position in geometry. But can there be diversity of position in an entirely empty space or one filled with an undifferentiated universal medium (such as the ether of modern physicists)? "So long as we leave matter out of account," Bertrand Russell very aptly says, "one position is perfectly indistinguishable from another .... Indeed, before spatial relations can arise at all, the homogeneity of empty space must be destroyed, and this destruction must be effected by matter."[76] But it is possible to be even more explicit. In fact, Henri Poincaré said that we would have no geometry if there were no unchanging solid bodies moving in space (Ch. 1, p. 25). Now to explain the existence of solid bodies we shall have to go into the constitution of matter, whereupon will be revealed to us, first the *given* of Perrin's absolute magnitudes of atoms and then the anomalies encountered by the Council of Brussels. Finally it is obvious that the irrationals of sensation and transitive action appear only as creators of diversity in time and space; they are therefore linked to the concept of diversity as well. Thus everything in the irrational seems to be connected, and it appears extremely difficult to establish differences and classifications, even ephemeral ones (for, at best, such a classification would obviously be liable to be overturned at any moment by a new discovery). Let us simply say then that science, unlike Hegel, attributes an indeterminate number of aspects to the irrational.

It is precisely because he had previously stripped the irrational (or at least that part of it he meant to include in the domain of concrete reason) of all the complexity this concept actually includes, that Hegel conceived the idea of subjecting it directly to this reason. Now that was to attempt the impossible. Of course the goal of reason must be to make everything that does not come from within subject to itself – this is its proper function, since this is what we call *reasoning*. Furthermore, we saw in our earlier book that explanatory science as a whole is nothing more than an operation carried on according to this schema. But the understanding sought by science comes slowly and gradually. If one tries to rush things, as all those who construct systems of global deduction obviously must do, one is sure to end in failure. What the failure demonstrates is that there is no possible solution, that the irrational does not let itself be dictated to, anticipated, totally constructed by deduction. If one tries to constrain it, it avenges itself by driving the deduction off its course into paths that have nothing in common with those followed by nature; and certainly it never avenged itself more cruelly than in the case of Hegel.

NOTES

1. Pierre Duhem, *La Science allemande* (Paris: A. Hermann et Fils, 1915), p. 22.
2. See for example *Wiss. der Logik*, 3:39 [Miller 52]: The content of traditional logic is "spiritless"; similarly, *Wiss. der Logik*, 5:143 [Miller 682], on the "worthlessness" of "syllogistic wisdom." On the other hand, Hegel celebrated Aristotle's "infinite merit," which consists in being the first to have undertaken a description "of the phenomena of thinking just as they occur"; this merit "must fill us with the highest admiration for the powers of that genius." But "it is necessary ... to go further" (5:31 [595]). Boutroux was certainly right in insisting on the fact that Hegel did not entirely abandon Aristotelian logic and that in particular he did not deny the principle of contradiction, contrary to what is sometimes claimed (cf. p. 301, note 11, below). It nevertheless remains true that, as Boutroux concedes, he made "an audacious and novel application" of this logic (cf. Berthelot, 'Sur la nécessité' 141) and that his system, according to Berthelot's apt remark, "constitutes from beginning to end a *reductio ad absurdum* ... of the postulates peculiar to Aristotelian logic" ('Sur la nécessité' 166).
3. Cf. Berthelot, 'Sur la nécessité' 16.
4. Johann Friedrich Herbart, *Sämmtliche Werke* (Langensalza: Veit / Hermann Beyer & Söhne, 1882), 4:188; cf. 2:193, 8:77, on the contradiction also presented by the concept of substance [8:77 erroneous].
5. See Berthelot, 'Sur la nécessité' 142–145. On the *coincidentia oppositorum* in Nicholas of Cusa and on the concept of identity and its negation in Jakob Böhme, see Karl Rosenkranz, *Hegel als deut. Nat.* 10, 299. Cf. also what Croce says on this subject, *Ce qui est vivant* 30 ff. [Ainslie 36 ff.]. On J. J. Rousseau, who also has sometimes been cited as one of the precursors of Hegelian *dialectic*, see Léopold Leseine, *L'Influence de Hegel sur Marx* (Paris: Bonvalot-Jouve, 1907), p. 22.
6. Schelling, *Zur Geschichte*, I, 10:96. Cf. *Jugement de M. de Schelling sur la philosophie de M. Cousin*, trans. Willm (Paris: F. G. Levrault, 1835), p. 17: "... this same philosophy to which he owed the principle of his own method, ... it was the simplest method to seize what it contained that was most proper and original."
7. See for example *Einleitung zu dem Entwurf*, I, 3:308 ff. The *Einleitung* dates from 1799, therefore well before any of Hegel's publications. On the question of how much Schelling may, in his turn, have owed to Fichte, indeed whether Hegel might not even have borrowed the elements in question directly from the latter without passing through Schelling (questions that do not interest us here), cf. Kuno Fischer, *Geschichte* 7:205 ff. Wilhelm Dilthey, who believes he can trace the invention of the dialectical method back to Fichte, nevertheless recognizes that for him "contradiction takes on a different meaning than for Hegel" (*Die Jugendgeschichte Hegel's*, Berlin: G. Reimer, 1905, p. 53).
8. Schelling, *Zur Geschichte*, I, 10:126. Cf. Ch. 12, p. 335 below. On this question

Hartmann obviously treats Schelling unjustly: "The fact that Hegel's dialectic requires contradiction and includes it, while Schelling's dialectic excludes it, a fact that makes an important difference, was not noted by Schelling, clearly due to his excessive zeal in claiming Hegel's method also as his property" (*Schelling's philosophisches System*, Leipzig: Hermann Haacke, 1897, p. 26). Our citations in the text show that Schelling, on the contrary, was perfectly aware that the transposition of the conflict onto the terrain of logic belonged exclusively to his emulator. But the transposition itself seems to him to be entirely unjustified (cf. below Ch. 12, pp. 311 ff., and Ch. 18, p. 521).

9. Cf. Karl Rosenkranz, *Hegel als deut. Nat.* 31 ff. Schelling himself recognizes that the idea of this triadic arrangement came to him from Kant (thesis, antithesis, synthesis) and claims for himself only the honor of having "won recognition for this type in its most extensive application" (*Einleitung in die Philosophie der Mythologie, Werke*, II, 1:312).
10. Croce rightly emphasizes the fact that for Hegel this "first triad" made up of the terms *being, nothingness, becoming* includes all the others (*Ce qui est vivant* 19 [Ainslie 22–23]).
11. Hegel, *Wiss. der Logik*, 3:78–79 [Miller 82–83]; cf. also 3:105–106 [103]. McTaggart, on precisely this subject of the exceedingly important notion of becoming in Hegel, endeavors to show (just as Boutroux does; cf. p. 300, n. 2) that Hegelian dialectic, contrary to what ill-informed adversaries have sometimes claimed, in no wise rejects the principle of contradiction. "An unreconciled predication of two contrary categories, for instance Being and not-Being, of the same thing, would lead in the dialectic, as it would lead elsewhere, to scepticism, if it were not for the reconciliation in Becoming" (*Studies* 9; cf. also p. 24 on the importance of the idea of becoming in Hegelian deduction).
12. Ferdinand Lassalle (*Die Philosophie Herakleitos des Dunkeln von Ephesos*, Berlin: F. Duncker, 1858, pp. 7 ff.) has strongly emphasized this concept of the identity of being and nonbeing in Heraclitus. As a good Hegelian, he has somewhat exaggerated its significance, but in general his view appears quite accurate on this point.
13. Charles Bernard Renouvier, '*Les sept Énigmes du monde* de M. du Bois-Reymond,' *La Critique philosophique* **11** (1882) 1:188.
14. Hegel, *Enc., Logik*, 6:14 [Wallace 15]. Cf. also 6:47 [47]: Thought is "the basis of everything," and 6:316 [288]: "The notion is a true [*schlechthin*] concrete" [Meyerson's brackets].
15. *Enc., Logik*, 6:49 [Wallace 50]. Karl Michelet, in his foreword to the *Naturphilosophie* (7_1:xiii [Petry 1:183]), also stressed the point that "the Hegelian method ... develops the Idea of space, time, motion, and matter etc., out of the logical Idea." In Boutroux's words, nature for Hegel is "the externalized spirit, or the idea in the form of something other than itself" (Berthelot, 'Sur la nécessité' 146). Lucien Herr, in his fine résumé in the *Grande Encyclopédie* ('Hegel,' 19:1000), correctly points out that in Hegel "the idea ... becomes nature." On the essential unity of the subjective and objective worlds in Hegel, cf. also William Wallace, *Prolegomena* 270, and p. 82: "Nature is spirit in disguise." See Appendix 8.

16. Andrew Seth, *Hegelianism and Personality*, 2nd ed. (Edinburgh: William Blackwood and Sons, 1893), pp. 117–118.
17. Edward Caird, *Essays on Literature and Philosophy* (Glasgow: James Maclehose and Sons, 1892), 2:534.
18. René Berthelot, 'Sur la nécessité' 115; cf. pp. 117, 133, 148, 157, 162. Rosenkranz had already made the point that this philosophy should instead be called *philosophy of mind*, given that for Hegel "the concept of mind alone makes possible the concepts of nature and of the idea qua logic" (*Hegel's Leben* 100), and William Wallace has explicitly protested against the term *panlogism* (*Prolegomena* 80), objecting that within the limits of Hegelian deduction there is room for many things that we customarily designate as irrational.
19. See among others Kuno Fischer, *Geschichte* 8:1162, and Eduard von Hartmann, *Geschichte der Metaphysik II*, *Ausgewählte Werke* (Leipzig: Hermann Haacke, 1900), 12:214 et passim.
20. Hegel, *Naturphilosophie*, 7_1:6 [Miller 2]. On Hegel's *panlogism*, see Appendix 8.
21. "Logic exhibits the elevation of the Idea to that level from which it becomes the creator of nature ..." (*Wiss. der Logik*, 5:26 [Miller 592]).
22. ["I am that which was, is, and will be, and my veil no mortal hath lifted" (Wallace 10).]
23. Hegel, *Enc.*, *Logik*, 6:xl, and *Naturphilosophie*, 7_1:17, 696 [Miller 10, 444–445]. Cf. *Phil. des Geistes*, 7_2:290 [Miller 180]: "When people assert that man cannot know the truth, they are uttering the worst form of blasphemy. They are not aware of what they are saying. Were they aware of it they would deserve that the truth should be taken away from them. The modern despair of truth being knowable is alien to all speculative philosophy as it is to all genuine religiosity." Moreover, as far back as his first published work, his doctoral thesis *De orbitis planetarum* (published in 1800), Hegel could not be more explicit on this point: *"Verum mensura et numerus naturae a ratione alieni esse nequeunt: neque studium et cognitio legum naturae alia re nituntur, quam quod naturam a ratione conformatam esse credamus, et de identitate omnium legum naturae nobis persuasum sit"* ["But in truth the measure and number of nature cannot be alien to reason (*ratio*): the study and knowledge of the laws of nature rest on nothing other than our believing that nature has been formed by reason (*ratio*) and our being convinced of the identity of all laws of nature"] (*De Orbitis*, 16:28 [Adler 301]).
24. [Spinoza, *Ethics*, Pt. 2, Prop. 7: "The order and connection of ideas is the same as the order and connection of things."]
25. See Harald Höffding, *Histoire de la philosophie moderne* [hereafter *Histoire*], trans. P. Bordier (Paris: Librairies Félix Alcan et Guillaumin Réunies, 1906), 2:175 ff. [*A History of Modern Philosophy*, trans. B. E. Meyer (USA: Dover, 1955), 2:174 ff.]. On the identity of thought and reality in Schelling, see Appendix 16.
26. John McTaggart Ellis McTaggart, *A Commentary on Hegel's Logic*, (Cambridge: Cambridge University Press, 1910), p. 1. Let us note that when they do consent to set forth the true foundations of the Hegelian doctrine, historians of

philosophy and even generally the authors of monographs are content, as Croce so aptly points out (*Ce qui est vivant* 5–6 [Ainslie 5]), merely to repeat, often almost literally, the content of one of the master's works chapter by chapter (abridging insofar as possible): this is what was done (as the Italian philosopher points out) by Kuno Fischer in the two fat volumes he devoted to Hegel (*Hegel's Leben, Werke und Lehre*, Vol. 8, Parts 1 and 2, of the *Geschichte der neueren Philosophie*) and, let us add, by Georges Noël in his *Logique de Hegel* (Paris: Félix Alcan, 1897), with the sole difference that the first followed the *Wissenschaft der Logik* and the second the *Logik* of the *Encyclopädie*. Neither of them tried to consolidate the two works nor, in general, allowed himself the least divergence from the model.

27. *Enc., Logik*, 6:50 [Wallace 50; Meyerson's brackets]. On this subject, as in many other concepts in his *Naturphilosophie*, Hegel is not entirely independent of Schelling. To the latter belongs in particular the idea of considering magnetism as the fundamental phenomenon in nature, as well as that of "constructing" it by means of abstract categories. See especially *Ideen*, I, 2:164 [Harris 128]: "Magnetism is the general act of animation, the implanting of unity into multiplicity, of the concept into difference." It is "therefore determinant of pure length, and since this is manifested in body by absolute cohesion, of absolute cohesion." Likewise *Darlegung*, I, 7:64: Philosophy "sees in the magnet nothing less than the living law of identity, the A=A expressed in space but not at all affected by it."
28. *Enc., Logik*, 6:50 [Wallace 50]. Cf. also *Naturphilosophie*, 7_1:38 [Miller 24] on this "impotence of Nature," which is incapable of maintaining the species, whose concept is marred by the birth of monsters; likewise 7_1:265, 651, 695 [Miller 175, 416, 444–445], and *Wiss. der Logik*, 5:45 [Miller 607]. Hegelians have not completely abandoned this position, as we can see in Wallace who, after insisting on the fact that nature "is all a unity of development and has a life-history written in its organism for intelligence to read and to reconstitute," adds that this can be done only by supposing that "all its accident and irregularity is but the inevitable imperfection of reality as given in parts and successions" (*Prolegomena* 477). Furthermore this "impotence of nature" is also a legacy from Schelling's philosophy. Cf. for example *Transc. Idealismus*, I, 3:341 [Heath 6]: "The dead and unconscious products of nature are merely abortive attempts that she makes to reflect herself."
29. *Naturphilosophie*, 7_1:246, 371 [Miller 163, 239]. Cf. also 7_1:401 [258], where the role of water and air in the fabrication of sulfuric acid is analogously explained by the fact that since the entire process has the form of a syllogism, three agents are needed; 7_1:413 [266] where the chemical process appears as a "real syllogism"; 7_1:500 [321] where the development of the plant "splits up into three syllogisms." But the State is also a syllogism, as likewise is God, just as He was earlier for Abelard (cf. Victor Cousin, Introduction, *Ouvrages inédits d'Abélard*, Paris: Imprimerie royale, 1836, p. cxcviii).
30. As early as 1840 Trendelenburg notes (*Log. Untersuch.* 1:90) that "there is no known example" of the application of the dialectical method "in the physical and natural sciences." Moreover, see below (p. 276) for Karl Michelet's significant

admissions in this regard.

31. At Jena, in the summer of 1803, of 52 professors (including the rank of *Privatdozent*) in the faculties taken as a whole, twelve taught pure philosophy; in the winter of 1803–1804, twelve out of only 48 (Kuno Fischer, *Geschichte* 8:67), and the craze certainly did nothing but grow during the whole of the period that followed. On the favors enjoyed by Hegelian philosophy in Prussia, see Appendix 5.
32. Justus von Liebig, *Aus Justus Liebig's und Friedrich Wöhler's Briefwechsel in dem Jahren 1829–1873* (Brunswick, F. Vieweg, 1888).
33. *Enc.*, *Logik*, 6:358 [Wallace 326–327; Meyerson's brackets]. Cf. *Naturphilosophie*, 7_1:3, 196, 511, 520, 536, 606 [Miller 1, 130–131, 328, 333, 343, 388].
34. *Phänomenologie*, 2:40 [Baillie 109]. In 1840 Karl Michelet, the faithful disciple of the master, again renews this anathema against the "so-called philosophers of nature" and approves of "the forthright way" in which physicists have rejected the "turbid mixture of thought and empiricism" that was being offered to them. He makes clear that the blame is addressed in particular to productions that appeared in the first decade of the nineteenth century and explicitly commends the botanist Link who, speaking of the facile explanations of these philosophers and the eternal tripartition they introduced, declares: "Nature has never been so abused" (Foreword to *Naturphilosophie*, 7_1:viii-ix [Petry 180–181]).
35. [Juvenal, *Satires* II, 24. *Quis tulerit Gracchos de seditione querentes*?: Who would endure complaints of sedition from the Gracchi (who were archetypal revolutionaries)?]
36. Cf. below, Ch. 12, p. 320, on the relation between Hegel's *Philosophy of Nature* and Schelling's.
37. *Naturphilosophie*, 7_1:163 [Miller 109]. Hegel maintains the existence of the four elements of Empedocles and Aristotle; they are "physical Elements," which are of another order than chemical elements but have more generality (161 [107]). See Appendix 11.
38. Descartes, *Discourse on Method*, Elizabeth S. Haldane and G.R.T. Ross, *The Philosophical Works of Descartes* (Cambridge: Cambridge University Press, 1931; reprint New York: Dover, n.d.), 1:121.
39. On the place of experiment in Descartes, cf. Ch. 14, p. 363, below.
40. *Phil. der Geschichte*, 9:vii [Sibree xiii-xiv]. The preface is by Gans, but one cannot doubt that he faithfully expresses Hegel's thought. Cf. also *Naturphilosophie*, 7_1:37 [Miller 23], which formulates more or less the same idea, attributing the shortcoming constituted by this intervention of chance to "the impotence" of nature. Quite characteristically, Gans is not content to state that the tendency to "construe" everything is "alien to intelligence" and "ridiculous" and that "philosophy is degraded by this mechanical application of its noblest organs," but also adds by way of argument that "a reconciliation with those who occupy themselves with its empirical details is thereby rendered impossible."
41. Croce is not entirely wrong when he states that Hegel's hostility toward mathematicians and scientists [*physiciens*] "did not arise from contempt for those

disciplines; it came rather from an excess of love" (*Ce qui est vivant* 135 [Ainslie 167]; we have permitted ourself to replace the term "*naturalistes*" used in Buriot's otherwise fine translation with the word "*physiciens*," since the Italian word seems to us to designate here those who are concerned with science in general, excluding mathematicians).

42. Rosenkranz (*Hegel's Leben* 199) maintained the contrary, noting that the extracts left behind by Hegel were taken "not only from German books but also from French or English authors," and this assertion has been widely reproduced, even by English language commentators (cf. for example James Hutchison Stirling, *The Secret of Hegel*, Edinburgh: Oliver & Boyd, 1896, p. xix). But here as elsewhere, Rosenkranz is a rather unreliable witness as to what his master knew. On the other hand, it is possible that his statement is materially correct and that Hegel, who was deeply interested in the English political movement, acquired sufficient command of the language late in life to understand the undemanding text of a political sheet, as is suggested in particular by his article on the English Reform Bill which appeared shortly before his death (*Über die englische Reform-Bill*, *Werke*, 17:425–470 ['The English Reform Bill,' *Hegel's Political Writings*, trans. T. M. Knox (Oxford: Oxford University Press, 1964), pp. 295–330)]). But neither in the three volumes of the *Logik* nor in the three of the *Encyclopädie*, nor in any other work of Hegel, where the citations of French authors are exceedingly numerous (there must be hundreds of them in the *Naturphilosophie*), do we recall ever having come across any citation from an English text. Newton is quoted exclusively from Latin texts.
43. See for example *Naturphilosophie*, 7_1:660–661 [Miller 422]. Likewise, in the *Phil. des Geistes*, 7_2:201 [Miller 124], Pinel's work is declared to be the best of its kind.
44. James Hutchison Stirling says (*The Secret of Hegel*, Edinburgh: Oliver & Boyd, 1896, p. xix) that Hegel was "grounded to the full" in mathematics and the physical sciences. That is a clear exaggeration in the case of the physical sciences, where his knowledge seems to have been more broad than really deep. As to his mathematical knowledge, cf. p. 283 above. Rosenkranz reports that the very copious extracts left by Hegel demonstrate that "almost no work of any repute on mathematics, physics or physiology had been omitted from his studies" (*Hegel's Leben* 152–153), but of course does not inform us whether these extracts demonstrate that the philosopher had actually assimilated the content of his readings (which, as specific examples make clear, his disciple and biographer would have been quite incapable of judging).
45. Here again Hegel has only followed the example of Schelling who, after initially holding a different opinion, came out in favor of Goethe's optics in the *Darstellung meines Systems der Philosophie*. Cf. Rosenkranz, *Schelling* 170.
46. Hegel was bound to Goethe by gratitude and personal friendship. It was Goethe who had had him named professor (without stipend) at Jena in 1805 and who the following year had procured for him an annual salary of 100 thalers. Their friendship did not suffer from the fact that Hegel left Jena shortly thereafter; furthermore he later went back to visit Goethe (Kuno Fischer, *Geschichte* 8:64 ff., 606). See Appendix 13.

47. Eduard von Hartmann has quite appropriately insisted on this important aspect of the Hegelian theory (*Geschichte der Metaphysik II*, *Ausgewählte Werke*, Leipzig: Hermann Haacke, 1900, 12:218).
48. Cf. his reproaches on this account with regard to "abstract and classifying reason [*der tabellarische Verstand*]" (*Phänomenologie*, 2:42 [Baillie 111 – we have translated Meyerson's French; Baillie translates this as "The pigeon-holing process of understanding"]). This curious disdain for spatial relations is once again borrowed from Schelling. Cf. *Darlegung*, I, 7:64: "The philosophy of nature represents what is directly positive in nature without taking into account any other element, for example space and so on, which are of no importance [*den Raum und das übrige Nichtige*]." [Meyerson's brackets.] Likewise in the *Aphorismen über die Naturphilosophie* (*Aus den Jahrbüchern*, I, 7:224): "the mechanistic conception of nature is consequently a conception based on abstraction, arising when one leaves aside all that is real and positive in nature and considers that which is of no value." These two works date from 1806 and Hegel has thus adopted here the opinions Schelling professed in a relatively late phase of his career. Earlier, in fact, Schelling's position seems to have been less clear-cut. For example, in the *Ideen*, which dates from 1797, he seems to take for granted that all differences between bodies must be reducible to their relations with the three dimensions of space (I, 2:175 [Harris 136–137]).
49. *Wiss. der Logik*, 3:326 [Miller 273; Miller's brackets]; cf. also 3:456 [Miller 375].
50. *Naturphilosophie*, 7_1:85–88 [Miller 56–59]. D. F. Strauss, writing in 1854, that is, at a time when the antimetaphysical reaction in Germany was at its height (Strauss himself declares that he "is aware that philosophy's day has passed for the time being and that the day of empirical science has dawned"), and attempting to defend his master Hegel, "the dead lion," from "a few kicks," sets out to show in particular that the famous error in *De orbitis* – where Hegel had thought he could demonstrate the necessity of a gap in the planetary system between Mars and Jupiter, some time after Piazzi's discovery of the first asteroid, Ceres – was only a passing thought and did not in any way follow from the essential principles of the Hegelian philosophy (*Die Asteroiden und die Philosophen*, *Werke*, Bonn: Emil Strauss, 1876, 2:335, 336). It is true that Hegel did not take up the demonstration in his later works, probably because he had subsequently learned of the discovery. But the conviction underlying the planetary theory of *De orbitis*, namely that of the eminent dignity of the concept of power, is, on the contrary, altogether fundamental for Hegel and he constantly maintained it. *Quae progressio quum arithmetica sit, et ne numerorum quidem ex se ipsi procreationem i, e. potentias, sequatur, ad philosophiam nullomodo pertinet* [As this progression is arithmetical and does not follow the generation of numbers out of themselves – i. e., it does not follow the powers – it in no way pertains to philosophy] (*De orbitis*, 16:28 [Adler 302]), he says, in speaking of Bode's law. On this concept cf. also the quotation in our note 51 of this chapter.
51. *De orbitis*, 16:4–5 [Adler 281]: "*De qua cum Mathesi Physices conjunctione praecipue monendum est, ut caveamus, ne rationes pure mathematicas cum rationibus physicis confundamus .... Verum ab ipsa totius ratione sejungendae*

sunt ejusdem analysis atque explicatio .... Quapropter ne quae ad rationes cognoscendi mathesi proprias atque formales pertinent, cum physicis rationibus confundamus, iis quibus mathematica tantum est realitas, physicam realitatem tribuentes [With respect to such a conjunction of physics and mathematics, one must above all be warned to be careful not to confuse the purely mathematical relations with the physical ones (*rationes*) .... But the analysis and the explication of the whole ... must be distinguished from the very reason (*ratio*) of the whole .... For that reason let us not confuse what belongs to the proper and formal relations (*rationes*) of mathematical knowledge with physical relations (*rationes*) by attributing a physical reality to those relations the reality of which is only mathematical]." Similarly, 16:7 [Adler 284]: "*... neque lineis, in quas ex illo postulato directio vis per lineam exposita resolvitur, ob commodum mathematicum physicam significationem tribuendam esse* [... no physical meaning, on account of mathematical advantages alone, must be attributed to the lines into which the direction of a force exhibited by a line is decomposed in virtue of this postulate]." In connection with this conception we already find lively attacks against Newton: "*Illam autem, quae a Newtone incepta est, mathematices et physices confusionem* ... [Nonetheless, ... the confusion initiated by Newton between mathematics and physics ...]" (16:17 [292]; cf. also 16:5–12 [282–288]), as well as praise for Kepler (cf. Ch. 13, p. 350, below).

52. Rosenkranz observes that as a student Hegel studied mathematics on his own rather than in school (*Hegel's Leben* 10).
53. Kuno Fischer, *Geschichte* 8:63; Karl Rosenkranz, *Hegel's Leben* 161.
54. Cf. for example *Wiss. der Logik*, 3:215, 283 ff., 338, 355 ff. [Miller 189, 240 ff., 282, 387 ff.].
55. *Enc.*, *Logik*, 6:148, 235 [Wallace 144, 217–218]. Cf. *Naturphilosophie*, 7_1:58 [Miller 38]. Also *Wiss. der Logik*, 5:282 [Miller 789–790]: "Progress consists in the reduction of the unequal to an ever greater equality." On this point Hegel challenges Kant's concept of synthetic *a priori* judgments in mathematics (cf. also *Wiss. der Logik*, 3:236–237, 238–240 [Miller 205–206, 207–208]). Although this is an interesting issue, it strays too far from our subject.
56. *Wiss. der Logik*, 3:40 [Miller 53]; cf. 5:315 ff. [814 ff.]. It should be noted that Schelling, on the other hand, had celebrated the merits of Spinoza's geometric method as applied to philosophy, calling it a "great example" (*Über die Construktion in der Philosophie*, *Werke*, I, 5:126–127).
57. To characterize Rosenkranz's career it will suffice to recall that he himself declared in the preface of one of his works that he was resigned to the fact that everything he had produced would fall into oblivion but that his biography of Hegel would live on and that the historical figure of the master, as he had presented it, "would survive the centuries" (*Hegel als deut. Nat.* v).
58. Rosenkranz, *Hegel als deut. Nat.* 113–114 [Hall 157]. It is also interesting to note that Taine, despite his clear view of Hegel's philosophy as a whole, does not seem to have understood his particular attitude toward mathematics. "Hegel finds a method of construction and conceives a new idea of the universe: he applies this method to mathematics, to the physical sciences ..." (*Les Philosophes classiques* 141). Now we have just seen that one of the characteristic traits of the

method is precisely the fact that it must not be applied to mathematics. But this is so clearly an anomaly that Taine, in recounting the program of the system, set it right, no doubt unconsciously.

59. *Naturphilosophie*, 7_1:164, 171, 621 [Miller 109, 114, 398]. This is why the chemical process "represents the dialectic through which all the particular properties of bodies are brought to destruction," or to *transitoriness, die Vergänglichkeit* (*Naturphilosophie*, 7_1:421 [Miller 271]).
60. *Naturphilosophie*, 7_1:175 [Miller 117]. This is yet another idea Hegel shares with Schelling. Cf. *Ideen*, I, 2:119 [Harris 93].
61. *Naturphilosophie*, 7_1:176, 184, 358, 459, 529–530, 623–624 [Miller 117, 122, 231, 296, 339, 399–400]. Perhaps one ought not impute all these glaring errors to Hegel alone; at least some of the German biologists of this era seem to have professed extremely odd opinions concerning the chemical action of the organism. For instance a citation from Autenrieth (*Naturphilosophie*, 7_1:620–621 [397–398]) establishes that this scientist, who appears to have enjoyed great authority, actually believed that the animal organism could transform bodies not containing nitrogen into nitrogenous substances, just as it sometimes seems to have been supposed that the element carbon was formed in plants themselves (cf. Schelling, *Weltseele*, I, 2:513).
62. *Wiss. der Logik*, 4:92 [Miller 457–458; we have substituted "reason" and "consequence" for "ground" and "grounded" to conform to Meyerson's translation].
63. *Wiss. der Logik*, 4:93–95 [Miller 458–460]. This could also be translated: "treat as a basis that which has no basis [*Grunloses als Grundlage*]." We see here the device of allusions through puns, which Hegel particularly fancied. [We have substituted "reason" for "ground" to conform to Meyerson's translation; Meyerson's brackets.]
64. *Wiss. der Logik*, 4:95–96 [Miller 460–461]. See also 3:451 [Miller 370], following the passage on gradual coming-to-be and ceasing-to-be quoted in Chapter 8 (p. 235, note 5), for the affirmation of the "tautology" and the "intellectual difficulty" entailed in any attempt to make things conceivable in this way, in conformity with the demands of abstract reason. Furthermore "what Socrates and Plato call sophistry is nothing more than argumentation from reasons" (*Wiss. der Logik*, 4:103 [466]). Cf. also *Phänomenologie*, 2:119 [Baillie 201]: "In this tautological process understanding ... holds fast to the changeless unity of its object, and the process takes effect solely within understanding itself, not in the object. It is an explanation that not only explains nothing, but is so plain that, while it makes as if it would say something different from what is already said, it really says nothing at all, but merely repeats the same thing over again."
65. *Wiss. der Logik*, 4:226–227 [Miller 559–560; Meyerson's brackets]. Cf. a completely analogous passage in *Enc.*, *Logik*, 6:304 [Wallace 277], which proves that in the meantime (that is, from 1813 to the end of his life) Hegel's opinions on the subject remained unchanged.
66. Nothing in either Rosenkranz or Haym shows that Hegel clearly conceived his logical system prior to *De orbitis*. In the *Theologische Jugendschriften* we were

able to discover only a sort of allusion to the synthesis of opposites, rather remote and, what is more, belonging to a fragment dating from 1800 (Herman Nohl, *Hegels theologische Jugendschriften nach den Handschriften der Kgl. Bibliothek in Berlin*, Tübingen: J. C. B. Mohr, 1907, p. 348 [*Early Theological Writings*, trans. T. M. Knox (Chicago: University of Chicago Press, 1948), pp. 312–313]).

67. Harald Höffding, *Histoire de la philosophie moderne*, trans. P. Bordier (Paris: Librairies Félix Alcan et Guillaumin Réunies, 1906), 2:184 [*A History of Modern Philosophy*, trans. B. E. Meyer (USA: Dover, 1955), 2:182].
68. *Wiss. der Logik*, 3:120, 147; 5:340, 343 [Miller 115, 137, 834, 835–836]. We can see in Friedrich Engels (*Herrn Eugen Dühring's Umwälzung der Wissenschaft*, Leipzig: Genossenschaftsbuchdruckerei, 1878, p. 116 [*Herr Eugen Dühring's Revolution in Science* (New York: Charles H. Kerr, 1935), p. 143]) what importance the Hegelians attributed to this concept of the "negation of the negation." According to Engels, Jean Jacques Rousseau had also used it.
69. Benedetto Croce, *Ce qui est vivant* 93 [Ainslie 114]. Baillie, a devout Hegelian, while declaring that Hegel's logic holds the key without which the whole system remains an insoluble mystery, nevertheless admits that for the reader the logic remains "the most forbidding and impossible part" of the doctrine (*The Origin and Significance of Hegel's Logic*, London: Macmillan, 1901, p. vii).
70. McTaggart, imbued though he is with the validity and importance of Hegelian logic, nevertheless acknowledges that explanation by means of abstract reason (*understanding, Verstand*), that is, by the concept of causality alone, "will be found to rationalize the event sufficiently for the needs of the moment" in "the work of ordinary thought" and that, on the other hand, "justification of the Reason at the bar of the Understanding" is necessary, whereas it is impossible to provide this justification, to "prove the validity of the Reason to the Understanding in a positive manner" (*Studies* 12, 15, 88). [The parenthetical "understanding" is provided by Meyerson in English.]
71. Friedrich Engels, *Herrn Eugen Dühring's Umwälzung der Wissenschaft* (Leipzig: Genossenschaftsbuchdruckerei, 1878), p. 6 [*Herr Eugen Dühring's Revolution in Science* (New York: Charles H. Kerr, 1935), p. 20; Meyerson's brackets].
72. Adam Żółtowski, *Metoda Hegla I Zasady Filozofii Spekulatywnej* (Cracow: Nakładem Akademii Umiejętności, 1910), p. 268.
73. Adolf Trendelenburg, *Log. Untersuch.* 1:25. Victor Cousin, as enthusiastic as he is about the Hegelian *system*, also feels this to be an especially weak point: "Hegel sees quite clearly that pure being without determination is a true nonbeing; but how from this identity of being and nonbeing can he draw becoming, *das Werden*, I mean a real and effective becoming?" (*Fragments et souvenirs*, 3rd ed., Paris: Didier, 1857, p. 183). Paul Janet develops this objection further, altogether in the same sense as Trendelenburg (but apparently independently of him), pointing out that becoming is not the identity of being and nonbeing, but the passage from one to the other, "which is quite a different thing," and that this passage requires the intervention of the ideas of time and motion. This is why the idea of becoming is actually borrowed from experience

(*Etudes sur la dialectique dans Platon et dans Hegel*, Paris: Ladrange, 1861, pp. 352 ff.).

74. McTaggart, *Studies* 160–170; cf. also p. 198 for the objection of F.C.S. Schiller: "*Sub specie aeternitatis*, the temporal process is not, as such, real, and can produce nothing new."

75. Andrew Seth notes that for Hegel, when he is "in his Platonising mood," the real world is that "of timeless forms, ... not the world of existing things and persons" [Meyerson inserts "in his Platonising mood" in English, though Seth in fact speaks of a "Platonizing strain"], and that in spite of the great importance the *Philosophy of History* attributes to development in time, one often has the impression "that it is permissible to treat time as an unessential factor, which virtually disappears when the necessity of the evolution has been grasped" (Seth, *Hegelianism and Personality*, 2nd ed., Edinburgh: William Blackwood and Sons, 1893, pp. 216, 182). Furthermore, this affirmation will not come as too great a surprise to the reader who recalls what we set forth in the preceding chapter, namely what infinite pains Hegel takes, whenever he is to *explain* a manifestation, a development in time, to connect it to the *explanation* of something preexistent, that is to say, basically, to a more or less disguised preformation. Obviously, deep down inside, he does not reason on such occasions according to his logic, for which becoming, the passage from nonbeing to being, is something in direct conformity with reason, but according to everyday logic for which there is no rational explanation except the identity of the antecedent and the consequent. It should be noted that already the *philosophers of nature* of Schelling's school sometimes seem to play with this idea of a nontemporal nature: "What we call nature is not the nature that presents itself to sensible perception and which is limited to what is finite, but the one which is eternal and immutable in itself" (Henrich Steffens, *Grundzüge der philosophischen Naturwissenschaft*, Berlin: Realschulbuchhandlung, 1806, p. 16).

76. Bertrand Russell, *Essais sur le fondement de la géométrie*, trans. Cadenat (Paris: Gauthier-Villars, 1901), p. 97 [*An Essay on the Foundations of Geometry* (Cambridge: The University Press, 1897), p. 77]. D'Alembert earlier followed an analogous line of thought when he stated in the *Preliminary Discourse* of the *Encyclopedia* ('Discours préliminaire des éditeurs,' *Encyclopédie*, Paris: Gauthier-Villars, 1751, 1:v) that "extension in which we did not distinguish shaped parts would be only a distant and obscure vision where everything would elude us because we would be unable to discern anything clearly" [*Preliminary Discourse to the Encyclopedia of Diderot*, trans. Richard N. Schwab (Indianapolis: Bobbs-Merrill, 1963), p. 18]. Curiously enough, Hegel expressed the opposite opinion (*Naturphilosophie*, 7_1:47 [Miller 30]).

CHAPTER 12

SCHELLING'S OBJECTIONS

Hegel's doctrine is a monism. It may well be the most perfect, or at least the most extreme, monism recorded in the history of human thought. In the words of Trendelenburg, a highly competent critic almost contemporary with the epoch in which the system was at its height, the Hegelian method "undertakes the audacious task of developing thought and being in their oneness." "If such a dialectic exists," the philosopher goes on to say, "through which thought, in developing itself, develops by its own force the inner nature of things, we shall possess thereby, at one fell swoop, the fullness of truth and certainty" (*Log. Untersuch.* 1:23).

How chimerical this enterprise is by its very essence has never been shown more clearly and convincingly than it was by Hegel's most immediate predecessor in the evolution of German philosophy, a fellow student five years his junior who for years, due to the anomaly of his precocious maturity, had given the impression of being his teacher – Schelling. From almost the exact moment Hegel had declared his independence in the *Phenomenology*, Schelling had remained nearly completely silent. Four years after Hegel's death he broke his silence in a very short and apparently incidental publication which nevertheless had an enormous impact; what is more, it deserved this honor, for it is in many ways a highly valuable work in the history of philosophic thought.[1] It is the preface to the German translation of Victor Cousin's *Fragments philosophiques*.[2]

After briefly characterizing both his own philosophy and that of Hegel (in the process making the accusation of quasi-plagiarism to which we referred above, Ch. 11, pp. 265 ff.) and pointing out Hegel's originality in applying the process to the logical idea – "this is an invention for which he alone deserves credit and which intellectual poverty has justly admired" – Schelling shows that "the logical movement proper to the notion sustains itself, as could be anticipated, ... within the limit of what is purely logical; but as soon as one must leave it to set foot on the terrain of reality, the dialectical movement stops and is broken." The transformation which then takes place "is no longer a dialectical movement but something quite different to which it would be hard to give a name, for which there is no category in a pure rational system, and for which the inventor himself has no name in his own system." It is, in short, an "attempt to regress ... toward scholastic dogmatism, and to found metaphysics on a purely rational principle exclusive of all reality, a vain attempt for the simple reason that the *empirical* element or the reality, rejected at the outset, is reintroduced into the system as if through a back door." Therefore Hegel's attempt shows "once again that it is impossible to arrive at reality through the rational." For

> one can quite well begin a system of philosophy with an *a priori* principle, a purely rational principle, by simply prefixing a few preliminary considerations. That is not the problem. But just as all these forms termed aprioristic express only the negative side of all knowledge, that without which no knowledge is possible, and not the positive side, that by which it is born, and just as, consequently, their character of universality and necessity is only a negative character, so also in this absolute *prius* which in its universality and its necessity is nothing other than what cannot be conceived anywhere at all, that is to say being in itself (ἀυτὸ τὸ ὄν), one can recognize only the universal negative character, that without which nothing is, but not that by which something exists. (*Jugement* 16–19)

In his later work *Zur Geschichte der neueren Philosophie*,[3] Schelling, in the chapter consecrated to Hegel, completes this masterful criticism. He first repeats (what in fact constitutes the kernel of his argument) that "logic presents itself as the purely negative element of existence, as that without which nothing could exist but from which it does not at all follow that everything exists only thereby" (*Zur Geschichte*, I, 10:143). But he undertakes, on the other hand, to show how Hegel came to choose *pure being* as the starting point for his development: it too is a notion suggested to him by "the earlier philosophy" (that is, by Schelling's own system). But for Hegel it is a simple notion of a scholastic sort, an entirely empty notion; the identity of being and nonbeing is a proposition "which gives us nothing" (like all Hegelian philosophy, for that matter). As for the process by which Hegel means to advance, starting from this notion, its alleged necessity rests only on the fact that

> thought is already accustomed to a concrete being having more content, and that consequently it cannot be satisfied by this meager fare of pure being where what is thought is a content in general, but is no particular content .... It is thus not, finally, a necessity that lies in the empty concept itself, it is rather a necessity that resides in the thinker and is imposed on him by his memories, which do not allow him to stop at this empty abstraction,

and what actually constitutes "the implicit driving force of this progression is still the *terminus ad quem*, the real world, at which knowledge must ultimately arrive" (*Zur Geschichte*, I, 10:129 ff.).

However, the weakest point of Hegelian deduction remains the starting point itself: the transition between knowing and being, between the logic and the philosophy of nature.

> At the end of the *Logic* the idea is both subject and object; conscious of itself, it constitutes, in being ideal, also the real, consequently it has not the slightest need of becoming real in any way other than it already is. If then one nevertheless assumes that something of this sort occurs, one does not suppose it because of a necessity lodged in the idea itself, but simply because nature does after all exist. (*Zur Geschichte*, I, 10:152)

Hegel says that "the idea expels [*entlässt*] nature from within itself." What can be meant by this term?

> What is clear is that it is flattery to call this explanation theosophical .... It is a pretty pass the Hegelian philosophy has reached here ....it is a wide and nasty ditch; by pointing it out (I first mentioned it in the preface to Cousin) I have provoked many hostile reactions, but no valid response that is not completely misleading. (*Zur Geschichte*, I, 10:153–154 [Meyerson's brackets])

Was Hegel unaware of this difficulty? In order to be persuaded that he was not, one need only examine the way in which the passage from the logical to the real takes place for him. These deductions are exceedingly obscure and, what is more, not at all identical in the *Science of Logic* and the *Encyclopedia*. They make use of all the resources, all the artifices of Hegelian metaphysics – one might even say, but for the respect due the serious efforts of such a mind, all the acrobatics and all the prestidigitation of which this art is capable.[4] If Hegel had recourse to such subterfuges, if, in spite of everything and despite the fact that he himself sometimes seems to deny it, he tried to persuade us that, to use Seth's vivid expression, "logical abstractions can *thicken*, as it were, into real existences," that they are able to "take flesh and blood and walk into the air," so that "the whole frame of nature is no more than a duplicate or reflection of the thought-determinations of the Logic,"[5] it is surely because he found such a transition necessary. Since the idea creates reality, it was *necessary* that there be a passage from logic to the philosophy of nature. The essential structure of the system unrelentingly demanded it. To see this, moreover, one need only observe that this transition is found in Plato. As is well known, the great ancestor of all idealism avoids the problem (for he cannot do otherwise) by appealing to mythology (it is apparently to this procedure that Schelling is referring when he declares that one would actually flatter Hegel's procedure by calling it theosophical). No doubt Plato had a better excuse than Hegel in this case. Still, he came after Parmenides, whose philosophy had already clearly shown that rational being, the only being whose existence can really be drawn from our reason, is being that is universal and undifferentiated, which is to say, as Renouvier pointed out, fundamentally unreal.[6] But since that time, the knowledge acquired in more than twenty centuries of philosophic labor had come to be added to it, knowledge which had constantly tended to reinforce the awe-inspiring intuition of the venerable Eleatic. The thinkers of the Middle Ages in particular – Arabs, Jews or Latins – had made it sufficiently clear that the being which, in St. Anselm's terminology, is *causa sui*, the one whose "essence involves existence"[7] can be only a concept of the divinity with indistinct attributes, a divinity which, if one penetrates behind the mysterious veil in which it is wrapped by theological artifices, is the very image of the permanent being of Parmenides. We saw above that Maimonides already affirms that God can have only negative attributes (Ch. 6, p. 153). And if Spinoza believed he could revive St. Anselm's undertaking and tried to demonstrate the existence of Being as logically necessary, Kant seems to have destroyed this "ontological proof" once and for all by establishing that "the concept of an absolutely necessary being is a concept of pure reason, that is, a mere idea the objective reality of which is very far from being proved by the fact that reason requires it." Indeed, "the unconditioned necessity of judgments is not the same as an absolute necessity of things." Thus the proposition that a triangle has three angles "does not declare that three angles are absolutely necessary, but that, under the condition that there is a triangle (that is, that a triangle is given), three angles will necessarily be found in it." Any proposition affirming existence is synthetic; "how can we profess to maintain that the predicate of existence cannot be rejected without contradiction? This is a feature which is found only in analytic propositions, and is indeed precisely what constitutes their analytic character." Thus, "whatever ..., and however much, our concept of an object may contain, we must go outside it, if we are to ascribe existence to the object," and our consciousness of any existence whatever "belongs exclusively to the unity of experience."[8]

We see that Hegel's attempt has as its indispensable prior condition a return, on a vaster scale, to the ontological proof of St. Anselm and Spinoza. And if one looks at it closely, Schelling's refutation basically amounts to an amplification of Kant's.

As the reader will understand, it would be entirely outside the compass of the present work to try to make an actual summary of the sequence of Hegel's arguments (which would be possible, moreover, only by plunging deeply into his metaphysics). Suffice it then to say that his principal device consists in impoverishing the category of Being beyond all measure, in pouring scorn upon it (to use the expression of the English philosopher whose enlightening expositions we have used more than once on this subject),[9] at the same time enriching the category of knowing almost as extravagantly. Clearly his aim is to create the impression that a complete knowledge of all the relationships constituting the concept of the thing can succeed in being equivalent to the thing itself, in *creating* it, so to speak. If he were successful, the category of essential Being would be more or less abolished; it would be identical with that of Knowing. But it is evident that the undertaking is altogether futile: as Kant told us, there is no more identity between Knowing and Being with an infinity of relationships than with a single relationship.

It is worth our while to dwell at more length on this subject, for we have clearly arrived at one of the most essential points of the Hegelian position. From this vantage point we can at least get an idea of what constitutes at the same time its lasting reputation and its irremediable weakness. Indeed, they both have a common source. The almost irresistible attraction which the doctrine certainly exerted in its own time and which, as we have seen, is not completely without effect today, arises from the fact that Hegel actually attempted an undertaking that constitutes an eternal postulate of the human mind, a secret postulate if there ever was one, for while we aspire to it with all the fibers of our intellect, we dare not admit this aspiration to ourselves, because at the same time we have the feeling it cannot be satisfied. But Hegel did not hesitate to try his luck, and if he is not the only one who dared to do so, he is certainly the one who put the most insistent effort into the attempt. He announced that he had succeeded and it cannot be doubted for a single instant that he was sincere in his claim: well he might make this mistake, since, as we are quite aware, a whole people and a whole epoch made the same mistake along with him and since, what is more, the profound sincerity of his faith was certainly one of the most important elements in his success.

However, Hegel himself was not totally impervious to the state of mind to which we have just alluded, one that moves us to affirm and deny, by one and the same movement so to speak, that the world is intelligible.

Before the enormity entailed by an unqualified affirmation, and above all in the face of attacks by adversaries who had completely laid bare this foundation of his doctrine, he sometimes hesitated, equivocated, even retreated. These are what have been called, in an expression which may not in fact be too strong, his *palinodes*. For example, in the preface to his *Philosophy of Law*, he loudly proclaims as the guiding principle that what exists is in conformity with reason, *reasonable* (*vernünftig*),[10] and, quite logically, in his *Philosophy of History* he sketches a rough picture of the entire evolution of humanity, applying all the force of his genius to demonstrating that this evolution is in essence necessary, that great men have been only its blind instruments, themselves unaware of the ends they served, to such an extent that one ought to be able to tell the history of the world without proper names – this is one of the aspects of his philosophy that had the most powerful effect on the imagination and the strongest, most lasting repercussions.[11] But in the face of the violent protests this affirmation evoked in all who were inclined to criticize the political and social situation in Germany at the time (and certainly it was open to criticism), Hegel quickly contradicted himself, stating that his claim was not to be understood as referring to all that appears as really existing, but only to that part of this same reality which is reasonable in itself (*Enc.*, *Logik*, 6:10 [Wallace 10–11]) – thereby reducing the resounding apothegm to a common and yet pretentious tautology. And his disciples in their turn perpetuate the mistake when, in the face of one of the master's particularly audacious claims, they try to show that it is not to be taken literally, that Hegel did not pretend everything in the universe was deducible, that he merely asserted that there was reason in nature; for that is also, if not a tautology, at least a truism, since no human knowledge would be possible if we were perchance inclined to disregard it.[12] The truth is that if Hegel did not intend to deduce *everything*, if he admitted that the universe contained fortuitousness due to chance and to the impotence of nature, he nevertheless certainly thought he could deduce everything truly essential. Once again, that is what constitutes his strength. And we must not be surprised to see that to some extent the charm continues to operate, that the most distinguished minds are seduced by this prospect of a solution – however remote and obscure it may be and however forbidding the access routes may seem – which at least promises to penetrate into the essence of things, by finally identifying Being with Knowing.

But there can be no doubt that the promise is a false one. No philosophy can accomplish that task, by nature impossible, and Hegel's was no exception. Nevertheless he is greatly to be admired for having gone so far and so openly along this path the human mind has constantly in view, a path wide open and at the same time barred by impassable barriers. Surely all those who try to fathom these eternal problems will draw lessons of inestimable value from his attempt.

However, the observations we have just made seem to give rise to a peculiar sort of difficulty. If it is true that Hegel only followed to its final conclusion an eternal aspiration of the human mind and that this is the problem idealistic philosophy in particular unfailingly runs up against when it seeks to include nature in its deductions, how does it happen that the refutation, the clear and incontestable demonstration of the inherent impossibility, of the fundamentally chimerical nature of the whole enterprise should have come from Schelling of all people? For at bottom Schelling's position was without question just as idealistic as Hegel's and, what is more, the affirmation of the rationality of nature was one of its most essential foundations and, indeed, its actual point of departure. Schelling began with Fichte, whose disciple he was early in his career and whose philosophy he initially adopted, ostensibly proposing only to complete it on one important point, namely the theory of nature.[13] Indeed, Fichte, who in turn began with Kant, had suppressed the Kantian thing-in-itself, by decree so to speak; for him the self posits for itself the nonself. Thus for Fichte nature (as Schelling shows in the writing where he breaks definitively with his master) becomes a simple limit, in short something without real content, that is, something properly nonexistent.[14]

First of all, Schelling wanted to show how the self, that is to say its reason, manifests itself in nature, which could indeed be considered a simple extension of the Fichtean system (Fichte himself seems to have understood it in this way at the beginning). But the disciple immediately established his independence. He demanded, in fact, that this "philosophy of nature" he was establishing serve as a real base for philosophy in general, that philosophy, after descending "into the depths of nature, ... rise to the heights of the spirit," that is to say that it construct the whole system by means of conceptions the philosophy of nature had provided for it. Thus the philosophy of nature assumed a preponderant place in the system, which explains why the system as a whole has sometimes been given the name *Naturphilosophie*; toward the end of his life, Schelling himself, while protesting that it was a simple *denominatio a potiori*, did not seem to find the appellation completely inappropriate.[15]

Thus Schelling was obviously engaged in an enterprise quite analogous to that of Hegel. And one can well understand that a thinker of this quality, once he had made the decision to combat the movement he himself had been so instrumental in creating, should have immediately dealt it mortal blows by exposing, with incomparable and implacable mastery, just where the inherent weakness of the position lay. But, having said this, one is all the more surprised that he decided to take such a position. Could he have labored under a delusion as to the true implications of his attacks, could he have failed to understand that what he sought to undermine – or rather to ruin utterly – was the very foundations of the idealistic conception of nature of which he was one of the protagonists? To answer this question, we must look more closely at Schelling's true position with regard to this "philosophy of nature" and consider in particular what distinguished it from that of Hegel. We believe this study will help the reader better fathom the essence of the Hegelian doctrine of nature by permitting him to understand to some extent the spiritual antecedents of this philosophy and the milieu in which it developed.

Schelling, in the very writings where he attacks his rival, took care to inform us, at least partially, why (at least in his opinion) the arguments he had put forward were in no way damaging to his own philosophy, that is, in other words, how his position differed from Hegel's.

One point on which Schelling insists strongly (and rightly so) is one which becomes immediately obvious if we simply examine the hierarchical way in which the subjects are arranged in the two systems: in passing from one system to the other, the *philosophy of nature* changes position completely. For Schelling it is placed at the very beginning; the mind, as we have just seen, as soon as it approaches the philosophic domain, must enter at once into the examination of the phenomena offered by nature in order to seek there the materials for all subsequent construction. For Hegel, on the contrary, the penetration takes place by a sort of transcendent psychology, the "phenomenology of mind," or else, according to his later conception (for he does seem to have changed his mind on this point), by the logic itself. In any case, for him the *philosophy of nature* is placed only after the logic, as an explanation or an emanation of the latter, on the same level as the *philosophy of mind*.

Therefore, as Schelling himself points out in attempting to define the divergences between his philosophy and that of his rival, in his own system the mind, "starting from the abstract subject, from the subject in its abstraction," enters "into nature from the very first step, and there is thus no need for a later explanation in order to make the transition from the logical to the real," so that the theory "concerns itself with the pure *how* of things without pronouncing itself on their real existence"[16] or rather, as Schelling explains in the same work, so that this existence is considered as due to chance. "The *primum existens*, as I have called it, is the primitive fortuitousness (chance insofar as it is primordial).[17] All this construction begins with the emergence of the primitive fortuitousness – which is unlike itself – it begins by a dissonance and undoubtedly must begin that way," which means that "for the concept of matter one must first think of nothing more than something in general, which is no longer a nothing, that is, is no longer pure freedom" (*Zur Geschichte*, I, 10:101, 104). In other words, what exists, the material, must first and foremost be considered as *given* – which makes it unnecessary to deduce it by starting from purely rational or reasonable notions, as Hegel did.

In this given, that is to say in the whole of reality, the *philosophy of nature* seeks what is in conformity with reason. Its very name is indicative of this task, although one cannot say it is its defining characteristic, for theoretical science has the same goal. Schelling himself is well aware of the similarity. In one of his first works, *Ideas for a Philosophy of Nature*, he points out the analogy between his system of explanations and that of mechanistic physics in general and of Le Sage and Prévôt in particular.[18] Although he declares that this theory, "like all atomistic theories, is a tissue of empirical fictions and arbitrary assumptions" (*Ideen*, I, 2:70 [Harris 52]), he nevertheless treats it with decided favor. He contrasts the purity of this system with the impure mixture presented by physical theories in general. Physicists ought to have applied themselves to perfecting this theory, which gives way only before higher considerations. Furthermore, atomism is "the only consistent system of empiricism."[19]

In a somewhat later work, he similarly states that the

> tendency of all natural science is ... to move from nature to intelligence [*auf das Intelligente*]. This and nothing else is at the bottom of the urge to bring *theory* into the phenomena of nature. – The highest consummation of natural science would be the complete spiritualizing of all natural laws into laws of perception [*des Anschauens*] and thought. The phenomena (the matter) must wholly disappear, and only the laws (the form) remain.[20]

The continuation of the passage leaves no doubt that Schelling is speaking of the action of genuine science and not (as one might almost be led to believe) of the role attributed to his *philosophy of nature*.

Hence it is, that the more lawfulness emerges in nature itself, the more the husk disappears, the phenomena themselves become more mental, and at length vanish entirely. The phenomena of optics are nothing but a geometry whose lines are drawn by light, and this light itself is already of doubtful materiality. In the phenomena of magnetism all material traces are already disappearing, and in those of gravitation, which even scientists have thought it possible to conceive of merely as an immediate spiritual influence, nothing remains but its law, whose large-scale execution is the mechanism of the heavenly motions.[21]

It is almost superfluous to point out how much this statement by Schelling, with its idea of the gradual spiritualization (we would say rationalization) of nature by means of laws and theories, is consistent in some ways with the true credo of modern science[22] and how much this attitude differs from that of Hegel, particularly in the respect he shows for theoretical science, although not without a number of reservations.

In short, for Schelling the *philosophy of nature* is not something fundamentally different from science as the scientists understand it; it is simply science pushed further because it has been better equipped by the resources that philosophic speculation provides:[23] it is "speculative physics."[24] Because of this, it is without doubt immensely superior to ordinary science, whose conceptions are only a "tissue of empirical fictions and arbitrary assumptions devoid of any philosophy" and "revolve" in an "everlasting circle" (*Ideen*, I, 2:70 [Harris 52–53]). Nonetheless theoretical science is already, as it were, a transition between science and the *philosophy of nature*: following the passage from the *System of Transcendental Idealism* cited earlier, after repeating that science has a "tendency to render nature intelligent," he concludes with the statement that "through this very tendency it becomes *nature-philosophy*, which is one of the necessary basic sciences of philosophy" (*Transc. Idealismus*, I, 3:341 [Heath 6]).

This continuity between the *philosophy of nature* and science and the fact that the former, just like the latter, relies on reality as science understands it explain why Schelling behaves quite differently toward experience than Hegel. No doubt we sometimes find declarations in Schelling in which he displays a confidence in the results of his scientific deductions which appears to be almost the equal of that shown by Hegel in analogous cases: the position occupied by phenomena in the system of the philosophy of nature is "the only explanation of them that there is" and furthermore "all doubt is ... removed" concerning this explanation

(*Ideen*, I, 2:71 [Harris 53]). But it should be noted that here, as always in Schelling's writings in the philosophy of nature proper, it is a question of deductions which themselves begin with experimental data and that the feeling of their certainty can thus have no other source than the perfect agreement (real or imagined) between these deductions and experience. Schelling explicitly states, in the case of the principle of "general duality" – although he believes it "to be as necessary a principle of all explanation in nature as the concept of nature itself" – that an assumption of this sort must be confirmed by experience.

> This absolute presupposition must bear its necessity within itself, but it must also be subjected to empirical verification, for if all the phenomena of nature do not allow themselves to be deduced from this presupposition, if in all of nature there exists a single phenomenon that is not necessary according to this principle, or even goes so far as to contradict it, the presupposition is thereby declared erroneous and must cease forthwith to be valid as a principle.[25]

In fact, "no procedure of idealistic explanation is to be admitted into the philosophy of nature." This manner of explanation would be as

> senseless for physics and for our science, which is placed on the same terrain as physics, as the old procedures of theological explanation or the introduction of a general finality into science, which is disfigured thereby. For any procedure of idealistic explanation removed from its proper domain and dragged into that of the explanation of nature degenerates into the most extravagant nonsense, examples of which are not hard to find. (*Einleitung zu dem Entwurf*, I, 3:273; this work is dated 1799)

It is most remarkable, incidentally, that this passage, and especially the last sentence, which seems to be directed specifically at Hegel's *Naturphilosophie*, was actually written before the author of the *Phenomenology* began to express independent opinions. So we see that Schelling, in many passages, appeals explicitly to scientific experimentation[26] and openly avails himself of confirmations (real or imagined) that it had afforded him.[27]

Thus the *philosophy of nature* of Schelling (and his disciples) is really something quite different in essence from that which bears the same name in Hegel. People have most often maintained the contrary, but they have allowed themselves to be deceived by a purely outward resemblance. The resemblance does in fact exist, and it is quite often rather striking. Hegel, displaying little inventiveness in this regard, and even a certain lack of imagination, it seems, is usually content to utilize the "constructions" of Schelling's school, and we have called attention to some of these borrowings in the course of the present work (cf. in particular Ch. 11, pp. 303–306, notes 27, 45, 48). But if one digs a bit more deeply, the dissimilarity between the two theories becomes glaringly obvious. For Hegel's *Naturphilosophie* is sustained by a sort of internal framework of coherent propositions which constitute a rigid and entirely aprioristic system, based, or at least claiming to be based, solely on logical propositions, whereas there is nothing similar in the other "philosophers of nature" (except for certain writings of Schelling with which we shall deal shortly). These philosophers are really only doing "speculative physics," that is, a science one might say is obsessed with hypotheses, which constructs them at every turn – whether they are suitable or not – on the narrowest and shakiest bases, sometimes even on purely imaginary bases, worrying very little about testing the solidity of these edifices and, on the contrary, proclaiming its chimeras as certainties. These extravagances, for which experimental data furnished the point of departure (or at least the pretext), nevertheless maintain some contact with theoretical science, even though they wander far into the limitless domain of entirely arbitrary assumptions. On both these counts, then, Hegel was right not to want to be identified with these rivals: as we have seen, he abhorred theoretical science, and he was not entirely mistaken in calling the efforts of other *philosophers of nature* vague and in protesting against any attempt to assimilate them with his work of systematic deduction.

Let us recall, however, that Schelling, coming as he does between Fichte and Hegel, is, or at least pretends to be, as firmly idealistic as either of them. Therefore the result of the investigation at which the mind will arrive in entering *freely* into the world of natural phenomena and science will be predetermined: this examination can only reveal that the phenomenon, the scientific *given* is entirely consistent with reason. From the beginning of his philosophic career, while he is still ostensibly only a disciple of Fichte, Schelling is most explicit on this point. "The infinite world," he says in one of his first works, "is nothing other than our creative spirit itself in innumerable productions and reproductions." He likewise declares that "our representation is simultaneously representation and thing" and "that we have no alternative but to affirm that it is not matter that engenders spirit, but spirit that engenders matter."[28] These are opinions to which he firmly adheres, since the demonstration of the rationality of nature is, as we have seen, the very purpose of his *philosophy of nature* and the justification of its place in the system. "Nature," says Schelling in his Introduction to *Ideas for a Philosophy of*

Nature (1797),

> should be Mind made visible, Mind the invisible Nature. Here then, in the absolute identity of Mind *in us* and Nature *outside us*, the problem of the possibility of a Nature external to us must be resolved. The final goal of our further research is, therefore, this idea of Nature. (*Ideen*, I, 2:56 [Harris 42])

Spinoza, "with complete clarity, saw mind and matter as one, thought and extension simply as modifications of the same principle." Of course his philosophy is "the most unintelligible that ever existed," which is why it "has lain unrecognized for over a hundred years." But Schelling means to renew it by bringing about a reconciliation between these conceptions and those of Leibniz (*Ideen*, I, 2:20, 36 ff., 71 [Harris 15, 27–28, 53]). Furthermore he has no doubt that his philosophy is equal to the task. The *philosophy of nature*, he says in the body of the same work,

> up to the present time ... is the most fully worked out endeavour to set forth the theory of Ideas and the identity of Nature with the world of Ideas .... What had not perhaps for long been so much as suspected, or at best been considered impossible, ... has in part been actually achieved already by the *Philosophy of Nature*, while in part it is on the way to doing so. (*Ideen*, I, 2:69 [Harris 52]; cf. Appendix 16)

The *philosophy of nature*, he declares a little later, "is the Spinozism of physics" (*Einleitung zu dem Entwurf*, I, 3:273).

But if this philosophy, after "descending into the depths of nature," (cf. p. 317 above), really succeeds in demonstrating the "Spinozistic"[29] identity of idea and nature, if it succeeds in actually breaking nature down into elements belonging exclusively to the intellect, in showing it to be composed of purely rational elements, why would it not go on to the inverse operation and construct nature by means of such elements? This is indeed – at least up to a point – the opinion of Schelling himself. The *philosophy of nature*, for which "the objective is made primary, and the question is: how a subjective is annexed thereto, which coincides with it," is only the first part of the system; it must be followed by a second part in which "the subjective is made primary, and the problem is: how an objective supervenes, which coincides with it" (*Transc. Idealismus*, I, 3:340–341 [Heath 5–6]). This, as it were, synthetic counterpart to the *philosophy of nature* is what Schelling presents in the *System of Transcendental Idealism* published in 1800 and in the work entitled *An Exposition of my System of Philosophy* published the following year, to which he subsequently always refers as the most complete résumé of his opinions in this area. In the first pages of this *Exposition* he states his intention of finally bringing together there, in a single system, what he had presented under two different aspects, namely as *philosophy of nature* and as *transcendental philosophy*. But in reality the *Exposition* presents only the second standpoint, that of the *System of Transcendental Idealism*; the work is, moreover, only a fragment, as the author explicitly acknowledges.[30] Another work similarly conceived but much longer and more complete, although it too seems to have remained unfinished, is written a few years later but remains in manuscript form and comes to light only in the posthumous edition.[31]

All these works, which from the standpoint of form thus resemble Hegel's subsequently published *Naturphilosophie*, nevertheless differ from it and also from the writings Schelling himself classified as *philosophy of nature* by their for the most part completely general tenor. In them we find the deduction of the three dimensions of space and also, simultaneously, that of the existence of magnetism, electricity, and the chemical process, as well as of gravitation and cohesion.[32] But all the wealth of detail that characterizes what is properly called the *philosophy of nature* is lacking. It is not surprising that these writings, at least as far as their scientific content is concerned, have attracted less attention than Schelling's earlier works; it is almost exclusively the latter that are imitated in his disciples' *philosophy of nature*, and from this point of view Hegel's attempt to excuse himself for the attacks in the *Phenomenology* by drawing a distinction between the master and his imitators (cf. Appendix 19, p. 596), is not entirely vain.

Due to the difference in actual content between the works of Schelling belonging to these two categories, as well as to the lack of systematic rigor that characterizes his work in general, it is impossible to settle in any clear way this fundamental problem of the true nature of the scientific *given* that forms the basis for the *philosophy of nature*. Indeed, if Schelling's Spinozistic statements are to be taken literally (as he unquestionably appears to understand them in the passage just cited), it is certain that in the synthetic exposition, not only must all *given*, any element that is not strictly rational, disappear, but also that this exposition, form aside, must coincide perfectly with the analytic exposition. We have just seen that in practice one cannot judge whether or not this is the case. But we do have a few theoretical statements by the philosopher on the subject. For example, from the very first pages of the *System of Transcendental Idealism*, he declares that neither *transcendental philosophy* nor the *philosophy of nature* alone can completely represent the parallelism between nature and the intelligent element; it is necessary that it be the combination of the two sciences, which, "on that very account ... must forever be opposed to one another, and can never merge into one (*Transc. Idealismus*, I, 3:331 [Heath 2]).[33] Similarly, in the *General Deduction of the Dynamic Process*, which appeared the same year, he again explains that "the sole task of science is to *construct* matter. The task is feasible, although the use that is made of this general solution will never be complete .... It is a task beyond all finite forces, one that in nature itself could be accomplished only by means of an unconscious production." Thus the scientific given of the *philosophy of nature* is not a genuine irrational (to use our nomenclature); but to prove it rational would require something like Leibniz's "infinite analysis," which is why the *philosophy of nature* and *transcendental philosophy* can never be joined.[34]

Was that really Schelling's final position, the definitive solution that satisfied him once and for all? From the vague and incomplete way in which this master writer explains himself on this subject and from the fragmentary state of his works in this area in general, we are inclined to believe, rather, that he felt real hesitation and that his opinion, far from being fixed, oscillated between a total "Spinozism" and the recognition of an irreducible scientific given.

Such an interpretation is suggested by the frankness, one might almost say the naïveté, with which the two contrary points of view are frequently put forth in the same work, hardly a few pages apart and without the reader's being informed that this constitutes a contradiction which, at the very least, must be resolved by means of an auxiliary hypothesis. For instance, in *Ideas for a Philosophy of Nature*, the definition of *philosophy of nature* as the application of philosophy to science and the claim that the whole of his philosophy "is nothing else than natural science" is found on page 6, while beginning on page 20 "Spinozism," the identity of mind and matter, is solemnly proclaimed (*Ideen*, I, 2:6, 20 [Harris 5, 20]). Similarly, in the *Introduction to the Sketch of a System of the Philosophy of Nature*, it is stated that this philosophy "is the Spinozism of physics," and a little later on the same page the author protests against any attempt to introduce idealistic explanations, adding on the following page that the *philosophy of nature* is realistic "through and through" and is consequently only speculative physics (*Einleitung zu dem Entwurf*, I, 3:273, 274).

Later in this work we even find an explicit attempt to justify maintaining these two points of view simultaneously. The passage is extremely interesting and merits being quoted in its entirety. Schelling begins by protesting against the fact that false intentions are imputed to the *philosophy of nature*.

The statement that the science of nature ought to be able to deduce all of its propositions *a priori* has sometimes been understood to mean that science ought to do completely without experiment and draw its propositions from itself without any intervention from experience; this assertion is so absurd that even the objections raised against it deserve only pity. Not only do we not know this or that thing in particular, but at the beginning and in general we do not know anything except through experience or through its intermediary and consequently the whole of our knowledge is composed of propositions about experience.

But after this profession of faith, which could not be faulted by the most dedicated disciple of John Stuart Mill, comes a quick reversion to the *a priori*.

These propositions are transformed into aprioristic propositions only by the fact that one becomes aware of them as necessary, and thus any proposition, regardless of its content, can be elevated to this dignity, given that the difference between *a priori* and *a posteriori* propositions is not at all, contrary to what some may have imagined, a difference originally inherent in the propositions themselves, but, on the contrary, a difference depending on the relation of these propositions to our knowledge and to our manner of knowing, so that any proposition that is for me simply historical, in other words, a proposition of experience, becomes an *a priori* proposition as soon as I directly or indirectly fathom its inner necessity.

Therefore "any primitive natural phenomenon" is "simply necessary ..., since there is no chance in nature." All this is crowned by the following twofold affirmation: "It is thus not we who know nature as *a priori*, it is nature that *is a priori* .... But if nature is *a priori*, it must also be possible to recognize it as something that is *a priori*" (*Einleitung zu dem Entwurf*, I, 3:278).

The deduction contained in the last sentence is certainly irreproachable and, what is more, altogether essential for Schelling, since, as we know, it is the *philosophy of nature* that is to undertake this task. However, we do seem to have reasoned in a circle, for here we are back at our starting point. If nature is recognized as entirely aprioristic, what need will there be for experiments? How would one even insert them into such an exposition of nature (always, of course, *after* nature had been recognized as aprioristic)? So this naive manner of understanding the proposition concerning the *a priori* character of nature, a manner Schelling could not ridicule enough at the beginning of the quoted passage, was not so absurd after all. It simply pushed to its final conclusion one aspect of the ambiguity Schelling's doctrine certainly contains.

Would it be disrespectful to so illustrious a name to suppose that Schelling only belatedly and gradually became fully aware of this ambiguity? It is well-known how blind the most highly honored thinkers have been to the lacunae in their own work, lacunae that minds of incomparably less breadth have no difficulty in discovering.

What is more, the supposition we have just made would have the advantage of furnishing the basis for an intelligible solution to the troubling problem presented by Schelling's life and philosophic production. We beg leave to devote a few pages to this subject which, as the reader will see, is linked to questions of great interest. Indeed it seems to us that the peculiar dispositions of this profound mind and even his changeable personal inclinations are likely to cast some light on the way the human intellect in general behaves in this domain not easily penetrated by direct introspection.

Schelling, five years younger than his fellow student Hegel, is a rare example of intellectual precocity (especially in the domain of philosophy, which usually seems to be reserved for more mature efforts). His first publications date from 1795, when he is barely twenty; they immediately attract the attention of the philosophic public, and Fichte waxes enthusiastic about them. Let us add that these writings, both in content and in form, undoubtedly merit this reception. The ensuing years witness an uninterrupted flow of important works. *Ideas for a Philosophy of Nature* appears in 1797, when Schelling is twenty-two, *On the World-Soul* the following year, the *First Sketch of a System of the Philosophy of Nature* in 1799, the extremely interesting *Introduction* to this *Sketch* a bit later in the same year and the *System of Transcendental Idealism* in 1800, along with the *General Deduction of the Dynamic Process* (we are omitting works of lesser importance). At that moment Schelling is just twenty-five, and we really cannot fail to be amazed at the volume and above all the quality of his production. But it continues at almost the same pace in the early years of the new century. And then, a little before the end of the first decade of the nineteenth century it comes to an abrupt halt. From then on, Schelling, formerly so productive, seems struck with sudden sterility. He is almost completely silent or produces only things of little significance. This is not a voluntary halt in his production, he does not remain silent because he believes he has said everything he had to say and that his system is henceforth complete; on the contrary, he acknowledges, at least implicitly, that many essential things are lacking and promises to fill these gaps in new publications. But these promises are never fulfilled. As once the succession of his publications flowed uninterruptedly, so now flows the succession of his projects; the booksellers of Leipzig or elsewhere continually announce publications by the great philosopher, but these publications never actually see the light of day. They are not all aborted at the same stage: for some the author does not seem to have gone much beyond the title and a very vague outline of the contents; others, on the contrary, were begun quite seriously and interrupted when one part was already completely written and even, in some cases, set in type and a few copies printed (see Appendix 17). But once again, nothing reaches the public which, after long and eager anticipation – an eagerness sustained by the booksellers' announcements – finally tires and grows accustomed to the silence of this philosopher who formerly appeared so inexhaustible in the expression of his thoughts. In Schelling's last twenty years there appear a few writings (of which the *Preface* to Cousin is undoubtedly the most remarkable) which show beyond any question that his mind has lost none of its vigor and that his means of expression have remained as brilliant as ever, but which add little to our knowledge of his philosophy. Finally, after his death, several works on which he had worked diligently during his very last years, that is from his sixty-seventh to his eightieth years, are published in an unfinished form; they are systematic works, but rather theosophical in content, showing that at this stage the philosopher's interest has shifted somewhat. Religious concerns, already very pronounced in the fragment of *The Ages of the World*,[35] had entirely prevailed, which is understandable, since Schelling was in fact called to Berlin as a philosopher of religion, to serve as a sort of counterweight to the anti-Christian movement initiated by Hegelians of the far left, a movement whose most brilliant expression – along with the writings of Feuerbach, Bruno Bauer, etc. – was Strauss's *Life of Jesus*, published in 1835.

An anomaly as striking as this philosopher's long and obstinate silence has understandably not escaped the notice of biographers and historians of philosophy. It has frequently been explained by a circumstance in Schelling's family life, namely the death of his first wife in 1809, which, indeed, took place at almost the same time the philosopher underwent the metamorphosis of which we have spoken (see for example Kuno Fischer, *Geschichte* 7:151). But for many reasons this hypothesis seems a bit forced (see Appendix 18).

A very different explanation has been advanced by Hartmann. He believes that Schelling generally lacked independence of mind (he was *ohne eigenes Rückgrat*, literally "without a backbone of his own") and that the first part of his philosophic career was more fruitful for the simple reason that he had just come to know the great thinkers who had preceded him – Bruno, Spinoza and Hume in particular – and had endeavored to follow their example.[36] Apart from the fact that this completely ignores the real value of Schelling's own philosophy (Hartmann himself treats the philosopher much more fairly in the rest of his book), one does not quite see how in so few years Schelling could have entirely exhausted the vein presented to him by the whole of philosophy as it existed in his epoch, to the point of running out of "productive models." This hypothesis would explain even less why Schelling himself constantly had the conviction that, far from having said everything, he still, on the contrary, had essential things to reveal (although he did not actually succeed in doing so).

Bréhier in his turn has applied himself to the same problem. He believes that Schelling, being "above all a professor and journalist" and "a conversationalist as well as a marvelously evocative orator" (which is certainly an extremely accurate appraisal), needed the speaker's rostrum in order to produce, and it is for this reason that the long period of tranquility at Munich was so completely sterile in terms of philosophic production.[37] However, from 1822 to 1829 Schelling teaches at Erlangen without this resumption of professorial activity marking an analogous resumption of his philosophic production.

Finally, Hegel has also been implicated, and it is indeed most remarkable that the moment Schelling ceases to write almost coincides with the moment Hegel, in his *Phenomenology*, emerges as an independent thinker and quite rapidly attains a celebrity eclipsing that of his emulator. It is, so to speak, out of disgust for the attitude of the philosophic public, out of resentment at his rival's success, which he considers unjustified – or at least greatly exaggerated and too easily won – that Schelling is supposed to have retired to his tent. There is certainly some truth in this supposition and Schelling himself, in speaking, after Hegel's death, of his philosophy and his success, shows this disgust and spite quite strongly.[38] But one might wonder whether these declarations are entirely sincere, that is, whether they tell the *whole* truth. For if that had been the true reason, the sole reason, one would think that Schelling would have said so immediately, at the very beginning of this period of silence, or at least sometime during it, in order to justify the strange change of pace that baffled the public at large and even his supporters, as he must have sensed. Now, on the contrary, by his incessant promises and announcements he implicitly acknowledges that he has an obligation to explain himself; and yet he does not do so. His attitude thus resembles that of a man from whom everyone expects a revelation, a man who would himself like to make this revelation, but whose tongue is tied by a definite circumstance having to do with the very nature of the revelation to be made.

If we try to take in at a single glance the whole of Schelling's abundant production during his most active period, one fact stands out immediately, a fact even more striking if one compares these writings to those of Hegel. For the latter everything is system, one would almost say nothing but system; all the works are classified (or at least claim to be classified) in a rigid and preconceived plan, and when there are deviations (as for the *Logic*, which actually replaces the *Phenomenology* as an introduction to the system), the modifications are concealed as carefully as possible. For Schelling, on the other hand, everything is fragmentary; one would say that these were only improvised, provisional, occasional works. As the titles frequently indicate, they are *Ideas*, *Sketches*, designed simply to pave the way for a later definitive work. But that definitive work he never manages to provide – no doubt because the works he begins with this in mind do not satisfy him.[39] It is quite characteristic that the most voluminous and systematic work Schelling produced during this whole period, a work dealing for the most part precisely with science, should have remained shelved, although it was almost finished, and in spite of the fact that, during the period that followed, the philosopher was undoubtedly tempted more than once to publish it.

It is just as significant that he appears rather annoyed when others make very sincere attempts to state his position. The annoyance is expressed quite strongly in the case of Cousin, who certainly did not lack good will toward him. He does not reproach his interpreter so much for having insufficiently understood him as he does in general for having given his compatriots "imperfectly conceived ideas of a philosophy that is thus far completed only in the mind of its original author." It would be better to wait for "his [Schelling's] works that are going to be published in the course of the winter" and "will once and for all put an end to petty discussions" (Plitt, *Aus Schelling's Leben* 3:42–43).[40] We recognize here the eternal promises of future works, and it goes without saying that this one remained as empty as the others and that the winter of 1828–1829 (the letter is dated 27 November 1828) did not see the appearance of any Schelling work. That does not prevent him from writing to Cousin in July 1833, with reference to the publication of the *Fragments philosophiques*: "I would have preferred that you wait for my own explanations which are ready for the press and will leave nothing to be desired" (Plitt 3:71). Again, five years later, in 1838, he tells him that "German philosophy is on the point of and even in need of undergoing yet one final crisis, and that one cannot judge the beginning, nor the middle, nor even the *beginning of the end* until a scientific movement such as that of German philosophy is entirely over and arrives at its true goal." He adds: "My current ideas ... (and I have never ceased working on them) are not meant to be presented by one of my disciples; I am the only one who can develop them." He goes so far as to express his displeasure that "German philosophy" was announced as the subject of a Paris competition; it ought to have been "put off another few years" (Plitt 3:136–137),[41] that is to say, apparently, the time Schelling believed necessary to come to terms with himself.

Is it too venturesome, then, to suppose that at bottom things were really just what appearances suggest, and that a serious difficulty inherent in the philosophic stance he had taken and in the relations of this philosophy with that of Hegel is what prevented him from giving the definitive presentation of his ideas that the public awaited?[42] Indeed, the ambiguity referred to above became more difficult to maintain as soon as Hegel asserted himself. For in Hegel the ambiguity has disappeared. He has made his choice. Nature is essentially *a priori* and, what is more, he really deduces it – or at least (what here amounts to the same thing) claims to have done so. Of course there still remains something irrational, or rather nonreasonable. But the irrational elements are minor characteristics, due to the *play* of nature, to its *impotence*; on the other hand, everything essential in nature is declared to be in complete conformity with reason, composed solely of elements borrowed from reason. It is clear that such a doctrine was truly, at least in what it claimed to have attained, the culmination of the movement German philosophy had followed from Kant to Fichte and from Fichte to Schelling. The abolition of the thing-in-itself, which led to the identity of the self and the nonself, the former creating, *positing for itself*, the latter, and the penetration of nature by philosophic thought, showing directly that nature itself was only thought – here these two mutually complementary conceptions merged, as it were, into a third, which seemed to be the definitive consummation for both of them. They had been only two links of an uninterrupted chain of thought; but here is the third link, which is in fact the ultimate one, for this time the identity of being and thought no longer appears in a distant fog, as for Fichte, it is no longer stated in equivocal theoretical utterances only to be immediately denied in the practice of the *philosophy of nature*; it is real, immediate, *palpable*, the master affirms it, he shows it in detail, covering the entire field of human knowledge, the sciences of nature as well as those of the spirit, physics and history alike, jurisprudence and theology alike. You have only to cover all these domains with him and absorb his explanations (granted it is sometimes a bit difficult to grasp their meaning) in order to be as convinced of it as he is.

There can be no better example of this than the attitude of Victor Cousin. Cousin, hampered by his limited knowledge of German, frankly admits he understood almost nothing of Hegel's philosophy, either in 1817 when he was becoming acquainted with the master or later when he studied the *Encyclopedia.* "Not everything was entirely unintelligible to me,"[43] he says in speaking of the Hegelian ideas, and as for the *Encyclopedia*, he declares that "it was a book all bristling with formulas of a sufficiently scholastic appearance, and written in a style by no means lucid, particularly for me."[44] He adds elsewhere: "I plunged into it, but it resisted me completely and I did not get much out of it. Upon my return to Heidelberg, I asked Hegel himself for the explanations I needed, but to no avail; he always evaded my questions, not perceiving that he was answering them sufficiently by not answering them" (*Fragments et souvenirs* 180). This did not prevent Cousin from immediately forming the deepest admiration for the author of the book.

> From our first conversation I divined what he was, I comprehended his whole reach, I felt that I was in the presence of a superior man; and when I continued my Journey from Heidelberg into other parts of Germany, I proclaimed him wherever I went, I prophesied him, as it were; and upon my return to France, I said to my friends; gentlemen, I have seen a man of genius. (*Fragments philosophiques* xxxviii [Ripley 90–91])

He never revised this first impression, as is demonstrated by the fact that, after Hegel's death, he published these recollections in the preface to the *Fragments philosophiques*, where, at the risk of displeasing Schelling, he stated that Hegel had not only developed and enriched the system his predecessor had left "filled with all manner of imperfections and defects," but had given it "in many respects a new aspect" (*Fragments *philosophiques* xl-xli [Ripley 92–93]; cf. Appendix 15, p. 588).

That no doubt appears contradictory. Sensing this, Cousin concisely explains, with perfect insight and precision, what impelled him toward the Hegelian doctrine. "Hegel was dogmatic; and although I could not yet get my bearings in his dogmatism, it attracted me" (*Fragments et souvenirs* 80).[45] Cousin *believed* in idealism; the philosophy of nature filled him with enthusiasm, he "saw nothing else in Germany" (cf. Appendix 15, p. 586). Basically, however, he undoubtedly could not realize how fragmentary all this was, and thus how deprived of all real efficacy. But now at last the system was arising, set forth coherently, *dogmatically*. Even tainted with a little obscurity, was not such a doctrine irresistible?

There can be no doubt, moreover, that in addition to the great attraction bound to be exerted by the announcement of a complete resolution of reality (as we showed, p. 331), the groundwork laid in Germany by Schelling, Fichte, and even in a certain sense Kant, had much to do with the overwhelming success of Hegel's philosophy. Schelling himself certainly understood this, and he also perceived that in many respects Hegel was essentially only drawing the logical conclusions of the doctrines that he, Schelling, had taught. That is unquestionably the true source of his incessant accusations of plagiarism against Hegel and the Hegelians. Schelling himself recognizes this state of affairs more or less explicitly. His frankest statements to this effect are found in a document which has not been included in his *Works*. It is the clandestine publication of a course taught at the University of Berlin, instigated by an enemy who wanted to attack him. Schelling never denied that the text was an exact reproduction of his words; he did not even do so before the bar of justice, where he simply sued his adversaries for unauthorized publication, in effect countersigning the words that were attributed to him.[46]

Here then is how Schelling speaks of Hegel:

> The fact that I mention Hegel shows how much I esteem him. I see that Hegel alone has saved the fundamental thought of my philosophy for posterity, and, as I recognized in his courses on the history of philosophy in particular, he maintained it until the last and preserved it in its pure form. While others of us let ourselves be taken in by the material element of the acquired conception, he admirably maintained the *method* in its purity. No one else could have completed the preceding philosophy more ably than he. He made a positive philosophy out of the philosophy of identity, and in so doing raised it in general to the level of an absolute philosophy, leaving nothing outside it.

Nevertheless Schelling criticizes the fact that for Hegel logic forms only a part of philosophy, whereas it ought to have been "absolute logic, the ideal of the pure science of reason." "However," he adds, "the fact that all this science must be dissolved into logic was something I recognized only later and not independently of Hegel."[47]

At first glance, we have the impression that Schelling has completely revised his judgment since his *Preface* of 1834. Upon closer examination, the change appears much less significant, at least as far as content is concerned. The form is indeed entirely different: there is praise where in the former work there were only attacks. No doubt this is because the resentment that seems to have been very deep immediately after Hegel's death – perhaps in the face of the unanimous expressions of sorrow at his passing, the sometimes hyperbolic eulogies lavished upon the deceased, etc. – has softened in the meantime. It is perhaps also because he almost has the feeling of having triumphed over his adversary, since he succeeds him in his professorship. Finally it is because he is speaking before an audience he knows admires Hegel. But that he has not been completely mollified we know not only through the declarations of the *Zur Geschichte* cited above, which, since they are much later than the *Preface*, must date from approximately the beginning of his career at Berlin, but also through certain letters written by Schelling at this time and through numerous passages from the posthumous works, which are certainly more recent in origin. In these works Hegel is treated with no consideration at all, his dogmatism is "all the more repugnant because of its narrowness," and Schelling applies to him these words of Terence: "*Haec si tu postules certa ratione facere, nihilo plus agas, quam si das operam ut cum ratione insanias*."[48] He remains just as firm in his condemnation of the Hegelian transition from the logic to the philosophy of nature. By this transition, by the way in which, for him, the idea "expels nature from within itself" (cf. above p. 313), Hegel resembles Jakob Böhme, whom he pretends to despise. Hegel is as theosophical as Böhme, except that the intoxication of the latter is real, whereas Hegel would merely like to appear intoxicated (*Phil. der Offenbar.*, II, 3:121–122, 82). Hegel was entirely devoid of artistic sensibility and of any sense of the whole, or else "he would have felt the movement break down between the logic and the philosophy of nature and would have understood, by the very way the latter is coupled to the former, that he had strayed from the straight path" (*Phil. der Offenbar.*, II, 3:88).

These posthumous works also show us how praise and blame are reconciled in Schelling's mind. Hegel was right to reduce everything in his philosophy to logic. This philosophy, and all that preceded it (including his own philosophy, as Schelling explicitly states)[49] are only a part of the true philosophy, the part he calls *negative* philosophy. But the latter is incomplete without the second part, *positive* philosophy, which is precisely what Schelling feels called upon to construct and which alone will be able to deal with what exists, which is beyond the reach of negative philosophy.[50] Hegel's error was to try his hand at this, and his system, in the long run, is not even a true negative philosophy, precisely because he professes to include the positive in it (*Phil. der Offenbar.*, II, 3:80).

Exactly what these two philosophies – negative and positive – of Schelling's final period are, we of course do not pretend to set forth here; we can only refer the reader desiring to know the essence of this system (which clearly had very little influence on the course of philosophic thought, especially compared to that exerted by the writings of the philosopher's early period) to the short work by Delbos,[51] which we consider the best résumé of this complex philosophy ever published in any language. Here, as the reader will no doubt understand, the subject does not interest us except insofar as it can serve to clarify Schelling's attitude toward the Hegelian system.

From this perspective, we can begin by asking ourselves what status this new system gives to the *philosophy of nature*, which occupied such a considerable place in the former system and which, if it was perhaps not Schelling's most original creation, was certainly the one by which he had most strongly influenced the minds of his contemporaries. Strangely enough, it is a question for which it is difficult to find a clear-cut answer. Hartmann has said that in his first period Schelling had a philosophy of nature but no nature, whereas in his final period he had nature but no philosophy of nature.[52] That is not literally true, for among the works of this period of Schelling's life can be found a treatise entitled *Exposition of the Process of Nature*[53] which came to light only after the death of the philosopher but which reproduces a course taught at Berlin in 1843–1844, thus making it roughly contemporaneous with the works the editor has placed in the Second Series. Now this *Exposition*, also only a fragment to be sure, is certainly *philosophy of nature*, almost in the manner of the *Transcendental Idealism* and the *World-Soul*, except that it is still more general and imprecise than these works of the first period. Thus on occasion Schelling reverted to his old ways. But when one considers the masterworks of this epoch, beginning with the *Introduction to the*

Philosophy of Mythology, one certainly does not quite see how the *philosophy of nature* could be linked to it. Of course Schelling does not pass over it in complete silence and sometimes he even appears to want to defend it, but in a rather strange fashion. For instance, in violently attacking Fichte, whom he accuses of having "destroyed all rational or intelligible connection between things," he adds:

> after such an unfounded [*bodenlos*] idealism, the first thing one could still accomplish in order simply to return to the ways of philosophy was obviously to bring to light the immanent reason that resides in the things themselves and to find their intelligible interconnections. (*Einleitung in die Philosophie der Mythologie*, *Werke*, II, 1:465)

Here then the *philosophy of nature* is justified, or rather excused, simply because the opportunity presented itself, as a salutary reaction against the excesses of Fichtean idealism. In short, when we read these works, we have the impression that the *philosophy of nature* no longer interests its author – no doubt because he has recognized that (contrary to what he had formerly thought) it does not lead to the complete rationalization of reality and therefore one will be unable to achieve a complete deduction of nature from the principles it has revealed.

But this conviction of the impossibility of a total deduction (and that is the essential point to retain from the passages we have quoted) came to Schelling through Hegel, or at least not independently of him. It is because he saw that Hegel's philosophy was in certain respects a logical extension, a systematization of his own that he clearly perceived where this path was leading and consequently, after serious consideration, became convinced of the inadequacy of his own work and the necessity of setting out in search of a new solution. This solution is elaborated only slowly, undoubtedly all the more so because, with the maturity of age and experience, this time he insists on attaining complete internal consistency and on reducing his thought to a system. Now nature, which endowed him with flashes of genius, seems to have been rather parsimonious when it came to the gift of systematization, which Hegel possessed to the highest degree.

To his great credit, Schelling fights stubbornly against this inadequacy during most of his life, imposing on himself a silence that withstands the pleas of his friends and the attacks of his enemies. That would also explain why he appears so impatient toward those who, with the best faith in the world, expound his ideas: his past philosophy no longer satisfies him (one might almost say exasperates him, although in many passages of his posthumous works he seems to defend it more or less) ever since Hegel showed him where it was leading.

Indeed, if one looks closely, one has the impression that even during his early period Schelling was fundamentally much less idealistic than the resounding formulas we have quoted might suggest, and that the realistic element, far from being accidental (as Schelling would have us think and as he apparently tries to convince himself) expresses, on the contrary, an intrinsic and important aspect of his true inclinations: while in theory professing the most intransigent "Spinozism," it appears that when the time came to apply it directly, to translate it into sensible reality, he was shocked by the extreme solutions this doctrine imposed and took refuge in a realism, which was no doubt less logical but was also less repugnant to his immediate feelings.[54] Consequently, as soon as Hegel's first independent works appeared, Schelling could only be strongly repelled by their extravagances, which he obviously found facile and crude. And he may have immediately, if not clearly understood, at least strongly sensed the decisive value of those objections he was to expose with incomparable vigor and mastery in the *Preface* to Cousin's *Fragments*.

But certainly he also developed spontaneously in this direction. It is, as a matter of fact, in 1806, a year before Hegel's *Phenomenology*, that there appears, as an introduction to the second edition of the treatise *On the World-Soul*, the small work entitled *On the Relation between the Real and the Ideal in Nature*,[55] and it is in this work that the concept of *will* first assumes an important place in Schelling's thought.[56] However, we find the germs of the conception going back to the very beginning of his philosophic career. Indeed, in the *Treatises for the Elucidation of the Idealism of the Science of Knowledge*, dating from 1797, he writes: "This self-determination of the spirit is called *will*. The spirit wills and it is free. No other reason can be shown for the fact that it *wills* .... The spirit is a primitive will. This will must thus be as infinite as spirit itself."[57] He also states in the same work that "what can be calculated *a priori*, what is accomplished in conformity with necessary laws, is not an object of history, and, inversely, what is an object of history must not be calculable *a priori*."[58]

During the next decade, which includes the period of Schelling's greatest philosophic activity, this aspect of his philosophy tends to become rather blurred; however, in the *Transcendental Idealism*, which dates from 1803 [sic; actually 1800], is to be found a passage on history and determinism almost identical to the one we have just cited (*Transc.*

Idealismus, I, 3:584 [Heath 196]).

But the true evolution of Schelling's philosophy toward voluntarism certainly starts with the little work from 1806. In the *Philosophical Inquiries into the Nature of Human Freedom* (1809), Schelling proclaims that

the irrational and accidental element which reveals itself as connected with what is necessary in the formation of all beings, especially organic ones, proves that it was not merely a geometric necessity which operated here, but that freedom, spirit and self-will played their part too.[59]

From then on, will assumes a preponderant place for him, so much so that in the *Philosophy of Revelation* it becomes "the foundation of nature in its entirety,"[60] and this aspect of Schelling's later philosophy (which is very much concerned with the problem of freedom)[61] obviously contributes to separating him from Hegel and also, eventually, from the philosophy of nature – although he tried to introduce the concept of will into the latter, as we have just seen.

It must be added, however, that a more personal sentiment could have played some role in Schelling's unnatural silence. In the evolution of German philosophy, Schelling came between Fichte and Hegel. Now the former had already proclaimed an allegedly integral idealism. Granted this doctrine turned out to be incomplete in its treatment of nature; but Schelling and his disciples had announced that they had succeeded in correcting the deficiency. Hegel and his followers, for their part, had gone this affirmation one better. And now Schelling was faced with the necessity of deserting the flag.[62] It is understandable that he hesitated for a long time before making public his change in position.

He resigned himself to it in the works that were published posthumously. We saw above how he attacks Fichte in the *Introduction to the Philosophy of Mythology*. In the *Introduction to the Philosophy of Revelation* he opposes both Fichte and Hegel even more decisively, if that is possible.

Everything encountered in our experience can be produced *a priori* in thought alone, but then it is *only* in thought. If we meant to transform that into an objective statement, to say that everything in itself is only in thought, we would have to return to the standpoint of Fichtean idealism. If we want something that is outside thought, we must start from a being that is absolutely independent of all thought, prior to all thought. The Hegelian philosophy knows nothing of such a manner of being; there is no place for such a concept in that philosophy. (*Phil. der Offenbar.*, II, 3:164)

All that, however, was already implicitly contained in the declarations of the *Preface*. It is thus not surprising that the Hegelians cried treason. But the master had died, he whose defense and above all whose counterattack would surely have been far more formidable. Is that merely a coincidence? Or can we not suppose that Schelling, more or less aware of the weakness of his position and instinctively fearing the retort of a polemicist as vigorous as Hegel, in some sense felt his courage mount as a result of the latter's death? What is clear is that if Schelling spoke of the works of his rival in his courses at Munich or Erlangen while Hegel was still alive, he could not have used the language he did in 1834. For word would have gotten around and Hegel surely would have taken up the attack and, what is more, would not have been able to appear so cordial on the occasion of their chance interview in 1829 (see Appendix 19).

Schelling's attack understandably created a great stir. It tolled the death knell for the Hegelian philosophy as an "unassailable" metaphysical system – although certain disciples a bit blinded by their devotion tried to perpetuate the illusion of sacrosanctity long after this date.[63] And if it was certainly not the determining cause of the formidable antiphilosophic reaction that broke out in Germany shortly thereafter (since the reaction had deeper roots), it probably played a large role. For Schelling might conceivably have failed to recognize that his attack hit home when directed against himself as well: since he had not gone all the way to the logical conclusion, since he had retained realistic elements in his philosophy, he probably had the illusion that he could take shelter behind them and watch imperturbably as Hegelian idealism collapsed. But for the philosophic public, Hegelianism, not without reason, represented the whole of idealistic philosophy, the philosophy of which Hegel was the final flowering. Everything tumbled down at one blow, so to speak. What followed was, as we have seen, a most violent reaction, the reign of the most extreme empiricism and materialism in Germany. And when there finally came a philosophic revival a few decades later, it was not in the name of Hegel, nor even in the names of Schelling or Fichte that it took place; on the contrary, thinkers loudly proclaimed their intention to ignore this phase of philosophic thought entirely and go directly back to Kant.[64]

We have already pointed out that Hegel rendered an immense service to reflective thought by attempting an almost superhuman enterprise, one that this thought ardently desires to see carried out, even while recognizing it to be superhuman: this enterprise is the direct rationalization of the real – not as a nebulous and indeterminate theme, as many an earlier theory had presented it, nor as a remote goal, as science makes it seem while at the same time immediately limiting this asymptotic progress by the admission of an irrational, so that one no longer draws nearer to the goal indefinitely, but only nearer to a limit that is infinitely distant from it – but as a thing to be realized in an immediate way, *in fact*. His example makes us see clearly where this process is leading, and consequently Schelling can, through his objections, make us fully realize the enormity of the effort that idealism, pushed to its logical conclusions, demands of our imagination; he can show us all the dreadful expanse of the abyss that, in this case, our reason must cross at a single bound.

Need we warn the reader that, as was true for Hegel, our exposition can in no way pretend to *do justice* to Schelling's thought? As we pointed out earlier in connection with the conflict we believe we have uncovered there, we have been exclusively interested in a particular aspect of his thought. We must add that the simplification entailed by this method may have even more drawbacks in the case of Schelling, whose position fluctuates and sometimes even actually changes entirely, than in the case of Hegel, which is much more of a piece. But we have found it impossible to avoid this shortcoming, and can only refer the reader who would like to complete, or even rectify, the very partial and imperfect image of Schelling's philosophy that emerges from the preceding pages, to the fine study by Bréhier.

Nor should we even wish to give the impression that Schelling's chief merit in the area with which we are concerned consisted in the refutation contained in the 1834 document. The *philosophy of nature* of Schelling and his disciples is unquestionably an aberration from our current perspective; but one will not have to delve very deeply into the history of scientific thought to become convinced that theories that had dominated whole generations were also called aberrations by the proponents of the theories that superseded them.[65] Of course it can be argued that this is a far more serious aberration than the one involved in those theories; but it is nevertheless only a question of degree and not one of principle, and the *philosophy of nature*, as we have seen, did not at all remain sterile from the standpoint of scientific progress; on the contrary it furnished science with general ideas and, indirectly, with discoveries of great value. Furthermore we have shown how inferior the fundamental conception of Schelling's philosophy is to that of Hegel from the standpoint of rigorous logic; still we must admit that it is by that very fact less forbidding, less foreign to what actually constitutes the essence of our mind. This ambiguity, this hesitation between irreconcilable points of view, this mixture of "Spinozism," of general formulas radiating extreme idealism and of more precise claims deriving from realism and calculated to serve, in a manner of speaking, as a refuge for the mind as soon as the pitiless logic of the idealistic conception shocks it by consequences altogether too extreme – this is the very essence of our intellect. In that respect Schelling's doctrine is indeed more adequate for the intellect: it is, if we may venture to say so, more *human* than Hegel's. The fact that the *philosophy of nature* of the former seems immeasurably less shocking in relation to current ideas than that of the latter is only one particular aspect of this observation.

NOTES

1. This is not the opinion of Kuno Fischer, who believes, on the contrary, that the fact that it was Schelling's first philosophic publication in more than twenty years and that it was also the first time the latter had brought his disagreement with Hegel out into the open in writing gave this work an importance it would not otherwise have had (*Geschichte* 7:227). We hope the reader will disagree after examining with us the content of this short work.
2. A good French translation of this preface was published shortly after its appearance in Germany under the title *Jugement de M. de Schelling sur la philosophie de M. Cousin, traduit de l'allemand et précédé d'un Essai sur la nationalité des philosophies* (Paris: F. G. Levrault, 1835) by Joseph Willm [hereafter *Jugement*]. The original German is found in *Werke*, I, 10:201–224. Paul Grimblot inserted a translation of the same preface as an appendix to his translation of *System of Transcendental Idealism* (*Système de l'idéalisme transcendental*, Paris: Ladrange, 1842, pp. 377 ff.); his translation seems to be an almost unmodified reproduction of that of Willm, from which Grimblot even borrows the title (the German text reads simply: *nebst einer beurtheilenden Vorrede des Hn. Geh. Raths v. Schelling*). On Cousin's personal relationships with Hegel and Schelling, see Appendix 15.
3. *Werke*, I, 10:1–200. According to the editor, this posthumous work is composed of notebooks prepared for the courses taught at the University of Munich. He dates it 1827, except for one part which would be earlier, since it was taken from a manuscript dated Erlangen, 1822 (cf. the note on 10:96). The whole thing would therefore considerably antedate the *Preface* to Cousin (which is placed after the *Zur Geschichte* in this chronologically arranged collection). But one need only glance at the chapter devoted to Hegel to see that this supposition is inadmissible, and that at least this part of the work must be much later than the *Preface*, which is explicitly cited in it.
4. Even Rosenkranz, the most faithful of the faithful, admits that "the transition from the idea qua logic to the idea qua nature has always remained affected with a certain obscurity in Hegel's philosophy" (*Hegel als deut. Nat.* 46–47). Later

Hegelians have not advanced much beyond their predecessors on this point, and Grubich no doubt expresses their general opinion when he acknowledges that Hegel "constantly wrapped the description of the transition from the idea to nature in something of a mystic veil" (*Über das Verhältnis Hartmanns zu Hegel und Schopenhauer*, Leipzig: Fritz Eckhardt, 1908, p. 15). But on the other hand the most authoritative interpreters of the Hegelian position fully admit that the establishment of a transition between thought and reality in this doctrine is an inescapable necessity. For example, Caird, while protesting that Hegelian metaphysics "does not, as is often supposed, supersede science by an *a priori* construction of the universe," nevertheless states that "if to find thought in things be more than an empty word, then the movement or process, which thought *is*, must explain at once the transition from thought to what in opposition we call '*things*,' and must give us the means of reconciling that opposition" (*Essays on Literature and Philosophy*, Glasgow: James Maclehose and Sons, 1892, 2:437, 534).

5. Andrew Seth, *Hegelianism and Personality*, 2nd ed. (Edinburgh: William Blackwood and Sons, 1897), pp. 132 ff.; cf. also pp. 111 ff., 117.
6. See Ch. 11, pp. 266–267 above, where, moreover, we saw to what extent Hegel himself seems to be confounding the being of Parmenides with nonbeing.
7. Spinoza, *Ethics*, Pt. 1, Def. 1: *Per causam sui intelligo id, cujus essentia involvit existentiam, sive id, cujus natura non potest concipi nisi existens* [By that which is *self-caused*, I mean that of which the essence involves existence, or that of which the nature is only conceivable as existent].
8. Kant, *Kritik der reinen Vernunft*, *Werke*, ed. Cassirer (Berlin, 1913), 3:411–415 [Norman Kemp Smith trans., B620–630].
9. Andrew Seth, *Hegelianism and Personality*, 2nd ed. (Edinburgh: William Blackwood and Sons, 1897), pp. 122, 133.
10. *Phil. des Rechts*, 8:17 [Knox 10]. The dominant position attributed to the principle is indicated by the double form in which it is expressed – *was vernünftig ist, das ist wirklich, und was wirklich ist, das ist vernünftig* [Knox: "*What is rational is actual and what is actual is rational*"] – as well as by the typographical arrangement: it is printed in italics (*Sperrschrift*) and centered on two lines in the middle of the page.
11. Here again Hegel had been preceded by Schelling. "History as a whole is a progressive, gradually self-disclosing revelation of the absolute" (*Transc. Idealismus*, I, 3:603 [Heath 211]).
12. One might say that the polemic surrounding this famous apothegm concerning the real and the reasonable has never ceased. In his preface to the *Philosophie des Rechts* Gans takes great pains to reduce it to a tautology (7:x ff.), and Jean Jaurès also seeks to defend, or at least to explain it (*De primis socialismi germanici lineamentis apud Lutherum, Kant, Fichte, et Hegel*, Toulouse: A. Chauvin et fils, 1891, p. 80), while, quite recently, Paul Roques declares it entirely tautological (*Hegel, sa vie et ses oeuvres*, Paris: Félix Alcan, 1912, p. 230).
13. Schelling's adversaries accused him of having plagiarized Fichte at the beginning of his career, but Fichte himself does not appear to have ever made any such

accusation (Kuno Fischer, *Geschichte* 7:143). Schelling's *Philosophy of Nature* seems so much an obvious complement to Fichte's philosophy that Kuno Fischer could say that Fichte had "by-passed the solution" of this problem (7:5). Xavier Léon, in his fine work entitled 'Fichte contre Schelling' (*Rev. de méta.* **12** [1904] 949–976), followed the disagreement between the two philosophers through all its phases and cleared up many points the German biographers had left obscure. We see there in particular how carefully Fichte applies himself to refuting his young rival after 1801. He is still combatting him in his last important work, the *Theory of the State* [*Die Staatslehre*], which appeared hardly a year before his death (Léon 972, 974). Bréhier has drawn attention to Metzger's doubts as to the preponderant influence of Fichte, and it is easy to believe that the young Schelling had fathomed the philosophy of his elder less than he himself imagined. However, Bréhier himself points out over and over again the connection between the positions of the two philosophers (*Schelling*, Paris: Félix Alcan, 1912, pp. 4, 8, 234).

14. Schelling, *Darlegung*, I, 7:10 ff. One cannot say that Schelling is actually unfair to Fichte on this matter. For example, Xavier Léon, a great authority on Fichte's doctrine, spells out the relation between his position and Schiller's as follows: "For Schiller nature conserves a reality outside spirit; it still has an existence in itself; for Fichte nature is entirely the unconscious work of spirit; it is the limit spirit opposes to itself in order to determine itself by reflecting itself" ('Schiller et Fichte,' in Charles Schmidt et al., *Etudes sur Schiller*, Bibliothèque de philologie et de littérature modernes, Paris: Félix Alcan, 1905, p. 88).
15. *Zur Geschichte*, I, 10:107 ff. Furthermore, it would have been unseemly to protest too vigorously, for in one of his first works, the *Ideen zu einer Philosophie der Natur* (published in 1797), he had proclaimed his philosophy to be "nothing else than natural science" (I, 2:6 [Harris 5]). In the same vein, his disciple Steffens declares: "The philosophy of nature is not a part of philosophy; it is itself all of philosophy, subject only to the potency [that is, the determination] of the form" (*Grundzüge der philosophischen Naturwissenschaft*, Berlin: Realschulbuchhandlung, 1806, p. 16 [Meyerson's brackets]).
16. *Zur Geschichte*, I, 10:146 ff. The expression we translate as "the pure *how* of things" is "*das reine* Was *der Dinge.*" Cf. *Philo. der Offenbar.*, II, 3:58: "Only sensation [*der Sinn*] can teach that something *is*, or in other words that it exists, even in the case of something that reason has grasped" [the bracketed insertion is Meyerson's, but the German actually reads "*die Erfahrung*" (experience)]. Moreover, in the continuation of this passage he indicates that he means to oppose himself to Hegel on this point. Cf. also II, 3:63, 127, 163, 171. This is the distinction between essence and existence, which of course goes back much further than Schelling; cf. for example Leibniz, *Opuscles* 9 [Parkinson 134].
17. The original has "*das erste Zufällige* (*Urzufall*)."
18. On this theory cf. IR 79–80 [Loewenberg 79–80, 107].
19. *Ideen*, I, 2:209, 212 ff. [Harris 167, 170 ff.]. Early in his career Schelling had stated that all the sciences would ultimately be reduced to a "universal mathematics" (*Allgemeinen Uebersicht der neuesten philosophische Literatur*, *Werke*, I, 1:463). He declared himself a supporter of dynamism; a chapter of the *Ideas* (I,

2:178 [Harris 143]), is entitled 'On Attraction and Repulsion in General, as Principles of a System of Nature,' and later in the same work he asserts that "all quality of matter rests wholly and solely on the intensity of its basic forces" (3:272 [Harris 216]). He also attempts to link his own work to that of atomism by claiming kinship with "dynamic atomism." But with his characteristic fickleness, as early as 1801 he abandons this point of view (which had been attacked by Eschenmayer), declaring that he had wanted to use it only "as a way of entering into the system." He adds: "I gladly leave this whole atomistic conception to Eschenmayer and all those who wish to practice with it" (*Über den wahren Begriff der Naturphilosophie*, *Werke*, I, 4:94).

20. *Transc. Idealismus*, I, 3:340 [Heath 6; Meyerson's brackets – we have changed "intuition" to "perception" to conform to Meyerson's translation].
21. *Transc. Idealismus*, I, 3:340–341 [Heath 6]. In a much later work, the *Darlegung des wahren Verhältnisses* (published in 1806), he writes along the same lines: "In this intention directed toward the knowledge of being, philosophy agrees entirely and immediately with physics ... and it is a futile enterprise to try to provoke strife or discord between them. Furthermore, it is a ridiculous enterprise when entered into with such obvious ignorance of the essence of these two sciences as there is on the part of Fichte" (I, 7:100).
22. Cf., however, our comments on this subject in Ch. 15, p. 406.
23. Cf. *Ideen*, I, 2:6 [Harris 5]: "It remains for philosophy to interpret what is read" in the book of nature by physics, chemistry and mathematics. Moreover, though Schelling was the uncontested master of the *philosophy of nature*, he did not invent this way of considering phenomena. Franz von Baader, whose work is in many respects inspired by very analogous tendencies, is considerably earlier (cf. Émile Bréhier, *Schelling*, Paris: Félix Alcan, 1912, pp. 184 ff.), while simultaneously with Schelling, but independently of him, another German thinker, Eschenmayer, actually attempted, in a limited field, a *philosophy of nature* that he links directly to Kant (we can see that this idea was really "in the air" at the time, which explains its rapid success). Eschenmayer dealt chiefly with magnetic phenomena, which he of course seeks to construct *a priori*. He explicitly claimed priority with respect to the philosophy of nature (Rosenkranz, *Schelling* 151 ff.), stressed the mathematical side of the theories and criticized Schelling for having neglected that aspect, but constantly displayed great admiration for the latter, on whom, as a matter of fact, his works appear to have exercised significant influence (Kuno Fisher, *Geschichte* 7:44–45). The correspondence of Schelling published by Gustav Leopold Plitt under the title *Aus Schelling's Leben in Briefen* (Leipzig: G. Hirzel, 1869–70) contains a great many letters between him and Eschenmayer.
24. *Ideen*, I, 2:70 [Harris 52]. In a work published shortly thereafter, the term "speculative physics" even appears in the title: *Introduction to the Sketch of a System of the Philosophy of Nature, or On the Concept of Speculative Physics and the Internal Organization of a System of this Science*. The work is dated 1799 (*Werke*, I, 3:269–326) and paragraph 3 is entitled: "The Philosophy of Nature is Speculative Physics" (I, 3:274).
25. *Einleitung zu dem Entwurf*, I, 3:277. It should be noted that Schelling was quite

consciously making an important statement of principle here. Indeed, the passage we quote is almost entirely italicized in the original, a fact all the more noteworthy because Schelling appears much less inclined to abuse this device than German writers in general.

26. *Ideen*, I, 2:5, 114, 116, 164, 313 [Harris 4–5, 89, 90, 127, 248]; *Weltseele*, I, 2:501.
27. In response to Fichte's attacks, cf. especially *Darlegung*, I, 7:108, where Coulomb's discovery of diamagnetism is explained as a confirmation of Schelling's "discovery" that magnetism is a general property of matter. In 1832, on the occasion of Faraday's discovery of magnetoelectric induction, Schelling, who does not yet know this work except through a brief and partially erroneous paragraph in the daily newspaper, nevertheless claims publicly in a session of the Academy of Munich that he had affirmed the unity of electricity, magnetism and chemical action thirty years before ('Ueber Faraday's neueste Entdeckung,' *Werke*, I, 9:446 ff.). Still later, in the *Darstellung des Naturprocesses*, a posthumous work reproducing a course taught at Berlin in 1843–44, he points out that assertions of the philosophy of nature in this area preceded the discovery of the Voltaic pile (*Werke*, I, 10:358). On the other hand, in this document he is forced to admit a considerable disappointment in an analogous case. In the *Darlegung* he had prided himself on having earlier affirmed that two different bodies could always be considered as the two sides of a magnet and that a German scientist (Ritter) had subsequently constructed a genuine magnetic needle by soldering a piece of copper to a piece of silver. The experiment was not confirmed, however (I, 10:350).
28. *Abhandlungen zur Erläuterung des Idealismus der Wissenschaftslehre*, *Werke*, I, 1:360, 362, 373–374.
29. It is hardly necessary to warn the reader that we use the terms "Spinozistic" and "Spinozism" in the sense they were given by Schelling and especially by those connected with him, with no claim whatsoever as to the true meaning of Spinoza's philosophy, which does not concern us here. Fichte uses the terms in almost the same sense when, in opposing himself to Schelling, he speaks of "new systems in the manner of Spinoza [*spinozisirende neuere Systeme*]" (Xavier Léon, 'Fichte contre Schelling,' *Rev. de méta.* **12** (1904) 952, n. 8). It is chiefly during his late period when he is reacting violently to Hegelian monism that Schelling declares that "the philosophy of Spinoza is not an empty unitary doctrine. Spinoza is not a mere successor to the Eleatics, his oneness [*sein Eins*] is not the abstract oneness of Parmenides, but a oneness genuinely embracing everything .... A mind like Spinoza's ... could not go back to those indigent elements whose poverty had already been exposed by the Socratic dialectic and in which only our contemporary anti-Socratic dialectic could see an exalted wisdom. Its substance is not simply an empty oneness; he conceived it as substance both extended and thinking" (*Philosophie der Mythologie*, *Werke*, II, 2:71). [Meyerson's brackets.]
30. *Darstellung meines Systems der Philosophie*, *Werke*, I, 4:105.
31. *System der gesammten Philosophie und der Naturphilosophie insbesondere*, *Werke*, I, 6. In this edition the work extends to 440 pages, as compared to little

more than a hundred for the *Darstellung*. The editor dates it 1804.

32. On these deductions see Ch. 14, p. 388, n. 3.
33. This idea of a bipartite philosophy whose two aspects are opposed to one another, while mutually evoking and completing one another, appears to have haunted Schelling throughout his career. It entirely dominates the philosophy of his late period. However, in that period Schelling places the opposition between his two philosophies, which he then terms negative and positive, on a quite different terrain than he had in the *System of Transcendental Idealism*. Cf. p. 335 above.
34. *Allgemeine Deduktion des dynamischen Processes*, *Werke*, I, 4:3.
35. On the evolution of this work, see Appendix 17.
36. Eduard von Hartmann, *Schelling's philosophisches System* (Leipzig: Hermann Haacke, 1897), p. 3.
37. Émile Bréhier, *Schelling* (Paris: Félix Alcan, 1912), p. 244.
38. See for example *Erste Vorlesung in Berlin*, *Werke*, II, 4:359–60, and *Phil. der Offenbar.*, II, 3:86–87.
39. Bréhier does no more than speak for all those who have more or less studied Schelling's work in noting that the latter "never could arrive at the complete exposition he had in mind" (*Schelling*, Paris: Félix Alcan, 1912, p. 83). Schelling, during the final period of his career when he consciously opposes Hegel, knows quite well where the shoe pinches. He takes great pains to defend his philosophy against the reproach of "not being a system." This philosophy "was, on the contrary, a born system, its singularity consisted precisely in the fact that it was a system. The question of whether the external presentation conformed more or less to the Scholastic [*schulmässig*] precepts could be considered irrelevant; the system was included in the thing itself, and he who possessed the thing thereby possessed the system." But this passage is preceded by another which candidly recognizes Hegel's merit in this respect. Hegel, "while others almost without exception were stumbling, at least firmly maintained the method, and the energy with which he established a false system, but nevertheless a system, could, if well directed, have been of inestimable value for science. That is exactly what made him so compelling. In fact, I have seen that those who expounded him the most ardently spoke very little (except for a few apothegms and aphorisms) of the particulars of his doctrine, whereas they always pointed out the fact that his philosophy is a *system*, and a complete one" (*Phil. der Offenbar.*, II, 3:87 ff. [Meyerson's brackets]). In response to the violent attacks against their master, the Hegelians take malicious pleasure in pointing out how Schelling's conceptions vary as to the very foundations of his *philosophy of nature*. For example, Rosenkranz observes that the primitive force is gravitation in the *Ideas*, light in the *Weltseele* and magnetism in the *Entwurf* (*Schelling* 95).
40. Cf. the continuation of the same passage in Appendix 15, p. 588.
41. Rosenkranz notes in 1843 that Schelling disavows and censures all published accounts of his teachings (*Schelling* v).
42. In a response to friends who wondered at his silence, Schelling himself alluded to the difficulties he was encountering in seeking to express his ideas in "a whole that conformed completely" to what he wanted. This time he means "to consider

only his own satisfaction" and "to express the whole of himself." His work contains "slag" and "it is difficult to extricate himself from everything that is creating an obstacle" and "thus free [himself] completely." It is true that he announces to his correspondent that he has finally succeeded, but we have seen how he deluded himself on this point (Kuno Fischer, *Geschichte* 7:165). It is thus, as Fischer realized, an interior conflict, a "lack of harmony between himself and his work," that was at the root of the philosopher's silence.

43. Victor Cousin, *Fragments et souvenirs*, 3rd ed. (Paris: Didier, 1857), p. 79.
44. Victor Cousin, *Fragments philosophiques*, 2nd ed. (Paris: Ladrange, 1833), p. xxxvii [*Philosophical Essays*, trans. George Ripley (Edinburgh: Thomas Clark, 1839), p. 90].
45. We saw above (p. 346, n. 39) that Schelling, for his part, was perfectly aware of the attraction exercised by Hegel's philosophy precisely because it was a *system*.
46. The author of the publication was Paulus, a professor at Heidelberg, a former friend of Schelling's but for many years his bitterest enemy. The friendship dated from the very beginning of Schelling's literary career (see the very flattering letter from Paulus dated 1793 – Schelling is barely eighteen! – in Plitt, *Aus Schelling's Leben* 1:37). The rupture occurred in 1803 or 1804, as we see by a very friendly reference to Paulus in a letter from Schelling to Hegel on 11 July 1803 and a hostile passage in a letter to Eschenmayer on 22 December 1804 (Plitt 1:467 and 2:45). Another remarkable facet of this quarrel, in which both sides were extraordinarily relentless, is the age of the two adversaries. At that time Schelling was in fact seventy-two and Paulus eighty. Schelling lost his case. See Kuno Fischer, *Geschichte* 7:261 ff.
47. Kuno Fischer, *Geschichte* 7:266. Cf. similar statements in *Phil. der Offenbar.*, II, 3:86 ff. Cf. also *Erste Vorlesung in Berlin*, *Werke*, II, 4:365: "If a man has gone further astray, it is because he has dared more; if he has missed the goal, it is because he has followed a path his predecessors had not closed to him."
48. [Terence, *Eunuch* 61–63: If you should insist on doing this with a clear rationale, you would do no more than strive to be insane and rational simultaneously.]
49. *Phil. der Offenbar.*, II, 3:85, 86, 92, 93, 151, 171, 178.
50. *Phil. der Offenbar.*, II, 3:58, 59, 88–89, 127, 162–163, 171.
51. Victor Delbos, *De posteriore Schellingii philosophia, quatenus hegelianae doctrinae adversatur* (Paris: Félix Alcan, 1902).
52. Eduard von Hartmann, *Schelling's philosophisches System* (Leipzig: Hermann Haacke, 1897), p. 142.
53. *Darstellung des Naturprocesses*, *Werke*, I, 10:301–390.
54. Eduard von Hartmann (*Schelling's philosophisches System*, Leipzig: Hermann Haacke, 1897, pp. 19, 52 ff., 142) strongly insisted on Schelling's intrinsic realistic tendencies and on the opposition between these tendencies and his apparently extreme and at any rate loudly proclaimed idealism. But of course Hartmann (for whom the epistemological point of view is not paramount) views this conflict from a somewhat different perspective than we do.
55. *Über das Verhältnis des Realen und Idealen in der Natur oder Entwickelung der ersten Grundsätze der Naturphilosophie an den Principien der Schwere und des Lichts*, *Werke*, I, 2:356–378. From its placement in this edition, this work is

pushed back to the year 1798 (the date of the first edition of the *Weltseele*), while in reality it appeared only eight years later. Cf. Rosenkranz, *Schelling* 274.

56. *Über das Verhältnis des Realen und Idealen in der Natur*, *Werke*, I, 2:362; Rosenkranz insists with some justification on this fact (*Schelling* 275).
57. *Abhandlungen zur Erläuterung des Idealismus der Wissenschaftslehre*, *Werke*, I, 1:395.
58. *Aus der 'Allgemeinen Übersicht der neuesten philosophischen Literatur,'* *Werke*, I, 1:467; this work is actually an integral part of the preceding one (cf. I, 1:453, note).
59. *Philosophische Untersuchungen über das Wesen der menschlichen Freiheit und die damit zusammenhängenden Gegenstände*, *Werke*, I, 7:376 [*Of Human Freedom*, trans. James Gutmann (Chicago: Open Court, 1936), pp. 52–53]. See also 7:395–396 [Gutmann 75]: "All nature tells us that it is in no wise the product of mere geometric necessity; not sheer, pure reason, but personality and spirit are in it .... Otherwise geometric reasoning which has ruled so long must long since have fully penetrated nature and achieved its idol of universal and eternal laws of nature more fully than has yet occurred, since it must daily rather recognize more fully the irrational relationship between nature and itself."
60. *Phil. der Offenbar.*, II, 3:205: "*Wollen ist die Grundlage aller Natur.*" Cf. on the same subject, among others, *Einleitung in die Philosophie der Mythologie*, *Werke*, II, 1:464, 481; *Phil. der Offenbar.*, II, 3:93, 214. Cf. also Eduard von Hartmann, *Schelling's philosophisches System* (Leipzig: Hermann Haacke, 1897), pp. 59, 119. Furthermore, Hartmann believes that this voluntarist element in Schelling's philosophy was the source for Schopenhauer's opinions in this domain (pp. iii-iv; cf. also his *Schelling's positive Philosophie als Einheit von Hegel und Schopenhauer*, Berlin: Otto Loewenstein, 1869, passim). But Rosenkranz has already pointed out the analogy between the positions of the two philosophers (*Schelling* viii-ix).
61. Cf. Hartmann, *Schelling's philosophisches System* 213, where, as a matter of fact, this part of his philosophy is judged "a failure."
62. Heinrich Heine, who, whatever his own opinion, had only a rather limited competency in philosophic matters, sometimes expresses the opinion of the general literate public all the better for that. "The doctrine of Spinoza and the philosophy of nature, as explained by Schelling during his best period," he declares, "are essentially one and the same thing," and in another passage he explains that "the only merit of the modern Philosophy of Nature lies in demonstrating, in the clearest manner, the eternal parallelism that exists between spirit and matter. I say spirit and matter, and I employ these expressions as equivalents for what Spinoza calls thought and extension" (*De l'Allemagne*, *Oeuvres de Henri Heine*, Paris: Michel Lévy Frères, 1855, 5:220, 94–95 [*Religion and Philosophy in Germany*, trans. John Snodgrass (Boston: Beacon Press, 1959), pp. 150, 73]). One sees that what captivated the public about Schelling's ideas was indeed the fact that they presented themselves as progress toward the demonstration of the rationality of nature, that is, toward exactly the goal Hegel claimed to have achieved. Moreover, this point of view is one that orthodox Hegelians constantly stressed, and Karl Hegel, the editor of his father's

correspondence, speaks as if it were taken for granted that Schelling's merit is to have first "posed the principle of the identity of thought and being, of spirit and nature, a truth whose implementation as a philosophic system, according to the dialectical method, was reserved for Hegelian philosophy" (Hegel, *Briefe*, 19_2:389). Cf. also Rosenkranz, *Schelling* xxiii, xxix, 7.

63. Cf. for example the title of Rosenkranz's [sic] work, *Hegel: der unwiderlegte Weltphilosoph* (the unrefuted world philosopher). [The author of the work is actually Karl Ludwig Michelet (Leipzig: Duncker & Humblot, 1870.]

64. "*Zurück zu Kant!*" (Back to Kant!) was the veritable rallying cry which, as it were, summed up the program of German philosophy for an entire generation. It was launched by Otto Liebmann in 1865 (*Kant und die Epigonen*, 2nd ed., Berlin: Reuther & Reichard, 1912, p. 216). [Liebmann's text actually reads "*Es muss auf Kant zurückgegangen werden.*"]

65. The author of these lines remembers quite well that in his younger days, when "structural chemistry" was beginning to carry the day, the "theory of types" was the object of violent attacks of this sort and that the term *aberration* was quite generally applied to this theory in discussions among laboratory chemists. And yet the ideas of Dumas, Laurent and Gerhardt paved the way for those of Kekulé to such an extent, and moreover bear such a resemblance to them, that it is almost impossible to imagine the latter without the former.

CHAPTER 13

HEGEL AND COMTE

From the scientific point of view, as we have said, Hegel's attempt appears curiously aberrant. We are surprised that it occurred so late in time, since barely a century separates it from us; from the standpoint of logic it looks rather as if it could have been produced well before Descartes: it is something of an anachronism. What made it possible is obviously Hegel's attitude toward science. Granted he says he wants to respect science and profit from its teachings, but only on the condition that he can be selective about it, accepting part of it and abandoning the rest. What he accepts is only the empirical part: facts and generalizations based directly on these facts. Everything theoretical, explanatory, leading more or less to assumptions about the inner nature of things, he believes must be rejected out of hand.

This bias seems quite astonishing to us today, for the prestige of science has grown enormously in the meantime and we are accustomed to treating it with incomparably more respect. To temper our disapproval, however, it will suffice to observe how much this attitude in some respects fundamentally resembles that of August Comte.

This affirmation may seem somewhat surprising at first glance, given that we customarily represent the positions of these two thinkers, and especially their attitude toward science, as being at opposite poles: Comte would seem to be "the antithesis" of Hegel from all points of view.[1] There is certainly much truth in this suggestion, and we ourself shall have occasion to insist on this aspect of the two doctrines a little later. Nevertheless, the analogy between the two thinkers is not purely specious. In particular, nothing is easier than finding theoretical declarations of a frankly positivistic cast in Hegel.

For example, he states that Kepler demonstrated his laws "in the sense that he found for the empirical data their general expression." Kepler discovered his laws by induction, based on Tycho's observations; "to elicit the universal law from these isolated phenomena is the work of genius in this field." It has since been claimed that the proofs of these laws had been discovered only by Newton. "Seldom has fame been more unjustly transferred from a first discoverer to another person." Newton's method can offer advantages of convenience from the standpoint of the analytic process; but, as we saw above (p. 282), the distinctions and determinations introduced through mathematical analysis are to be kept entirely distinct from what must have a physical reality. It must be recognized that, because this distinction has not been preserved, "physical mechanics is steeped in an unspeakable metaphysics" (*Naturphilosophie*, 7_1:98–107 [Miller 65–72]).[2]

Of course Hegel does not for a moment dream of forbidding aprioristic constructions, since his effort in the scientific domain amounts precisely to an attempt at such a deduction. But in his system the revelations that must result from the deduction arrive in science from outside, as it were; as for "empirical science" itself, it must limit itself to gathering observations and fashioning them into laws. Its proper task is accomplished as soon as these laws are formulated, since law must suffice for scientific analysis, as Hegel puts it in the case of Kepler. We need hardly point out how consistent this point of view is with Comte's (aside, of course, from the metaphysical construction which Hegel would superimpose on this purely legalistic science [see Ch. 1, n. 39, p. 30]). Furthermore, the example of Kepler's laws and Newtonian reduction is particularly well chosen in this connection. In fact, Hegel was perfectly correct in maintaining that from Kepler to Newton there is not merely progress in the more precise determination of the paths of the planets as a result of the perturbations they exert on one another (what is more, this is a part of Newton's work whose merit Hegel recognizes; see *Naturphilosophie*, 7_1:101 [Miller 67–68]), but that there is also the introduction of the concept of mutual attraction, a concept that, whatever formal reservations Newton himself may have had, certainly ends up transforming what in Kepler was only a collection of numerical relations, into an ontological being: the force of gravitation. For our imagination would seem to be incapable of effecting such a transformation so long as the relation in question remains too complicated, and for that reason the resistance of Hegel (who is horrified by such "metaphysics" born directly of mathematical deduction) to the Newtonian simplification itself is perfectly understandable. We ourself – in seeking to imagine what a truly positivistic or phenomenalistic science could be like, one that dissolved the common sense object but replaced it with no new reality, nothing objective, neither atom nor mass nor force nor energy nor hypostasized quality – have had recourse to this image of a science coming to a halt, so to speak, at Kepler's laws (IR 504 ff. [Loewenberg 433–434]), and it is obvious that the coincidence is not at all fortuitous.

We could multiply examples and citations here almost at will. Let us merely add one particular point. We saw that Hegel combats the fundamental conception of the chemical element. What disturbs him especially is the assumption that the element persists in compounds, an assumption in which he sees (and rightly so) a manifestation of the tendency toward causal explanation. Now this opinion that the elements do not persist in chemical compounds is not as extravagant as one would at first think. Not of course in the form it assumes for Hegel – no chemist worthy of the name has believed, since Lavoisier, that one element could be transformed into another during a chemical reaction or a biological process – but in a less crude form, it has sometimes enlisted distinguished supporters. Henri Sainte-Claire Deville held such an opinion, and quite recently the position was defended by Ostwald (cf. Ch. 10, p. 253, and IR 264 [Loewenberg 237]). For both these scientists this conception is certainly tied to their hostility toward atomism and, at least for Ostwald, to a positivistic or pseudopositivistic profession of faith (cf. Ch. 6, p. 164).

The undeniable analogy between Hegel and August Comte insofar as their attitude toward science is concerned manifests itself in an external but altogether typical way: neither one, in speaking of *science*, really means to designate what his contemporaries understood by the term. What Auguste Comte has in mind is a science greatly purified according to his prescriptions, having forsworn, for example, overly close scrutiny of phenomena by means of "equivocal" research instruments such as the microscope [*Cours* 3:369–370], strictly abstaining from all research capable of shaking an established law such as Mariotte's principle, and also forbidding any investigation whose present or future utility (according to some indefinite criterion) is not previously established, as, for example, research into the chemical constitution of the celestial bodies, or even, in general, all astronomical research extending beyond the limits of the solar system. And of course it is also a science freed from theories or at least granting them only an entirely subordinate position, not allowing them to concern themselves with the inner nature of phenomena, nor even, in general, with the way phenomena are produced. This entirely *positive* science – which is the only one the "true speculative regime" of the future will tolerate, dealing severely if need be with all those whose works would threaten it with "active disorganization" [*Cours* 6:638–639] – is what Comte means when he speaks of *science* without qualification. It is what allows him to vituperate against Regnault [2:453], to ridicule Leverrier, whose "alleged discovery ... if it could have been real ought to have truly interested only the inhabitants of Uranus," to deny that Fresnel's admirable optical theories have any interest [2:453] and to accuse the great chemists of his time "of a metaphysical turn of mind."[3] For Hegel it is much worse: his "science" is exclusively the one that deduces everything from the concept, or, in other words, his own *Naturphilosophie*; the other, the science of the scientists, he calls *raw empiricism*. In many passages of his works he opposes the two terms "scientific" and "empirical," always in this same sense.

That is a bias so strange that at first glance it might seem incomprehensible. But one must realize that it is not without analogy in the history of human thought. On the contrary, here as on many other points, Hegel is closely connected to Kant.

It is true that Kant professes opinions on the essence of scientific knowledge that one would at first judge diametrically opposed to those of Hegel. Whereas the latter, as we have seen, means to banish mathematics from physics, Kant states that "in every special doctrine of nature only so much science *proper* can be found as there is mathematics in it." But we must note that in order to arrive at this conclusion, he nevertheless started from premises exactly like Hegel's. For Kant as well, nothing is scientific except what can be deduced. "Only that whose certainty is apodeictic can be called science *proper*; cognition that can contain merely empirical certainty is only improperly called science." A theoretical knowledge "deserves the name of natural science only when the natural laws that underlie it are cognized *a priori* and are not mere laws of experience." Applying these rules to the chemistry of his time, Kant declares that even "the most complete explication of certain phenomena by chemical principles always leaves dissatisfaction in its wake, inasmuch as through these contingent laws learned by mere experience no *a priori* grounds can be adduced."[4] Thus, since nature must be partially, but essentially, rational, anything that is not rational cannot, by that very fact, belong to this *essence*. Hegel shares this opinion, except that for him deduction is no longer mathematical, but logical. Anything that cannot be deduced is not science and thus everything Comte understands under that name remains strictly excluded from Hegel's science. The "margin" that is left by Hegelian deduction and that includes a body of empirical facts and laws with which it formally refuses to concern itself, that it disdains, so to speak, is exactly broad enough to include the whole of "positive" science.

In this respect, then, the Hegelian system can, strictly speaking, be reconciled with positivism, which explains why this coupling of the two has been, implicitly or explicitly, attempted time and time again.[5]

Obviously Comte and Hegel were wrong. Not, of course, in judging that in principle science could be criticized, for in creating science, the human mind obeys a sort of obscure instinct, obeys tendencies whose true nature it cannot itself succeed in knowing except through a process of analysis. Now it is not to be taken for granted that the scientific instinct must always be infallible. As a matter of fact, assuming that the science of today represents a perfect model – which can only be a postulate – we are forced to observe that the science of the past has not always resembled it, and that consequently mankind has frequently gone astray on the path leading to its present state. It considered as sciences things we can no longer believe worthy of the name, such as the art of the haruspices, magic, judiciary astrology and alchemy. Similarly the path to which scientific thought was committed during a long succession of centuries, and not too long ago at that – the path of Peripatetic science – now appears to us to be a dead end. Those who criticized these errors fought the good fight, and humanity owes them endless gratitude for having put it back on the right track. Indeed, both Hegel and Comte thought that is what they were doing. Their mistake is thus not one of principle but one of *degree*. They treated science too lightly; they were too ready to believe that they had completely laid bare its foundations. This is how Comte was led to mistake a part of science for the whole; and Hegel, who had gone more deeply than Comte from the theoretical point of view, used his analysis only as an instrument of war against science itself. If, on the contrary, we truly take the trouble to scrutinize all of modern science, we see that it constitutes a harmonious whole, that in fact almost *everything* in this edifice is perfectly justified and that there is no part which must be rejected, nor even considered a sort of temporary appendage; and we likewise arrive at the conviction that despite their paradoxical appearance, the principles that guide it are the right ones and even, insofar as we can tell at the present time, the only possible ones, since they indicate the unique way by which man has thus far been able to fathom the secret of nature.

Comte's first positivistic work, the *Considérations philosophiques sur la science et les savants*, dates from 1825, and the *Cours de philosophie positive* began to appear in 1830, the year before Hegel died at just over sixty years of age. We can thus consider these philosophers contemporaries and ask ourselves if the spirit of the times did not play a role; that seems at least probable. Indeed, on the one hand, the practical results of science did not command as much admiration as they do today and, on the other hand, the time was not forgotten when science had occupied only a subordinate rank, when theology had claimed the right to dominate it, to call it to task, to dictate the way it must follow. Since theology had to renounce this dominant role, was not philosophy (which claimed to have brought about this revolution) destined to take its place? And why would philosophy respect the authority of the scientists any more than it had that of the priests? Hegel, to be sure, was profoundly impressed by the spectacle of the overthrow of prevailing ideas and by that of the ensuing political revolution (see Appendix 5, p. 568). And as for Auguste Comte, it is quite well-known that the fact that the spiritual foundations of the old prerevolutionary world could have been shaken was a veritable obsession with him and that a type of government based on the unity of sentiments and beliefs which had characterized the society of the Middle Ages appeared to him to be the supreme ideal.

Because he recognized the impossibility of reconstituting this unity on the terrain of Catholicism, Comte had conceived the idea of replacing religion by science, but by a science that, at least in its form – the immutability of its dogmas, a strongly constituted spiritual hierarchy armed with the secular sword by the very fact that it was to dominate the whole of the social edifice – would bear a close resemblance to its ecclesiastical model.

Hegel certainly never formulated so grandiose a plan. But that he likewise aspired to establish an official doctrine exercising control over the minds of the populace (his own doctrine, of course), is made quite clear in a plan he submitted to the minister Altenstein for instituting a publication that would bear the official government stamp and formulate what was henceforth to be considered orthodox.[6] The minister, as devoted as he was to Hegel, was astute enough not to go along completely with these visions of organization run wild. But he did put at the philosopher's disposal everything the already existing organs of the State could provide, and Hegel thus experienced the joy of at least partially putting into practice the intellectual constraint of which Comte could only dream (see Appendix 5, pp. 566–567).

That Comte and Hegel were only sons of their time can be seen through the example of a man who was far from being their intellectual equal and yet, like them, conceived (and partially implemented) a plan for an official doctrine to control minds and to force them, by all the means at the State's disposal, into a single mold. We do no more than express the present consensus on the philosophy of Victor Cousin in pointing out his lack of originality, and even an absence of unity and precise contours in his thought. However, its author, profiting from his high official position and the prestige his teaching enjoyed at the time, meant to impose this philosophy, just as it was, on educated Frenchmen in general. We can see through Cousin's witty biographer how he went about it[7] and how, in 1848, at the very moment the July monarchy was crumbling, two of Cousin's disciples were busy writing a manual which included only passages from the writings of the master. This handbook, says Jules Simon, "would have been officially authorized and officiously prescribed." And he adds: "Philosophy would have had its catechism. It already had its bishop" (*Victor Cousin* 117 [Masson 120]). We should not be surprised then that Auguste Comte dreamed of an analogous constraint on behalf of a doctrine whose pretensions were certainly a great deal more justified than those of Cousinian eclecticism, and that Hegel, without a qualm, used the support lent him by the secular arm to bring about the actual domination of his philosophy in the State that officially adopted it.

We need hardly point out that it would nevertheless be committing a great injustice to try to put Comte's scientific opinions on the same plane as Hegel's. However bizarre some of the arbitrary demands formulated by the author of positivism, however inconsistent with the principles that truly direct the research of scientists, it is certain that he had an incomparably more accurate understanding of the whole of science and that, what is more, he possessed to a very high degree the scientific instinct Hegel so sorely lacks. Comte's epistemology is only partially correct, in the sense that it defines only a part of science, the part that makes predictions and states laws, which we have called purely legalistic science. But even in denying theoretical and rational science (and he was undoubtedly wrong to do so), Comte nevertheless obeyed a deep-seated inclination that in fact characterizes science today, as we shall see in our final chapter (pp. 523 ff.). Furthermore, Comte proclaimed his partial truth with incomparable vigor, making it the core of a doctrine that embraces all the forms of activity of the human mind. His influence on his contemporaries, and especially on the succeeding generations, was immense; it still endures, and we have seen its repercussions in science, where, moreover, its effect is prolonged by Mach's doctrine, which derives directly from it. There can be no greater contrast than that between Comte's extremely powerful and persistent influence and the complete scorn the scientists of Hegel's time exhibited toward his scientific doctrine, and the scientists of the following generations have shown not the slightest interest in reconsidering this severe verdict.

In seeking to understand Hegel's attitude toward science, we have most assuredly found sufficient cause to justify this scorn. We must, however, add one more trait, one whose role is perhaps not the least important here. Indeed, perhaps no peculiarity of Hegel's personality shocks the inmost feelings of the man schooled in modern science as deeply as his scorn for nature. It has been said that Hegel scorned science, which is true in some sense, as we have just seen, for he treats it quite cavalierly; he calls it to task and means to dictate to it, and everything in it having to do with explanation certainly appears empty to him; however, he respects experimental science, readily admits the importance of its conquests and finds high praise for their authors. On the other hand his contempt for nature is absolute. We do not believe that in the entire immense Hegelian corpus, although it pretends to embrace the whole of man's spiritual activity, can be found one sentence, one expression testifying that a spectacle of nature moved him or provoked his admiration in the slightest. That was, surely, an innate predisposition, and his correspondence reveals that already, at the age of twenty-five, when he visited the Bernese Alps, the spectacle left him indifferent. The sight of the glaciers "has nothing of interest about it," it offers "nothing great or pleasing." The traveler finds in it "no satisfaction except that of having approached such a glacier" and judges that the bottom of the glacier resembles a very muddy street. In general, he observes that "neither the eye nor the imagination discovers in these shapeless masses any point whatever where the former could alight with pleasure and the latter find a subject of occupation or of play." Reason perceives in them "nothing awe inspiring, nothing that imposes astonishment or admiration." The sight of these eternally dead masses, he adds, gave me "nothing more than the monotonous and, in the long run, boring impression: it is so [*es ist so*]."[8]

But perhaps even more characteristic of his attitude is the way he speaks of the starry sky. He finds the admiration Kant professed for this sublime spectacle to be foolish; it is a subject of constant derision for him, one to which he returns on several occasions. The immensity of the celestial spaces is an example of "bad infinity"; one must beware of any admiration or even surprise on its account, and as for the stars, they are quite simply comparable to an eruption on the skin.[9]

It is understandable that with this disposition he does not hesitate to consider the immense mass of natural phenomena to be almost negligible. Scientists may concern themselves with them, but this work has no real interest except insofar as it prepares materials for true science, the one that seeks, in these scattered materials, the warp and weft of the spirit. But everything not related to this deductive network, everything that remains outside it, all this is simply an effect of the "play" of nature, evidence of its spirit of arbitrariness and its inability to realize the concept and, on these grounds, unworthy of arresting the attention of the serious thinker.

Trendelenburg testifies that already in his day it was not unusual to speak of the "logical arrogance" of the Hegelians (*Log. Untersuch.* 1:88). The expression was fully justified, but here we have, without doubt, the most surprising and extreme manifestation of this overweening logic which scorns everything that does not emanate from it. How foreign all that is to the spirit of our times is eloquently demonstrated by the attitude of Hegel's disciples. Not only did they completely dissociate themselves from the Hegelian *philosophy of nature*, but they generally sought to soften the more shocking aspects of the master's intransigence in this domain. McTaggart offers us a characteristic example of this tendency. For him, "we can no more deny that there are signs of rationality in the universe, than we can deny that there are signs of irrationality." But the latter tell "nothing against the rationality of the universe." Indeed, such a conclusion is due quite simply to the fact that our intelligence persists "in demanding what cannot and should not be obtained." This is why "taken by itself, philosophy proclaims its own inadequacy. For it must assert things to be completely rational, and therefore completely explicable, which, all the same, it cannot succeed in completely explaining" (McTaggart, *Studies* 184, 225, 221, 208). Thus it is not the universe that is at fault, it is our reason, which the philosopher declares to be irrational, so to speak, apparently in the name of a superior intelligence, an *intellectus angelicus* for whom the irrationalities we perceive in the external world would be harmoniously resolved. The solution at first appears contradictory – for what is rationality if not conformity to reason, our reason, since we can know no other (McTaggart is not entirely oblivious to this objection; see *Studies* 225) – but basically what it amounts to is quite simply observing that our intelligence is antinomic and contradicts itself. Be that as it may, this attitude certainly involves a kind of humility of mind which could not be further from "the logical arrogance" Trendelenburg mentions.

On the other hand, McTaggart's attitude is perfectly consistent with that of contemporary science. Since the scientist constantly submits his hypothetical deductions to empirical verification and is resigned to overthrowing them entirely if necessary, that is to say, if a better and above all more comprehensive deduction came along, he obviously cannot consider his reason with any of the absolute confidence that inspired Hegel's thought. Therefore nothing is further from his most basic intentions than the idea of abasing nature before his reason. He finds no phenomenon unworthy of his attention, however minute and insignificant it might appear, and even the most trifling of them seems likely to him to reveal something of the mystery he probes unceasingly even while sensing that it is unfathomable. If a difficulty stops him, if the phenomenon completely contradicts the predictions he formulated on the basis of suppositions that nevertheless appeared entirely consistent with everything nature had revealed to him up to that point, he will certainly not dream of accusing nature of revealing its impotence, as Hegel did; he will blame himself and his own inadequacy; he will think that he has *not yet* fathomed the mystery on this point, not for a single instant doubting that if he succeeded in penetrating it, the apparent chaos would be revealed to him as a cosmos full of harmony and beauty. At the very place where science has celebrated its most incontestable triumphs, he feels that what has been accomplished is infinitely insignificant compared with what remains to be done. When he has, by dint of complicated operations, calculated a few points of the catenary curve, he knows quite well that if he simply hangs a cord by its two ends he will immediately see the complete solution of this "transcendental" equation with his own eyes, and when, through successive approximations by means of arduous operations sometimes taking months and years, he has succeeded in more or less determining a part of the trajectory a planet will follow as a consequence of numerous perturbations, again he does not doubt for a single second that nature resolves this problem instantaneously, completely and without the slightest difficulty. Nature "pays no attention to analytic difficulties," said Fresnel, who was nevertheless very nearly Hegel's contemporary. And when he is forced to observe the existence of a permanent barrier to his comprehension, of an irrational, the scientist, rather than passing the same judgment as Hegel, which to him would seem tantamount to blasphemy, will recognize his own impotence with regard to this nature which inspires in him an almost religious awe.

To be sure, that is not quite Comte's attitude. One could perhaps go so far as to say that in accusing nature of not obeying its own laws below certain limits, he was manifesting at least something of the state of mind Hegel reveals with so much arrogance. But as we noted (Ch. 4, p. 77), it would seem that he did not consider these overly minute phenomena which he would close to scientific research as truly arbitrary, but only as too complicated for us ever to succeed in knowing the laws that govern them. In that case, then, it is more a question of too much humility on the part of reason. Moreover, the very foundation of the positivistic doctrine, with its renunciation of all penetration into the domain of the true essence of phenomena and of the way in which they are produced, certainly implies a spirit of humility of this sort and is, in this sense, diametrically opposed to the Hegelian conception.

It remains no less true that the analogies we have pointed out bear on an essential side of the two doctrines. Not only do both Hegel and Comte mean to impose harsh controls on scientific and philosophic thought (which can be explained by the spirit of the times, as we have shown), but they mutilate science in an entirely similar manner, claiming to confine it to the domain of strictly experimental rules. Both of them (although for quite different reasons) completely disregard science's search for rationality, or rather forbid it to make the search, and if this interdiction constitutes the very foundation of positivism (as its name reveals), it is also (as we have noted in Ch. 11, p. 289) the starting point, or at least one of the principal starting points, for the ideas of Hegel.

NOTES

1. Kuno Fischer, *Geschichte* 8:1176. Wundt, for whom Hegel's philosophy and positivism are "the two most remarkable phenomena of the last century, phenomena that, in spite of all their innate defects, contain the significant seeds of a subsequent development," treats them as absolutely antagonistic positions (*Einleitung in die Philosophie*, 5th ed., Leipzig: W. Engelmann, 1909, pp. 266 ff.).
2. Cf. *Wiss. der Logik*, 3:462 [Miller 380]: "The quantitative side of this fact [of planetary motion] has been accurately ascertained by the untiring diligence of observation, and further, it has been reduced to its simple law and formula. Hence all that can properly be required of a theory has been accomplished" [Meyerson's brackets]. Hegel's strange understanding of the respective roles of Kepler and Newton was formed quite early. Indeed, it is found in *De orbitis planetarum* (*Werke*, 16:17): "*Patet inde, quanto purius fuerit Kepleri ingenium et indoles*" ["This shows how much purer the talent and natural inclination of

Kepler were" (Adler 292)]. Alluding to the anecdote according to which Newton's ideas on gravitation were inspired by a falling apple, Hegel jokingly declares that this is consequently the third time the apple has brought misfortune to mankind, having first been the source of original sin, then of the Trojan War and finally of the misfortune of the sciences and philosophy. Cf. Chapter 11, p. 306, note 51, above.

3. Cf. Ch. 4, pp. 73–74, and IR 6, 9, 11, 446 [Loewenberg 20–21, 22, 23–24, 390].
4. Kant, *Premiers principes métaphysiques de la science de la nature*, trans. Andler and Chavannes (Paris: Félix Alcan, 1892), pp. 4–5 [*Metaphysical Foundations of Natural Science* (Indianapolis: Bobbs-Merrill, 1970), trans. James Ellington, pp. 4–6]. Furthermore, we know that for Kant "the construction of concepts," the "presentation of the object in *a priori* intuition," takes place only by way of mathematics (p. 5 [5]).
5. In truth, it does not seem that this side of the Hegelian logic and *philosophy of nature* – although it is the one that least separates it from the modern scientific conception – has been sufficiently considered by the commentators. We have seen that they generally tend to ignore everything having to do with Hegel's properly scientific conceptions (Ch. 11, p. 269), but even when they do speak of them, only rarely do they mention these restrictions. McTaggart is, we believe, the only one to lay any stress on them – to our mind insufficiently so, for if Hegel, as this commentator very aptly says, does not attempt "to reduce everything to a manifestation of the idea" (*Studies* 29; Ch. 11, p. 267, above), it is really only because he conceives the intervention of this *irrational*, which is due to the play of chance, to nature's impotence to realize the concept. Therefore, in declaring that philosophy, although incapable of showing *how* things are rational, "must be contented with an abstract demonstration that things must be rational," McTaggart concedes much more than Hegel would have on this point (*Studies* 207–208).
6. Hegel, *Über die Einrichtung einer kritischen Zeitschrift der Literatur* (*An das Ministerium des Unterrichts eingesandt*), *Werke*, 17:368–390. Unlike the other official documents published in the same volume, this one is undated, but must be from Hegel's final period. The author stresses from the outset that the organ he seeks to found will be an integral part of the administration (17:373, 380, 381); the editorial committee "will at the same time have the status of a governmental organ [*Behörde*]" (17:383 [Meyerson's brackets]). The committee members will of course be appointed by the government and must include officials of the higher administration of public instruction. It will therefore be "really one of the monarchy's central institutions" (17:384) and it will not be possible for any Prussian writer to remain indifferent to the fact of being recognized by the publication, "under the eyes of the higher administration, so to speak" (17:378–379). Professors will thus be led to teach above all what has been really established and to be wary of misguided originality. Here Hegel cites Voltaire's remark: "Germany is a country rich in misguided originals" (17:372).
7. Jules Simon, *Victor Cousin*, 4th ed. (Paris: Hachette, 1910), passim, esp. Ch. 3, pp. 76 ff., 'Le régiment' [*Victor Cousin*, trans. Gustave Masson (London: G. Routledge & Sons, 1888), pp. 79 ff., 'The Regiment'].

8. Cf. the extracts from Hegel's correspondence included in Rosenkranz, *Hegel's Leben* 475–483, and Kuno Fischer, *Geschichte* 8:22.
9. Hegel, *Naturphilosophie*, 7_1:92, 461 [Miller 62, 297]. William Wallace recognizes that Hegel in general "looked down upon the mere natural world" ('Hegel,' *Encyclopedia Britannica*, 11th ed., 13:206). That was certainly a quite characteristic and very individual peculiarity of Hegel's personality, one for which neither his time nor his entourage can be held responsible; German romanticism, following the impetus given it by Jean Jacques Rousseau, one of Hegel's favored intellectual sources in his youth (cf. Rosenkranz, *Hegel's Leben* 17), professed a boundless admiration for nature, particularly its untamed aspects, and Hegel's most intimate friend, the poet Hölderlin, arrived at a veritable deification of nature in his cult of Greek classicism (*Hegel's Leben* iv). Moreover, Schelling had already underlined this distinctive feature of his rival's philosophy: "Jacobi can hardly treat nature worse than Hegel does ..." (*Zur Geschichte*, I, 10:152).

CHAPTER 14

HEGEL, DESCARTES AND KANT

We have seen that Hegel, in spite of his solemn declaration that "the hidden essence of the Universe" is unable "to resist the courage to know" (Ch. 11, p. 268 [*Enc.*, *Logik*, 6:xl]), that is, that nature is entirely intelligible, was nevertheless far from truly wishing to deduce everything, that, on the contrary, he left a *margin* to experimental science (we showed in the preceding chapter how extremely broad this margin is). That is a characteristic he shared with all analogous systems imagined since remotest antiquity. Indeed, we noted in Chapter 4 (p. 92) that the systems of the Ionian thinkers at least implicitly contained reservations of the same sort, and in both Plato and Aristotle these restrictions are formulated explicitly. They are found again in Descartes. To be convinced of this, one need only read the continuation of the passage from the *Discourse on Method* cited in Chapter 11 (p. 274) where he so proudly and confidently proclaims the complete triumph of deduction in the case of knowledge of things "which are the most common and simple of any that exist" (such as "the heavens, the stars, an earth," etc.). Descartes immediately adds:

> Then, when I wished to descend to those which were more particular, so many objects of various kinds presented themselves to me, that I did not think it was possible for the human mind to distinguish the forms or species of bodies which are on the earth from an infinitude of others which might have been so if it had been the will of God to place them there, or consequently to apply them to our use, if it were not that we arrive at the causes by the effects, and avail ourselves of many particular experiments.[1]

Thus the great deductive chain is perfectly continuous; it is, to use the Kantian term, entirely *a priori*, self-sustaining and appealing to nothing that does not come out of reason itself. It is this chain that furnishes everything truly essential in nature. But alongside it there are secondary phenomena. These "more particular" things can no doubt be connected to the deductive chain by going back from the effects to the causes. But they cannot really be deduced from these causes, at least not unequivocally, since an analogous deduction would enable us to affirm the existence of things that do not actually exist. In other words, these phenomena, which might be called adventitious, can be known only *a posteriori*. Around the domain reserved for pure deduction they form a sort of fringe or margin (as we termed it in Hegel) conceded to experience.

Appearances to the contrary, this way of thinking has much in common with that of a great many of our contemporaries, scientists or not. What, after all, is the belief in the possibility of reducing the universe to a mechanism? Let us suppose this deduction to have been accomplished: everything will be intelligible, that is, again, deducible *a priori* – except for a margin which (this is, as we have seen, the presupposition inevitably implied by mechanism) will include, most importantly, qualitative sensation. But this irrational element is ruled to be accidental, as it were; it is assumed that it will not enter into the deductive chain, which for its part will include everything essential, as it did for Descartes and Hegel. Moreover, this is not mere coincidence, since Descartes's system is in some sense a mechanism and even, if one will, the most thorough and logical form of mechanism ever imagined.

Thus what is truly characteristic in systems of global deduction is the continuity of the deduction they suppose.

Furthermore, the historical kinship of these conceptions is undeniable. And on the subject of this kinship – leaving aside corpuscular mechanism for the moment (we spoke in Chapter 5, pp. 129 ff., of the way it is linked to the Eleatic doctrine of the permanence of being and we shall come back to this question later in Chapter 17, pp. 496 ff.) – we can point out that, interestingly enough, a circumstance we would at first judge to be of prime importance, namely the form of the deduction, the question of knowing whether it is logical or mathematical, instead appears to be altogether secondary. For example, we saw (Ch. 4, p. 97) how directly Aristotle's theory follows from Plato's, in spite of the fact that the latter is a panmathematicism and the former a panlogism. Descartes returns to mathematicism, and this time the break at first glance seems complete, for Descartes is one of those men who incarnate most completely the spirit of the Renaissance, which destroys all that characterized the period of the Middle Ages, and Peripatetic philosophy in particular. But we know that this is only an appearance and that this great founder of modern thought has the closest relationship with the Scholastics.[2] As for his attempt at deduction in particular, it is undoubtedly founded on the eternal aspirations of the human spirit. Nevertheless, given the state of scientific knowledge in his time (which Descartes epitomized in the most complete fashion and which he himself had helped to extend as much as any other creative mind of any epoch in the evolution of humanity) and the importance experiment was quite obviously beginning to assume in it, one can only wonder at the negligible role he attributes to experiment and, in general, at the unshakable faith he professes in the certainty of his deductions. But this is because humanity had just lived through long centuries of domination by a doctrine for which the deducibility of the real world was a veritable dogma. Descartes was retraveling the same road Aristotle had taken in departing from Plato – but in the opposite direction: from panlogism to mathematicism. However, once again the fundamental postulate of the complete rationality of the real remained standing.

The parentage of the Hegelian theory is just as clear. Hegel claims to derive all his *philosophy of nature* – all that series of attempts to "construct" physical reality, to arrive at phenomena through logical deduction – from Kant (*Wiss. der Logik*, 3:202 [Miller 179–180]), and we have seen (Ch. 13, p. 353) that there is indeed an analogy as to the role of deduction in the epistemological doctrines of the two philosophers. We have also seen that the analogy is masked by the fact that the deduction is not the same in the two cases, being mathematical by preference for Kant and logical or pseudological for Hegel. This is because, here again, from our present point of view, the form of the deduction was basically less important than we would at first think and because the evolution which had been accomplished from Plato to Aristotle and from the Peripatetics to Descartes was taking place once again; Hegel's claim in itself constitutes an excellent proof of this. Furthermore, one need only look closely to see the contours of this analogy become clear.

As a matter of fact, what Kant believed he could establish is that in each science there is a *pure* part, that is, one that is purely aprioristic, depending solely on the inherent structure of our mind. In his *Metaphysical Foundations of Natural Science* and in the posthumous work entitled *Transition from the Metaphysical Foundations of Natural Science to Physics*, he undertook above all to develop this purely aprioristic science, which includes not only what we call kinematics (and which Kant called *phoronomy*), but also a good part of dynamics, that is, in short, the most essential parts of what we include under the name *rational mechanics*. Moreover, Kant deduces entirely *a priori* many things outside mechanics as well, such as, for example, the conservation of the weight of matter in phenomena where the matter undergoes quite obvious changes in properties, that is, in chemical phenomena. Indeed, he clearly perceived (and this observation is extremely important from the standpoint of the epistemological analysis of the principles of conservation in general) that in this particular case the proposition is composed of two elements of somewhat different significance. For example he says in the *Foundations*: "In universal metaphysics there is laid down the proposition that with regard to all changes of nature, no substance either arises or perishes, and here [in mechanics] there is only set forth what is substance in matter." But this determination of the nature of a substance is also aprioristic, since for him the concept of matter includes not only that of mass but also that of weight.[3]

It is from this aspect of Kantian theory that the *philosophy of nature* and Hegel's position are derived. Indeed, what Kant's successors found worthy of imitation is not the prudent way in which the master of critical philosophy had abstracted these general propositions following a study of the masterworks of mathematical physics, especially those of Newton, and of course even less is it the role mathematical deduction plays for Kant: it is solely the fact of the deduction itself. Kant had believed he could deduce a part of science, a sort of skeleton, which he called his "metaphysical principles." But this is because Kant was a dualist, because he posited the existence in nature of something not depending on our mind. But if one rejected this last element, this *thing-in-itself*, if the mind resorbed the world into itself, was it not logical to declare that Kant had stopped halfway, that *all* science had to be deducible, even if by a route very different from Kant's? Or at least philosophy was to deduce *a priori* not only a skeleton of science, but everything essential in it, *all* its dominant traits, leaving to "empiricism," to scientists properly speaking, only the task of determining its secondary features, the traits by which nature, because of its powerlessness to realize the notion, spilled over into forms of chance (Ch. 11, pp. 271 and 275). Despite the inordinately exaggerated pretensions of aprioristic deduction in this theory, it is nevertheless impossible not to recognize what this entire way of understanding the role of the deductive process has in common with the Kantian position.[4]

What cannot seriously be denied is that the Kantian deduction in very large part derives directly or indirectly from the Cartesian deduction. Of course, as far as the details of science are concerned, Kant followed Newton rather than Descartes, but without speaking of the general direction the latter had given to mathematical physics and which had been largely retained by Newton, it is well-known what enormous prestige the Cartesian conceptions had enjoyed throughout Europe for a very long period ending not long before the time Kant was writing. However, we must recognize that the scientific climate had changed considerably in the meantime, and the change concerned precisely the role attributed to deduction, whose prestige had been badly damaged. The Baconian ideas, after triumphing in England, had, especially through the intermediary of French thinkers and scientists, made inroads on the continent, which they were in the process of conquering in its turn. The immense authority of Newton had much to do with this evolution. Although he was not as *Baconian* as one would sometimes like to think, and in any case more Baconian in his epistemological formulas than in the practice of science, it is certain that he places experience on an entirely different level than Descartes; and his resounding declarations, above all the famous *hypotheses non fingo*, contribute greatly to the blossoming of a new spirit. Henceforth it is no longer deduction that occupies the first place in science, it is experience. This current, without being as powerful as it later became, was certainly already manifesting itself with great vigor among scientists at the moment the *Metaphysical Foundations* was published, which no doubt explains why Kant's work exercised so little real influence on the course of science, even in Germany. "Newtonian physicist" though he was, and however strong the influence of Newton's arguments in each particular case, Kant reacted against the spirit prevailing in the science of his time by effecting a return toward the Cartesian tradition. And it is hardly necessary to point out how much the place occupied by experimental knowledge in Kant – coming *after* deductive knowledge and not interfering with its development – conforms to the model furnished by Descartes.[5]

By that fact, by the continuity of the deduction of the metaphysical foundations it would impose on science, Kant's undertaking – however modest its pretensions compared to those of the immense constructions of Descartes and Hegel – is indeed to be classed with them. And we understand how Kant, deriving from Cartesianism, was able to engender Hegelianism.

Thus Hegel's attempt at a global deduction of nature is connected, through Kant, to Descartes and, through Descartes, to the great deductive current and to Plato. It is therefore not a sort of monstrosity without precedent in the history of human thought, as it has been judged, for example, by scientists (such as Liebig) who had to combat it. To be sure it seems anachronistic, as we have pointed out. But even here we see that the author of the *Naturphilosophie* was not unrelated to Kant, that in a sense he was only following (at a great distance, to be sure) an inspiration given by the author of the *Metaphysical Foundations*. Whatever its merit, Kant's movement away from the science of his time back toward Cartesian deduction certainly constituted a regression, and it is this regression that Hegel pursued and pushed to its logical conclusion, by abandoning mathematical deduction in favor of deduction by the notion, or, in other words, by taking for his model not Descartes but Aristotle – indeed, as we have noted, he himself explicitly recognized this kinship (Ch. 11, p. 264 above).

If, however, we consider the limits of what was meant to be deduced in this fashion, we see that Hegel fell far short of his model. Granted that Aristotle and his followers in the Middle Ages limited deduction to the *universal*, as we saw in the case of Gersonides, these universals comprised all that makes up science, since science can treat only of genus. Hegel declares only certain very general aspects of science to be deducible, all the rest having arisen out of the arbitrariness of nature and being subject to empirical knowledge alone. In spite of all its "logical arrogance," the Hegelian philosophy is obliged to take into account the fact that an enormous body of scientific knowledge has been acquired since the time of the last practitioners of Peripatetic physics, and this evolution prevents it from pushing its regression beyond certain limits.

What we have just recognized about the role of deduction in Descartes and in Kant can help us elucidate a particular side of the Hegelian theory which has perhaps not always been fully understood or judged with sufficient fairness. We saw that the author of the dialectical method has been criticized for not confining himself to a rigorously aprioristic development of thought, but for surreptitiously introducing elements coming from outside, borrowed principally from experience, and we have acknowledged that this objection is well-founded. But some critics seem to have pushed the argument too far. For instance Trendelenburg, after observing (probably quite correctly) that Hegel's logic, far from having arisen spontaneously and only later been applied to nature, is on the contrary tailored to fit the *Naturphilosophie*, intends, as we have seen, to use this observation to demonstrate that this logic can therefore not be a product "of thought relying strictly on itself" (Ch. 11, p. 290), that is to say that it does not constitute purely aprioristic knowledge. We do not find this manner of argument entirely legitimate. What we object to is thc connection he means to establish between these two sorts of considerations, for we are in agreement both with where the argument begins –

with the observation that Hegel's logic came later than his epistemology – and where it ends – with the introduction of empirical elements into dialectical reasoning.

Trendelenburg's criticism is linked to another he makes which also bears on the relations of Hegel's theory to experience. Experience, Trendelenburg claims, can in no way enter into the system, given that it "is never shown how the dialectical method absorbs the material acquired by the empirical sciences, since nowhere is a door left open" through which it can enter and that "the dialectical method is self-contained, without a gap." Now it is clear from what we have just seen that this criticism rests on false premises. Hegel did not have to show how empirical knowledge was to enter into his "science" of nature, precisely because for him this science was entirely aprioristic. But it does not follow that he denied the role of experience in science. As a matter of fact, he did not, as we have seen (Ch. 11, p. 275), and, Trendelenburg to the contrary, there is no inconsistency here. Hegel thought – just like Descartes and Kant – that the aprioristic development had to be continuous; but once this chain had been forged, empirical knowledge could then attach to it the results of its observations and experiments, summed up in laws and embracing a body of phenomena as vast and as minute as one would like – provided, however, that one did not pretend to use these laws to explain anything at all in the phenomena, nor above all to attach to them rational scientific theories involving mathematical elaborations, explanation being the exclusive prerogative of aprioristic deduction by the notion.

However, did not Hegel himself acknowledge that philosophy had to make use of the content of the empirical sciences? Is that not inconsistent with the assertion of the absolute aprioricity of the deductive chain? We believe not. And, to start with, we can note that the objection, if true, would apply almost equally well to Descartes and Kant. There is no doubt that Descartes's *Principles* are saturated with all the empirical knowledge of the epoch. Would one challenge that the development found there is entirely deductive and aprioristic? And in the same way Kant certainly claims to establish the *Metaphysical* (that is, *a priori*) *Foundations of Natural Science*. But one need only open the work of that name to see that experimental science is involved at every step. Would these two great minds have fallen into a quite obvious trap that a layman would be able to detect at a glance?

Clearly they would not. Neither Descartes nor Kant believed he was compromising the aprioristic purity of his deductions because he did not disregard available scientific knowledge. Descartes, who went furthest along the deductive route, would certainly have readily admitted that the existence of a perceived external world had been useful to him in this work of *pure* reasoning (to use the Kantian nomenclature), from which he claims to have drawn the heavens, the stars, an earth, water, fire, minerals, etc. (see Ch. 11, p. 274 above) – not only for checking its results and establishing their conformity with what exists, but also for giving an impetus to this work of reason and guiding it, by showing it the goal to be attained. And as for Kant, his position on this point is clear. Of course his *metaphysical foundations* are, by definition, aprioristic and by that very fact presuppose that reason can arrive at them by closing its eyes, so to speak, to the perception of the external world; but he by no means claims to have arrived at them in this way himself; he claims, on the contrary, to have obviously *separated them out* from existing science, that is from a knowledge in the constitution of which experience had played a considerable role (which Kant certainly did not repudiate). There is no contradiction there, or at least only the one resulting from the existence, on the one hand, of our reason and, on the other, of a perceived world on which it acts. For our reason obviously needs this perceived world upon which to act. But just as it does not at all follow from the simple observation that, in the words of the famous medieval dictum, *nihil est in intellectu quod non fuerit prius in sensu*, since one must add the Leibnizian corrective *nisi intellectus ipse*,[6] so is it not contradictory to try to separate out this *intellectus ipse* and its procedures, by analyzing the work our reason accomplishes, using our perceptions as raw material. In a certain sense it can be said that there is no other way to learn anything about the functioning of our reason. Indeed, if anything is certain, it is that, since reason cannot observe itself, we cannot know its work directly, that only its end results, in the form of decrees promulgated by reason, arrive at the level of our consciousness. It is true that we frequently make the opposite claim and say we know the *reasons* for our decisions; but it is clear, supposing that we really do know them, that we were able to do so only after the fact. Pascal makes this amply clear:

M. de Roannez used to say: 'The reasons occur to me afterwards, but first of all the thing pleases or shocks me without my knowing why, and yet it shocks me for reasons I only discover later.' But I do not think that it shocks for reasons we discover afterwards, but that we only discover the reasons because it does shock. (*Pensées* 457 [Krailsheimer 351])

If our reason could know itself, it would be through an immediate awareness free of all doubt; we would say, "this is how I reason" with the same certainty with which we say, "I'm cold." Now the very existence of logic shows that this is not the case; if a science was needed to teach us how we reason, if research was required to recognize it and if the results of this research could be debated, it is because all that work is accomplished unconsciously, in spite of ourselves as it were, almost like the work of our digestion or our heart. Furthermore, in this logical research, the substratum, that upon which the reasoning we are attempting to analyze is operating, is composed and can only be composed of objects of our perception. Hegel, who certainly cannot be suspected of trying to diminish the part played by the *a priori* in our reasoning, praised Aristotle as having been the first to undertake a description "of the phenomena of thinking just as they occur."[7] Nothing could be more accurate, for it is only by observing, from the outside so to speak, how we have reasoned, that we succeed in separating out a small fragment of this *intellectus ipse*, namely the role our reason must have played in the reasoning being analyzed. Thus, in order to know this indubitable *a priori*, we had to use reasoning in which perceptions, empirical knowledge, played a part. Kant therefore did not depart from his program when he used a similar procedure to abstract, from experimental science, procedures he nevertheless considered to be entirely *a priori*. And Hegel in his turn only followed Kant's example, here as elsewhere, in asking experimental science to prepare the content of the particular in such a way that it can be included in philosophy (Ch. 11, p. 275). By that he obviously means that what is specific in phenomena, once science has raised it to a high degree of generality, will thus be rendered suitable for becoming the object, the result of an aprioristic deduction,[8] and this hope is, for him (as it also is for Kant), all the less unreasonable, illogical, because he does acknowledge that a whole part of the physical sciences must remain exclusively empirical, since it is fundamentally unsuitable for entering into this deductive chain. Thus a whole preliminary work is needed in order to separate out the empirical knowledge from this slag (as one would be tempted to call it), which is only the expression of nature's powerlessness to realize the notion. This work of refining, as it were, is what Hegel sees as the proper task of experimental science.

But we can go still further. We saw above (p. 363) that the foundations of the mechanistic faith are analogous to the principles that guided Descartes and Kant with respect to the perfect continuity of the deductive chain that constitutes the essential thread of phenomena. Of course the mechanist does not believe, as did these two thinkers, that he can immediately separate out this chain by an operation on the results acquired by science. But he judges that such an operation will be possible some day and that the task of science is to bring that day closer and closer. Obviously, when that goal has been reached, nature will have become intelligible – the hope is chimerical, but there is no doubt that mechanism actually does imply it. Now this intelligible or rational component which, as such, can only be aprioristic, will have been separated out by scientific reasoning, that is, by deductions starting from experimental data. It too then is a path analogous to the one followed by Hegel. And in the same way positivism based on deductions, for which the pattern was set by Sophie Germain and Goblot (Ch. 3, pp. 65 ff.) and which postulates that the laws of nature will ultimately be reduced to a single proposition which itself will be recognized as necessary, assumes that we shall attain this ultimate generalization, this state where everything will be "completely illuminated,"[9] by following the path of experiment.

Thus only rigid positivism on the model of August Comte or Mach, where deduction serves solely to abridge the formulas and from which all intelligibility is excluded, repudiates a notion of this type. But as soon as it is a question of truly explaining phenomena, one is forced to appeal to it. Moreover, that is only a reiteration of our observation that in claiming to explain a fact, we are thereby asserting that we can conceive of it as necessary, as following from the very foundations of our reason. Consequently the process criticized by Trendelenburg, which consists in abstracting from reasoning about phenomena something that will then be recognized as aprioristic, is implied by the very concept of explanation.

Trendelenburg is equally wrong to believe that if he could establish that Hegel's logic were intentionally constructed with an eye to the use he meant to make of it later in his philosophy of nature, it would demonstrate that this logic is not an aprioristic construction. The Hegelian logic could perfectly well have been only a "purified perception," an "anticipated abstraction of nature,"[10] and not be any less an emanation of the intellect. It would quite simply have proved that the philosopher had really completed the task to which he had avowedly committed himself, that he had succeeded in making the order of thought coincide with the order of things. Furthermore, this is the belief of the followers Hegel still continues to hold today, and these Hegelians, as we know, form a not inconsiderable school. We, on the contrary, remain convinced that Hegel's logic is, in Hartmann's words, "the imposing attempt of a great mind to accomplish an impossible task." But Trendelenburg's alleged demonstration – which is at bottom only a statement of the problem to be resolved – certainly does not suffice to establish this impossibility. What establishes it for us, apart from the theoretical reasons we pointed to in Chapters 11 and 12, is the absolute sterility of the doctrine from the standpoint of the growth of scientific knowledge.

We noted a few pages ago (p. 367) that Kant's ideas had exercised very little influence on the course of science. Moreover, Hegel himself realized as much; he believed this lack of success stemmed from the fact that Kant's scientific construction lacked breadth (cf. Ch. 11, p. 276), and his own construction certainly leaves nothing to be desired in that respect. But clearly he was entirely mistaken in this belief: in broadening his theory he widened the moat separating him from science, particularly since science in its turn had evolved, and evolved in a diametrically opposed sense, becoming still more addicted to experiment and, as a consequence, less inclined toward global deduction than it had been in Kant's day.

This evolution which, at the moment Hegel was writing, was going to be consecrated as dogma by the appearance of positivism, is so obvious that its scope has most often been exaggerated. Comte and his spiritual ancestor Bacon set the example by insisting, very grandiloquently, on the importance of the reforms to be accomplished or already accomplished and the total transformation science as a whole was going to undergo or had undergone on that account.

It is obvious that these assertions are not only highly exaggerated, but also materially incorrect, in the sense that by brushing deduction aside, or at least assigning it a completely subordinate role, they failed to recognize the true essence of scientific research: the impetus toward the comprehension, the rationalization of reality. Is the empiricist claim to be totally disallowed, however? On the contrary, it is clear, after what we have just seen, that it must contain some truth, since in current science the role of experience and, consequently, that of deduction are no longer exactly what they still were, for example, in the time of Descartes.

What has changed from this point of view can be recognized through our analysis above. As a matter of fact, it is quite simply the characteristic trait of continuity in deduction. For Descartes it was obviously a sort of unshakable dogma. But even for Kant this faith had not notably weakened. We have spoken of the way in which he establishes the framework of his "first principles" ["foundations"] and of his belief in the particular importance of the role of mathematics in science (Ch. 13, p. 353 and this chapter, p. 365). Indeed, mathematical deduction is something more for him than it can be for us. No doubt Kant is just as aware as any scientist today that the vast majority of these deductions are only fragmentary and cannot be connected directly to his skeleton of *a priori* "principles"; but for him that appears to be a temporary situation: these fragments are destined to be welded together later through the progress of knowledge, at which time the deductive principles, which constitute the true framework of the edifice, will stand out all by themselves, so to speak, just as, for example, the study of the motion of the celestial bodies has, by way of the Copernican system, the laws of Kepler and the Newtonian simplification, led to the discovery of the law of universal gravitation – for Kant this law is aprioristic, however, since he considers attraction to be part of the concept of matter itself.

Now it is quite certain that here the basic credo of contemporary science is essentially different from Descartes's or Kant's. It is different in that, little by little and most often without a clear awareness of the change taking place, the scientist has ceased to consider the possibility of a continuous deduction, has ceased to hope that he would ever grasp even a fragment of this definitive chain. It can no doubt be said that the science of all epochs, in its everyday practice – that is, apart from the moments when, prompted by a powerful mind such as Aristotle or Descartes, it sought to include the whole of phenomena in a single construction – has tacitly accepted the fact of a deduction that is essentially discontinuous. Indeed, deduction in contemporary science does not pretend to start from propositions drawn exclusively from our own intellect; it clings anywhere at all, more or less by chance, to anything that seems, for any reason whatever, to be *established*, and consequently also to propositions of empirical origin or to others whose origin science has not even bothered to elucidate – which is the case with the principles of conservation in particular, as we saw (Ch. 5, pp. 119 ff.; cf. Ch. 17, pp. 500 ff. below). Furthermore, that is only one specific example of the general observation formulated in Chapter 3 (pp. 56 ff.) concerning the underlying weakness of many scientific deductions due to the fundamental circumstance that our reason derives its satisfaction from the fact of the deduction itself, independently of the solidity of the starting point; we see that, in this form, the modern conception was implicitly postulated by the science of all times. This is because mankind, throughout its history, as we have pointed out many times in the course of this work, while *wanting* nature to be rational and while being convinced that it would to some extent yield to this requirement, at the same time felt that nature also sometimes resisted it.

But in modern times it is the latter feeling, that of reality's resistance to the attempt at rationalization, that has gradually gained strength. As long as the idea of an ultimate complete reduction to mechanism remained standing, however, the scientist, while in practice accepting the discontinuity of deduction, necessarily had to consider the discontinuity as destined to be replaced eventually by a construction perfectly logical in its essential parts, where the irrational would manifest itself only incidentally, finding its place in the *margin* (cf. pp. 353 and 364 above). At the present time, on the contrary, the scientist has a distinct feeling that the discontinuity of his deductions is definitive and irremediable, or, in other words, that the irrational does not allow itself to be dismissed as a by-product of the deductive chain, but must be included in the chain itself, of which it must always serve as the point of departure. And, similarly, science does not believe that any mathematical theory, however firmly established it may appear for the moment, is really *definitive*, that it can be considered as a necessary part of a system, of a theory of the world that will one day be established for all time. On the contrary, science is quite ready to overthrow its theoretical edifice, at any time and down to its deepest foundations.

Of course that is an implicit rather than an expressly avowed conviction, even for the contemporary scientist; furthermore it has had difficulty enough becoming established. It is true that Galileo already seems to have had at least an inkling of the truth in this domain. Indeed, he has quite recently been quoted as declaring that it is not true we can fathom reality: on the contrary, it is "impossible, even for the most speculative minds, to arrive at complete knowledge of even a single thing in nature, be it ever so slight."[11] If we allow this statement of principle its full scope, we see that it applies not only to the Peripatetic definition, but also to the mathematical formula serving, like the former, to replace the physical phenomenon. But the question is whether Galileo meant to apply his statement with such generality. We must always be a bit wary of these resounding affirmations, at once dogmatic and negative, which are meant to serve as weapons against adversaries but which the author reserves the right to apply exclusively to them. "*Hypotheses non fingo*," Newton proclaims, and less than twenty years after the master's death the Newtonian Musschenbroek cries: "All hypotheses are banished!"[12] Which did not prevent Newton from formulating, in the famous Thirty-first Query of his *Opticks*, the complete program of a theoretical physics based on the hypothesis of central forces (that is, action at a distance, to which physicists of all epochs have shown so much aversion), nor did it deter his disciples from working conscientiously to develop this conception (cf. IR 46, 53 [Loewenberg 49, 53]). What master and disciples meant to designate as "hypotheses" was, as a matter of fact, only the Cartesian theories.

In the same way, Lavoisier, attacking phlogiston, declares himself to be hostile to hypotheses. As he says in the preface to the *Traité élémentaire de chimie*,

> It is not surprising that in the physical sciences in general one has often made assumptions rather than reached conclusions, nor that suppositions transmitted from one age to another have become more and more imposing by the weight of the authority they have acquired, to the point that they have finally been adopted and regarded as fundamental truths, even by very fine minds.[13]

But Priestley turns the same argument against Lavoisier. "Speculation," he declares in his turn, "is a commodity of little value in physics; new and important facts are ... a great deal more precious."[14] Perhaps, as Berthelot did well to point out, Priestley is exhibiting the eternal bitterness of a man who has a gift for discovering new facts (a gift Priestley indeed possessed in the highest degree) against a man endowed with the even greater gift of coordinating such facts into a coherent theoretical conception. But clearly there is also the fact that what Lavoisier considers a hypothesis is phlogiston theory, whereas the one by which he means to replace it, the theory of chemical elements persisting in compounds, appears to him as the pure and simple expression of reality; Priestley, on the other hand, judges that his observations, stated in the language of the phlogiston theorists, a language to which everyone was accustomed at that time, are quite simply facts, and he protests against an interpretation which, by translating them into the language of the new conception, seems to him to introduce an inadmissible portion of speculation.

Even Schelling – the creator of the *philosophy of nature*, the author of so many hypotheses that, to say the least, amaze us by the singular audacity with which they have been constructed on very narrow bases having no solidity whatever – affirmed that the construction of physical reality *a priori* "provides the basis for a general theory of natural phenomena, in which there is hope of being able to dispense with all the hypotheses and fictions ..." (*Transc. Idealismus*, I, 3:453 [Heath 92]), just as his disciple Steffens declared that their philosophy meant to show how one could understand nature "by itself" and "without the external aid of hypotheses and principles."[15]

And it is likewise possible that Galileo was referring only to the Peripatetics and would not have been willing, for example, to consider his own propositions and mechanical explanations as ephemeral. We are led to this conclusion by observing how long it took for the conviction that physical fact is impenetrable to become established, even in scientific minds of the highest order. This can be illustrated by a story dealing with an epoch two centuries later than that of Galileo and having as its protagonist a scientist who, although perhaps not attaining the stature of the great Florentine, nevertheless bears one of the most illustrious names of the nineteenth century.

> One of the late lamented masters of mathematical physics, Monsieur Sarrau, told me a typical anecdote about Cauchy, which he had heard from Father Gratry. Cauchy was walking with Father Gratry in the Luxembourg Gardens. They were talking of the next life and of how happy the elect would be to know at last, unveiled and without restriction, truths lengthily and painfully pursued in this world. Alluding to Cauchy's research on the mechanical theory for the reflection of light, Gratry suggested that one of the great intellectual joys of the illustrious geometer in the next life would doubtless be to fathom the secret of the nature of light and to know the last word on the problems of optics he had been pondering. Cauchy protested. On this point he could not admit that it was possible ever to learn anything more than he already knew; he could not imagine that the most perfect intelligence could understand the mechanism of reflection any differently from the way he had explained it. He had given a *mechanical theory* for the phenomena; his piety did not extend to the belief that God Himself could possibly do anything else and do it better.[16]

There is an undeniable analogy between Cauchy's way of thinking and that of the early seventeenth century Peripatetics about whom we spoke in our Chapter 4 (pp. 94 ff.). Just like them, he believed that there was something definitive in physical science which would remain unchanged in the centuries to come (since it was in conformity with the very nature of things). Furthermore Cauchy's ideas are obviously close to Kant's, and there is no doubt that they were, perhaps in a somewhat less decisive form, those shared by most men of science in the first part of the nineteenth century. As a matter of fact, the belief in universal mechanism was then fairly general. In 1900 Cornu was still proclaiming: "In the world there is nothing except matter and motion."[17] Already at that time he seemed to be only the somewhat tardy spokesman for convictions that had inspired a generation dead and gone. But some fifteen or twenty years earlier we dare say no physicist would have contested this principle. The affirmation of universal mechanism, as we pointed out (pp. 364 and 375), certainly includes the belief in the fundamental rationality of phenomena. No doubt it does not absolutely follow that wherever one has arrived at a mechanical explanation, it must be definitive; one can, on the contrary, think (as is suggested by mathematical deduction) that wherever we find one figurative hypothesis we can find an infinity of them and that consequently the way remains open for future upheavals. But that is a rather recent belief among physicists; hardly a generation ago the immense majority of them certainly interpreted the affirmation of mechanism as Cauchy did. What we have here is an evolution in the basic credo of science, not readily apparent but fundamentally extremely important, which has taken place during the course of the last few generations. On the one hand, it has been recognized that any hope of ultimate reduction must be abandoned, and this conviction has become so strong that, on the contrary, the mechanical phenomenon is being deduced from the electrical phenomenon (cf. Ch. 3, p. 60 and Ch. 6, p. 151), and, on the other hand, we have been forced to admit, in the case of phenomena believed to be perfectly well-known, that new observations were capable of entirely overthrowing the received conceptions, indeed, even of suggesting, as has occurred in the case of black body radiation (cf. Ch. 2, pp. 36 ff., and Ch. 6, p. 168), that irrationals existed where they were least expected. It is thus that the notion of the ephemeral character of all theories has finally triumphed over that of Kant. However, we cannot disregard the fact that Bacon's empiricism and Comte's positivism also played quite an important role in this evolution. Just as the work of Francis Bacon had done science an immense service as an antidote to Peripatetic deduction, the work of Auguste Comte was in its turn extremely useful as a counterbalance to the excessive tendency to reduce everything to mathematics and mechanics – and especially the tendency to consider this reduction as complete, or at least easily realizable – which was certainly the characteristic of the physical sciences for many long years. By proclaiming the sovereignty of experience, by eliminating deduction or assigning it an entirely subordinate role, empiricist or positivistic theory has, as it were, depreciated deduction in the eyes of the scientist and has thereby prepared him to understand better the true nature of the theoretical reasoning of science. On the other hand, it hardly bears repeating that this epistemological position, as prestigious as it was, was entirely powerless to modify the inner nature of scientific reasoning: explanatory science has continued to flourish.

It has also continued to register successes, which is a point not to be overlooked. It has been said that science is a work which succeeds, and that is glaringly obvious to all. But one must understand that this success is not limited (contrary to what usually seems to be implied) to mere *prediction*, that is, to purely legalistic science, but extends to explanatory science as well. The success of scientific explanation becomes especially evident if we compare it to the failure of Hegelian deduction. Of course science is incapable of *completely* explaining a phenomenon, even the most insignificant one; on that point Hegel is perfectly right. But Hegel went on to affirm that since the work of scientific explanation is by that very fact vitiated in its essence, it must therefore be absolutely futile. It is, in short, a *reductio ad absurdum*. Is it valid? It cannot be, since explanatory science exists and celebrates new triumphs each day. And yet, as we have admitted, the starting point of this demonstration is perfectly correct. Hegel really succeeded in *dismantling* scientific thought, which is just what he says it is: a continuous chain of "abstract" identities which in the final analysis can only end up in an indistinct whole in time and space. Is there truly any contradiction, then, between his demonstration and this result, and does "rational science" triumph in opposition to reason?

In Chapter 17 the reader will see that in a certain sense this is indeed the case, and we shall attempt there to make the nature of this "epistemological paradox" somewhat more clear. But insofar as Hegel is concerned, it is not difficult to see where his reasoning breaks down. The demonstration would be irrefutable if it were established that nature must be rational, adequate to our intelligence. Of this Hegel was firmly convinced. It is a postulate: *it is not possible* for nature to resist the mind; the mind *must* be able to understand it, or at least understand its main outlines (while what remains outside this understanding can only be totally unimportant, irrelevant, unworthy of serious study). Now if carried out by means of mathematical mechanism, this understanding results in an impasse, since it leads straight to acosmism, to the negation of nature itself. This path therefore cannot be the right one; there *must* be another one, and it is this other path we must seek, and must follow when it has been found – as Hegel believed he had done.

But what the results acquired by science show us is that nature is not entirely rational. Nor, to be sure, is it entirely irrational, for then we could not reason about it at all. It is rational and irrational by turns, by fits and starts we might say, without our being able to know in advance where it is or is not in conformity with the requirements of our reason. This is why mankind has sometimes been able to demonstrate a genuine prescience with regard to nature (as for atomism and the laws of conservation) and, more often, to be grossly mistaken in trying to predict it. But the overwhelmingly important point, the point that constitutes a peremptory refutation of the Hegelian demonstration, is that the successful anticipations have taken place almost exclusively by the route he condemns. Some confirmations have even exceeded expectations. The most adamant supporters of the atomistic conceptions would surely not have dared suppose, before the work of the last few decades, that we would succeed in directly observing the immediate effects of molecular motion through the microscope, and in exactly determining the absolute dimensions of the molecules themselves, and the early results acquired in this domain were met with rather marked skepticism (Ch. 6, pp. 163 ff.). But today there can no longer be any reasonable doubt about them, and it consequently has been accepted that the theory has been capable of furnishing a marvelously precise spatial image – one that is therefore adequate to the requirements of our imagination – of at least one aspect of this irrational concept of diversity in space. But perhaps even more marvelous is the success obtained in the case of the temporal irrational by the kinetic theory of continuous change. Indeed, our reason, our causal tendency demands the preservation, the permanence of everything; but the fact of becoming belies it, and it is this fact that Carnot's principle affirms and specifies. The contrast seems complete. But through the device of the calculation of probabilities, mechanism actually installs itself in the very heart of the domain ruled by this principle and succeeds, here again, in providing a spatial image. Now mechanism is unquestionably an eminently causal conception, framed for the sole purpose of maintaining the permanence of the essential being through changing appearances. That it has to some extent succeeded in rationalizing even the continuous and infinite change of Carnot's principle is certainly a result that could almost be qualified as paradoxical.

To be sure, science does not succeed as well in every area. At present the irrationals of sensation and transitive action are at just about the same point they were long centuries ago, at the dawn of human thought: neither the efforts of physiologists, nor those of the authors of the different corpuscular or dynamic theories have been able to accomplish much toward their rationalization. So far, the phenomena treated at the *Council* of Brussels also seem to resist any attempt to furnish a model, a spatial image at all satisfying to the mind; the great scientists called together in that assembly have verified this, as we have seen. But there it is a question of recent things differing so widely from anything previously known that it is perhaps allowable to maintain a hope for the future: we can suppose that scientific ingenuity has not had its final say, indeed even that new discoveries will be forthcoming to light a way through these shadows.

Needless to say, none of the rationalizations accomplished in this way are complete, nor can they be. Perrin's absolute dimensions of molecules are a *given*, in which reason, deduction has no part, and at present we see clearly that, even were we to succeed in deducing these numbers from others characterizing matter in an even more fundamental way, there will always have to remain a final *given* relative to spatial dimensions, which will be impossible to eliminate. And likewise the *explanation* of Carnot's principle by the calculation of probabilities necessarily leads, as Boltzmann established, to positing a past improbable, and therefore inexplicable state. Furthermore, these rationalizations have not been carried out according to the models furnished by Descartes and Hegel, by attempts at global deduction. To return to a mathematical image we used earlier with respect to laws [p. 88 above], physicists have not sought to draw a single circle connecting all the points of the curve, but have been content to study the radius of curvature at different points, to retrace the curve by means of more or less discontinuous arcs of circles, leaving spaces in between for the appearance of the irrational. And it is unquestionably to this prudent procedure, in conjunction with the use of mathematical deduction, that they owe all their successes.

These successes of the mathematical and mechanical method are nonetheless worthy of our highest admiration. For – just think of it – science has truly succeeded where Hegel's thought had failed so dismally. It has succeeded in penetrating the realm of the irrational; it has been able (only partially, to be sure, and trying not to force it too much) to discipline, to dominate this irrational.

Thus Hegel was entirely mistaken in believing that because mathematical deduction does not proceed from the notion, it can only be an operation "external" to the thing (Ch. 11, p. 275). The new element that it is able to generate, by the very fact that it does not start from the notion, offers, on the contrary, that precious quality of agreeing with the inner being of nature. Not only is the agreement absolute and unlimited in the realm of the mathematical sciences themselves – on this point everyone obviously agrees – but it goes beyond this realm; it manifests itself, incompletely and sporadically it is true, but nevertheless clearly and tangibly, in the realm of the physical sciences. Mathematical deduction, whatever Hegel may have said, is not an artificial operation, but something in profound conformity with the working of our intellect. Moreover, it is obviously, and again contrary to Hegel, something that closely and mysteriously conforms to the order of things themselves.

Recalling what we established in Chapter 4 (pp. 82 ff.) about the *structure* that must be attributed to reality because we succeed in formulating laws which it apparently obeys, we can say that, in the same way, the success of our theories reveals a certain structure of this same reality.

Need we point out that this is a completely *a posteriori* observation? Mathematical propositions enjoy, in comparison with physical laws, the astonishing privilege of being directly and unreservedly applicable to reality. The problem posed by this observation is immediately resolved if we suppose that reality is only a body of mathematical concepts. That is undoubtedly the point of departure for the ancient panmathematicists, and probably for Galileo as well. But Descartes drew from it the impeccably logical conclusion that if everything is mathematical, everything must be deducible, and he consecrated the immense resources of his genius to the attempt. He failed, but the attempt was not at all absurd in itself. Today, to be sure, such an attempt appears condemned in advance, because contemporary science, by its general attitude, instills in us a distrust of pure deduction and, consequently, of panmathematicism. But philosophy had not waited for the testimony of science, and Aristotle, as we shall see (Ch. 15, p. 414), combatted the panmathematicism of his master Plato with extremely solid arguments; Newton's profound remark [concerning the impossibility of deducing temporal and spatial diversity], which we quoted in Chapter 6 (p. 152) and which is purely philosophic in nature, can be understood in the same sense. Descartes was certainly familiar with Aristotle's objections, and most assuredly he could have arrived at a deduction analogous to Newton's. But no doubt arguments of this sort would not have seemed to him to preclude his enterprise; they have quite a different force for us since we can support them with precise scientific theories. But if there could, at that time, be no truly convincing argument against the possibility of an undertaking such as the one Descartes was attempting, there could also be no question, either then or later, of any sort of argument in favor of universal mathematicism. And as for this partial mathematicism married with empiricism which characterizes the *modus operandi* of contemporary science, it is obviously the result of much trial and error, ranging from pure mathematicism to more or less radical empiricism, trial and error in which Descartes's attempt certainly loomed large. Even less, then, could it be a question of demonstrating *a priori* that this mitigated mathematicism constituted the only path leading to the understanding of phenomena. Thus Hegel's attempt was not absurd either.

Similarly, Hegel's failure, as complete as it is, must not prevent us from recognizing how well his attempt conformed to the inmost tendencies of the human mind, which *wants* nature to be rational. His undertaking is on common ground with Descartes's enterprise and the other attempts at global deduction to which thought and science owe so much.

Were those who have called the Hegelian conception of science a monstrosity therefore entirely wrong? On the contrary, we have pointed out ourself in the preceding chapter how anachronistic it appears compared to that of Auguste Comte, which is more or less contemporary with it. It could no doubt be argued that this is due to the very fact that the *Naturphilosophie* constitutes an attempt at global explanation in an era which no longer admitted of such explanation. But why is it then that the doctrine still appears anachronistic when compared to the theory of Descartes? Hegel was destined to fail – granted – but what explains the enormity of this failure? It is surprising, to say the least, that such a powerful mind, quite conversant enough with the scientific knowledge of his time, was able, after Newton and Lavoisier, to produce a work alleging to treat science and yet so foreign to what we understand by the term that one might think it had fallen from another planet.

In order to see the reasons for this formidable anomaly, all we have to do is return to what we have just shown about the success of theoretical science. This success is due to the method it employs, that of mathematical deduction, which was also the method of Descartes, while Hegel deliberately sets this procedure aside (Ch. 11, pp. 282 ff.). This circumstance, which seemed secondary to us earlier when we were dealing with the kinship of the theories, is what played the decisive role insofar as the actual result of the undertaking is concerned.

We saw in our Chapter 11 (p. 268) that Hegel himself linked his *philosophy of nature* to Aristotle, and we acknowledged that there was indeed an analogy between the two conceptions in that they were both panlogisms (although the logic to which they appealed was not the same). We can see that the Peripatetic doctrine was also remarkably sterile from the point of view of scientific progress. To be sure, there was the difference that whereas the Hegelian theory was rejected out of hand by the scientists of its day, Aristotle's theory absolutely dominated science for many centuries. But its long reign only makes more apparent the futility of the efforts it inspired, of which one might say there is next to nothing remaining in present-day science. This statement may perhaps be judged too harsh, particularly given that many points remain obscure in our knowledge of the thought of those ages and that it is thus not at all impossible that later revelations will show that one current of scientific ideas or another which seems completely modern was prepared or at least anticipated by these thinkers. But it seems to us that what is currently known about this evolution is sufficient grounds for our conclusion. In this respect, Duhem's work in particular appears to us to contain all the elements of a formal demonstration. Duhem was, as is well-known, a firmly convinced and ardent supporter of the Peripatetic doctrine, to which he would have liked science to return. It is especially to this end, apparently, that he devoted himself to very thorough studies of the physics of the Middle Ages and the early Renaissance, studies whose results are recorded in many valuable works, most of which we have had occasion to cite either in the present work or in our earlier book. Now one need only glance through Duhem's publications with an open mind to see that the advances he points out, which are indeed sometimes extremely interesting (as, for example, the theory of *impetus impressus*, which was especially in favor at the University of Paris; cf. IR 120 [Loewenberg 117]) are in no way connected to the Peripatetic explanation by matter and form and are even elaborated in opposition to that system. We can make a similar observation about another branch of science, namely the chemistry of the alchemists and their successors. There progress is really accomplished only by a very marked deviation from the reigning Peripateticism. These chemical theories are certainly qualitative, as Aristotle's conception would have them, but apart from that, they are genuine scientific theories, and become more so as knowledge advances; the purely logical aspect tends more and more to disappear, sometimes not without resistance on the part of those intent upon maintaining the essence of the Peripatetic position, as becomes especially apparent at the time of the dissensions over mixtures (cf. IR 371 ff., 480 ff. [Loewenberg 327 ff., 415 ff.]). In qualitative chemical theory in its later form, that of phlogiston theory (as also in the qualitative theories of physics which are contemporaneous with it, or which did not die out until more recently, as for example Black's theory of caloric), there no longer remains any trace of Peripateticism to speak of, except for this qualitative character itself. And it is most instructive to note that this particularity also had to be abolished for science to be able to progress. As we have pointed out (Ch. 3, pp. 57 ff., Ch. 6, pp. 151 ff.), from the standpoint of our sensation a qualitative conception undoubtedly explains phenomena much better than a kinetic theory can. If it is caloric, the substratum and hypostasis of my sensation of heat, that passes from one body to another, I understand better why I am hot than if someone talks about movements that can have no immediate relation to my sensation of heat; and likewise, if it is phlogiston that passes, I think I understand why the body becomes combustible, while the reason for its affinity for oxygen escapes me. It has nevertheless been necessary in these two cases, as in so many others, to abandon the concept of quality in favor of that of quantity, and it is clear that this sacrifice was made for the sake of a better deduction of phenomena or, in other words (since in our science deduction is carried out by means of mathematics), for the sake of a more complete penetration of physical phenomena by the mathematical concept.

All resistance to the demands of science in this matter has been absolutely futile and will certainly always be so, and there is no better example of this incontestable truth than the case of Hegel.

Thus one cannot reasonably doubt that it is because he consistently tried to connect concepts directly, without going through the intermediary of mathematics, or, if one will, because of his exclusive use of logical processes (or pseudological ones, that is, processes conforming to his own particular logic), that his failure was so complete. And the anachronistic aspect of his attempt stems mainly from the fact that it was undertaken at a time when science had already reached a very advanced phase of development in the diametrically opposite direction.

In examining the work of Hegel we discovered the deep-seated reasons that prompted him to adopt such an extravagant attitude. But can one also appeal here again to an external influence from the milieu? Obviously we can no longer invoke the spirit of the times as we did in the preceding chapter to explain Hegel's claim to dictate to science, because Auguste

Comte's attitude, for example, is entirely different. Thus, since the time does not explain it, we must have recourse to the place, that is, in the present case, to particular features of the author's native land. That is indeed, we believe, the correct explanation. What contributed to distorting Hegel's thinking is the divorce between philosophy and the physicomathematical sciences that took place in Germany toward the end of the eighteenth century. In that country there has not been the sort of continuous current of "philosophy of science" that we see, for example, in France, where it extends from the philosophers of the second half of the eighteenth century, and in particular from d'Alembert, to Lazare Carnot and Ampère, then to Auguste Comte and Cournot, and finally to Renouvier, thus reaching our contemporaries. In Germany, after Kant (for whom, as we know, "the construction of concepts by means of the presentation of the object in *a priori* intuition" was carried out only by means of mathematics),[18] hostilities are declared between the two camps.

This was the era when the poet Schiller (who was also, of course, a notable philosopher)[19] expressed the general mood by proclaiming in a distich addressed to philosophers and scientists: "Let hostility reign between you! It is still too soon for an alliance; only by going your separate ways in your search will it be possible for the truth to be recognized."[20] This strange state of mind led to the even stranger consequence that, being unable to do without science and scorning the one they found being practiced by professional scientists, the philosophers tried to construct one to their own specifications. The result is the *philosophy of nature*, not only of Schelling and the disciples inspired by his principles, but also of Hegel himself – however careful the latter was to distance his attempt from theirs. But Hegel's effort, despite the signal failure to which it finally led, nevertheless teaches us a lesson, if only a negative one. Thanks to his incomparable vigor and the entire absence of any hesitation before patent absurdity (this is what could be called "philosophic courage" – Descartes and Spinoza also possess it to a high degree), Hegel has indeed demonstrated peremptorily, by the very completeness of this failure, how impracticable the way he had followed was.

The enormity of Hegel's failure also demonstrates how wrong science and philosophy would be to follow Schiller's advice to the letter, that is, to try to ignore one another. It serves no purpose to try to place each of them in a hermetically sealed compartment.

That is what positivism wanted above all to do, and we have seen how chimerical this project is: today's science is saturated with ontology, and scientists, despite what they themselves expressly claim, do metaphysics, just as their precursors have always done. But the positivists have not been the only ones who wanted to effect this separation; Hegelian or neo-Hegelian idealistic philosophers have sometimes followed in their footsteps. This attitude is no doubt contrary to that of Hegel himself, although it can be supported by certain specific aspects of his doctrine, in particular by the existence of the *margin* Hegel leaves to empirical science, which margin corresponds almost exactly to Comte's concept of science, an agreement that shows up in Hegel's numerous "positivistic" passages. The continuators have tried to systematize this agreement. For example, Bradley declares that for science

> matter, motion and force are but *working ideas*, used to understand the occurrence of certain events .... And for the metaphysician to urge that these ideas contradict themselves, is irrelevant and unfair. To object that in the end they are not true, is to mistake their pretensions. And thus when matter is treated of as a thing standing in its own right, continuous and identical, metaphysics is not concerned .... And thus, while metaphysics and natural science keep each to its own business, a collision is impossible. Neither needs defence against the other, except through misunderstanding.[21]

From the psychological point of view such an attitude is perfectly understandable: nothing is more natural than the fact that, after trying to invade and subjugate the scientific domain and after noting its complete failure to do so, Hegelianism is inclined to sign a sort of neutrality pact with science. But it is also quite certain that such an attitude is basically impossible. Bradley himself feels constrained to add: "But that misunderstandings on both sides have been too often provoked I think no one can deny." Now it is not a question here of more or less accidental misunderstandings, but of an organic phenomenon born out of a profound necessity: just as science cannot do without metaphysics, philosophy cannot do without science. How indeed could philosophy, whose proper task is to put us in agreement with ourselves, to present us with a fundamentally coherent image of reality, dare to neglect this formidable body of observations whose extent and whose ascendancy over our minds never stop growing? Furthermore, one need only consider without bias how thought has evolved during the last century to be convinced of the existence of this reciprocal relationship. Indeed, science and philosophy are so incapable of doing without each other that, if either one tries to do so, it immediately experiences an irresistible temptation to replace what it finds lacking by a surrogate of its own. And then the philosopher trying to manufacture a science is just as far removed from genuine science as the scientist putting together a "scientific" philosophy is from everything that can claim to be a part of genuine philosophy. Germany, as a consequence of the rupture between science and philosophy, is an excellent place to observe this double error. Each side, triumphing in its turn, seeks, as it were, to suppress the adversary. We have recounted the excesses of the *philosophy of nature*. But those of the other phase are just as instructive. Once the infatuation with metaphysics of Schelling's and Hegel's time had passed, Germany seemed to awaken from a long intoxication, or rather from an oppressive nightmare. This was the time when, even in the domain of philosophy proper, "a dog would not have wanted to accept a scrap of bread from Hegel's hand," to use Stein's picturesque expression.[22] And as for the men of science, anything that was philosophy or seemed to bear the slightest resemblance to it was an object of horror or scorn.[23] But the inevitable consequences of this attitude quickly made themselves felt, for the scientists began in their turn to do philosophy, proceeding immediately – it goes without saying – to the most extreme metaphysics. Need we add that this metaphysics, which was, moreover, only a crude yet insipid replica of eighteenth century French materialism, was no less useless from the standpoint of the development of philosophic thought than the *philosophy of nature* had been from the standpoint of the evolution of science.

On the other hand, the union of the very great philosopher with the very great scientist in the person of Descartes is precisely what enabled him to bear up under the crushing burden he had imposed on himself: the global deduction of physical reality. And if he failed – for to succeed was obviously beyond human capabilities – this failure brought human thought so many inestimable conquests, so many precious beginnings, that it merits the most resounding ovations.

NOTES

1. [Descartes, *Discourse on Method*, trans. Elizabeth S. Haldane and G.R.T. Ross, *The Philosophical Works of Descartes* (Cambridge: Cambridge University Press, 1931; reprint New York: Dover, n.d.), 1:121.]
2. See especially Etienne Gilson, *Index scolastico-cartésien* (Paris: Alcan, 1913), passim.
3. [Kant, *Metaphysical Foundations of Natural Science* (Indianapolis: Bobbs-Merrill, 1970), trans. James Ellington, p. 102.] Cf. IR 192–194, 461–462

[Loewenberg 175–177, 400–402]. It is interesting to note that in his first philosophic writing, *Thoughts on the True Estimation of Living Forces* (published in 1747), Kant tried to deduce *a priori* the tridimensionality of space (*Werke*, ed. E. Cassirer, Berlin: Bruno Cassirer, 1912, 1:21–23, § 9,10 [*Kant's Inaugural Dissertation and Early Writings on Space*, trans. John Handyside (Chicago: Open Court, 1929), pp. 9–12]). This somewhat naïve deduction (since the conception of three-dimensional space was surely implied in that of the expansion of force in conformity with the law of the square of the distance) disappeared in later works, where, given that space appears as an *a priori* form of our sensibility, there is no longer any need to deduce it. It is nevertheless possible (as Trendelenburg pointed out, *Log. Untersuch.* 1:187–188), that Hegel, in deducing the three dimensions from the three "moments" of the notion (universality, particularity, individuality), is connected, on this point as well, to an attempt Kant himself considered obsolete, so to speak. However, it is at least as likely that Kant's influence in this case was not direct and that Hegel, as in numerous similar cases we have cited, simply followed the example of Schelling, who deduces matter from the existence of three forces (instead of only two, as Kant had done) and then uses these three forces to explain the tridimensionality of space, magnetism creating the first dimension, electricity the second and the chemical process the third (*Transc. Idealismus*, I, 3:444 ff. [Heath 86 ff.]. Cf. an analogous but not identical deduction, *Stuttgarter Privatvorlesungen*, *Werke*, I, 7:447).

4. Hegel's immediate disciples never tire of repeating that the philosophy of Hegel is only the completion of the one Kant had had in mind (cf. for example Rosenkranz, in *Hegel's Leben* 317 and in *Hegel als deut. Nat.* 46, 337). Speaking of Kant's conception of matter, Trendelenburg says: "The essential aspect of this conception has remained and has continued to influence the new philosophy of nature" (*Log. Untersuch.* 1:212). Thus McTaggart is not entirely wrong, and in any case remains faithful to the spirit of the school, when he declares that Hegel's epistemology is fundamentally identical to Kant's (*Studies* 25), nor is James Hutchison Stirling's more general assertion that Hegel had "made *explicit* the *concrete* Universal that was *implicit* in Kant" (*The Secret of Hegel*, Edinburgh: Oliver & Boyd, 1896, p. xxii) unfounded from this point of view. Furthermore we know that in general the English-speaking neo-Hegelians claim almost as much kinship with Kant as with the author of the *Phenomenology* and are therefore sometimes called neo-Kantians.
5. William Wallace (*Prolegomena* 76) particularly stresses, and rightfully so, the fact that both Kant and his successors understood the deductions relative to nature as "consistent, continuous and coherent."
6. [Leibniz, *New Essays on Human Understanding*, Bk. 1, Ch. 1, § 2: "There is nothing in the intellect which was not previously in the senses ... except the intellect itself."]
7. [*Wiss. der Logik*, 5:31 (Miller 595), as quoted on p. 300, n. 2, above.]
8. Edward Caird (*Essays on Literature and Philosophy*, Glasgow: James Maclehose and Sons, 1892, 2:437–438) rightly emphasizes this aspect of the Hegelian doctrine.

9. [Edmond Goblot, *Traité de logique* (Paris: Armand Colin, 1918), p. 331, as quoted on p. 67 above.]
10. [Trendelenburg, *Log. Untersuch.* 1:68, as quoted on p. 290 above.]
11. Norero, 'Compte-rendu du IVe Congrès de philosophie,' *Rev. de Méta.* **19** (1911) 626.
12. [Ferdinand Rosenberger, *Die Geschichte der Physik* (Brunswick: Friedrich Vieweg, 1884), 3:4.]
13. Lavoisier, *Oeuvres* (Paris: Imprimerie Impériale, 1862), 1:3–4. The passage continues: "The only way to prevent these errors is to suppress or at least to simplify as much as possible the reasoning we supply, because it is the only thing that can lead us astray; to submit it continually to the test of experience; to keep only those facts that are pure givens of nature and cannot deceive us .... Convinced of these truths, I have imposed on myself the rule of never proceeding except from the known to the unknown, of deducing no consequence not following immediately from experiments and observations." But can Lavoisier, whose theory had caused science to accomplish such a prodigious leap forward on the eternal path toward the rationalization of reality, seriously be called to task for believing that his conception was identical with the facts, that he definitively held, to use our image, a fragment of the great deductive chain? Is it not natural, on the contrary, for him to write that "progress [in chemistry] today is so rapid, the facts fall into place so felicitously in the modern doctrine, that we can hope, even in our time, to see it come much closer to the degree of perfection it is capable of attaining" (*Oeuvres*, 1:5 [Meyerson's brackets]), which degree of perfection, as we see from the sentence preceding the one just quoted, is that possessed by geometry?
14. Marcellin Berthelot, *La Révolution chimique: Lavoisier* (Paris: Félix Alcan, 1902), p. 74.
15. Henrich Steffens, *Grundzüge der philosophischen Naturwissenschaft* (Berlin: Realschulbuchhandlung, 1803), p. iv.
16. Bernard Brunhes, *La Dégradation de l'énergie* (Paris: Ernest Flammarion, 1908), pp. 261–262.
17. E. de Baillehache, 'Les Unités électriques,' *Scientia* **8** (1910) 208.
18. Kant, *Premiers Principes métaphysiques de la science de la nature*, trans. Andler and Chavannes (Paris: Félix Alcan, 1892), p. 5 [*Metaphysical Foundations of Natural Science* (Indianapolis: Bobbs-Merrill, 1970), trans. James Ellington, p. 5]. Cf. also Ch. 13, p. 353, above.
19. See especially the very instructive study by Xavier Léon, 'Schiller et Fichte,' *Etudes sur Schiller*, Bibliothèque de philologie et de littérature modernes (Paris: Félix Alcan, 1905), p. 90.
20. Friedrich Schiller, 'Naturforscher und Transzendental-Philosophen,' *Werke*, ed. Boxberger (Berlin and Stuttgart: Spemann, n.d.), 2:176. Moreover, it is quite possible, as Boxberger indicates in a note, that Schiller's epigram was chiefly directed at the *philosophy of nature*, which was beginning to make itself heard at almost exactly this time; if so, he would have wished to protest solely against this bizarre amalgam of experimental knowledge and transcendental deduction (which mixture, on the other hand, appears to have pleased Goethe very much).

The form in which he expressed the protest and the way in which it was understood by the German public are nevertheless typical.

21. F. H. Bradley, *Appearance and Reality* (London: Swan Sonnenschein & Co., 1893), p. 284.
22. As to this prolonged disrepute of the Hegelian philosophy in Germany, cf. what D. Strauss, himself still a Hegelian at that time, felt obliged to admit in 1854 (Appendix 6, p. 573, and Ch. 11, p. 306, n. 50). Forty years later, René Berthelot ('Sur la nécessité' 119) likewise testifies that in Germany there is a tendency to consider Hegelian philosophy "as no longer possessing vitality and fecundity."
23. Here is just one of many possible examples. In 1877, Kolbe, a famous chemist, in attacking Van't Hoff's stereochemical conceptions in the most violent manner (as we related in Ch. 6, p. 175, n. 25) and also blaming Wislicenus, who had had the misfortune of approving the hypothesis of the young scientist, wrote: "Herewith Wislicenus makes it clear that he has gone over from the camp of the true investigators to that of the speculative philosophers of ominous memory, who are separated by only a thin 'medium' [untranslatable pun on the term *Medium* which in German designates both a *milieu* and a spiritualistic *medium*] from Spiritualism" (J. H. Van't Hoff, *Dix années dans l'histoire d'une théorie*, Rotterdam: P. M. Bazendijk, 1887, p. 21 [*Chemistry in Space*, trans. J. E. Marsh (Oxford: Clarendon Press, 1891), p. 18; Meyerson's brackets]).

BOOK FOUR

SCIENTIFIC AND PHILOSOPHIC REASON

CHAPTER 15

SCIENCE AND PHILOSOPHIC SYSTEMS

The necessity of a union between science and philosophy, which was the conclusion we reached in Book Three, seems at first glance to create an inextricable difficulty. If philosophy really needs only good will to accommodate itself to science and its changing aspects – the scientists of a given era always being more or less in agreement among themselves – science finds itself in an incomparably more difficult situation with regard to philosophy. For it is only too obvious that in philosophy there is only diversity and disputation; never has one particular metaphysical position truly reigned to the point of silencing all contradiction, let alone to the point of removing all doubt, and of course in our time this doubt more than ever reigns supreme. Now it may be argued that science cannot suspend all judgment in this matter. If it is true, as we have seen, that all scientific explanations are naturally, unconsciously, necessarily ontological, then science would seem to need this ontology *immediately*, not in the process of being worked out, but already completely worked out; in other words, it must have made a choice between the possible metaphysical systems.

Is this really the case and, if so, which metaphysics does it support?

Let us note, first of all, that it cannot be merely the metaphysics of common sense. This remark is not a superfluous one, for there has been no lack of attempts attributing a preponderant role to common sense and claiming to draw a complete philosophy from it, which philosophy was then seen as necessarily dominating all the conceptions our reason elaborates about reality and therefore, first and foremost, those of science. But perhaps the homage thus rendered to common sense was not always as sincere as it seemed. As a matter of fact, when philosophers appeal to common sense, they generally reserve the right to define it in their own way and often wind up replacing the primitive ontology, which is what it really is, with more or less idealistic conceptions – thus taking advantage of the tendency of all idealism to "turn around," to come back surreptitiously to the instinctive notions of naïve realism (cf. Ch. 10, p. 256). But it also does happen that common sense is actually set up as the supreme judge, that this or that philosophic or scientific conception is solemnly condemned in its name. In Chapter 11 (p. 263), we gave an example of this attitude when we spoke of Duhem's opinion of Hegelian metaphysics. In the name of these same principles, Duhem also rejected the new theories arising out of the consideration of electrical and luminiferous phenomena, and in particular the theory of relativity of Einstein and Minkowski. For Duhem, all that is needed in order to condemn this conception is the observation that "one cannot give a correct statement of it in ordinary language without having recourse to algebraic forms," and that it "upsets all the intuitions of good sense," for "in order to be true it is not sufficient that science be rigorous, it is necessary that it start from good sense to end up with good sense," and no judgment can be too harsh for this science that "goes its way, proud of its algebraic rigor, looking with scorn at the common sense with which all men are endowed."[1]

We need only recall what we have come to see during the course of the present work to realize how little assertions of this type can be founded on what science and the way it has evolved actually teach. Science undoubtedly starts with common sense; but how could it possibly return to it at the end of its labors? We saw in Chapter 1 (p. 26) that as soon as reason begins its work of scientific research, it feels compelled to note the inconsistency of the representation presented to it by simple perception in the first place, and consequently it has no alternative but to substitute a different representation, that of theories. Science does not do so lightly nor because it is so "proud of its algebraic rigor" that it scorns the common conceptions of humanity; on the contrary, the resistance consistently encountered at the outset by the opinions which we now consider the most justified and best-established bears eloquent testimony to the slowness with which reason abandons familiar conceptions. But reason cannot renounce its essential function, which is to reason, to reduce sensations to something coherent, and that is what forces it to transform, gradually and reluctantly, the reality it initially believed so firmly established. If it has been possible to make mistakes, to fail to recognize this process which actually dominates the entire course of scientific thought, it would seem to be solely due to a more or less willful confusion between the common sense representation and the one to which science in fact appeals quite often, namely, the representation provided by mechanism. Now in the first place we have been able to convince ourselves that the world of mechanism constitutes, for science, not a truly completed conception to which it adheres definitively, but rather an intermediate phase, since it reserves the right to resolve the atoms later into indistinct space (Ch. 5, pp. 135 ff.). And then, even leaving aside this ultimate outcome of scientific theories and thinking about the easiest representation to grasp, the most palpable representation of mechanism out of all those offered by science, namely corpuscular mechanism more or less as Boyle imagined it – a world of infinitely hard, infinitely elastic globules colliding with one another – one need only consider it carefully to recognize how completely such a representation overthrows our original ideas. Indeed, what does this greyness have in common with the reality of our immediate perception, riotous with color, heat and sound? And what an immense difference between these indivisible particles and the matter with which we are familiar! No doubt mechanism more or less unconsciously takes pains to create a sort of illusion here; Bergson has observed with remarkable perspicacity that the atoms said to be deprived of physical qualities are actually determined "only in relation to an eventual vision and an eventual contact."[2] But that is simply because reason, as we have seen, tenaciously maintains the representations to which it is accustomed. Thus one cannot appeal to the role played by the mechanistic conceptions in science in order to affirm the supremacy of common sense.

Furthermore, a mere glance at history shows that science today is filled with claims that violently shocked the common sense of those living at the moment they appeared. Let us consider the sphericity of the earth. It was already a firmly established scientific truth in antiquity, in Plato's time, a truth never seriously doubted by any genuine astronomer since. However poor an idea one may have of the scientific mentality of the Middle Ages, it is nonetheless certain that the cosmology commonly taught then, that of Ptolemy, firmly assumed sphericity. Now one need only examine the objections raised against Christopher Columbus to see how little common opinion at that time was aware of the most direct consequences of this doctrine, as for example the fact that gravitation must everywhere be directed toward the center of the terrestrial sphere: it was seriously argued that by sailing west one would slide down a mountain of water and find it impossible to climb back up again. Should this surprise us? And is it not sufficient to plumb the depths of our own consciousness to find an analogous sentiment? Ever since we were children, I dare say, we have so often heard sphericity affirmed, and everything we learn, the trips around the world, the existence of climates, etc., so presupposes this fundamental idea – which is reconfirmed by each glance at a geographical map with its system of meridians and parallels – that it is difficult for us to entertain a formal doubt in this regard. However, who among us can picture the antipodes in any other way than with their heads down and their feet up? Here we see that the opposition of common sense, in spite of the uncontested domination of the scientific theory for more than twenty centuries, has not yet been reduced to silence, since, for our immediate feeling, the direction of gravitation still remains a privileged direction parallel to itself in absolute space. One could cite many other cases almost as flagrant, and it would be particularly easy to find, until nearly the end of the eighteenth or indeed even the beginning of the nineteenth century, passages in which what we consider the most firmly established teachings of science, such as the theory of inertial motion or the theory of atmospheric pressure, are cruelly ridiculed in the name of the common ideas of mankind, of this good sense "with which all men are endowed."

Thus – it has been said many times, but this truth cannot be repeated too often or learned too well – in the realm of science there is not and cannot be any domination by one common doctrine, one unshakable catholic doctrine formulating, according to the well-known definition, "what has been believed everywhere, always and by everyone."[3] Here everyone is not necessarily right against Voltaire; on the contrary, it is possible that the innovator, however paradoxical, however unheard of his ideas appear to his contemporaries, will nevertheless end up making them prevail, that humanity will, after lengthy resistance, adopt ideas it at first saw as absurd chimeras. That is what finally convinces us that if science contains a particular metaphysics, it cannot be that of common sense, which, by virtue of our constitution, is elaborated in everyone in exactly the same way.

But if science abandons common sense, what conception does it put in its place? It is surely not the Hegelian metaphysics which, as we have seen, was never able to gain any acceptance in science. Is it then another system dreamed up by the philosophers? Or does science end up, in spite of everything, with a system of its own, a system the philosophers had not thought of?

In order to resolve this problem, we shall not take the direct route. We shall not try, at least not initially, to examine science itself, because we would undoubtedly run the risk of being led astray by preconceived ideas we might have without being aware of them. We shall instead examine the solutions that have been proposed by philosophers or scientists, especially those still being proposed today.

These solutions are many and varied. This statement might seem surprising, at least insofar as scientists are concerned. Have we not asserted that they generally admit to being positivists and that positivism is an altogether well-defined doctrine consisting precisely in the exclusion of metaphysics? Indeed we have, but we have also seen that the positivistic doctrine, in all its rigor, is impossible to apply. One cannot do science or talk science without including as a substratum a body of presuppositions about being. That is why scientists, although they often profess a rigid positivism in theory, in reality imply a whole metaphysics when it comes to putting forth their positions (sometimes a rather inconsistent metaphysics, it is true – we shall shortly see the reason for this). But precisely because of the clarity and basic simplicity of positivism, it is relatively easy to separate out, to lay bare the elements that are fundamentally foreign to it.

For the sake of convenience in the examination we are about to undertake, we can divide the most distinct of these solutions or attempted solutions into four major classes: 1) mechanistic or atomistic theory, 2) energetic theory, 3) a more or less refined philosophic realism (such as Hartmann's "transcendental realism"), and finally 4) mathematical idealism. It goes without saying that this classification lays no claim to rigor, and even less does it aspire actually to cover in its totality the vast field of possibilities upon which we are touching here; but perhaps by examining these few types of contemporary thought that appear particularly characteristic, we shall have a somewhat clearer idea of the general solution.

Mechanism (which is also termed materialism when one means to stress the fact that the system ultimately tends to include in its explanations the whole body of psychic phenomena as well, and atomism when one wishes to emphasize that it generally assumes the discontinuity of matter) was for a long time considered the philosophy *par excellence* of the physical sciences. This practice is still perpetuated in many popular accounts – it is easy to see that many of those who pretend to be positivists are in reality imbued with the purest mechanistic spirit. If we turn to real science, we at first have the impression that it follows the same current; certainly as soon as it is a question of molecules and atoms, they are treated as real things, as the ontological reality forming the substratum for the reality of common sense and thus serving as a substitute for it. One could even believe that this conviction has become even stronger of late, since we have now succeeded in knowing the absolute magnitudes of atoms.

We noted in Chapter 5 (pp. 128 ff.) that, contrary to widespread belief, the mechanical or atomic theories are not of truly scientific origin. No doubt science was able to give them the particular form in which they now appear, but their foundation goes back much further than science or, at least, arose simultaneously with it. Atomism is found entirely formed in Democritus at a time when Greek science is barely in its infancy, and also among the Hindus in very remote antiquity; and we easily see that in spite of what appear to be such complete changes brought about by electronic theory, for example, certain essential traits remain perfectly immutable, as modern scientists have been somewhat surprised to discover. For example, Gouy's and Perrin's demonstration concerning the reality of the invisible molecular motion that is the source of Brownian motion offers an incontestable family resemblance to the demonstration by which Lucretius, in immortal verse, establishes the materiality of air.[4] That, as we have seen, is due to the fact that atomism is in reality an aprioristic construction, or at least one framed to satisfy an innate tendency of our mind, the causal tendency, which makes us seek something stable in what is changing, thus leading us to explain the appearance of the changing body by the supposition of a change in the combination of eternal particles. The atom of Democritus, as Aristotle testifies (Ch. 5, p. 129), is only the permanent being of the Eleatics, and Descartes showed us (Ch. 5, p. 126) that the only intelligible change is change of place. Science certainly confirms atomism and in large measure completes it. Is not science the work of man, and is it not natural for it to express a tendency that is thus its own? But it must be recognized that this confirmation sometimes goes beyond what one could legitimately anticipate; here and there reality unexpectedly shows itself to be in agreement with our mind.

No doubt the modern physicist needs a reality detached from the self – an ontology – just as much as the physicist of any other epoch did, and since by his observations he has destroyed the ontology of common sense from which he started, he is more or less forced to believe in the beings he creates, in molecules, atoms, electrons. However, he comes too close to these beings to be able to overlook the difficulties inherent in these conceptions, and consequently in the concept of mechanism as well. Not only does he realize that he is in no position to come up with any truly satisfying representation of the reality of things, but he is too aware of what there is in phenomena that is irrational, radically irreducible to the categories of our understanding, to be able to be mechanistic in the philosophic sense of the term, as his predecessors frequently were, that is, to affirm that the essence of the real must be constituted by a material and intelligible whole (Ch. 14, pp. 375 ff.).

Unlike mechanism, energeticism is a doctrine of genuinely scientific origin. Its author is Ostwald, and so we shall deal here principally with the theory as propounded by this learned chemist. That is not an idle warning. As a matter of fact, not all those considered energeticists actually profess the same opinions. Many see no more in the position than its opposition to mechanism, and see the energetic formulas only as rules which would make it possible to do without any assumptions about the substratum, about the reality of things. Those who are of this persuasion can be classified quite simply as positivists.[5] Ostwald's position is basically quite different; energy for him is a veritable ontological being, a thing in itself, whose variations are the sole cause of the infinite multiplicity of phenomena. In this respect, then, energeticism resembles mechanism. It is easy to see that it also resembles mechanism in that it follows from the same tendency of our intellect, since it too seeks to explain all variation by the displacement of something that remains constant: here it is energy, as it was the atom in the earlier case. From the philosophic point of view, energeticism is thus in no way superior to mechanism. One might think that as a result of its origin it would be superior from the scientific point of view. Just the opposite seems to be the case. The most immediate utility of scientific theories is, first of all, to allow a great number of facts to be classified from a single viewpoint, then to guide the researcher by enabling him to predict new facts. Now in terms of the body of facts known at the very moment energeticism was born, the theory was already, if anything, inferior to atomism: to mention only one particular point, Carnot's principle is certainly an anomaly in both systems, but its statistical explanation seems much less contrived than the attempt to assimilate mass with entropy, even though the conservation of mass is obviously of the same order as that of energy (cf. IR 396 ff. [Loewenberg 347 ff.]). As for predictions, the services mechanism has rendered are obviously countless; they have sometimes surprised the researchers themselves, who at first believed they had found only a quite superficial analogy, whereas subsequent discoveries established the existence of a marvelous agreement. We shall mention only two examples: the earlier is Fresnel's deductions for the wave motion of light,[6] and the more recent is the way in which facts linked to Le Bel's and Van't Hoff's conception of the asymmetrical carbon atom complete and clarify Kekulé's structural theory, despite the fact that the latter had originally appeared almost intolerably crude. Energeticism can be credited with nothing comparable. It could no doubt be argued that this is only because the theory is quite recent. But as fate would have it, since the formulation of the energetic doctrine important discoveries have been made casting new light on the constitution of matter. Now these discoveries, which are those that have ensured the triumph of kinetics over thermodynamics (cf. Ch. 1, pp. 22 ff., and Ch. 6, pp. 164 ff.), clearly run counter to the energetic position. In fact, probably the most profound distinction that could be established between energeticism and contemporary mechanism (from the standpoint of the consequences each of the two theories entails for the conceptions of physics) is the following: the former is a doctrine of the continuous, while the latter is a doctrine of the discontinuous. Now the discoveries in question establish the discontinuity of matter, and work such as Perrin's that has led to establishing the absolute size of atoms, underscores the triumph of this point of view. We cannot see how any attentive observer could fail to be struck by the extreme vigor of the thrust of science toward atomism at this time,[7] and it is perhaps not too early to say that from this perspective philosophic energeticism already belongs to the past. If it still has supporters, it is apparently only because of the more or less conscious confusion mentioned above which assimilates it with a theory of purely *legalistic* relations, allowing it either to proclaim the positivistic abstention from all metaphysics or even to fit in with atomism.

With transcendental realism we return to the realm of conceptions of philosophic origin. We have chosen to present this system in the form given it by Eduard von Hartmann, because this philosopher strove to support his position by considerations based on the theories of the physical sciences. The way in which Hartmann connects his system to science can be summed up very briefly. He points out that science, by means of observation, though starting from the world of common sense, nevertheless ends up substituting for it an entirely different conception, namely mechanism. But, on the other hand, it cannot establish that mechanism exists in itself, since the last term of the reduction remains mysterious. Finally, throughout its work, in the course of which directly perceived reality is destroyed, science maintains the notions of time and space. Thus it ultimately arrives at a noumenon subject to the conditions of time and space, that is, a particular metaphysical system which is precisely the one Hartmann designates by the term transcendental realism (cf. IR 483 ff. [Loewenberg 417 ff.]).

Hartmann's system does not appear to have attracted much attention among scientists. It nevertheless seems in some respects to be quite representative of the convictions (mostly unconscious, to be sure) of the average experimental physicist. The latter, while inclined, on the one hand, to believe in the reality of the external world he is exploring, has to be aware, on the other hand, of the inextricable difficulties inherent in the representation that science offers him in its mechanical theories. As a result, the body of inner convictions this physicist holds concerning the reality of the sensible world will actually most often fluctuate somewhere in the vicinity of a "noumenon subject to the conditions of time and space," and in this sense Hartmann does not err in seeing this as the current philosophic theory of the modern physicist. It is certainly implied in any number of scientific expositions of our day. If it is not asserted more clearly, it is not simply because the scientist does not dare venture into the realm of philosophy, nor even because he is obedient to positivistic suggestions and sincerely believes he can abstain from metaphysics, but surely also because he feels deeply, if vaguely, that this is not a truly coherent and final conception from the scientific point of view, but rather an intermediate step toward something even less concrete.

Indeed, when we examined the way science proceeds in its work of explanation (Ch. 5, pp. 135 ff., Ch. 6, p. 166), we recognized that science by no means stops at the atom but, on the contrary, tends to dissolve it into a universal medium, the ether, which is and can only be a hypostasis of space. Thus what it finally ends up with is an indistinct whole in time and space, the sphere of Parmenides, acosmism (to use the term – and an accurate one – invented by Hegel, who applied it, as we know, to Spinoza).

And thus Hartmann's claim to deduce his realism directly from the conceptions of science breaks down.

These considerations taken as a whole appear to push science clearly toward idealism, and this is the tendency behind the last of the conceptions we have mentioned, namely mathematical idealism. What we meant to designate by this term is a whole body of current theories tending to renew and elaborate one of the loftiest and most seductive philosophies known to mankind, namely Platonism, which, according to the Greeks themselves, had "mathematicized" nature.

We referred to Plato's panmathematicism in Chapter 8 (pp. 216 ff.) in our discussion of explanation by geometric figures, pointing out how consistent it is with the spirit of science today.[8] Modern pan-mathematicism also avails itself of one aspect of this consistency. We believe that an exposition of the principles of this position will convince the reader that a metaphysics of this sort is quite often implicit in many works dealing with the theory of science and its relation with thought. But one school in particular has given it a more clearly defined form in a series of most remarkable works: we are referring to the Marburg school. It goes without saying that we in no way pretend to criticize the whole of this vast edifice here, nor do we aspire to give those unfamiliar with these works any idea of the great richness of detail amassed by the scholarship of Hermann Cohen and his disciples and of the vitality of the thinking that has reduced this detail into a whole marked throughout by a strong unity. We shall be concerned with only one particular aspect of this philosophy, namely its attitude toward science. This attitude can be summed up briefly as follows: from the fact that mathematical physics, as one of the thinkers of this school, Cassirer, puts it, "turns away from the essence of things and their internal nature in order to turn toward their numerical order and connection and their functional and mathematical structure," one tends to conclude that the physical sciences reduce the sensible world to a body of relations of function. It is therefore a world of mathematical concepts, a purely ideal world which is substituted for the common sense world of sensible realities, and it is in this idealistic conception that science and philosophy must meet and unite.

First of all, from a general and abstract point of view, we can see what the philosopher-physicist finds tempting in this particular form of the idealistic position. We saw in Chapter 1 that science, as it advances, moves further and further away from considerations involving the person of the observer. Clearly, therefore, when one wants to pass legitimately from the world of scientific reasonings to that of a subjectivism based on immediate sensation, it will be necessary to backtrack over the whole route science has had us travel, that is, to proceed again by the way of the common sense conception.

That is a consequence idealistic philosophy evidently has some difficulty admitting. The immense scientific or pseudoscientific construction that Hegel attempted to erect, considered from this point of view, is actually nothing other than what can be done with science if one begins, not with the ontology of common sense, but *after* destroying that ontology in favor of a world of concepts. No doubt the contemporary supporters of Hegel would be little inclined to renew this adventure; they would be more likely to admit with Bradley that science and philosophy must ignore one another (an attitude we have shown to be equally impossible). But from the standpoint of theory, their demands have remained the same. Thus F.C.S. Schiller, a contemporary philosopher who is not a Hegelian, but whose thought must have been considerably influenced by the Hegelian doctrine – as is clearly the case for all the English thinkers – demands that "these entities of physics [be construed] as pragmatic constructions which are relative to, and justified by, the scientific success which attends their working, but *should not* be construed as metaphysical assertions."[9] There is some truth in this precept insofar as it is addressed to the philosopher: he must not consider the existence of the "entities of physics" to be a result definitively imposed by physics (which, as we shall soon see, it in fact is not). But if one is to see this, at the same time, as a program to be followed by science, it is certain that it is an entirely chimerical view. Indeed, we saw in Chapter 1 that all areas of current physics – even those like thermodynamics that at first seem more or less consistent with the positivistic program – imply the affirmation of the metaphysical existence of *things* independent of our sensation. We also pointed out how unlikely it seems that a true science can ever be constructed on any essentially different basis. Here we can press the question a bit harder by asking how one would go about developing an area of physics according to Schiller's model. What would become of atoms, for example?

Obviously the atom or the molecule of pure mechanistic theory can be assimilated to a billiard ball and henceforth treated, according to John Stuart Mill's or Bertrand Russell's precept, as a "possibility of sensation" or "sensibilium" – but of a sensation, it should be noted, that is already transposed, since it will be a question not of the sensations of a human being, but of those of a being made entirely in our image who would have sufficiently refined senses or would himself be sufficiently reduced in size to perceive these miniature billiard balls. But let us think of the chemical atom, the carbon atom for example. How do we go about defining it by "the success which attends [its] working,"[10] that is, as a function of the experiences that caused us to imagine it, at the same time being very careful not to suppose its existence *ipso facto*? No doubt if there were a good *theory* of the chemical atom, if we could reduce all the properties of carbon (as has successfully been done for the "dissymmetry" of its valences) to the fact of the spatial arrangement of particles having only properties, if not intelligible (which we know is impossible), at least not more unintelligible than those we admit in mechanics, we would assimilate it again, with more or less effort, to a common sense object and would treat it as a "possibility of sensation." But such a mechanical theory does not exist, and therefore the transition from the atom to experience – even if we mean to limit ourselves to fundamental experiences such as combustion – proves to be too complicated for us to be able to preserve the tenuous thread of the possibility of sensation; whether we want to or not, if we do not wish the thread to break, we shall be obliged to strengthen it, which can obviously be accomplished only by positing the ontological existence of the concept, as science does. Clearly if we hoped to see science eventually reduce the entire material world to a mechanism, we could consider the current state as ephemeral and suppose that there will come a day, no matter how remote, when science will be able, after a fashion, to speak the language of the idealistic philosopher. But, we are well aware, this hope is no longer possible. And the electrical phenomenon, which is currently the fundamental phenomenon, has nothing to do with our direct sensation. If we wish to arrive at direct sensation, to succeed in defining the electron as "possibility of sensation," it will be possible to do so, as for the chemical atom, only by backtracking over the course science has followed starting from common sense perception, and it will be a task even more complicated, if that is possible.

We saw in Chapter 12 (p. 320) how Schelling stressed the fact that as science progresses "the phenomena themselves become more mental, and at length vanish entirely" (*Trans. Idealismus*, I, 3:340 [Heath 6]). For him this "spritualization" of the real paves the way for the introduction of purely idealistic notions into science, and certainly conceptions of this sort are more or less clearly implied in many of the epistemological arguments of our day. But that is simple equivocation, a more or less conscious confusion between two sorts of metaphysical doctrines brought together in the term idealism (as we shall try to explain shortly). If, indeed, what is meant by the word is all the various theories generally included under this name, with the exception of mathematical idealism (which was surely, at bottom, the way Schelling understood it), it is certain that the course of science, rather than coming closer to it, is moving further from it and that, contrary to what seems to be suggested, the more the world of the things of our immediate perception is destroyed, the more difficult it becomes to introduce such an idealism into science.

Thus idealistic concepts have no place in science and, considered from this angle, Bradley's precept is fully justified: philosophic criticism must not seek to interfere with the deductive chains of science; it must let the scientist follow his deductions and develop his theoretical conceptions by positing the objectivity of phenomena.

This observation unquestionably has a paradoxical side. For in the end scientific theory tends to destroy the reality of the world of our perception just as much as the most idealistic philosophy does, since it finally resolves it into an indistinct whole. The two tendencies are therefore in agreement here, or rather they are obviously one and the same fundamental tendency of the human mind, the tendency to want the world to be intelligible and to be satisfied only in destroying it. This tendency manifests itself in two ways. Why then cannot a closer relation be established between the two aspects of this one effort?

In order to understand why, we need only observe how idealistic philosophy and science respectively go about destroying perceived reality. Science begins by disregarding, that is, by abolishing the singular in favor of the general, of universals. Having thus impoverished reality in order to create legalistic science, it creates theoretical science by stripping reality as much as possible of any qualitative element, transforming it into a collection of beings such as the atom, force, mass, velocity, energy: abstract and quantitative beings. Finally it dissolves these substances in their turn into the undifferentiated ether. But it should be noted that science accomplishes this dissolution only at the very end. In the interval between the creation of the genus necessitated by law and the disappearance of matter into the ether – an interval that in fact embraces science in its entirety – the quantitative being of the theory remains as real as the common sense objects out of which it came and which it replaced: if we grasp atoms or electrons, "singular points in the ether," just as they are about to blend into it, we shall find them as independent of our sensation, as *substantial* as the things we believe we perceive. In a certain sense they are even more substantial (as we showed in Chapter 1, p. 24), for they are assumed to be actual substances, and science, by taking away their qualitative aspect, has also added to their perdurability.

In sum, we can say that in starting from the world of our direct perception which, if one wishes to characterize it in two words, is essentially a *sensible reality*, science attacks the first of these terms; as its work progresses, its concepts move further and further from sensation, becoming less and less sensible, whereas their reality is reinforced.

Idealistic philosophy follows exactly the opposite course: it sacrifices the second term to the first. We have only to contemplate the whole of the admirable work of Bergson to ascertain that this is so. From the very beginning, by an analysis whose profundity has surely never been surpassed, he probes the "immediate data" of our consciousness: perception is dissociated for the sake of pure sensation. Here again the world of common sense is dissolved, but the process has nothing in common with that of science (unless it be the end result), for reality does not dissolve into space; on the contrary, the spatial form is what disappears first. What was a sensible reality certainly remains sensible, but it is no longer a reality in the sense in which science understands the term: it is no longer separable from the *self*.

The reasons for this diversity between science and philosophy can also be understood from a slightly different pont of view. Indeed, what we rediscover here is the distinction between mathematical deduction and purely logical deduction (using the latter term to designate all the various processes by which we mean to connect our ideas *without* the intermediary of mathematical concepts). Science uses the first type of deduction and philosophy the second. Aristotle and Hegel wanted to make logical deduction prevail in science, and the conceptions of the former (if not of the latter) triumphed for centuries, to the point that science really did attempt to conform to them. But finally all that belongs to the past, no doubt superseded forever. In spite of all the efforts undertaken on its behalf by certain eminent thinkers and in particular, most recently, by Duhem (cf. Ch. 14, pp. 384 ff.), scientific Peripateticism seems to have no chance of rising again; the indications to the contrary that some have wished to emphasize appear to be founded on purely superficial analogies, and as to the fundamental aspect of Aristotle's doctrine, the purely logical linking of concepts, it is certainly impossible to discover the slightest trace of it in contemporary science. Furthermore science is at present too solidly entrenched, its dominion over the human mind too firmly established, to have to fear a renewed attack by Hegelianism or, in general, a real encroachment on the part of any idealistic philosophy whatsoever.

However, as we have seen (Ch. 14, pp. 387 ff.), science and philosophy must not combat one another, and it is even dangerous for them to remain ignorant of each other. For they are both emanations of our reason, and it is thus possible that in spite of the fundamental diversity whose basis we have just noted, they approach each other in this or that particular bit of reasoning. For instance, to cite only the most recent example, the fact that

Einstein's and Minkowski's theory of relativity seeks to impose new notions of time and space on us certainly makes those notions that form the basis for immediate perception appear to be purely subjective, to be incapable of belonging to what is assumed to exist outside us. Now this is a conception that, in a certain sense, resembles that of many a philosopher in this domain, in particular (as supporters of relativity theory have not failed to call to our attention) the Kantian notion of space as the form of our perception.[11]

Moreover, insofar as this diversity between science and philosophy is concerned, an important reservation must be made, a reservation bearing on the distinction to which we just alluded (p. 407). We spoke of "idealistic philosophy" above only for lack of a term to designate the whole class of different philosophic systems that, in contrast with realism, form concepts based on the idea and proceeding by logical deduction, for the observations we have just made refer exclusively to these systems. But these are not the only systems possible; on the contrary, philosophy can, at least partially, follow a route analogous to that of science, forming its concepts by means of mathematical deduction. It is precisely this mathematical idealism that concerns us here, and it is easy to understand how, given the fundamental agreement between this philosophy and science, mathematical idealism is particularly well-suited to the essential concepts of science. Indeed, in the domain of mathematical concepts the agreement between our reason and reality is complete. Therefore, if philosophy bases itself on these concepts, its arguments can easily link up again with those of theoretical science.

Furthermore, let us recall what we observed about the independence of the entities created by theory with respect to our sensation and also about the unlimited perdurability of these entities. Indeed, it follows from this that science, in destroying reality, leaves its mathematical aspect intact, or at least would like to: at the very moment the "singular points" are about to dissolve into the ether (p. 407), they are still characterized by the figures in which they are arranged, and one would have us believe that this trait persists, even though the *points* have already lost all distinctive existence, are no longer anything but purely geometric points.

What is more, this observation is obviously implied by what we showed in Book Two, notably in Chapters 5 (pp. 135 ff.) and 8. If it has been possible for science to try to explain phenomena by space and, ultimately, to reduce the diversity of things to space, it is certainly necessary not only that there be complete agreement between geometry and reality but also that geometrical figures, in spite of everything, retain, in our imagination, a mysteriously substantial quality, a certain corporeity, so to speak; indeed, without this the enterprise would have appeared absurd from the beginning, and the devices used in the reduction to space would not have been able to deceive us for a single moment. Now we saw that such is not the case: the illusion is real, as attested to not only by Descartes, but by modern ether theorists as well. That is also the profound source of the sentiment that allows Kant to contrast the "pure philosophy, or metaphysics" which is "pure rational cognition from mere concepts" with that philosophy which undertakes "*the construction of concepts* by means of the presentation of the object in *a priori* intuition" and which "is called mathematics." In fact, from mere concepts

> the possibility of the thought (that it does not contradict itself) can indeed be cognized, but not the possibility of the object as a natural thing, which can be given (as existing) outside of the thought. Therefore, in order to cognize the possibility of determinate natural things, and hence to cognize them *a priori*, there is further required that the *intuition* corresponding to the concept be given *a priori*, i. e., that the concept be constructed. Now, rational cognition through the construction of concepts is mathematical.[12]

It is this feeling that obviously forms the deepest foundations of panmathematicism.

It should be noted, however, that, as we have pointed out (Ch. 6, pp. 143 ff.), geometry itself cannot be considered entirely rational. Without speaking of the concept of *diversity* with which our mind must come to terms in all its reasonings without exception (and which we also treated in connection with Hegel in Ch. 11, pp. 277 ff.), it is clear that the notion of space exhibits particular traits which we must accept as given. For this very reason, the term panmathematicism we have just used presents a certain imprecision. One can in fact consider excluding this geometric irrational by creating a pure *panalgebrism*. We shall not dwell on this alternative, for the simple reason that what interests us here is the explanation of the physical phenomenon, which, as we believe we have established in this work (especially in Ch. 8), is carried out by means of spatial functions. Therefore our panmathematicism also includes the whole of geometry, and could thus be called *pangeometrism* (as opposed to the above-mentioned panalgebrism). Consequently it includes the geometric given, and in particular the tridimensionality of space – we dealt with this in Chapter 5 and noted in Chapter 14 (p. 388, n. 3) the sterility of efforts by philosophers to deduce this property of our extension *a priori*. But what we must recall here is that if empirical elements enter into mathematics they are to be found contained in what forms the premises, as it were, of this science, in the initial definitions, axioms and intuitions, and that once these foundations have been laid, the rest is developed, or at least is capable of being developed by the pure work of reason (Ch. 4, pp. 96 ff.; Ch. 5, pp. 109 ff.). Supposing, indeed, that experience does enter into mathematics, as has been claimed and as appears likely, its role there certainly seems to be quite different than in the physical sciences, since it is apparently possible to eliminate it afterwards, to establish a proof by deduction, leaving no room for doubt or future corrections. And it is particularly significant that this demonstration seems to us to be *always* possible, for even in the extremely rare cases where experience has not yet been eliminated, we remain persuaded that it *will* be, that only the perspicacity of the mathematicians was wanting. Moreover, this deduction is complete, not merely approximate as for the physical phenomenon: for a mathematical theorem one could not conceive of an evolution analogous to the one undergone by Kepler's laws and Mariotte's law (subject to considerations that we set forth in Appendix 21).

Nor can we imagine that, as a result of a later evolution, a theorem of geometry might come to appear to us as nothing more than a probable proposition founded on statistics, even though this sort of evolution does not seem to be at all precluded in the case of the law of universal gravitation or even the principle of inertia. This is because, unlike the physical concept, the mathematical concept appears to be entirely determined by its definition, to be transparent, as it were, to our intellect, and if interpretation is sometimes involved, as happens, for example, when we apply arithmetic to geometry, it is in virtue of a convention made once and for all, one which can leave no hesitation in our minds. To be sure – and this is an important point – the mathematical concept is not transparent in the sense that our mind is able to grasp all of its *implications* immediately. The properties of conic sections as revealed by analytic geometry were obviously contained without exception in the definitions of these curves given by Greek mathematicians; but the point is that they were found there only implicitly, and an effort was necessary to bring them out, to make them explicit. What we can say, however, is that whatever knowledge about these curves may be acquired in the future, there will not be any that cannot be deduced from the definitions. And, *a fortiori*, no subsequent discovery will be able to lead to a result involving a contradiction between the new proposition and the original definition. The transparency of the mathematical concept must then be extended in the sense that our intellect feels itself *capable* of bringing out all the implications by its work alone, that it knows this concept to be entirely *open* to its operations, concealing no element that can resist it.

We have seen that the problem posed by the recognition of the fundamental agreement between mathematical deduction and reality, a problem that consists – to quote Brunschvicg – in "justifying *a priori* the mathematical form assumed by scientific knowledge of the universe," has always greatly preoccupied philosophers. It is resolved immediately if one adopts the philosophic position of which we have just spoken. If reality is essentially only a collection of mathematical concepts, it is not at all surprising that, as Kant puts it, in science "only so much science proper can be found as there is mathematics in it."[13]

There is no question, moreover, that the agreement between our reason and our perception in this domain lends a high degree of reality to anything of a mathematical nature offered by our perception. Sophie Germain even believed she found there a demonstration of the reality of the external world in general:

> Will one doubt that a type of being has an absolute reality when one sees the language of calculation, starting from a single reality it has grasped, give rise to all the realities related to the first by a common nature? If the only thing such relationships had to commend them was the fact that our intellect is able to conceive them, how could it happen that the observation of facts comes, by such a different way, to show a structure outside human thought similar to that whose model man finds within himself?[14]

Others, of a different turn of mind, have seen here a peremptory argument in favor of idealism. Even the Hegelians used this argument, although their doctrine made such a point of belittling the importance of the role ordinarily attributed to mathematics. "Arithmetic, geometry, stereometry show that there exists in nature an idealism of proportions and forms that could in no way be explained by the fact of a pure and simple aggregation of eternal atoms," writes Rosenkranz (*Hegel als deut. Nat.* 329). In reality these two deductions obviously are indistinguishable, for they are both based on the recognition of the agreement of which we have spoken. Let us also note that this feeling that mathematical concepts are of a higher reality is clearly linked to the distinction between primary and secondary qualities and, thereby, to the whole edifice of universal mechanism, which is, in this respect as well, only a step in the direction of panmathematicism.

But we can approach the question of the relationship of mathematical idealism to the sciences from still another angle, by considering mathematical physics. In this way the agreement we have just noted will be confirmed and made more explicit; however, the difficulties inherent in the solution will at the same time become quite apparent.

Let us open a treatise on mathematical physics and examine any proposition whatsoever. It will appear in the form of two terms connected by an equal-sign or, more rarely, a symbol for inequality. The terms themselves are composed of symbols each of which represents a datum expressed numerically. However, these numbers are not *abstract*; each time we express a number, we append to the numeral an appellation indicating the particular nature of the number in question; for example, T will represent time and we shall state it in seconds, V a velocity that we shall express in meters per second, W a weight in kilograms, etc. And what is characteristic of these statements is that as a general rule they are not composed of symbols of one and the same kind or combined with purely abstract numbers, but above all that they are symbols of different kinds which are combined with one another by different mathematical operations, such as multiplication or division.

It would seem to be clear, first of all, that operations of this sort are not, *ipso facto*, legitimate. How can we conceive of a weight (in kilograms) multiplied by a time (in seconds)? Is this not something like multiplying meters of cloth by liters of milk?

But in order to see the nature of the operation in question even more clearly, let us leave the terrain of mathematical physics for a moment and move to that of pure mathematics. We have seen (Ch. 1, pp. 25 ff.) that geometry itself seems to be founded on substantialistic conceptions. Now such conceptions are clearly found there in their simplest form, and it is this very simplicity that will help us in this case.

An area in geometry is understood as having two dimensions; they are two lines, two lengths, expressed in meters for example, that are combined by multiplication. That is obviously an altogether exceptional operation; even in geometry one cannot combine one area with another area in this way, nor one solid with another solid. For lengths themselves this privilege is limited; beyond the third power the symbol can no longer be translated into our reality. Just as obviously, the operation we perform on lengths is not of the same nature as those we use for abstract numbers: when we multiply one number by another, we never obtain anything except a number like the first two, whereas here two lengths give us an area, that is, something essentially different from the two factors.

Thus an arithmetical operation is being applied to a geometrical construction; when one knows how the operation is performed, it allows one to calculate the geometrical construction and foresee the result, but it does not actually exhaust the content. Geometric concepts are not pure quantities. One can apply the category of quantity to them, but in their essence there is something that escapes this category.

These objections are extremely ancient. Indeed, their essential features (taking into account, of course, the very different state of the sciences at the time) can be found in the polemic Aristotle directs against his master in the *Metaphysics*, notably when he points out that numbers cannot be assimilated with ideas, since they are devoid of all quality. For example, a single number can always be formed from several numbers; but how can a single idea be formed from several ideas? Likewise a line cannot be derived from an area, nor an area from a solid.[15] That is the selfsame geometrical argument we have just used – the fact that it happens to be presented inversely, so to speak, in Aristotle is a simple consequence of the fact that for him the concept of line, insofar as it is more abstract, is considered superior to that of area, as is the concept of area to that of volume.

The process whose functioning we have just noted in geometry can be seen to continue and develop in the physical sciences. A mass multiplied by an acceleration yields a force; multiplied by half the square of velocity, kinetic energy; a thermal energy divided by the temperature gives an entropy – and so on: mathematical physics abounds with examples of this kind, one could almost say is made up of nothing but such examples. For everywhere there is a calculation one uses concrete numbers, in each case one moves from one class of these magnitudes into the other by multiplication and division, and in each case the calculation has meaning only because one interprets the result according to a well-defined method. To be sure, calculation furnishes a numerical datum, but the latter in itself does not constitute the result of the operation; to it must be added interpretation, which alone allows the result to be stated in the form of a concrete number.

What is added in this way by interpretation, in transforming what was only an abstract number into an acceleration, a thermal energy or an entropy, can be designated by the term *quality*: these numbers are no longer pure quantities; they are magnitudes belonging to a specific class.

And, most importantly, what distinguishes these magnitudes from pure quantity is that they refer, as the term "*concrete* numbers" clearly indicates, to real concrete being; they are indissolubly linked to it and cannot be defined or even imagined distinctly without first positing the concept of a spatial being. We pointed this out in Chapter 1 (p. 23) in the case of the concept of temperature as it is encountered in thermodynamics. But the observation is valid for all concrete magnitudes used by physics. One obviously cannot imagine a thermal energy, an entropy or an acceleration without thinking about *things*. In other words, what interpretation adds to mathematical data is in fact ontology.

If we now wish to analyze more closely what this quality is that is superimposed on pure quantity in this fashion, in order to understand what constitutes the specific nature of the element that characterizes these classes of magnitudes, what makes a temperature different from a thermal energy and a thermal energy from a mass, we shall undoubtedly arrive at the conviction that these distinctions are more felt by us than clearly discerned – which, needless to say, is due precisely to the fact that it is a question of physical being, by its essence opaque, imperfectly fathomable by our understanding. However, in certain cases science manages to get a better grasp of this *quid proprium*, this quality, to isolate and specify in certain aspects what it contains that is inaccessible to all further reduction.

We are already acquainted with this irreducible element: it is, in fact, nothing other than the irrational we treated in Chapters 6 and 7. To make it immediately obvious how this concept is related to the question at issue here, let us suppose for a moment that the impossible has occurred and the physical world has been completely reduced, explained, without there remaining anything opaque, anything that resembles a substance to which we might attach a *why*. To use Leibniz's formula, supposing that we had come to recognize that bodies are formed of particles in the shape of globules, we would be able to explain *why* they are globules and not cubes; that is to say that we would have succeeded, according to the ideal Descartes thought he had achieved, in understanding the entire world as *necessary*, as conditioned in its existence by the sole fact of the existence of one unique matter, which we have moreover seen to be a simple hypostasis of space. As a result all independent *data*, and in particular all unattached numerical data, all coefficients will clearly have disappeared. Furthermore, one can verify in detail that this is what science actually tries to do, that it strives to reduce the coefficients to one another, to deduce them from one another. All new theories, all discoveries establishing a link between previously unconnected phenomena lead to this result. To take only one of many examples, by the establishment of the electromagnetic theory of light all numerical data peculiar to optics cease to exist as such and must henceforth be derived from those characterizing the physics of electricity.

That obviously stems from the fact that, due to this theory, all *specificity* of light has disappeared. Not, of course, as a sensation: sensation is a primordial fact with which physics does not concern itself, which it eliminates, apparently by relegating it to physiology, but actually by realizing that it cannot have any hold on it, that what is involved is an irrational of a higher order (Ch. 6, pp. 144 ff.). But as a distinct physical fact, that is, from the standpoint of what happens in space, light has, in a manner of speaking, ceased to exist; it is no longer anything more than one manifestation of electricity among many others, and not only ultraviolet or infrared vibrations, which have no effect on our retina at all, but Hertzian waves as well are completely assimilated with light by the physicist. Optics has ceased to constitute a separate branch of physics, and one cannot even say that it truly forms a subdivision of the branch devoted to electricity.

On the other hand, wherever there is an irrational, specificity continues to exist, and the irrational is precisely the expression of what there is that is specific in the order of phenomena in question; it makes clear what, in the whole of their properties, in their *quality*, is irreducible to pure quantity. Thus, to take the irrational at its origin as it were, in geometry, the tridimensionality of space is certainly an irrational, a fact, which we must be content merely to acknowledge and which *characterizes* our space, which expresses a property, a qualitative element lying hidden within this concept.

The same is true of the other irrationals we have discussed. The dimensions of the atoms are the expression of diversity in space, that is to say of the fact that there are parts of space that are occupied by matter, or in other words of the fact that not everything in this matter allows itself to be reduced to the properties of continuous extension. On the other hand, Carnot's principle or "the improbable initial state" that must be presupposed in order to explain this principle, to rationalize it, are the expressions of diversity in time, that is, of a particular character of the phenomena of the world around us, a character that makes *everything flow*, as Heraclitus said.

This is why wherever the specificity of phenomena seems very pronounced we can suspect the existence of a particular irrational and why, for example, a hypothesis of this sort can be formulated about chemical phenomena (Ch. 6, pp. 167 ff.).

Moreover, a vague supposition of this sort is all that can be attained under such circumstances, even in the most favorable case, prior to the definitive discovery. In particular, it is completely impossible, as we saw (Ch. 6, p. 169), to predict what form this future irrational will take. It might even be maintained that in some respects it is not desirable for scientists to speculate too much about things of this nature, to make assumptions which, unless the author is some sort of superhuman genius (as rare in the sciences as in any other branch of intellectual activity), could lead research astray. Indeed, since the form in which the irrational will appear is unpredictable, the safest course is to ignore the possibility that it will appear, to pretend, even if one does assume its existence, that it does not exist, that everything is rational, explicable. For then, in the process of seeking to explain everything, the irrational will emerge all by itself, as it were. Therefore researchers in physical chemistry do well to speak of physicochemistry, to work as if they believed there could be no distinction between chemical and physical phenomena. If this distinction does exist (as can also be supposed), this sort of work tending to merge the two orders of phenomena is the very thing that will inevitably bring it out and a new irrational will consequently be identified.

In this regard, the situation is altogether analogous for the biologist, and no doubt the more or less vague awareness of these ultimate aims of science influences his mental attitude; it is also what explains the curious attraction mechanistic conceptions seem to hold for these scientists. We have seen that even supposing the most brilliant success (almost inconceivable, we must admit, in the light of what has been accomplished up to now) of the efforts of explanatory science in the realm of life, it appears extremely probable that a whole series of new irrationals will have to be added to those that are offered by inorganic nature (or will be offered at that time, for nothing assures us that Planck's *quanta* must close the series). Certainly that is a state of affairs which at first sight would seem unfavorable to the thesis in question: one would be more inclined to think that if he had the slightest awareness of the existence of these irrationals, the biologist ought not believe in the possibility of explaining everything. But this awareness is precisely what he cannot have – one would almost dare to say must not have. Indeed, as a matter of fact the properly physical irrationals do not interest the biologist, who is inclined to envisage physics more as a completed whole devoid of mystery so that he can use it as the starting point for the explanation of the phenomena in his own field, phenomena whose complexity rightly seems infinitely more formidable to him. There remain the hypothetical irrationals that will characterize living matter. Now, insofar as this domain is concerned, the researcher is above all aware that a large number of the phenomena must be reducible to those of nonliving nature. Will there be something irreducible, something irrational? Even if he believes so, it will serve no purpose for him to dwell upon the idea in the course of his work. In fact, in the present state of our knowledge it does not seem that any power of deduction can permit him to indicate, even summarily, the nature of this irreducible element. He thus does not know where the irreducible is lodged, if we may be allowed this image, which part of his field of research contains it. And consequently he must logically extend his attempts at reduction over the field in its entirety, pretending that this irrational does not exist. In other words, the mental attitude of this scientist must be exactly that of a man who believes that everything in the living being is reducible to the phenomena of inorganic nature. As a result one can understand that this attitude, which is suggested to him by his work itself, should come to be part and parcel of his mentality, that is, that he should become resolutely mechanistic to the point of ignoring the physicists' indications to the contrary, and remaining deaf to their objections, however obvious and irrefutable they may be. And, *a fortiori*, the arguments of the philosophers remain without effect as far as he is concerned.

But, needless to say, in transforming what must be essentially only a provisional attitude, a working hypothesis, into an article of philosophic faith, one alters the very foundation of the hypothesis in the least legitimate way and, what is more, without any advantage for science. For the specificity of a great number of phenomena exhibited by organisms appears, on the whole, quite difficult to deny, and the research biologist has no need to deny them, but simply needs to be allowed to put them aside, to forget them for the time being.

If we now return from the consideration of these entirely hypothetical future irrationals to those whose existence has already been determined by science, we shall see that what we have termed irrationals are precise data sometimes capable of assuming a mathematical form. This circumstance alone is therefore not enough to guarantee their conformity to universal mathematicism. The reason is simple: they are not abstract but concrete numbers, *physical data*, and what distinguishes them from pure numbers is, as we have just seen, ontological in nature. Since ontology ultimately takes refuge in the irrational, this irrational can in one respect result from mathematicism, and it then serves as the starting point for deduction which proceeds mathematically and aims to explain reality; but it is not itself mathematically deducible: if we go back from explanation to explanation, the chain will stop there. Therefore the scientific irrational in some respects resembles the one which, according to Renouvier, would be constituted by an act of free will; it also represents, in a totally different order of considerations it is true, an "absolute beginning." In order to see that this analogy is not completely specious, one need only consider that the believer will be able to suppose that the absolute dimensions of the atoms, like the improbable state at the origin of the irreversible universe, are the result of a decree, an act of will on the part of the divinity. Moreover, that is obviously one and the same observation in two different forms. "Can one discover," writes Joseph de Maistre, "that the planets are retained and moved in their orbits by two forces that balance one another (whatever may be true of these two forces) without discovering at the same time that they were established in the beginning for this noble purpose?" And he likewise supposes that a Christian, "discovering that the leaf on the tree has the property of absorbing a great quantity of noxious air ... will cry out: Oh Providence! I worship you and thank you!"[16] But the connection can clearly be established only on the condition that we remain ignorant. If the mechanism by which carbon dioxide is absorbed by the green parts of the plant were completely known, the believer would be forced to push the assumption of an arbitrary divine act further back. And similarly, if we assume that a cosmogonic theory of our planetary system has been firmly established, divine intervention to explain the position of the planets and the stability of their orbits (de Maistre speaks as a Newtonian, as did his whole era) becomes unnecessary and thus inadmissible. That is surely what Laplace meant by his famous response to the emperor Napoleon, who reproached him for not having mentioned God in his cosmogony: "Sire, I had no need of that hypothesis." It is more than likely, given what is known about the character of Laplace (who was a perfect courtier), that the remark is apocryphal. Nevertheless, like many other historical anecdotes, it constitutes a "legendary truth," identifying precisely what distinguished Laplace's position from that of Newton (whom the emperor probably had in mind, and who needed divine intervention to maintain the stability of the orbits). However, what one sees as a manifestation of almost superhuman pride on the part of the astronomer is only a statement of a fact, namely the obvious circumstance that, as a result of the causal explanation of a series of phenomena (which Laplace believed he had carried out), the believer is obliged to push the supposed act of divine free will further back.

This pushing back stops at the irrational. That which is irrational and therefore essentially inexplicable in causal terms will always be conceived as being of divine initiation, which initiation, of course, would have an end in view. Only two conditions are necessary. First of all, the irrational must be genuinely definitive. That is not in fact the case for the irrational concerning the absolute dimensions of the atoms, as we saw, and therefore if tomorrow someone succeeded in deducing these dimensions from the general constitution of a universal medium or from the general properties of electricity, etc., the theological assumption would once again have to be pushed further back. Secondly, the end must appear to be a worthy goal from the strictly human point of view (Ch. 7, pp. 195 ff.). That is obviously why no one has ever attempted to represent Carnot's principle as resulting from divine free will, despite the fact that it points to a very clear irrational. Indeed, in order to consider as reasonable the end toward which everything is necessarily proceeding according to this proposition – namely the destruction of the universe by the disappearance of all diversity – it would be necessary to assume that this destruction, or at least the disappearance of all consciousness in absolute uniformity, was an ideal to be attained; perhaps a Buddhist might admit this, but the conception would not fit in very well with religion as Europeans generally understand it. On the other hand, the kinetic explanation of the principle clearly improves the situation from this point of view. Progress toward a more probable state becomes of necessity logical and thus finds itself beyond the reach of divine intervention, even though the divinity can intervene at the origin in order to create the improbable state and thus ensure, under given *compossible* conditions (as in Leibniz), the existence of our changing world. And since this improbable state constitutes a definitive irrational, as we have seen, there will no longer be any need to push our assumptions further back.

Thus science itself, in its effort at rationalization, succeeds in certain cases in defining the limits of the power of our explanatory reason, and will no doubt do so increasingly in the future.

But it is only natural that the physicist absorbed in the manipulation of formulas, and perhaps even more so the mathematician who furnishes him procedures and the philosopher who has chosen to study their work, come to lose sight of these considerations to some extent. This is the illusion whose source we have sought. It consists in believing (more or less implicitly, to be sure) that by carrying out mathematical operations on concrete numbers one would be able to break physical reality down into its component parts without allowing any of it to be lost, for it apparently follows that one would also be able to reconstitute it by the inverse process, to construct it, so to speak, out of pure mathematical formulas, out of pure quantities no longer tainted with any qualitative element – or perhaps we should say tainted with nothing physical, nothing referring to existence in space. Indeed, as can easily be seen, the term quality is apt to take on a somewhat troublesome meaning here. We have used it to designate what goes beyond determinations of pure quantity, which is certainly consistent with common usage. In that sense, then, we can speak of quality even in geometry, for determinations in space offer an aspect which escapes pure quantity and, to cite a specific example, "the qualitative study" of curves undertaken by Henri Poincaré[17] clearly refers to this aspect. However, one can also speak of the quality of a pure quantity and in particular of that of a number; one will say, for example, that this or that number has the quality of being even or of being divisible by nine. This extension of the term, which does not, strictly speaking, appear illegitimate, nevertheless offers serious disadvantages in the present case, in that it is apt to create a sort of ambiguity. Indeed, as soon as pure quantity can bear quality, can present a qualitative aspect, does that not tend to suggest the existence of quality in general within pure quantity? Panmathematicism appears to us to succumb to this temptation, taking advantage of the ambiguity to move from pure quantity to spatial quantity, and from there to physical quantity, that is – if we may put it this way – by skipping over the irrational, which, moreover, is consistent with its program. Rarely, perhaps, will one find a clear and peremptory affirmation of a truly complete panmathematicism, outside of certain formulas of the Marburg school. But there would not seem to be any doubt that it forms the basis for many current conceptions among the scientists and thinkers of our time.

A comparison that is called for here and that the reader himself will no doubt already have made is the one between the attitude of this extreme idealism and that of positivism. Before tackling this subject, however, it seems appropriate to glance at the relationship of positivism, which we treated in the first chapter only as a function of materialism, to other philosophic positions in general.

As we have seen, positivism means to leave metaphysics aside; but we have also seen that this attitude is extremely difficult to sustain, that as soon as one engages in science one runs into ontology, that it is impossible to do science without implicitly affirming the ontological existence of *objects*. The positivist would like science to be only a system of *relations*. Now it is also a system of *relata*, and scientific relations themselves, as it is easy to convince oneself, come into play only between relata, and can do so in no other way. This is why the formula *relations without relata* [*rapports sans supports*], which, if it could be fully and immediately applied, would indeed be the most complete expression of the positivistic program, necessarily becomes a design for the future, a goal the positivist more or less consciously strives to attain. Now, needless to say, ridding science, scientific reality, of all *reality*, reducing it to a body of rational formulas, is the program *par excellence* of idealism. It is therefore understandable that in spite of himself the positivist feels unconsciously attracted by this philosophic position – unless, of course, he has previously adopted the materialistic stance, consciously or not. One can boldly state that any positivist who is not resolutely materialistic is a latent idealist at heart. Thus one finds these would-be positivists in all the various camps of metaphysical idealism. At a time when the great popularity of Hegelianism in Germany had barely subsided and positivism was only beginning to become popular in France, Taine claimed to reconcile the two positions, loudly proclaiming to be the disciple of both Auguste Comte and Hegel, and Lionel Dauriac, an excellent judge in this matter, assures us that "Renouvier undoubtedly owes his phenomenalism to Auguste Comte."[18] This syncretism is clearly favored by the fact that, as we pointed out in Chapter 12, the "margin" Hegel leaves to "the empirical," that is, to experimental science, can be understood, with a little good will, as including all of science as Auguste Comte defined it. No doubt many neo-Hegelians today (and there is nothing to prevent there being a few genuine scientists in their ranks who started out as the most orthodox positivists and are convinced they have remained so) still dream of a reconciliation of this kind. But if, as a general principle, science and philosophy cannot exist side by side as if they were in separate sealed compartments, it is obviously impossible to fit two doctrines into the framework of a single conception, when one of the doctrines would embrace science alone and the other would apply to all the rest. If Hegelianism is taken at all seriously, it will not be able to forgo the development of a theory of nature, a *Naturphilosophie*, which will end up, like it or not, in an attempt to introduce idealistic concepts into science, and we have already seen how chimerical any such enterprise must be.

Of course many other formulas are possible: they all run into the same obstacles to varying degrees. Two of them, however, two ways of going from positivism to idealism, seem to merit particular attention.

In Chapter 3 (pp. 66 ff.) we spoke of that form of positivistic theory which, although maintaining that science is uniquely composed of laws, lays particular stress on the fact that the progress of our knowledge must consist in deduction by means of which laws are derived from one another. The ideal goal would be to reduce them all to an ultimate formula, which formula, as is explicitly stipulated (and, what is more, required by the logic of this whole system founded on deduction), must appear *necessary* – which, as is readily understood, will at the same time necessitate all the formulas that will have been reduced to it, that is to say, according to the premises, all the rules of science without exception.

Now, how are we to imagine the ultimate proposition or propositions (for nowhere does Goblot actually assure us that there must be only one unique proposition comparable to Sophie Germain's "unique fact" [p. 66 above]) that are to crown the edifice? What is at issue here is not the fact that we cannot at the present time form an idea of a content or contents so fashioned that all the other propositions of science would derive from it or them. This is only natural, since it is understood that it is not a question of *a priori* knowledge. Of course, once the construction has been completed, everything will be *illuminated* [p. 372 above] and the entire world will be deducible; but until then we must painfully climb up the chain of deductions, feeling our way. Thus from this side or, if one prefers, from below, from what must connect these propositions to the phenomenal world, the conception is justified, or at least offers no more difficulties than any other system of the same sort.

What is literally unimaginable, on the other hand, is (to make use of the same image) the connection of the ultimate propositions from above, the fact that they will have to be seen as logically necessary, that is to say drawn from the inmost recesses of our reason. For reason, as we know, can only provide us with empty frameworks.

One can therefore state in advance that it will be essentially impossible to carry out this operation successfully. If our deductions go up to a single proposition, this proposition will certainly appear impossible to deduce from the content of our reason alone, and if there are several propositions, they will likewise all be of this same type.

That this is the case can be verified in part by examining certain areas of science, especially thermodynamics, where, as we have seen (Ch. 1, pp. 21 ff.), the deduction is actually carried out starting from two abstract principles, the principle of the conservation of energy and Carnot's principle. The conservation of energy, by the very fact that it involves *conserving* something, can to a certain extent – to the extent that it fits into the general framework of the sphere of Parmenides, the thing immutable in time – be conceived as rational, deducible. But Carnot's principle is certainly contrary to the requirements of our reason: it is rational for things to persist and not for them to change. Or at least if one wishes to find a reason for the principle, one is obliged to seek it in a domain quite different from that of strictly legalistic science. Indeed, we saw (Ch. 6, pp. 156 ff.) that there has been no lack of attempts to do so and that the procedures utilized (which in part have certainly succeeded to a surprising degree) have nothing in common with those of a truly positive science, being, as a matter of fact, the very procedures positivism forbids, procedures of a quite unmistakable kinetic theory, presupposing the existence of things in space and having no meaning without this "metaphysics."

In the same context, we can note that it is not impossible to find (setting aside the principles of conservation, the essence of which is patent in this regard) scientific propositions having a basis that is more or less implicitly and partially *a priori*, thereby resembling those ultimate propositions of which we spoke above. For example, the constant 2, which appears as an exponent in the statement of the Newtonian law of attraction, is termed *essential* by Henri Poincaré, in contrast with all the others, which appear as *accidental*. Now if one asks oneself why this given has come to assume a particular dignity, there can be no doubt as to the reply: it is because it is founded on the spatial image which shows us the "central force" emanating from its point of origin and spreading out, so to speak, over spherical surfaces greater and greater in diameter. Which proves that, in order to find an aprioristic foundation for something that concerns reality, science is obliged to call upon what is already implicitly contained in this concept of external reality.

Furthermore, we need only examine things a little more closely in order to see the true nature of the obstacle that stands in the way of the theory.

It introduces a completely foreign element of rationality into the positivistic schema. But what is more, this concept of rationality dominates the theory to such an extent that what it ultimately seeks is a veritable rationalization, which it expects to be complete, thus scorning science's teaching on the subject of irrationals. This characteristic obviously arises from the fact that, being concerned only with laws and disregarding explanatory theories, it fails to take note of the fact that at a certain point scientific explanation encounters an impassable barrier. Thus, all things considered, this is an absolutely idealistic conception, an attempt to pass from the positivistic doctrine to a philosophy conceiving nature as logically necessary, to reduce nature to *idea.* In fact, the ultimate proposition which is to be the apex of the deductive process will clearly have to be drawn from the depths of our reason. We have just convinced ourselves, by precise examples, that everything in science demonstrates the futility of such an enterprise. But, considering it from a more general point of view, we recognize without difficulty that at bottom what we have here is something identical to Hegel's audacious attempt. For, contrary to what positivism claims, laws do not contain relations alone, as we are well aware: they contain, or at least imply, substantive relata, assumptions about being, and it is altogether chimerical to wish to separate them from this inherent element. Consequently the ultimate proposition must be revealed, like all others, as a proposition concerning being. Either explicitly or implicitly, this proposition must stipulate being and, as a result, since we claim to extract this proposition from our reason, we believe ontological being can be deduced from reason: that is in fact what Hegel claimed he had succeeded in doing and what Schelling protested. As soon as we lay bare the true motivation behind this operation, it becomes perfectly clear how chimerical it is.

Mathematical idealism (to which we now return after a parenthesis which has not led us all that far afield) appears to reconcile positivism and idealism more easily, more naturally, as it were, than do the other formulas. Indeed, the concept of mathematical dependence is basically only the expression of dependence in general in its most precise form, the easiest form for reason to grasp: the function is the most concrete formulation of law. The world of pure mathematical functions of which we have just spoken is thus only the clearest image of a world of laws without substances or, as has been said, of relations without relata. Is that not quite simply what the positivist is calling for? Of course if one looks more closely the divergence is profound. The positivist believes that as soon as one engages in science one can get rid of all considerations having to do with being, whereas the supporter of mathematical idealism postulates this elimination only as the result of a long effort, as an ideal that science tries to approximate. By a curious sort of reversal, the positivistic program becomes the more ambitious of the two in the practice of science, since it insists on the immediate application of what at first appears to be a program common to the two doctrines. From the philosophic point of view, however, mathematical idealism has incomparably more ambitious aims than positivism. The latter, indeed, simply declares that science can abstain from knowing the essence of things, whereas mathematical idealism claims to penetrate into this very essence, to know it through and through. Obviously its doctrine is much more consistent with the true nature of science than is that of positivism; but it also appears less simple, less direct, at least at first sight. And since the resemblance between the two formulas is not completely obvious, it may happen that the scientist or thinker who started out as a positivist and believes he still is one, but at the same time has discovered the impossibility of truly applying positivism to science, sometimes enters (most of the time unconsciously) into the paths of the most abstract idealism.

Thus, as Höffding aptly remarks, "Platonism and positivism attempt to enter into the closest possible relations with each other."[19]

But perhaps we shall see the implications of the doctrine even more clearly if, having examined it from the strictly scientific point of view, we now compare it to analogous metaphysical conceptions.

Since mathematical idealism is above all an *idealism*, we must recall what Schelling, in his polemic against Hegel, has taught us about the difficulty of moving from the idea to reality. What becomes here of this transition from the abstract to the concrete, this "wide and nasty ditch" (Ch. 12, p. 313) that we have just rediscovered, as hard to cross as ever, in the positivistic position which tries to deduce laws from a unique and necessary formula?

In appearance the obstacle is incomparably smaller this time. Whereas for Hegel the transformation from rational categories to beings requires an intolerable effort of imagination, for Descartes the supposition of the existence of geometrical figures in space seems to be almost a matter of course. It is clear that this advantage is due precisely to the peculiar nature of mathematical knowledge. Geometrical truths belong to our thought as much as to nature; thought and nature seem to merge. And as a result, when we have reduced external reality to collections of elementary triangles, as Plato did, or to configurations of "singular points" in the undifferentiated ether, as modern theoreticians would have it, we are convinced that they are made up solely of elements belonging to our intellect. However the obstacle, if it appears less impassable, has not disappeared for all that. And first of all it is clear that it is necessary from the outset to posit the existence of space at the same time we posit the existence of thought. This is what Cartesianism did, and we are aware of the difficulties it suffered as a result of the existence of this primordial dualism. But if one intends to pursue with rigor the synthesis, the reconstruction of nature by means of mathematical concepts, this difficulty of "concretization" is going to manifest itself immediately in still another way.

We have pointed out (pp. 409–410) that in our imagination geometrical figures seem, in spite of everything, to possess a mysterious sort of *corporeity* allowing them, up to a certain point, to persist in space outside our consciousness. Obviously that is still only a shadowy existence, however, and if we wish to transform it into concrete existence, something will have to be added to the figure to differentiate it from the surrounding space, whether this distinction has to do with the tridimensional content of the entire figure or only with the points that mark its corners: we are in fact speaking of the transition from the abstract to the concrete.

Or rather, we are still speaking of only one phase of it. In connection with the properly scientific aspect of panmathematicism, we saw (p. 421) how, through the mathematical operations of physical theory, the material concept is gradually divested of its reality, as it were, even though in each case the physicist retains the illusion of having performed only a simple operation of calculation. If we now mean to go on to the inverse operation, to reconstruct reality out of simple formulas, it will be necessary each time to add what the physicist was unable to take into account during his operations, or rather what he consciously disregarded, what in a manner of speaking he lost, namely the qualitative element that is precisely what distinguishes reality from our abstract concepts; in each case there will be a transition from the abstract to the concrete or, to be more exact, from a less concrete concept to a more concrete one, until the material body, with all its properties, is reconstituted. In short, in terms of Schelling's image, instead of jumping the ditch all at once, we shall have crossed a specific number of ditches one at a time, each of which, taken separately, appeared to constitute a much less formidable obstacle. And this circumstance would seem to add greatly to the facility of the whole operation, whose beginning, as we have seen, is rendered plausible by the perfect agreement between the mind and reality insofar as mathematical propositions are concerned.

But basically the difficulty is the same, for what is really involved is the conflict, the irreconcilable opposition between the *inevitable* belief in external reality and the *irresistible* belief in the rationality of this same reality. Both beliefs, as we have seen, manifest themselves with great vigor in science, and, more particularly, its theoretical part is in fact their common work, reason ordering science to seek explanations by deducing phenomena (Ch. 2) and the feeling of reality imprinting its explanations with an indelible ontological character (Ch. 1).

Furthermore, we can see that romantic idealism attempted something similar to this breaking up of the obstacle. Hegel in particular, with his different categories of being (*Sein*, *Dasein*, *Fürsichsein*, etc.) evidently seeks to create gradations in the transition between the abstract and the concrete. But that is because in Hegel, since he appeals to concepts directly suggested to us by our sensation, the first of these three concepts, that of *being*, however abstract one would like to think it, whatever care one has taken to strip it of everything that actually constitutes reality, nevertheless immediately invokes reality in its entirety. Indeed, this is why the transition appears so shocking to us. There is nothing of the sort to be feared in the case of the concepts of mathematical physics, which are concepts created by science, and this is what, once again, makes us understand the particular attraction mathematical idealism holds for the physicist, who finds familiar images there.

Thus each of the three metaphysical conceptions we have discussed (we can leave aside energeticism, which no longer seems to be anything more than a shadow of its former self) can legitimately claim to have arisen out of science. It is certain that science, since it starts from common sense and only transforms it gradually and reluctantly, speaks the language of mechanistic ontology most of the time; it is equally true that, in probing the inner nature of matter, science little by little destroys the concept of matter, although initially leaving the determinations of time and space intact; and that finally pushing the rationality of our image of the world to its extreme limit, it tends everywhere to substitute the concept of mathematical magnitude for that of quality. But we must add that science, if it seems to impose each of these three metaphysics one after the other, also invalidates them one after the other. It destroys the first two in its march toward the third and, when it has reached the third, it points out its limits itself, by imposing on us the given, the irrational, which we find recalcitrant to it.

Thus if it is, strictly speaking, *extravagant* to claim that science ignores ontology, if it is patent, on the contrary, that it maintains and confirms the concept of an external thing, that it is forced to seek untiringly to create a more and more coherent representation of this external reality, it is nonetheless true that it is not and surely never will be able to succeed in making this representation entirely acceptable to the mind. Incapable of making a definitive choice between the different systems of philosophy, science, although unable to do without metaphysics, basically finds itself in something like a state of indifference or, if one will, ataraxia, insofar as metaphysics is concerned.

As for the individual scientist, if he insists upon conscientiously adopting any kind of consistent conception of reality, he will no doubt often choose the one best suited to the part of science he has chosen to study. In these terms we earlier attributed mechanistic convictions to the biologist and Hartmann's transcendental realism to the experimental physicist, and in the same way mathematical idealism could be called the characteristic doctrine of the theoretical physicist. Obviously there is something arbitrary about such a schema, and, needless to say, many of the best scientific minds do not fit into these categories. But the important thing is that whatever the philosophic convictions of the scientist and however carefully considered and firmly established they seem, they intervene and become truly active only when he engages in genuine philosophic speculation. When he *does science*, on the contrary, they are silent, they become at least temporarily inoperative. In the practice of science, even if in theory he advocates the most extreme idealism, the scientist follows only his scientific instinct, which is obliged to accommodate the imperious predisposition for ontology. What hides the strength of this predisposition from us is the fact that it does not appear (to use a physical image) *rigid*. On the contrary, it seems to yield to the slightest pressure. Nothing seems easier than to strip it of its whole domain piece by piece: from the moment we "mend" the stick that the water shows us to be broken, we slide imperceptibly and apparently without serious resistance toward a conception that shows us this stick broken down into a whirl of atoms, after which we decompose these atoms into subatoms or electrons and finally dissolve them into the abyss of the undifferentiated whole. But this is because the predisposition for ontology replaces this lack of rigidity (to continue our image) by a truly unlimited elasticity. The slightest relaxation of the pressure, be it ever so momentary, is enough for all or part of the lost ground to be regained immediately. It is a commonplace in philosophy to recognize that the most stubborn solipsist *sees* matter when he opens his eyes in the morning and *touches* it when he stretches out his hand. As d'Alembert says, these are truths "recognized even by the skeptics when they are not debating."[20] Bishop Huet, perhaps the most extreme Pyrrhonist in the history of modern philosophy, explicitly stated: "When it comes to leading one's life, doing one's duty, we cease to be philosophers, to be ... doubting, uncertain; we become idiotic, simple-minded, credulous, we call things by their names ...."[21] This "metaphysics" is formed in us instantaneously, irresistibly, to the point that we believe in a simple passive reception by our senses where there is actually a very complicated work of our brain. "I know," says Reid, "that this belief [in the reality of the world of common sense] is not the effect of argumentation and reasoning; it is the immediate effect of my constitution."[22] Furthermore, we have seen, in connection with the equivocation created by the conception of the state of potentiality (Ch. 10, pp. 256 ff.), how easily philosophers themselves find it to return to the notions of common sense, notions which Balfour rightly terms *inevitable*.

But the important thing is to understand that it is not a question here of an exclusive prerogative of common sense strictly speaking, as is sometime claimed. On the contrary, the phenomenon manifests itself vigorously at all stages of the belief in the external world whose evolution we have followed through science. The scientist *perceives* objects when he puts himself in front of the eyepiece of a telescope or a microscope; he *believes* in matter when he does biology, as he will *believe* in the existence of atoms in redoing Perrin's experiments, and in electrons if he repeats those of J. J. Thomson. And yet, when he reasons about these experiments, he will be very careful to attribute to the particles of matter no more properties than necessary, in order to make it possible, to some extent, for them to evolve toward something that can be merged with space. Thus the active philosophic convictions of the scientist, convictions whose bases he adopts and rejects by turns, move continually along a line stretching from the most immediate common sense to the most advanced idealism, depending on the field of study in which he is engaged at the moment.

That peculiarity does not set him apart from the rest of humankind.

Indeed, one can reasonably doubt, to use Roustan's apt expression, that anybody has the right to claim that all his beliefs are perfectly consistent.[23] William James even thought it possible to make an inventory of the subuniverses man generally believes to exist alongside the sensible universe; he finds seven.[23] At any rate, for the scientist, just as for the common man, all these convictions, like those of science itself, are *in flux*. And that is undoubtedly what contributes to the fact that the scientist sometimes becomes convinced (erroneously, to be sure) that he has been able to dispense with all metaphysics.

What also contributes in particular to this illusion is a characteristic and very important trait of the process of ontological creation we have just discussed, namely the fact that it is not only instantaneous and unconscious, but also *general*. Of course when he reasons consciously, every man, and even more so the scientist, can *direct* his reason up to a point, arbitrarily dictate what path it is to follow, which can lead to a multiplicity of sometimes widely diverging conclusions. But we are in the realm of the unconscious and thus, by preterition, of the involuntary. And on this level, we must note, the agreement between human beings is nearly perfect. In this realm, human intelligence, if it starts from the same principles, indeed seems to react in an almost identical fashion when confronted with the same phenomena, the same observations. That is why, when we are engaged in the same areas of science, we actually have many more ideas in common than we make explicit; and this is what allows us to speak of them, sometimes to treat them thoroughly, without appearing to become involved in ontology. This is not, as is often asserted and as Cournot among others apparently believed, because we remove, because we totally eliminate from our scientific statements all speculation about being and its relation to our intelligence,[24] since such a procedure is entirely chimerical, as we have seen. It is, on the contrary, because in similar circumstances we all put into them approximately the same ontological content. In other words we do not eliminate ontology on an individual basis; it is eliminated to some extent (if it is eliminated at all) because it is almost identical in all of us, in our relations among ourselves.

This state of affairs was pointed out, not without some surprise, in a quite recent publication by Urbain. "Perceptions are rarely discussed among laboratory scientists," says the famous chemist.

One can assume that contemporaries with essentially the same intellectual background have acquired common patterns of thought. Universal consensus on scientific facts is no doubt largely based on this fact. The habit of calling a spade a spade permits men of science to avoid fruitless quarrels. It is wonderful to understand one another both as to the words and as to the perceptions they designate. This remarkable agreement creates an atmosphere of confidence among scientists, a solidarity from which they derive an assurance which is necessarily stronger than that of a single individual. There is perhaps no chemist who does not confound the reality of barium sulfate with his idea of it. I was curious enough to put the question to several of them. To all of them it appeared odd. From their startled looks I saw they all thought me mad to ask such a thing. This much is established: today's chemist makes bodies the absolute substratum of their properties, without worrying about the hypothetical nature of this conception.[25]

As one can see, these observations agree closely with those we have presented, both in Chapter 1 and in the present chapter, on the physicochemists' belief in the reality of the beings created by their science and on the foundations of this faith, as well as on the analogy between this faith and that of men guided by the common sense conceptions, of men who "call a spade a spade." The testimony is valuable, not only because of the authority of its author (whose name the reader has encountered many times in our pages), but because this scientist professes, in theory, a quite orthodox positivism and obviously finds the way of thinking he depicts with so much exactitude to be completely reprehensible.

What is more, the fact of the existence of an essential though merely implicit content characterizes our thought and its communication in general. "Language," Goblot aptly remarks, "is almost always quite elliptical: what is left unsaid is not the unimportant or the accidental, which could not be inferred, but what is so essential that it will not fail to be supplied."[26]

Finally, we should take into consideration here a particular process whose nature also seems to be made clear by the above analysis. We observed earlier that, according to current opinion, the abandonment of all philosophy works to the advantage of common sense conceptions, and realized that, on the contrary, the world of scientific theory is the object of an analogous act of faith. But we need only look a bit more closely to see that this opinion is nevertheless partially justified, or can at least be explained, by the fact that the faith, in these different cases, does not have the same intensity, if we may put it in those terms. In fact, the very different metaphysical conceptions between which the scientist finds himself torn do not all strike his imagination with the same degree of intensity; it is certain, on the contrary, that the ontology of common sense, multicolored and full of vigor, forms a complete contrast with the colorless beings of scientific hypotheses, which dissolve into the indistinct whole of Parmenides as soon as one tries to get closer to them. And since common sense ontology is also very convenient and, what is more, fully adequate for everyday life, it is possible for it to triumph to the point that it effaces all its rivals and leads to the conviction that it can be taken for granted. Despite its immense illogicality, that is the true state of mind of a great many of those who deny the ontological content of science: metaphysical reality – which forms the basis for their thought and which is essentially composed of common sense concepts, modified somewhat and above all completed by the most indispensable scientific notions, this modification taking place by an imperceptible process set in motion by experiences and reasoning (causal reasoning, just like that of common sense) – seems to them to have so certain an existence that they are almost incapable of seeing that what is involved is an ontological hypothesis like all the others. This explains why the physicist, although he pays lip service to the purest positivism and adjusts it, upon contact with scientific theories, in the direction of a sufficiently absolute mathematical idealism, upon occasion interprets this idealism as a simple "inverted realism," to use Hartmann's expression (Ch. 10, p. 256).

Science's attitude as we have described it unquestionably has something paradoxical about it, which is no doubt what makes it more difficult to accept. However, the discovery of this rather obvious paradox is nothing new. Consider this passage from Cournot:

> The intimate union and yet original independence of the philosophical element and of the positive or properly scientific element in the system of human knowledge manifest themselves here [in mathematics] through this very remarkable fact, that the mind cannot proceed in an orderly fashion in scientific construction without adopting some philosophical theory, and that nevertheless the progress and certainty of science do not depend upon the solution given to the philosophical question. (*Essai sur les fondements* 2:233, § 329 [Moore 474; Meyerson's brackets])

Clearly, Cournot's "remarkable fact" is at the very least quite closely related to the paradox with which we are concerned. Moreover Cournot, while he was thinking principally of mathematics, considered it only as a particular example of the entire system of "human knowledge"; and in other passages of the same *Essay* he has sufficiently pointed out the analogy, the continuity that he was establishing between the mathematical and the physical sciences in this respect (see esp. *Essai sur les fondements*

1:322 ff., §§ 150 ff. [Moore 229 ff.]). The fact that Cournot's starting point is quite different from our own – since, as we have seen, he admits the possibility of eliminating the transcendental element in scientific propositions – seems to add still more, if possible, to the credibility of the great thinker's testimony.

NOTES

1. Pierre Duhem, *La Science allemande* (Paris: A. Hermann et Fils, 1915), pp. 135, 136, 138–139, 143. As the title indicates, Duhem means to limit his condemnation to the science on the other side of the Rhine, but obviously the theory of relativity found champions as ardent as they were authoritative among scientists and philosophers outside Germany as well; we need only recall the important works of Langevin (known to philosophers especially through his brilliant exposition of the theory in the July 1911 *Revue de métaphysique et de morale* under the title 'L'Evolution de l'espace et du temps') and the very interesting comparisons made by Herbert Wildon Carr (*The Philosophy of Change, A Study of the Fundamental Principle of the Philosophy of Bergson*, London: Macmillan, 1914, pp. v, 10 ff., 38). It does not seem too soon to predict that the number of supporters of the theory will be considerably increased by the dazzling success it recently recorded in explaining the anomaly of Mercury, which had so long defied all efforts of astronomers, and in predicting with a truly surprising exactitude the deviation of light rays, verified during a solar eclipse in 1919.
2. Henri Bergson, *Matière et mémoire* (Paris: Félix Alcan, 1903), p. 22 [*Matter and Memory*, trans. Nancy Margaret Paul and W. Scott Palmer (London: George Allen & Unwin, 1911), p. 26].
3. St. Vincentii Lirinensis, *Commonitorium primum*, Ch. 2, *Patrologie Migne* (Paris, 1846), 50:639: *In ipsa item Catholica Ecclesia magnopere curandum est ut id teneamus quod ubique, quod semper, quod ab omnibus creditum est.*
4. Lucretius, *De rerum nat.* I, 269–299; cf. also II, 112–141. It is quite remarkable that Lucretius considered this indirect demonstration, by the effects that moving air is capable of exercising, to be so convincing that he neglected to make use of Empedocles' direct demonstration, which is undoubtedly one of the finest experiments coming down to us from antiquity (an illuminating exposition of this experiment is to be found in Burnet, *L'Aurore de la philosophie grecque*, trans. Reymond (Paris: Payot, 1919), pp. 251–252 [*Early Greek Philosophy*, 2nd ed. (London: Adam and Charles Black, 1908), pp. 253–254]). This is all the more significant because Lucretius knew Empedocles thoroughly, had consciously modeled the form of his poem on him, and had made him the subject of a dithyrambic eulogy (*De rerum nat.* I, 716 ff.).
5. Urbain's views on energetics would seem in particular to fit into this category ('La valeur des idées de A. Comte sur la chimie,' *Rev. de méta.* **27** [1920] 151–179, and 'Essai de discipline scientifique,' *La Grande Revue* **24** [1920] 47–74).
6. It will perhaps not be superfluous to note that the fecundity of this admirable

conception – Auguste Comte took pains to deny explicitly that it could ever offer "any real utility for guiding our mind in the effective study of optics" (*Cours* 2:453) – has not yet been exhausted even today after a century of important work it has inspired. Indeed, the theory by which Sagnac furnished the explanation of a curious phenomenon discovered by Gouy is directly connected to that of Fresnel (see G. Sagnac, *Comptes rendus de l'Académie des Sciences* **138** (1904) 479–481, 619–621, 678–680; **139** (1904) 186, and *Festschrift Ludwig Boltzmann*, Leipzig: J. A. Barth, 1904, pp. 529 ff.).

7. Hadamard, an extremely valuable witness both because of his high authority in everything having to do with mathematics and by the fact that, being a non-physicist, he is probably less apt to be carried away by something that might be only a passing phase, notes that an "evolution toward the discontinuous is taking shape at the present time" in physics (Jacques Hadamard, 'L'Oeuvre d'Henri Poincaré: le mathématicien,' *Rev. de méta.* **21** [1913] 620). Furthermore, we know that atomism today dominates not only the theory of electricity (as we showed in Ch. 6, p. 164), but also that of magnetism (see Pierre Weiss, 'Le Moment magnétique des atomes et le magnéton,' *Idées modernes* 335, 344).
8. Zeller quite rightly argued that "the theory of the elements of Philolaus and Plato is closely related to that of the atomists, given that they both set aside qualitative diversity of substances and allow only shape and size to subsist as the sole differences" (*Phil. der Griechen* 2:708 [erroneous citation]).
9. F.C.S. Schiller, 'Realism, Pragmatism and William James,' *Mind* **24** (1915) 521.
10. [Reading *application* for *explication* to conform to the Schiller quotation.]
11. Cf. Hermann Weyl, *Raum, Zeit, Materie*, 2nd ed. (Berlin: J. Springer, 1919), p. 3 [*Space – Time – Matter*, trans. Henry L. Brose (London: Methuen, 1921; reprint New York: Dover, 1952), p. 3].
12. Kant, *Premiers Principes métaphysiques de la science de la nature*, trans. Andler and Chavannes (Paris: Félix Alcan, 1892), pp. 5, 6 [*Metaphysical Foundations of Natural Science*, trans. James Ellington (Indianapolis: Bobbs-Merrill, 1970), pp. 5, 7]. We have already referred to this passage in Chapter 14, p. 386.
13. *Premiers principes métaphysiques de la science de la nature* 6 [Ellington 6], as quoted in Ch. 13, p. 353.
14. Sophie Germain, *Considérations générales sur l'état des sciences et des lettres aux différentes époques de leur culture, Oeuvres philosophiques* (Paris: Paul Ritti, 1879), pp. 157–158.
15. Aristotle, *Metaphysics* 991b21–27, 992a10–24. Zeller (*Phil. der Griechen* 2_2:297 ff. [Costelloe 1:319]) has admirably understood that Aristotle's objections refer to the qualitative aspect of geometric conceptions. Burnet (*L'Aurore de la philosophie grecque*, trans. Reymond, Paris: Payot, 1919, pp. 335 ff. [*Early Greek Philosophy*, 2nd ed. (London: Adam and Charles Black, 1908), pp. 337 ff.]) believes that these objections were addressed less to Plato than to the Pythagoreans.
16. Joseph de Maistre, *Examen de la philosophie de Bacon*, 3rd ed. (Paris: J. B. Pélagaud, 1855), 2:200–201.
17. See Paul Langevin, 'L'Oeuvre d'Henri Poincaré: Le physicien,' *Rev. de méta.* **21** (1913) 676.

18. Lionel Dauriac, 'Les Sources néocriticistes de la dialectique synthétique,' *Rev.de méta.* **17** (1909) 487.
19. Harald Höffding, *La Pensée humaine*, trans. Jacques de Coussanges (Paris: Félix Alcan, 1911), p. 129.
20. 'Discours préliminaire des éditeurs,' *Encyclopédie* (Paris: Gauthier-Villars, 1751), 1:ii [*Preliminary Discourse to the Encyclopedia of Diderot*, trans. Richard N. Schwab (Indianapolis: Bobbs-Merrill, 1963), p. 9].
21. Pierre Daniel Huet, *Traité philosophique de la foiblesse de l'esprit humain* (Amsterdam: Henri du Sauzet, 1723), p. 242.
22. Thomas Reid, *Of the Human Mind, Works*, ed. Hamilton (Edinburgh: Maclachlan, Stewart and Co., 1846), 1:183 [Meyerson's brackets].
23. Désiré-Auguste Roustan, *Leçons de philosophie*, Vol. 1: *Psychologie*, 3rd ed. (Paris: Ch. Delagrave, 1911), p. 374.
24. Antoine Augustin Cournot, *Essai sur les fondements de nos connaissances et sur les caractères de la critique philosophique* (Paris: Hachette, 1851), 2:21 ff., § 215 [*An Essay on the Foundations of our Knowledge*, trans. Merritt H. Moore (New York: The Liberal Arts Press, 1956), pp. 320 ff.].
25. Georges Urbain, 'Essai de discipline scientifique,' *La Grande Revue* [**24** (1920) 60], p. 16 of the offprint.
26. Edmond Goblot, *Traité de logique* (Paris: Armand Colin, 1918), p. 153.

CHAPTER 16

THE RATIONALITY OF THE REAL RECONSIDERED

In order that we may better understand the particular attitude of science as it takes possession of reality through reason – a work it shares with philosophy (Ch. 15, pp. 407 ff.) – we shall in this chapter reexamine the important problem of the agreement between thought and reality, in light of the results established in the course of the present work. We shall first use the Cartesian position as a basis for comparison.

On several occasions in Book Three we compared Hegel's scientific philosophy to Descartes's. In particular we noted in Chapter 14 (pp. 383 ff.) how anachronistic the former seems when placed beside the latter, and we endeavored to identify the specific traits of the doctrine that make it seem to step back into the past, so to speak. But we must admit that we do Hegel something of an injustice in making such a comparison, for in reality Descartes's doctrine is itself anachronistic, albeit in the opposite direction: it appears much closer to us than we would expect, given the epoch in which it arose. This is because, as we indicated in Chapter 11 (pp. 270 ff.), Descartes's work is valuable not only for its particular conquests in various areas of science, although these results are extremely important, but also and especially for its general spirit which – as any attentive reader cannot fail to notice – is surprisingly like that of present-day science. This point cannot be overemphasized, for the very reason that the current theory of science tends to obscure it.

The current theory actually grew out of the empiricist tradition, which we characterized in Chapter 14 (p. 378). These opinions, the origin of which goes back to Bacon, were, needless to say, definitively codified by Auguste Comte. For him experience is everything; it is actually the whole of science, since science is only a collection of laws, and since laws in their turn are only generalized experiences. As for mathematics, Comte certainly does not exclude it – he himself taught mathematics and in general had too powerful a scientific instinct not to sense the importance of mathematics in physics – but he reduces its role, completely subordinating it to experience. Indeed, mathematics is to be used only to give form to the laws, if necessary, and to establish connections between them. The latter purpose must be the exclusive concern of mathematical deductions, which can have no explanatory value in and of themselves, there being for Comte no explanation except through law, which is to say, basically, no explanation at all (Ch. 2, pp. 33 ff.). It is altogether typical, in this regard, that in spite of everything he comes to protest, like Bacon, against the "encroachments" of mathematicians into physics and to declare that "every attempt to refer chemical questions to mathematical doctrines must be considered, now and always, profoundly irrational, as being contrary to the nature of the phenomena ..." (*Cours* 2:281; 3:29 [Martineau 526]).

Now if that were truly the spirit that animates contemporary science, the striking resemblance it bears to the Cartesian construction would be an enigmatic phenomenon indeed. Of course it would not be an enigma with no possible explanation. The very fact that we seek to fathom nature and that we to some extent succeed certainly proves that there are points of similarity between nature and our mind. And therefore it would not be inconceivable that a system created by pure deduction might coincide, at least to some extent, with another system constructed by a simple generalization from facts, without any theoretical preoccupation. Granted. But it must of course be pointed out that the fundamental principle thus invoked, although certain and well-established, must nevertheless be called upon – given both the immense generality and the total imprecision of the formula (for we are, as we saw in Chapter 6, p. 172, entirely powerless to determine in advance where this coincidence between nature and the mind will manifest itself) – only as a last resort, where all other explanation is inconceivable. Now obviously this is not the case here. However true one supposes the "positive" procedures to be to the course of nature itself, however close to things themselves, however *natural* one considers this method, one still cannot seriously claim that the method and its results are part of nature. They, just like the deductive method and the science it created, can, in short, be no more than creations of human reason, whatever materials reason might bring into play. Consequently, it is evident that there is no need to have recourse to this "ultimate reason" of an agreement between nature and our intelligence in order to explain the resemblance between Cartesian science and contemporary science, and that the resemblance must instead be due to the analogy between the procedures utilized by the intelligence in the two cases. In other words, this resemblance alone would suffice, in the absence of any other proof, to make us suspect that, contrary to widespread opinion, science today does not conform to the Comtian model. Deduction in particular, which formed the basis for Cartesian science (and which is carried out, in our science as for Descartes, by means of mathematics, a fact that further tends to accentuate the resemblance between the two constructions), must play an incomparably more important role in science than Comte's program would allow.

We believe we have established in the course of this work how consistent this claim is with reality. By constantly exaggerating the role of experience and striving to obscure that of deduction as much as possible, Auguste Comte makes a travesty of the true tendencies of scientific research. It is not true that the gigantic effort by which humanity attempts to fathom nature is directed only toward action; if we do science, it is also – it is especially – in order to know, *to understand*. Now we cannot understand except by reasoning. To be sure, we need experiences. But we would like the results of these experiences to be of use to us in paving the way for arguments, deductions. Thus it is indeed deduction that constitutes the true goal of science.

However, as we noted in Chapter 14, Cartesian deduction differs from that practiced by the science of today in that it claims to be continuous. That is also a characteristic of Kantian deduction (Ch. 14, p. 367), and since Kant is closer to our own era, we believe there is some advantage in borrowing an example from the master of criticism in order to elucidate this subject. We choose the example of rational mechanics which, for Kant, appears to be entirely *a priori*.

As a matter of fact, one might think that, at least insofar as this branch of our knowledge is concerned, Kant's position is not very different from that of modern science. Indeed, does the adjective *rational* not suggest that we mean to derive this science solely from our reason? Nevertheless, it is easy to see that in the present case this is only an illusion.

We must note, first of all, that the term dates from an era when a conception analogous to Kant's really did prevail in science and when, as a result, this part of mechanics was considered (more or less implicitly) to be derived entirely from pure deduction. Yet even today the term does not shock us. This is because we no longer understand the rationality as absolute, but simply as relative. Indeed, among all the branches of our knowledge (with the obvious exception of *pure* mathematics) this is the one *most* consistent with our reason – which is exactly why it is also the furthest from reality, why it is founded for example on the concept of the reversible phenomenon, despite the fact that no real phenomenon can be reversible. It is also incontestably deductive, deriving its laws from a small number of principles. But in no way does it declare itself on the nature of these principles themselves. Therefore it leaves unresolved the question of knowing exactly what the principle of inertia is, whether it is a proposition capable of being demonstrated *a priori*, like a mathematical theorem, or simply an empirical law deduced from a great many experiments. Both opinions have had supporters, but it is not at all necessary to adopt either of them in order to embark upon the study of this science, and furthermore, no matter which one is adopted, nothing in the course of the arguments will be changed. As a matter of fact, as we have indicated above (Ch. 5, pp. 118 ff.) and believe we have demonstrated in our earlier book (IR, Ch. 3), neither of these two opinions is entirely justified, and the principle of inertia is merely *plausible*, that is, it contains an affirmation in which, by virtue of the particular structure of our understanding, we are disposed to believe on the basis of proofs we would under other circumstances judge insufficient. That stems from the fact that, from the philosophic point of view, the principle of inertia, like all the conservation principles, actually conceals a two-fold content: it is aprioristic in its general form, which stipulates the *conservation of something*, and *a posteriori* in its specification of what is conserved. As a matter of fact, the dividing line is drawn exactly where Kant perceived it, except that for him the second part of the principle appeared aprioristic as well, although the apriorism was in some sense inferior (Ch. 14, p. 366).

This intervention of the *a priori* in the constitution of the principles of conservation, the agreement – real or supposed – between nature and reason which these propositions seem to exhibit, gives them more weight than merely empirical laws, as we have seen (Ch. 5, pp. 120–121). Nevertheless, however much prestige they gain on this account, it is not enough to transform them into permanent immutable principles. This is proven by a mere glance at the history of science, which shows that propositions of this kind have arisen only to disappear again. For example, Descartes formulated the principle of the conservation of motion, which was at once almost universally accepted. This proposition undoubtedly contained the germs of our principle of the conservation of energy; nonetheless, it was actually *erroneous* in the form Descartes had given it, since it was inconsistent with mechanical phenomena already sufficiently recognized at that time. As soon as this was brought to light (barely a generation later), the Cartesian proposition was promptly abandoned. Similarly, Black's principle of heat ultimately could not hold up in the face of more and more frequent and precise discoveries leading to the conclusion that heat is capable of arising from the transformation of a different form of energy. In the same way, the hypothesis of phlogiston, which implicitly contained a genuine principle of conservation, the *conservation of the quality of combustibility* (as we saw in Ch. 5, p. 119), was discredited by Lavoisier's demonstrations founded on precise considerations concerning the *weight* of the materials involved in a chemical reaction.

Of course it can be pointed out that by encapsulating the history of each of these outdated principles in one brief sentence, we have omitted an important factor which accompanied the disappearance of the principle in each case. This is the fact that, although it is true the abandonment was finally motivated by refutations based on experiments, the opponents, those who succeeded in discrediting each of these principles, did not merely provide what might be called negative observations; they also proclaimed a new principle of conservation, one not only in better agreement with the facts, but in general also considerably broader in scope than the proposition they intended to abolish. Thus in the final analysis Descartes's conservation of motion only yielded in the face of Huygens' and Leibniz's conservation of *vis viva*, Black's conservation of heat in the face of Mayer's and Joule's conservation of energy, and, finally, Becher's and Stahl's conservation of combustibility in the face of Lavoisier's conservation of the weight of matter. It is most characteristic, precisely from the standpoint of the prestige of these conceptions of conservation and of the inmost need of our intelligence which they satisfy, that they succumbed only in the face of propositions of the same class: that is yet another way in which they resemble theories, which, as we saw, never disappear until other theories are available to replace them (Ch. 3, p. 64).

The fact remains, however, that they did succumb, and did so, it must be stressed, not by being merged into the proposition that was going to replace them, but by being refuted by it. Despite the fact that the conservation of energy is analogous to that of heat, not to mention being much more comprehensive, and despite the fact that the conservation of caloric was a step toward the conservation of energy, and probably even an indispensable intermediate phase of scientific thought on the way to the latter theory (Ch. 3, p. 61) – still, strictly speaking, Mayer's and Joule's principle *refutes* that of Black by establishing that heat cannot be conserved as such. Similarly, the conservation of *vis viva*, although it undoubtedly arose out of the conservation of motion, shows the impossibility of the Cartesian principle, just as the conservation of weight proves that phlogiston does not exist. So we see that the three early principles of conservation were actually *abolished*, and these examples clearly show that in fact such principles cannot in general be considered as necessarily forever immune to any attack, any attempt at modification or even refutation on the basis of experimental results. Thus, contrary to what Kant assumed, they cannot be part of the "metaphysical foundations of science." Between Kant's attitude and that of modern science, therefore, there is an essential divergence, the reason for which is perfectly clear. It lies wholly in the fact that for Kant these propositions were *pure*, that is, properly *a priori*, whereas for us they are a mixture of *a priori* and *a posteriori*, experimental science, of course, keeping complete control over what arises out of experience.

There remains, however, the important observation that this experimental element submits to the aprioristic form. In other words, what the existence of a principle of conservation shows is that in the domain encompassed by this principle there is agreement, up to a certain point, between our reason and nature. It is an agreement that reason has no doubt sought *a priori*, as is its wont, but that it has finally found, or should we say confirmed, *a posteriori*; through experience it has found something that certainly appears to it to be largely aprioristic. From this standpoint, then, the conception approaches that of Descartes and of Kant, both of whom, as we have seen, hoped to attain knowledge of the aprioristic framework of phenomena through scientific generalizations. And we see once again how unjustified was Trendelenburg's criticism of Hegel on this point (Ch. 14, pp. 368 ff.). At the same time, the success of science, not only with regard to the principles of conservation, but also from the standpoint of theoretical knowledge in general, makes it even more obvious that the Hegelian *Naturphilosophie* is completely futile.

We have seen (Ch. 14, pp. 379 ff.) how undeniably and brilliantly science has succeeded in this area. That is only a fact of experience, but it is obviously the most general purely empirical result that can be drawn from science and its history. Scientific knowledge in what we have come to see as its essential nature would not have been able to develop if the hope of seeing nature rationalized, intelligible, had not found there a real, if incomplete, satisfaction. Thus, from this point of view as well, there must be at least partial agreement between our intellect and the world around us.

That is no doubt an extremely vague result. It has to be, since, as we have seen, mankind's entire scientific effort can at bottom have no other aim than that of defining – in seeking to rationalize nature – the limits within which this agreement holds true, and since the notion of these limits is therefore necessarily fluid, being a function of the totality of human knowledge. The recognition of the partial agreement between reason and nature nonetheless seems very important to us. Indeed, we believe it allows us to understand better not only the way science operates (as we have just shown), but also, at least partially, the processes used by common sense.

In Chapter 1 (pp. 23 ff.) we called attention to the strong analogy that can be seen between the world of common sense and the world of scientific theories, both conceptions presupposing the existence of a series of beings independent of our sensation and thus situated outside the individual consciousness. We then noted the close connection between the two conceptions, science starting from common sense and transforming it only gradually and imperceptibly (pp. 26 ff.), a circumstance that also suggests we are dealing in these two cases with concepts quite similar in nature. Moreover we also pointed out that, compared to immediate sensation, the concepts of common sense, just like those of science, are characterized by a distinctive trait, their *perdurability*, although the latter possess this property to a higher degree, since they are assumed to be absolutely permanent, whereas the former are only relatively so.

It therefore seems difficult to avoid the conclusion that the body of suppositions we call the world of common sense owes its origin to a process analogous to the one which gives rise to the figurative theories of science, that it too is only a hypothesis constituted by reasoning using the schema of identification.[1] Between the two processes, however, there is the very important difference that the one to which we owe theories is played out before the conscious side of our intellect and as a result obeys our will, whereas the creation by common sense is carried out unconsciously and irresistibly. And this circumstance in its turn makes the problem seem incomparably more complex in the case of common sense than in the case of theories. Indeed, theories start from *perception*, that is to say from concepts in which the ontological character is already indissolubly inherent; therefore there is no need to ask how it happens that this character is imprinted on the beings they create. On the other hand, the raw material on which common sense operates is our *pure* sensations, which consequently, as such, contain by definition nothing that does not belong to our consciousness. How then does ontology arise here at the origin?

Of course, by the very fact that we are dealing in this case with something that happens outside the consciousness, all direct investigation is impossible. We cannot observe the process within ourselves: if we questioned ourselves about it, we could obviously not go beyond Reid's observation that the transformation occurs in virtue of our constitution, or Balfour's statement which qualifies it as inevitable (Ch. 15, p. 430). Nor can we study it outside ourselves: there is no human being, and probably no animal (at least so far as we can judge by the way animals react to their environment), in whom the perceptions of common sense do not take place in a manner analogous to the one we recognize in ourselves. But it is possible, we believe, to bring at least a little light into this darkness by referring to the commonplace that unconscious psychic processes are identical, or at least quite analogous, to those of conscious thought. No doubt that is only a *heuristic* principle. But we can see that in this case its use is in a manner of speaking inevitable, for we truly know no reasoning processes other than those of our conscious reason, and therefore if we hope to understand anything about the action of unconscious reason, we can do so only insofar as we have assimilated it to that of conscious reason.

Let us note here that for the creation of the objects of common sense to be possible, our sensations must to some extent lend themselves to it; they must be interconnected in a certain fashion. To be sure, they do not lend themselves to it completely: what our reason would demand is that the object be a substance immutable in time, which is not at all the case. But it does change slowly enough that we can indeed imagine it as an object, approximate it to a substance in designating it by the grammatical form we very appropriately term a *substantive*. When John Stuart Mill, and later Mach, admit that slowly evolving "groups of sensations" actually exist,[2] or when Henri Poincaré speaks of sensations "united to each other by I know not what indestructible cement and not by the hazard of a day,"[3] what they are affirming is the existence, exhibited by the common sense conception, of a particular structure of the whole body of our sensations, in the sense of Montaigne and Balfour (Ch. 4, p. 82).

Consequently we can to some extent recognize which particular features of our sensations allow, or suggest, this conception. First, there is the fact that our sensations obviously appear interconnected. A short while ago, as I was writing at a table, it afforded me certain tactile sensations; although I have moved away from it, I still see it and do not doubt for one moment – so interrelated are these sensations for me – that I need only make certain movements to experience the same tactile sensations, or analogous ones. The constancy of this link gives it such a hold over my mind that I am irresistibly forced to assume it even where it is impossible for me to verify its existence. I notice snow on a rock: this means that my visual sensation furnishes me a few imprecise but characteristic white and grey spots, reminding me that at an analogous distance and with analogous lighting I had a similar sensation and that later I could go closer and establish that this object was able to furnish me the whole range of tactile, etc., sensations implied by the concept of snow. However inaccessible the rock may be, I shall declare that it is covered with snow. In other words, I come to believe that these sensations, which can only be my personal sensations, are nevertheless linked together permanently, that is to say, directly.

My thought immediately goes further. For when *all* the sensations I presume to be caused by the object have disappeared, when I turn my head and no longer see the table, I nevertheless maintain that it still *is*. But clearly this is, at bottom, only an extension of the same process. Above, in supposing that sensations were interconnected, I was admitting that some of them could continue to subsist in themselves, independently of me, connected to me by the one sensation I actually experience (as in the case of objects revealed to me through sight). I now suppress this last connection by claiming that the entire group of sensations, though it has completely vanished from my consciousness, still continues to exist, to persist, and necessarily to persist outside myself – here (however prudent one might like to be in derivations of this sort) the etymology, with its *ex-sistere*, indeed seems to give some insight into the essence of the concept.

Furthermore there is no doubt that what we have here is really a particular quality of our sensations. They come and go, but above all they *recur*, if not in an absolutely identical form, at least in one similar enough that they can be incorporated into a system such as that of the world of objects, similar enough that it is to our advantage to construct this system in order to facilitate action designed to cause us pleasurable sensations, and especially to avoid painful ones. It is this experience and this expectation of its recurrence that, transformed into a fiction of persistence, create the object. However, we should explain what we mean by fiction. Surely it is fiction to pretend that what is made up of perceptions can continue to exist where there is no perception. But as chimerical and contradictory as this hypostasis is, it is not simply arbitrary, because it is based on a real or possible recurrence which is itself strictly experimental.

For instance, to use an example from the recent history of physics, it would be incorrect to say that the concept of potential energy was invented solely to make it possible to affirm the conservation of energy; it was also created because the energy of the motion which disappeared in the raising of a weight, etc., really *can* manifest itself experimentally, so that it is useful to consider it as mysteriously stored away.

But, in addition, a still more powerful factor sometimes intervenes. It is the fact that all sensations are presented to us indissolubly linked to one form, the spatial form, which particularly seems to govern their appearances, disappearances and reappearances, which form, moreover, lends itself in a very special way to the operations of our understanding.

This is yet another example of the agreement between reality and mathematics, more specifically geometry, which we discussed in the preceding chapter (esp. pp. 409 ff.) as a foundation for panmathematicism. But what must be pointed out here, where we are concerned with the concepts of common sense, is that there is not only agreement, but union, an immediate and fundamentally indissoluble union. All that our perception presents to us as really existing immediately assumes a spatial form, which cannot be stripped away without calling into question existence itself. Thus, we experience no great difficulty in stripping the perceived object of all the qualities properly so called that are undeniably attributed to it by common sense. With some effort we succeed in imagining the world of mechanical theory, a grey world without sound or light or heat. Yet this world, so different from the one provided by perception, can exist precisely because we picture it in space. If, on the other hand, we tried to take away its spatial characteristics, this whole world would immediately dissolve into nothingness, without there remaining anything whatsoever that truly exists. Here, then, existence and spatiality are synonymous, or at least inseparable, and that is yet another aspect of what we have recognized to be the superiority of panmathematicism as a metaphysical conception applicable to science. Indeed, if one sets aside the Cartesian dualism of thought and extension and tries to deduce reality from idea alone, it is at the moment when this reality must assume spatial form that the leap becomes most obvious. It will not help to make a distinction between existence and essence, as Schelling attempted to do in his later philosophy (Ch. 12, pp. 318–319), for essence will have to be limited to determinations from which everything relating to space is carefully excluded and whose degree of reality will consequently be almost nil from the scientific point of view. The accurate perception that herein lies the true difficulty of all nonmathematical idealism is obviously what caused Schelling and Hegel to go to great lengths to minimize the importance of spatial considerations (Ch. 11, p. 279).

Thus the creation of the world of common sense objects indeed seems to express the existence of a real agreement between reason and sensation. Unless, that is, one wishes, as certain philosophers have done, to deny altogether the existence of an understanding independent of sensation, to reject the Leibnizian amendment, *nisi intellectus ipse*, to Aristotle's famous statement, which the medieval philosophers had adopted. In such a case, the illusion of an agreement would spring from the fact that the understanding had simply drawn from sensation itself those elements in which the agreement appears to be expressed. Therefore our understanding would contain no autonomous preexisting principle of rationality as the source of the tendency to assume the conservation of all things. On the contrary, the tendency to consider objects permanent would be produced by the "structure" of sensations discussed above. We understand these objects to be perdurable because they recur under certain conditions, because they show themselves capable of constituting this system of common sense objects which facilitates prediction to such an extraordinary extent, and finally because the objects so constituted seem much more persistent than pure sensations.

We can see immediately that such a theory could not be applied to the scientific conceptions dealt with in this chapter (pp. 439–440). Here, indeed, the history of the sciences enables us to seize the very genesis of these ideas and to confirm the inadequacy of the empiricist theory. For empiricism to be true, atomism and the laws of conservation would have to be purely experimental in origin. Some have been so conscious of this that they have tried to affirm it explicitly, and even now it is frequently assumed more or less implicitly. But it is an untenable thesis; in order to understand anything about the history of these conceptions and their real situation in science, one is really obliged to have recourse to the causal tendency, a preexisting tendency to believe in permanence.

Of course, even admitting that the causal principle plays a role in scientific theories, we can deny that it does so in the conceptions of common sense. It will be objected that the analogy is misleading, that, on the contrary, the causal tendency observed in science is precisely the product of common sense, that the success of this system is what makes us conceive the permanence of all things and even the alleged principle of rational identity. The belief in the perfect rationality of nature could then be due to the fact that, since the data of the senses bear a form which allows our reason to proceed deductively, we are led to formulate an analogous requirement with regard to the rest of these data. In other words, panmathematicism is not engendered by the tendency toward the rational; on the contrary, the rational is the product of a more or less implicit and unconscious sort of panmathematicism.

That is not an impossible theory, but it is certainly a very difficult one to sustain. The break in continuity it necessitates between common sense and science is obviously already in itself a strong presumption to the contrary. But there are many other obstacles.

The causal tendency, we have seen, is manifested with great vigor in science in spite of often quite unfavorable circumstances over which it attempts, with some success, to prevail. For this to be true, for (in the words of Spencer, himself a confirmed opponent of this point of view) a "cognition ... that results from a long continued registry of experiences gradually organized into an irreversible mode of thought"[4] to be able to acquire such power over our understanding, it would seem that these experiences would at least have to confirm it in an absolutely uninterrupted manner. Now such is certainly not the case. Objects no doubt change much more slowly than our sensations, but they do nevertheless change, and the world of common sense offers an infinitely changing spectacle. Common sense itself displays a firm conviction that change is indeed the law of its real world, as is shown by maxims in all languages on the order of "All good things come to an end" ["*tout casse, tout lasse, tout passe*"]. So strong is this conviction that, in many cases, the absence of change appears to constitute an enigma requiring explanation. If someone showed me a table I had helped purchase twenty years earlier and its varnish was as fresh as the day it was bought, I would certainly conclude that it had been extraordinarily well taken care of, that it had been revarnished, or that it was another table of the same model. The last hypothesis is what would unfailingly occur to me in the case of a dog or a cat exactly like one I had known twenty years before. That, of course, does not mean that, in a different sense, all change does not seem to me to need explaining. But this very circumstance is in its turn enigmatic. If the conviction of the permanence of objects comes only from common sense, how can we account for the fact that the conviction goes so far beyond common sense, that the understanding finds the permanence of *perceived* objects insufficient and, in order to explain them, creates concepts like the atom, mass or energy, in which the permanence becomes absolute?

And, finally, how are we to understand the very birth of this conception of an external world? No doubt, once the system as a whole is in place, it facilitates prediction and action. But how was our understanding able to take the first step, how did it arrive at the paradoxical idea that sensation could subsist outside the self? If, as we assumed above, it was only obeying a preexisting general tendency whose effects we see elsewhere, if the belief in the permanence of an unperceived object is analogous to the existence of the element sulfur in sulfurous acid, I begin to understand. But if, on the contrary, the conviction of the permanence of sulphur must come from the recognition of the permanence of objects, the enigma of common sense remains undecipherable. Indeed, proponents of empiricism have constantly sought to demonstrate that there is no real problem here, that common sense does not posit the existence of objects, that common sense is, rather, perfectly idealistic. But that position is totally untenable, as we saw in Chapter 1. Common sense is unquestionably an ontology and, consequently, also an enigma, one which empiricist theory cannot even attempt to explain.

Thus we really seem to be forced to consider the intervention of the concept of identity as the expression, not of a simple action of the environment on the individual – an action in which the individual would play a purely passive role – but of a more complex process where the actions of the environment and the individual are combined: no doubt the environment does act, but the individual reacts and therefore contributes something that belongs to him in his own right.

This individual element can be understood in different ways, however. In particular one can attempt to connect the action of this principle to the general reaction of the individual to his environment. William James has said that "primarily then, and fundamentally, the mental life is for the sake of action of a preservative sort,"[5] and Roustan, in a well-constructed argument stated with exceptional clarity, has attempted to establish that this point of view is really sufficient to account for the functioning of scientific reason and that, as a result, one can completely disregard "pure understanding" as Descartes or Malebranche, for example, understood it.[6] Thus, from the present point of view, the individual finds both similarity and diversity in the physical reality around him. If he had to face this reality strictly as a passive observer, what would strike him most of all would undoubtedly be the fact of diversity, and he would then necessarily conclude in favor of the *panta rei* of Heraclitus, or rather the absolute skepticism of his disciple Cratylus, the impossibility of drawing anything like a coherent conception from the contemplation of this eternally puzzling labyrinth. But the individual cannot take such a disinterested attitude. He must live, which necessity entails for him the necessity of action and prediction – we saw in Chapter 2 (p. 32) that, according to Auguste Comte, this is the exclusive origin of science as a whole. Now clearly prediction is possible only if one admits the identical into reality. If the animal paid attention to what is particular about each blade of grass, it would never get around to grazing. Only because it makes the voluntary, intentional error of considering them identical does it arrive at the concept of grass in general and is it therefore able to make predictions about this grass. Nor does the carnivore question the existence or the identity of his prey as he pursues it, although the image frequently disappears from sight and constantly changes size and shape: the least doubt in this respect would render any pursuit impossible.

Unquestionably a theory of this kind is very appealing to a mind shaped by contemporary biological theories. For Locke, for Leibniz, for Kant, the human intellect was something independent of the rest of nature and also fully realized once and for all. We, on the other hand, because of the concept of evolution, would like to see it emerge from within nature itself, and any hypothesis that seems to lead in that direction is sure to find our understanding favorably predisposed. Such is the case here where the theory appeals to the very notions made familiar by Darwinism: this tendency to treat nascent diversity as identity, within animal life, within consciousness in its most primitive form, can be imagined to be, as it were, an accidental variation, yet one that persists insofar as it is useful to the life of the individual and of the species.

If we go a bit more deeply into this conception, however, we shall see that it entails consequences our reason has great difficulty accepting. In this regard we need only turn to our observations concerning empiricism proper. Of course the principal difficulty this hypothesis encountered no longer exists here: we do not have to explain how the idea of the identical arises out of the contemplation of diversity. But it is self-evident that the identity resulting from reaction to the environment can be only a limited one. Where, then, does the understanding acquire the concept of perfect, absolute identity, a concept that rules it to such an extent that it finds unsettling the fact that reality does not entirely correspond to it? Can it be said that we want perfect identity for the sake of more effective prediction? Is it not patent, on the contrary, that if there were no diversity, our understanding would not find anything on which to act? Obviously, then, if there were no variation in time, all prediction would be pointless: since nothing would ever happen, there would no longer be anything to predict.

Clearly the propensity of our understanding to go beyond what, according to this conception, must be its true and even its unique aim appears particularly enigmatic here. It will be said, with Roustan, that "the insatiable curiosity" displayed by science does not keep it from being "essentially a vital function," given that

> as soon as man has provided for his most pressing needs, it behooves him to forget himself, to pay more attention to external things than to his own needs, not remaining hypnotized by his desires, not being absorbed in them, but rather sympathizing unreservedly with this nature that is impenetrable to the overly egotistical animal. ('La science' 625)

True. But it is just as incontestable that this truth – the profound conviction that, since everything is interconnected, all knowledge, whatever it may be, must sooner or later become useful in the strictest sense of the term – is a recent acquisition of the human mind. Not that long ago, Auguste Comte proved to have only a very inadequate grasp of it (cf. Ch. 4, pp. 83 ff.), and Roustan rightly observes that "man reaps the greatest benefits from his scientific activity during the very periods when he forgets most completely the advice" of the founder of positivism ('La science' 627). But how does that fit into the framework of the theory? How can we conceive that, precisely in the case of the process of identification, the understanding, by unremitting and tireless effort, finally tends to reduce itself to the absurd? It would seem that the least we can do is credit the understanding with a particular faculty that would determine it to continue an activity in which it is engaged beyond the limits where it retains any sense, to continue it even when it turns against itself. But would that not be a properly logical faculty, and is it not therefore just as expedient to posit the principle of identity as an independent logical principle, that is, one not dependent on the *vital impetus*, not reducible to this impetus, although remaining closely related to it in many respects? It would be a factor properly characterizing the intellect and thus arising, in as primitive a form as one would wish, along with it, that is, with the consciousness. Whatever we do, and whatever advances science makes in this area, that is to say even if we were to succeed tomorrow in creating life by the juxtaposition of nonorganic materials, it is certain that the appearance of consciousness in such a collection of molecules would remain a miracle for us; it would be useless to attribute consciousness to inert matter itself, for by definition no manifestation of this consciousness appears there. But then why shrink from the eventuality of endowing this consciousness with a particular trait that, as we have seen, characterizes it always and everywhere?

Particularly since the supposition that would reduce the mind's penchant for logical identity to the vital impetus is perhaps not without risk from the standpoint of comprehending the paths followed by the human understanding. Indeed, it would seem to be because we tend to confuse the two tendencies in their respective actions that we have come to minimize the role of any properly rational element in science, to consider that science aims only at material action, true knowledge being the prerogative of other forms of activity of the human mind, especially of philosophy. Now there is no opposition between science and philosophy on this point. Although they proceed by different methods, neither has any true goal other than the penetration of the real by reason, "the honor of the human mind," as the mathematician Jacobi so aptly put it (Ch. 2, p. 33). Consequently, it would be utterly futile to try to reduce science to a sort of transcendent positivism by way of pragmatism, declaring that, since it gives up the possibility of attaining any definitive truth (Ch. 15, p. 405), the relative truth it attains can only be identified with the useful. Or else one must understand the useful not as what is profitable to life (of the individual or of the species), as does pragmatism, but in the sense of the quotation from Spinoza included in Chapter 2 (p. 32): "Neither does the mind, in so far as it makes use of reason, judge anything to be useful to it, save such things as are conducive to understanding." In other words, it is a question here of the rational, and it can only be to our advantage to eliminate all possibility of confusion between what is rational and what is merely useful.

Of course the error we have just pointed out need not be a consequence of the theory. One can quite easily assume that the principle has the quasi-Darwinian origin we have attempted to retrace, recognizing at the same time that in human reasoning, as we see it functioning from the most remote historical times, this factor functions independently. But that is nonetheless a new difficulty added to the other, or rather it is still the same difficulty recurring in a new and particularly tangible form, in the sense that these conceptions more or less bound up with the theory show us which way it is tending.

Thus the simplest as well as the safest point of view seems to be to consider the process of identification as part of the bedrock supporting the deepest layers of human knowledge, as a small fragment of the *intellectus ipse* which, according to Leibniz, must be added to what comes from the senses. As such, it is logically prior to all experience; according to the ideas set forth by Spencer in the continuation of the passage we cited on p. 448 (ideas this philosopher actually applies incorrectly to the conservation of matter), it constitutes "a cognition given in the form of all experiences whatever." Already the world of our perception is nothing other than a result of the intervention of this factor, with our sensations as the starting point.

Therefore, as we had already anticipated in Chapter 2 (p. 43), it is not strictly true that there is a direct opposition between the ontological tendency and the rational tendency, since the ontological tendency which creates the perception is itself already produced by the tendency to seek the reason, the explanation for our sensations. It is thus these sensations, insofar as they are stripped of all transcendence, it is the *immediate data of consciousness*, as Bergson calls them, that in fact ultimately constitute the element which is opposed to our reason and which our reason, with all the means at its disposal, seeks to assimilate – more or less successfully, as we have seen. This effort, which is essentially an attempt to impose the framework of the principle of identity upon sensation, is, from the standpoint of its final aims, a completely impossible enterprise. Nevertheless, the belief in the possibility of attaining this chimerical goal is indeed what is expressed by the postulate of the rationality of nature, a postulate which creates the whole of explanatory science (and consequently panmathematicism as well) and which also created the earlier attempts at global explanation. Experience – both that of common sense and that of science – in many cases corroborates our belief in rationality; it is the agreement between reason and nature, a partial agreement, for at other times nature seems to deny it. As a result, side by side with the ardent desire to understand nature as conforming to our reason there is created a profound belief that nature contains something that escapes this rationalization. We have seen that even in the most absolute systems of Ionian philosophy this belief is not entirely absent and that this limitation becomes stronger and stronger as knowledge advances (Ch. 4, p. 92; Ch. 14, pp. 363 ff., 368). But, on the other hand, even the most absolute empiricism could hardly go so far as to deny this fundamental postulate of the conformity of nature and our mind, for then it would become inconceivable for us to be able to fathom nature, and experience itself could no longer be justified. As a matter of fact (as we saw, Ch. 4, pp. 73 ff.), empiricism, especially as codified in Auguste Comte's positivism, already represents a sort of compromise, as does Hegelianism; it implicitly recognizes the agreement, but would limit it to the existence of laws. If our efforts throughout this work have not been entirely vain, the reader has grasped how inadequate this theory is, how badly it responds to the true rhythm of science. Indeed, science clearly proclaims the certainty of an agreement that goes far beyond simple lawfulness; it proclaims it quite as much by established results, such as the principles of conservation, the triumphs due to mechanism, etc. (Ch. 14, p. 380), as it does by its general attitude, by the very existence of theoretical, that is to say explanatory, physics, and its preponderant role in science, which can be founded only on the tenacious hope that even where this agreement seems lacking at the present time we shall be able to discover it by probing more deeply. Considered from this perspective, the positions of Comte and Hegel no longer appear to be opposed to one another in principle, but to represent one and the same understanding of the relations between our reason and nature, differing only in degree: they both allow – and this is consistent with the science of today – that the agreement exists, as well as the fact that it must end somewhere. But Comte fixes this limit short of the line drawn by science, while Hegel carries it well beyond. Hegel was undoubtedly mistaken, but one must not attempt to demonstrate his error, as Trendelenburg does, by considering the very fact that Hegel assumed this agreement to be grounds for a sort of *reductio ad absurdum* of his position. When the author of the *Logical Investigations*, after pointing out that all knowledge acquired by the dialectical method is *a priori* knowledge, asks triumphantly: "Do we then have, side by side, two sorts of science independent of one another and without any link between them?" (*Log. Untersuch.* 1:80), he is surely mistaken about Hegel's intent. What Hegel hoped is precisely that, since the order of ideas and the order of things necessarily coincide (Ch. 14, p. 372), the agreement would manifest itself at the point where the *a priori* would come to be fused with experience, that the results of deduction would be shown to form the fundamental web of nature, just as it is revealed to us by direct research, or, what amounts to the same thing, that the most general formulas of science could provide end points for chains of philosophic reasoning starting from the notion. The whole of Hegel's work is so obviously inspired at each step by this thought that one wonders how it would be possible to misinterpret it. And, once again, such a hope in itself can in no way be treated as absurd, for in the final analysis all science is forced to foster it to some extent. Trendelenburg himself acknowledges this a little further on. It has been claimed, he says, that Hegel "formulated clearly what others applied without being aware of it .... If what is meant by this is the fundamental idea that there is reason in things, we agree in our turn" (*Log. Untersuch.* 1:98). That is, literally, only Hermotimus's declaration, which is the basis of all philosophy and all explanatory science and even, as we have seen, of all science whatsoever, no matter how "positive" its formula appears to be.

However, this situation – the analogy that still holds between contemporary science and the most daring constructions of apriorism, insofar as the search for rationality is concerned – must not make us forget how essentially the attitude of this science differs, with respect to the deductions it uses, not only from the attitude of Hegel, but also from that of Descartes and of Kant. Today's scientist has the very clear feeling that in physics no theory can be complete. He knows that, unlike mathematical fact (Ch. 15, pp. 410 ff.), physical fact, regardless of our efforts and of how successful they are, can only be understood in part; all the rest remains inaccessible to our understanding, one might say *opaque*. Sophie Germain has captured this contrast extremely well. "To see our assurance," she says, "one would believe that, like the geometer, we have succeeded in expressing the nature of the [material] subject so precisely that all its properties are included in our definition." But this is only a deceptive appearance; in reality we feel very strongly that such is not the case, that "instead of an absolute equation which embraces the object of our research in its entirety, so that nothing belonging to it can be left out of this sort of characteristic definition, we know only a few properties relative to our senses," with the result that "we are completely ignorant of the essence" of matter.[7] Perfect knowledge of the real being, the *adæquatio rei et intellectus* according to Isaac Israeli's definition adopted by Saint Thomas,[8] is impossible.

In speaking of the evolution science has undergone in abandoning the hope of arriving at continuous deduction (Ch. 14, pp. 378 ff.), we have stressed the salutary role played by positivism. Let us add that some of Auguste Comte's utterances which appear the most shocking, the most contrary to the attitude of contemporary science, are explained to some extent if one takes into account the fact that he had to react against the tendency to believe in the possibility of the complete rationalization of physical fact. One need only open a physical chemistry textbook, for example, to see how wide of the mark Comte was in denying mathematical theories access to chemistry (p. 438 above). Yet the feeling underlying this prohibition, that of the specificity of the chemical phenomenon, of the resultant impossibility of reducing it entirely to the physical phenomenon – or as we have expressed it in Chapter 6 (pp. 167 ff.), of the existence of one or more chemical irrationals – was probably altogether correct. And, likewise, he was quite wrong in declaring that all assimilation between light and sound or motion would always be an "arbitrary supposition" and in condemning in general all tendencies to establish relations between what we now call the different forms of energy, stating that there are "six, and perhaps seven, irreducible branches" of physics.[9] But if he probably responded here above all to an awareness of necessary consequences that do in fact follow from the foundations of his doctrine (Ch. 1, p. 19 above), it is no less probable that, there again, his thought dimly envisaged the fact that the specificity of phenomena, where it appears especially clear-cut, is apt to indicate the existence of particular irrationals.

If we were to try to rationalize nature too much, and especially too soon, we would be certain to fail. Enriques, in a recent publication, has provided an excellent example of this. In mechanics, when we seek the resultant of two equal forces acting at an angle, it appears obvious to us that this resultant can only be the bisection of the angle in question, there being no reason for it to incline to one side or the other, as Poisson has already observed. But Enriques very correctly points out that this would no longer be true in the case of two forces determined by oppositely charged magnetic poles. This is because the geometrical representation would then in fact no longer be adequate to the phenomena, in the sense that "certain elements or certain given physical relations which cannot be substituted for one another would be expressed by means of geometrical elements and relations we consider equal."[10] Enriques's example is no doubt particularly striking. But what must be noted is that the restriction he establishes, far from applying only to exceptional cases, is, on the contrary, a general rule extending to mathematical physics as a whole. No mathematical representation, however faithful it seems to us, can actually express all the complexity of the physical phenomenon. In the best case it never expresses more than a single aspect of it, and the theory remains capable of being destroyed if we uncover a different aspect as we go more deeply into the essence of the phenomenon. Bouasse's catoptrics (Ch. 4, p. 98) will remain immutable so long as we preserve the mathematical bases, that is, as long as we schematize light surfaces and rays. But as soon as we go more deeply into their structures, the theory will have to be modified.

Thus, to the extent that science moves from the most abstract part of mathematics, that dealing only with figures, and moves toward physics, it gradually loses in pure rationality. D'Alembert expressed this with great clarity in his admirable *Preliminary Discourse* to the *Encyclopedia*:

The broader the object they [the sciences] embrace and the more it is considered in a general and abstract manner, the more also their principles are exempt from obscurities .... Our ideas become increasingly obscure as we examine more and more sensible properties in an object. Impenetrability, added to the idea of extension, seems to offer us an additional mystery; the nature of movement is an enigma for the philosophers; the metaphysical principle of the laws of percussion is no less concealed from them. In a word, the more they delve into their conception of matter and of the properties that represent it, the more this idea becomes obscure and seems to be trying to elude them.[11]

If we compare the mathematical sciences as a whole to the physical sciences as a whole in the light of what we have recognized above, we clearly see the source of our inner conviction of the higher rationality of the former, a conviction that makes the mathematicization of the real seem synonymous with its rationalization. This powerful and immediate feeling comes from the fact of the total and certain deduction of mathematics, as opposed to the – at best – partial and precarious deduction of the physical sciences.

For example, in the domain of physics, in spite of the fact that we no longer judge knowledge based on mathematical deduction to be entirely immutable, we still find it superior in nature to knowledge arising merely from generalized experience, and science retains much of the Cartesian spirit that, for Kant, lent a particular dignity to everything in physics that is connected with mathematical deduction (Ch. 13, p. 353).

This is because, as we saw (Ch. 4, pp. 85 ff.), contrary to what positivism would have us believe, laws do in fact appear to us to be further removed from things than theories are, to be external to them, so to speak, whereas theories are assumed to penetrate to the heart of things – their rationality, in virtue of our deep, ineradicable conviction of the fundamental rationality of nature, being proof that their formulas conform to the true essence of the real. This explains the fact – surprising at first glance – that in the case of laws man has always acted by successive approximation, whereas for theories it took centuries of disappointments and battles for us to realize that none of them could be true absolutely. In science there have never been any resounding battles over mere laws to compare with the battles that signaled the disappearance of the peripatetic theory of matter in physics, the geocentric hypothesis in astronomy or phlogiston theory in chemistry. Wherever history does seem to reveal rather heated discussions over laws, such as the battles over the law of multiple proportions, the principle of conservation of energy or the principle of maximum work, we see, on closer inspection, that it was basically theoretical conceptions that were involved or that there were (as for energy) *a priori* elements mixed in with the apparently empirical proposition. The battle over the Newtonian law of gravitation has (it is safe to say) never ceased; but (leaving aside, of course, Einstein's hypothesis, which is on a different plane) it dealt solely with the theoretical conception, the hypothesis concerning the means of production which this proposition inescapably suggests. We have seen (Ch. 4, p. 74) that Auguste Comte tried at one point to influence the attitude of physicists in this area, proclaiming that once certain laws, such as Mariotte's principle, were established, they were to remain forever untouchable. But the very way in which the passionate exhortations and vituperations of this powerful mind (and one which exerted such a strong influence on the thought of ensuing generations) fell on deaf ears, so to speak, shows how foreign all this was to true science: Mariotte's law did not find many defenders, at least in the absolute sense in which Comte wanted to preserve it, and the modern physicist uses this formula as a first approximation, applicable exclusively to the "ideal gas" he knows quite well he will never encounter in reality; in so doing he does not even have the impression that he is breaking with the tradition so ardently defended by the founder of positivism – so much does the process of successive approximation seem the only possibility in this case.

It is one of Kant's great merits to have recognized that, contrary to Cartesianism, which attributed everything to deduction and into which experience entered only surreptitiously, contrary also to the Baconian position for which science is reduced to experience and where deduction in its turn has great difficulty gaining entrance, science does not have a single origin, that it is a mixture of the *a priori* and the *a posteriori*: his mistake was in believing that the two domains could be delimited in advance. We know, on the contrary, that this is impossible, that experience and deduction are intermingled everywhere.

Indeed, that is also what distinguishes the method of contemporary science from that of Descartes. No doubt we would like to be able to deduce everything, as he did, and we certainly deduce everywhere it is in any way possible; science considers it an immense advance to be able to establish a deductive link between purely empirical observations, to be able to construct an explanatory theory: that is why this science bears so strong a resemblance to Cartesian science. But we also know that we cannot, we shall never be able to deduce everything, nor even everything essential, that it is impossible to reduce the science of nature to a coherent, rational conception: this is why we are forced to recognize that Descartes was mistaken. One might say that the Cartesian formula is an ideal for contemporary science; however, it must be understood that we are now convinced that we shall not continue to approach it indefinitely, the irrationals already discovered constituting impassible barriers to the realization of this ideal, barriers which are sure to be joined by others as science advances in its work.

Thus the true progress of science, which is the progress our understanding makes toward comprehending nature, must ultimately consist in determining the limits and the modalities of the agreement between nature and reason. That is the task of the scientist, one can safely say his only task, because the true scientist, whatever has been said of him, does not concern himself with practical applications that may result from his discoveries. Furthermore, he would quite often be incapable of predicting such applications, for they will result from other discoveries which, though probably less far-reaching than his, will nevertheless be genuine discoveries, and, what is more, due to minds otherwise disposed than his own.[12] But the principal credit for these conquests nonetheless belongs to the theoretician: however great a debt humanity owes Lister for establishing the method for the antiseptic treatment of wounds, we all feel that an infinitely greater gratitude is due Pasteur, whose theory made the method possible. And whatever the merit of the inventors of the internal combustion engine, it obviously does not even remotely compare to that of Carnot, despite the fact that he did not invent anything at all and was content to establish a principle concerning the maximum efficiency of any heat engine in general. Thus, even from the standpoint of utility, there is no doubt that theory is more valuable than any practical invention. So long as the work of understanding nature had not advanced very far, inventions could doubtless sometimes be the product of chance and of ingenious minds having only an imperfect acquaintance with the scientific knowledge of their time; but that has now become much more rare and will clearly become more and more so as science progresses. In the immense majority of cases, contemporary inventions are most certainly only simple applications of abstract scientific truths. Now, in order for there to be these applications, there must be science, there must be scientists whose sole preoccupation is the *honor of the human mind* (cf. above, p. 452). Jacobi, as we saw, was thinking in particular of pure mathematics, which is in fact the most abstract branch of human knowledge; however, this does not mean that its development has not exerted and must not always exert the most profound influence on the progress of the physical sciences and consequently on the development of applications having the most immediate practical significance. No one can seriously contest that the immense progress of physics in the last two centuries is a direct consequence of the invention of the infinitesimal calculus, that, as a matter of fact, such progress would have been impossible without this achievement in pure mathematics. And surely few physicists will be inclined to doubt that if mathematics today made a new advance analogous to that due to Newton and Leibniz, this fact would immediately have enormous repercussions for the progress of the physical sciences (as we showed in Ch. 5, p. 128).

Furthermore, one need only watch the scientist at work to realize the true nature of what he is trying to accomplish. He has examined a body of phenomena, a body of observations made by his predecessors or by himself, he has manipulated formulas, and the idea has come to him that these circumstances ought to entail this or that consequence, to bring about this or that phenomenon which has not yet been seen. Of course, in searching, he has taken pains to follow as closely as possible the course of nature itself, to take into account all the anomalies, all the irrationals that have been recorded. But, granting all this, he has *reasoned*, *deduced*, that is, he has assumed that, within the area embraced by his deduction, nature had to follow the very path of his reason. To be sure, once his reasoning is completed, he will verify it, and if nature decides against him, he will try a different path; but he will continue to form hypotheses and, in so doing, will obviously be doing nothing other than presupposing an agreement (be it ever so partial) between nature and the understanding.

This view of scientific work, a view according to which, in the words of Claude Bernard, "there can be no method for making discoveries" *(Méd. expér.* 57 [Greene 35]), runs directly counter to the theories proclaimed by Bacon. Indeed, Bacon believed that one could arrive at scientific discoveries by more or less mechanical inductive procedures; he went to great lengths to elaborate very detailed procedures, the use of which was to leave "but little to the acuteness and strength of wits," making them, on the contrary, "nearly on a level."[13] Now Bacon's procedures, we can safely say, have never been used systematically by any scientist worthy of the name, and no scientific discovery, great or small, is due to their application. Scientists themselves have sometimes recognized the sterility of work pretending to dispense with all anticipation of the results. On this subject we can cite the opinions of three illustrious men numbered among the creators of the eminently experimental science that was chemistry in the first half of the nineteenth century. "In order to undertake an experiment," says Berthollet, "one must have a purpose, be guided by a hypothesis."[14] Humphry Davy states that it is "only by forming theories, and then comparing them with facts, that we can hope to discover the true system of nature."[15] And Liebig, after having declared that between observations in the Baconian sense and genuine scientific research there is "the same relationship as between the noise of a child's rattle and music," points out that it is actually the scientific imagination which plays the greatest role in discoveries and that experience, like calculus, serves only to aid the thought process.[16]

Montaigne had earlier said something similar. In the passage from which we quoted a few sentences in Chapter 2 (p. 32), after stressing the "desire for knowledge" and the fact that "we try all the ways that can lead us to it," he adds: "When reason fails us, we use experience ... which is a weaker and less dignified means."[17]

But no one has better understood this essential aspect of scientific research, nor insisted upon it more vigorously, than Claude Bernard. For him the phrase we quoted above is not a casual remark, but is linked to opinions to which he returns time and time again. Bacon, he declares, "was not a man of science, and he did not understand the mechanism of the experimental method." This method develops nothing "except the ideas submitted to it." Indeed, "considered in itself, the experimental method is nothing but *reasoning* by whose help we methodically submit our ideas to experience, – the experience of facts" (*Méd. expér.* 82, 56, 7 [Greene 51, 34, 2]). It "is, in short, only logic applied to the coordination of the phenomena of nature, in order to discover their laws"; it "seeks to arrange in a logical order all the facts observed directly or provoked by experimentation in order to make them serve as *verification* of a preconceived idea, a preconceived idea which is, in reality, only our mind's logical anticipation of unknown phenomena." Moreover, it is impossible to devise an experiment without a preconceived idea; devising an experiment ... is putting a question; we never conceive a question without an idea which invites an answer. I consider it, therefore, an absolute principle that experiments must always be devised in view of a preconceived idea, no matter if the idea be not very clear nor very well defined.[18]

"The idea is what establishes ... the starting point or the *primum movens* of all scientific reasoning, and it is also the goal in the mind's aspiration *toward the unknown.*" "We must let ourselves go with it freely, provided that we observe the results of our experiment rigorously and fully," and the research scientist must "give free rein to [his] imagination," for "the idea is the essence of all reasoning and all invention. All progress depends on that."[19] On the other hand, "blind belief in fact, which dares to silence reason, is as dangerous to the experimental sciences as the *beliefs* of feeling or of faith which also force silence on reason." Indeed, a fact "is nothing in itself, it has value only through the idea connected with it or through the proof it supplies" and "if a phenomenon, in an experiment, had such a contradictory appearance that it did not necessarily connect itself with determinate causes, then reason should *reject the fact* as non-scientific. We should wait or by direct experiments seek the source of error which may have slipped into the observation" (*Méd. expér.* 85, 87 [Greene 53, 54]).

We beg the reader's indulgence for belaboring the ideas of the great physiologist. In spite of everything, the Baconian conception of science still enjoys great prestige, science textbooks present it as a matter of course, and it has not been superfluous to show how foreign it seemed to one of the most illustrious scientists of the nineteenth century. But we could just as easily evoke the testimony of contemporary scientists. For example, Soddy points out that if research scientists had had no theory to guide them, they would certainly have considered thorium the parent of radium, which would have been a serious error,[20] and Le Châtelier declares that "imagination is indispensable for leading the mind in new paths; without it the mind keeps going round and round in the same circle, repeating the same experiments or the same reasonings indefinitely."[21]

Everything conspires, then, to prove that the Baconian program is, if not entirely inapplicable, at least perfectly sterile. But what is the intrinsic reason for this? Why is it indispensable for the scientist to exercise his imagination in the search for laws, relationships? Why can he not interrogate nature directly and with no preconceived idea? Wherein lies the defect of this program, which at first sight certainly appears logical?

In order to answer these questions, we need only recall what we came to see about the "structure" of reality in Chapter 4 (pp. 82 ff.). If it is true that we can know laws only because the circumstances conditioning the phenomenon are subject to a certain hierarchy, are arranged, according to Arthur Balfour's felicitous image, in "fibers," then when we set out to search for a law, we must necessarily formulate an assumption as to how this structure will manifest itself at the point which interests us. As a matter of fact, the circumstances surrounding a phenomenon, we can be sure, are, strictly speaking, infinite in number. Which are the ones whose variations we shall study to see if we can, in any measure whatever, consider them as concomitant with those of the phenomenon itself? Obviously we must make a choice, in other words formulate a hypothesis about the existence of the "fiber."

Thus, to give an example, any phenomenon, since it takes place in time, is necessarily accompanied by a modification in the arrangement of the celestial bodies. But we shall certainly not try to link phenomena in this way, and a chemist who decided to study whether the outcome of the particular reaction he has in mind depends on the position of the planet Saturn would seem ridiculous to us today. But one must not forget that barely a few centuries ago alchemists believed in action of this sort, literally believed that certain phenomena we would term chemical could occur only if aided by a favorable constellation. And astrology also taught that the position of the stars exerted a direct and prevailing influence on political events and on the course of human affairs in general.

It would be difficult to exaggerate the hold these ideas had on humanity, both in antiquity and in the Middle Ages. In the Renaissance their influence becomes even greater, if that is possible; the astrologer becomes an important and quasi-official personage in the princely court. We know, moreover, that even some of the creators of modern astronomy were not proof against these beliefs we today consider pure superstitions. Tycho Brahe set himself up as a formal defender of astrology,[22] and Kepler, although he sometimes combats the abuse of astrology, believes in the influence of the conjunction of the planets on political events and censures Pico della Mirandola, who had declared himself the enemy of the astrologers.[23] As late as the eighteenth century, astrology numbered a distinguished advocate, the Count de Boulainvilliers,[24] and one need only glance at the numerous publications by the supporters of the "occult sciences" in almost all civilized countries[25] to be convinced that the belief is far from extinct and that among its defenders are those who in other respects must be considered educated men.

As for ordinary men, a quick examination of the language proves how deeply ingrained these beliefs had become. Expressions like *to be born under a lucky star*, to be *moon-struck*, words like *disaster* or *ascendant*, finally those very words, so simple and yet so full of meaning, *bonheur* and *malheur* ["happiness" and "unhappiness," which might appear to be derived from "good hour" and bad hour"], bear eloquent witness to this.[26]

Is it any wonder then that we witness the emergence of this false science, which Bailly aptly calls the "longest illness ever to afflict human reason"?[27] Not at all. In the first place, one must consider that *judicial* astrology, the one we usually have in mind when we pronounce the word "astrology," is based on so-called *natural* astrology, which affirms the influence of the celestial bodies on meteorological phenomena. It is only a slight exaggeration to say that in countries with very well-marked seasons (for example Chaldea, where astrology seems to have arisen), given that the appearance and cessation of the rains, of the great heats, etc., coincide with the appearance and disappearance of certain stars, the belief in a link between the two sorts of phenomena must have sprung up almost spontaneously. That the Hyades were *rain stars* and Sirius was a *heat star* appeared to the Greeks a sort of scientific commonplace needing no demonstration.[28] From this it was but a short step to a belief in the general influence of the stars, especially on important human affairs, wars, revolutions. This deduction no longer seems compelling today, because modern astronomy has accustomed us to thinking of the earth as a tiny planet lost, as it were, in the immensity of space and the multiplicity of systems. But at that time it was believed to be the center of everything there is, and it appeared absurd to suppose that phenomena as imposing as those observed in the sky should be without repercussions on earth; since God does nothing in vain, these phenomena had to have a purpose, be of some use, which, as we have seen, meant quite simply that they had to be useful to man, the "king of creation," as a present-day finalist has called him (Ch. 7, p. 196). That is precisely the argument put forth by Tycho in his defense of astrology.[29]

Thus we see that no phenomenon, even the most remote, can be considered as excluded *a priori* from our research. It is actually a question, therefore, of *divining* what we shall relate to what, and this is why it is indispensable for the scientific imagination to intervene and formulate a hypothesis, which can be nothing other than a prevision, a preconceived idea.

No doubt there are cases where, either as a result of theoretical conceptions that seem firmly established, or simply as a result of previously acquired knowledge, we take for granted a particular link between phenomena and certain of their conditions, and where, consequently, it is in fact possible to proceed according to a method which seems to resemble the Baconian program, studying the variation of the phenomenon in terms of one particular condition, with no preparation and no attempt at invention. A good many researchers who are conscientious but devoid of scientific imagination have quite often applied themselves to works of this sort, amassing data and measurements and drawing up methodically arranged tables that are included in voluminous publications. Bouasse has bitterly derided these laborious works,[30] not totally without justification. Their true scientific value is certainly slight, inferior to that offered by the least insight, the slightest theoretical progress, progress which teaches us something new, a relationship we did not suspect; as a matter of fact, their sole justification is the hope that they will one day be able to serve as a starting point for such a work of the mind. One must not even cherish the illusion that the meticulousness of these works guarantees their solidity. For any result acquired in this way obviously raises the question of the postulate underlying the whole body of measurements that have been carried out, namely the fact of a particular relation between the phenomenon and the condition. If, on the contrary, it were demonstrated that a condition other than those taken into account played an effective role, the whole thing would have to be begun again. Let us take a concrete example: a physicist has studied the variation in elasticity of a wire made of an iron and nickel alloy, as a function of its nickel content. But if it is later revealed that the manner in which the casting or cooling of the ingot was carried out, or that admixtures of minute quantities of manganese (which the experimenter was not able to take into account), etc., play an important role in the case, the table will become unusable. It will do no good to claim that one has made one's observations with *all other conditions being equal*, for one will never really be able to observe *all* the conditions.

But perhaps it will be helpful to turn to the science of the past rather than to that of the present, in order to bring to light this especially close relation between experimental science and hypothesis.

According to current thinking, hypothesis takes place after experimentation, is superposed on it, and at any rate is entirely distinct and clearly separable from it. It is also believed that, once this separation has been carried out, what remains is an unshakable scientific result: it is a *fact*, and it is considered something of a truism that facts remain while theories pass. Let us examine a specific example in order to try to see what is actually the case.

Here is a chemistry book signed with a name justly famous among succeeding generations: the *Digressions académiques* of Guyton de Morveau.[31] This book, which precedes the work of Lavoisier, is entirely imbued with phlogiston theory. Phlogiston is "pure elementary fire," the principle of volatility and dilatability, of color and odors and also "the agent of all dissolutions by acids" (pp. 107, 147, 157, 234, 245); the author does not entertain the slightest doubt as to its existence, it appears as a substance as real as any other, and Guyton even manages to draw up a table giving the phlogiston content of the various metals (p. 265). Of course it does not necessarily follow that the work lacks interest, in spite of the very strange form of the author's exposition when judged by present standards; on the contrary, it is easy to see that, precisely because of observations and considerations closely linked to the prevailing theory, this work constitutes an important moment in the progress of science. Guyton de Morveau in fact seeks to demonstrate that one of the elementary characteristics of phlogiston is that it is a *light* body.[32] To this end, he carefully collects all the known evidence on the increase in weight undergone by metals at the moment they are presumed to lose their phlogiston. He checks these data and makes them more precise by a series of experiments done with great exactitude, finally establishing that the phenomenon in question is altogether general and perfectly determined from the quantitative point of view. That was largely new, and it ran counter to the conception then commonly accepted among chemists, who generally treated quantitative considerations as something almost irrelevant from the theoretical standpoint; several years later still, after the first vigorous attacks by Lavoisier, Macquer declared himself to be completely reassured as to the fate of phlogiston, since the only arguments that could be advanced against the theory were quantitative ones (cf. IR 181–182 [Loewenberg 167–168], and Appendix 2, p. 551 below).

Let us now consider a whole series of experiments on Prussian blue described in this book. Guyton carefully examines the solubility of this body in acids, its magnetic properties as well as those of its products, the increase in weight iron undergoes in forming Prussian blue, its detonation with niter, its calcination in a retort, the products that result from its distillation, etc. These investigations serve to prove that (contrary to the claims of Macquer, whom, incidentally, Guyton does not name anywhere in this connection) "phlogiston is not the only substance combined with the metallic earth" in Prussian blue and that it "is not pure there," that is, that "the effect of calcination is not limited to taking away" from the Prussian blue "the *identical* phlogiston" involved in other reactions" (pp. 244–249).

Why does that appear so perfectly otiose today? The reader will kindly recall what we established in Chapter 1 (pp. 13 ff.) on the evolution laws can undergo as a result of the disappearance of the substance with which they are dealing. We have seen (precisely on the subject of elementary fire, of which phlogiston is only the last transformation) that this disappearance is not equally complete in all cases, that it allows of degrees, so to speak, that the disappearance of the substance can leave the *genus* intact, while at other times the genus itself dissolves. In this specific case, the phenomena of oxidation and reduction still constitute a genuine class today; our oxygen is, in short, simply a negative phlogiston, and Guyton's conclusions are thus, to a certain extent, translatable into the present language of science. On the other hand, for Prussian blue there no longer remains any trace of the genus itself: we have difficulty imagining that the formation of this substance, which we know to be a fairly complicated process, was likened to a simple reduction. And although Guyton was right, of course, in asserting that it was not a question here of the actual *identical phlogiston*, not only his demonstration, but his experiments themselves no longer interest us at all; the evolution of this scientific conception has been so profound that the scientific fact itself has disappeared. Not, certainly, in the sense that Guyton's results are shown to be materially inaccurate: Guyton was a careful experimenter and if a chemist today took the trouble to follow the same procedures, he would no doubt confirm the essential part of Guyton's data. But this is exactly what no chemist will care to do, and it is not difficult, after what we have just come to see, to understand why. The scientific fact is not just any observation whatever, it is an observation capable of being generalized, of leading us to formulate a law, for there is no science without law. Now law presupposes *genus*, and since the genus has vanished here, we know in advance that research on this point can lead nowhere, because there is no *fiber* there to be isolated.

This is obviously why it is generally so difficult to read scientific works belonging to the past: it often takes a considerable effort of imagination to succeed in seeing facts in the guise they assumed spontaneously for the observers of that time.

To realize how predisposed man is to delude himself by believing in the existence of a lawlike connection where today we find no trace of one, it suffices to note that there have existed sciences which we are forced to consider entirely fictitious. Judicial astrology, which we discussed above, offers us a case in point, but there are many others. The *alchemical* art properly speaking, that is to say the art of the gold-maker, as well as everything having to do with "occult" practices, certainly appears to the modern scientist to fit into such a category. And thus we see that in reality this past persists into the present, for one need only look around to discover fervent adherents of these doctrines. The case is altogether analogous for alleged sciences such as physiognomy or graphology where, for reasons easy to guess, it seems probable that the facts are connected, although no one has yet succeeded in establishing laws. Indeed, it is certain that in everyday life we all customarily judge people, at least to some extent, by their physiognomy. However, until now all attempts to establish any system of rules have ended only in ludicrous failures. Likewise, the uncertainty of the most elementary graphological verifications, those dealing only with the pure and simple identity of handwriting samples, has become proverbial. Believers in graphology will no doubt claim that there *must* be a connection; but that is not the question – strictly speaking, *all* the phenomena that we know must be interrelated – and what is at issue is to know whether this connection is so close and dominant that condition and phenomenon constitute a whole, a *fiber*, which can be isolated. Now of this we cannot be sure until the isolation has been carried out, that is, until actual rules have successfully been formulated. Thus it would be possible, to use a purely imaginary example, that the form of the human physiognomy depends primarily on the purely physiological activity of the lymphatic system in the embryo and the child, so that the influence of intellectual and moral particularities becomes truly knowable only if we are first able to eliminate what is due to the circulation of the lymph. And similarly, in the case of handwriting, some physiological influence or other of the nervous system or the muscular system could quite well be the most important factor, thus explaining all the failures of present-day attempts.

Indeed, in some cases the work directed toward isolating the *fiber* is so delicate that even the pure and simple verification of the results claimed to have been obtained can give rise to hesitations. In reality, any affirmation of a scientific fact is necessarily based on a calculation of probability, which in turn must be founded on a statistic.

Lavoisier, in criticizing the claims of the mesmerist Deslon, says that

> The art of inferring on the basis of experiments and observations consists in evaluating probabilities and judging whether they are strong enough or numerous enough to constitute proofs. This sort of calculation is more complicated and more difficult than one thinks; it requires great sagacity and is generally beyond the powers of the ordinary man.[33]

This is a truth of which we are not generally aware, because in well-executed scientific works the statistical data are presented as uniformly favorable to the formula to be established and because, consequently, probability appears to achieve certainty. All those who have done any laboratory work at all know that there is an element of fiction involved, that in reality, even in the hands of the most able experimenter, experiments sometimes do not succeed; in such cases, all sorts of explanations are customarily found for these failures and, just as frequently, the failures are not accounted for in the announcement of the results. Most of the time that is of no importance, since the phenomenon with which one is concerned is sufficiently clear-cut and depends on sufficiently simple conditions. However, in some particular cases, errors can result. Thus, to cite only one example, the chemists currently studying what are called *complexes* (cf. Ch. 3, pp. 56 ff., and Ch. 8, p. 228), using research methods science has recently put at their disposal, quite often, I have been told, have the impression that their predecessors described bodies which really do not exist at all.

The true nature of scientific claims becomes even more clear if we leave the terrain of established science and examine a branch of learning, or an alleged one, which, if it is not absolutely rejected by common scientific opinion, is certainly considered extremely dubious. Let us consider, for example, the art of the *diviner*. Can anyone really use a stick or an analogous instrument to indicate the nature of the subsoil, the springs, the minerals, etc., concealed in it? It is clear that in order to prove the affirmative, it would take a certain number of experiments carried out under conditions where the good faith of the operator was amply guaranteed, and which allowed a list of successes and failures to be drawn up, so that it might be established that the number of successes significantly exceeds the proportion that would result from a mere application of the laws of chance. And it is just as clear that until such a demonstration is made, science will be perfectly justified in treating claims fitting into this category as null and void.

In the case of the diviner, we are dealing, on the whole – at least so long as we refrain from appealing to the momentary predisposition, etc., of the principal agent – with a simple phenomenon, where success will be easy to define and where, as a result, a relatively limited number of experiments would suffice. The case becomes more difficult if it is a question of phenomena involving man's voluntary activity or only the functioning of his animal organism. Since success and failure are not very clear-cut, one will generally, with a bit of good will, be able to provide commentary, split hairs, and finally establish the result one was hoping for. That explains why beliefs such as that in the influence of the stars on human destinies could endure for centuries. They were certainly thought to be constantly confirmed by experience, whereas for us the vanity of all this bogus science appears beyond doubt. Thus good will alone, the will to believe – resulting here from the powerful desire to know the future, a desire that motivates every human being – was enough in this case to create the illusion, to make generations of observing scientists see a direct lawlike link where, according to our convictions today, there is not a trace of one.

Lavoisier made this clear in the continuation of the quotation begun above:

> Their errors [the errors of "ordinary men"] in this type of calculation are the basis for the success of the charlatans, the sorcerers, the alchemists, as they were earlier the basis for the successes of magicians, enchanters, and all those in general who delude themselves or seek to take advantage of public credulity. [Meyerson's brackets]

The case is altogether analogous for a good many medical practices – let us say past practices, to avoid awakening legitimate sensitivities. Surely we need only glance at a medical book dating from a not so remote past, for example one on iatrochemistry, in order to be astonished and quite often horrified at the remedies that were commonly prescribed, remedies without any possible effect on the organism, in enormous doses, indeed directly injurious remedies. And yet those who prescribed these remedies must have been convinced of their efficacy; in their eyes this efficacy was obviously an accomplished fact, confirmed by observations that could leave no room for doubt. Will it be said that they did not know how to observe properly? In one sense that is so, since the total result constituted an error, but it is not true in the ordinary sense of the word; Turquet de Mayerne, an iatrochemist and undoubtedly the discoverer of hydrogen, was certainly an uncommonly fine observer (we are thinking above all of his *Pharmacopoeia*).[34] If he went astray, it was not because he had observed badly in this or that particular case, but probably because he gathered only an insufficient number of observations, did not take the unfavorable cases sufficiently into account and therefore, in short, misjudged the probability resulting from the data at his disposal.

That is seen even more clearly, if possible, if we consider medical practices that now appear to be pure superstition. Renan relates that thousands of tablets from the faithful were found in Carthage thanking the goddess Tanit for her intervention in curing them. It would serve no purpose to argue that the priest, in addition to fulfilling religious functions, may have prescribed genuine remedies, for surely in the eyes of the patient, as well as in those of the healer no doubt, the really effective aspect of these prescriptions was that having to do with the action of the goddess. Renan is therefore perfectly right in stating that, in a certain sense, no experiment has been repeated more often. Must we believe in the goddess Tanit, then? Is it not certain, rather, that this is an infinitely more flagrant example of the illusion we saw in Turquet de Mayerne? Of course this illusion rests, just as judicial astrology does, on an aspiration deeply rooted in man's soul: the will to be cured, which motivates the patient and has an effect on his healer. Perhaps one will even discover a few traces of this state of mind – we dare say no more on this delicate subject – in contemporary medicine which, with its remedies extolled one day on the basis of clearly insufficient data and just as suddenly rejected the next, sometimes seems to forget the extreme complexity of the phenomena it is treating and does not seem sufficiently cognizant of the considerations of statistics and probability mentioned above.[35]

Thus all true scientific work implies a choice on the part of the researcher, the exercise of his faculties of imagination, judgment and rational understanding, and the intrinsic value of this work is, on the whole, dependent on the success crowning the exercise of his powers of divination.

What must again be noted in this context is the close connection between the search for a law and the search for a figurative theory. Indeed, how does a research scientist succeed in predicting a relationship of dependency between two series of facts except by implicitly assuming that there exists an internal link between these series? In other words, he has recourse – most of the time unconsciously, it goes without saying – to a hypothesis concerning the means of production. That is a type of reasoning Auguste Comte strictly forbade, as we have seen, although it seems to be indispensable to the very constitution of the laws of which, according to his system, science is composed.

We saw in Chapter 4 (pp. 84 ff.) that when the scientist moves from considerations having to do with relations, with laws, to suppositions concerning being, he obviously does not feel (contrary to what positivism ordains) that he is changing fields, and we have recognized that this is because everything we call science necessarily has an ontological character. But here is another aspect of the same thesis: not only does the physicist really never formulate relations except between relata [*supports*], but in seeking new relations, in reasoning, as he is constrained to do, about the possibility of these relations, he is forced to have recourse to the implications entailed, in the case in point, by the concept of the relatum [*support*]. Here again, therefore, his mental attitude has much in common with that of the scientist who formulates a figurative hypothesis, and we understand why the latter experiences no aversion to this sort of "metaphysics."

We also see the nature of what might be called the internal mechanism of our thought which, as soon as we have observed a constant order of phenomena, immediately makes us suspect that *behind* it lies a logically necessary connection between things (Ch. 3, pp. 50 ff.). At first glance, it looks as though in so doing we are superposing a foreign element upon the observations, and certainly from the purely logical standpoint this element is easy to eliminate: it is precisely this circumstance of which empiricism takes advantage so ingenuously. But actually one is only making manifest something the observations implicitly contained, for the supposition of an internal link must necessarily have preceded the discovery itself in the mind of the person who formulated them.

Despite the enormous difference between a philosopher of the Ionian schools and a contemporary research scientist with respect to both the starting point and the means of investigation used, it is clear that the latter closely resembles the former in that, exactly like him and in spite of all appearances to the contrary, he seeks to *divine* nature through the work of thought. Certainly he resembles him much more strongly than he does a research scientist who is radically empirical in the Baconian sense, who is supposed to study nature by direct procedures, excluding any preconceived idea, that is, any attempt at divination.

The scientist who does research starts from ground that has already been cleared. There he sees, or at least thinks he sees, where there is agreement between reason and reality and where the agreement ends. But in the region where he intends to go, everything is murky, and to guide his steps he has only the obscure glimmer of an insight, a more or less vague theory concerning the internal relatedness of things, resulting from bringing together some circumstances or other previously judged insignificant or too far apart to be connected. What inspires him, then, is above all the idea of analogy, in the broadest sense of the term.

We can further elucidate this concept. In letting himself be guided by analogy, the research scientist assumes that the phenomenon he is studying resembles this or that other phenomenon whose course is known to him or lends itself to being deduced. To take only the most familiar examples, Newton will assume that the component which combines with inertial motion to create the elliptical orbits of the heavenly bodies is the same as the one that causes the phenomena of the fall of heavy bodies on the earth. Fresnel will imagine that the propagation of light takes place by a mechanism resembling that of the motion of waves and from this conception will deduce a prodigious series of consequences that are so well verified by experience that this theory's potential still does not appear to be exhausted today, as we saw in Ch. 15 (p. 434, note 6). Gouy will think that the thermal agitation of molecules must move small visible masses suspended in a liquid, exactly as would occur in the case of bodies of molar dimension, and this idea will be confirmed to such an extent that this motion will be used by Perrin to determine the dimensions of molecules. Similarly, Kekulé, followed by Le Bel, Van't Hoff and Werner, will liken hitherto inexplicable chemical isomerisms to the properties of certain geometrical figures, such as the hexagon, the tetrahedron and the octahedron.

In each of these cases, needless to say, the supposition of an analogy encountered grave objections. For Newtonian gravitation, of course, the identity of the two forces, terrestrial and supraterrestrial, today appears complete. But in Newton's epoch, the time had not been forgotten when no assimilation seemed possible between what was happening on earth and what was observed in the "incorruptible" heavens, and weight was so far from appearing the necessary attribute of all matter that even a century later, on the eve of the triumph of Lavoisier's ideas, the concept of an absolutely light substance having what today would be called negative weight was not considered at all extravagant. Not to mention that the supposition of a force which both conforms to the nature of space in the way it spreads and at the same time ignores, as it were, this same space by the instantaneity of its transmission, was and very certainly remains extremely paradoxical (as is shown by the resistance of physicists contemporary with the discovery and the obstinacy of succeeding generations of physicists in their search for the *cause* of gravitation; cf. IR 77 ff., 180 [Loewenberg 78 ff., 167]).

The situation is even more apparent, if possible, insofar as the other theories are concerned. What could be stranger than to assume, as Fresnel did, that an extremely rare medium, as ether must be, does not vibrate longitudinally, like air, but transversally, like a solid body? And likewise, was it not singularly bold to suppose that one could *see*, directly, the immediate consequences of molecular motion? And is it not fundamentally bizarre to compare the most complex chemical reactions to simple geometrical figures?

By reasoning as they did, these men of science undoubtedly presupposed a conformity between nature and the human mind (as we pointed out on p. 460). But the basis for this presupposition was the belief in a similarity between series of phenomena that certainly appeared very different in nature to their contemporaries. In each case, the *discoverer* mentally dismisses these palpable differences, he more or less consciously forgets them, by making identical that which is diverse. And his discovery will appear all the more praiseworthy if the leap he has accomplished is more prodigious, that is, if what he has brought together hitherto seemed to us farther apart, or, in other words, if the identification is more forced. In this respect, therefore, the situation is exactly the same as the one we saw in the case of mathematical proof (Ch. 5, pp. 111 ff.).

Thus reasoning by analogy is, in its turn, nothing but a continuous effort to apply the schema of the identification of the diverse to the world of things.[36]

And this makes it even more clear that the process of identification is decidedly the process that creates science, since it is exclusively by this path that we try to achieve real understanding, that is, to explain reality to ourselves.

This observation will also help us better understand an aspect of the problem of explanation upon which we touched in Chapter 3 (p. 60). There we pointed out that according to common opinion explanation consists in reducing a less familiar phenomenon to a more familiar one and realized that this theory is untenable, that by introducing caloric, atoms, ether, electricity to explain everyday phenomena of our direct perception, science actually reduces something very familiar to elements that are unquestionably strange and mysterious. However, we must recognize that, in this case as in others of the same sort (cf. Ch. 1, pp. 59–60), our conviction can be complete only if we are in a position to realize how this error – a very general and deep-seated one, to be sure – could have arisen. Now the remark we just made about the role of reasoning by analogy allows us to understand this. Indeed, it is clear that the understanding, as it seeks analogy in its pursuit of rationalization, will first seize those analogies offered by familiar phenomena. It is not surprising, then, that science contains many explanations fitting this pattern, as for example the explanation of sound by undulations of the air or, from a more general standpoint, the reduction of phenomena of all sorts to mechanical phenomena. But we must note that this is not an indispensable condition, that, on the contrary, the dominant factor is solely the concern for rationalization, and that the understanding, if it finds nothing among familiar phenomena which can satisfy it in this respect, will push its search for analogies further; if need be, it will transform a known phenomenon, such as undulation, in such a way as to endow it with altogether strange and even contradictory properties (as Fresnel's theory does), or it will even seek analogies capable of reducing a familiar phenomenon to another infinitely less familiar one (as the electric theory of matter tries to do). These are not anomalies, but rather the normal conditions of the functioning of scientific reason.

This working of scientific reason is obviously of a much higher quality than that of ordinary reason. However, we must point out that the two are not *essentially* different, that they are, on the contrary, of an analogous order. In everyday life we are all obliged to judge, to anticipate or, more precisely, to guess the future; otherwise we could hardly act at all. Of course not all our actions have a determined, immediately assignable goal. On the contrary, most of our everyday gestures are those of habit or imitation: that is how things are done, which is good enough for us. We are like sheep, and with good reason, for we are well aware that we could not analyze the situation completely – since the nature that surrounds us is not completely rational – and we also feel that it would be pointless to try to unravel the tangle of reasonings and facts of experience that motivate a particular gesture: life is too short, and if I wanted to analyze everything in each of my actions, I would probably never manage to put my stockings on in the morning. We are thus quite happy that this work has been carried out by preceding generations and that we need only imitate them.

However, there are circumstances in which simple imitation of what I do every day or of what is ordinarily done in similar situations is not sufficient, obliging me to make my reason take a more active role, to reflect. The work reason accomplishes in such cases is, all things considered, quite analogous to the one we have just attributed to the scientist: following more or less vague and uncertain signs and symptoms, we try to recognize what things are and what course they will take in the future. And of course we can do so only by reasoning through analogy, by applying to the problem before us what we know about the course of events we recognize as similar, that is, by taking as identical that which nevertheless appears diverse to us. Thus the process of identification, as Stanley Jevons perceived, but to an even greater extent than he imagined, appears to be the most general pattern of all our reasoning without exception.[37]

Yet we must grant that this understanding of the process leading up to our reaction toward the external world is not the only one possible. Indeed, it rests on the postulate we have already used above (p. 444), which maintains that the action of our reason when we are not directly conscious of it is similar to that of our conscious reason; it is thus by a continuous chain of incomplete and rapid deductions and inductions that we arrive at the conclusions which dictate our behavior. The reader knows that we lean toward this position. But it is clear, since we are after all dealing with the unconscious, that no direct demonstration is possible, that the parallelism, on the contrary, may not be complete and that, at least in some individuals and in some particular cases, there may intervene an element foreign to conscious reasoning in the narrow sense of the word. This then might be something akin to animal instinct – instinct, of course, as understood by those who consider it a phenomenon *sui generis*, entirely irreducible to more elementary manifestations of organic matter, such as tropisms, etc. – a more direct penetration of the mind into nature, the establishment of an at least partial communion with nature.

To be sure, the idea of such a communion and the aspiration toward it are one of the eternal tendencies of the human spirit. In India, we can see that it has dominated since remotest antiquity, being expressed with infinite energy in certain aspects of Buddhism. But even in the Christian West, where religious conceptions seem rather to turn the mind away from nature, these tendencies appear very clearly in certain mystics, and in particular, with a singular beauty, in Saint Francis of Assisi. It is not surprising, therefore, that psychologists have thought of appealing more or less directly to this resource. The idea is so natural that novelists have found it a popular theme with the general public; among Russian storytellers, for example, the figure of the man who, be he game warden, hunter or simple tramp, has returned to the bosom of nature and finds himself in full communion with it to the point of being able to grasp it intuitively, has become almost a psychological commonplace.

Can an analogous assumption be made about the research scientist? Can we suppose that he too is "in communion with nature," at least under certain exceptional circumstances? Certainly a beginner watching the gestures of an old laboratory hand, seeing him unerringly carry out what seem to be the most complex manipulations and apply the appropriate remedy in each perplexing case with the certainty of success, initially has the impression that this man "is on the inside," that he sees these things from within. And he who writes these lines recalls quite distinctly that Bunsen, for example, who was not only a uniquely gifted laboratory chemist but also a great man of science, irresistibly gave such an impression to all those who saw him work. But one need only clearly picture what a supposition of this kind really implies to realize how inadmissible it is. Animals, trees, fields, and so on, are common sense objects which have always been familiar to man; the idea of entering into communion with them can thus, from a certain point of view, be made plausible to us. But to the scientist such communion would be of little use, for these objects of common sense are the very thing he is obliged to destroy. What would be necessary – at least for the contemporary scientist – is communion with electrical vibrations, atoms and electrons. But surely our imagination will refuse to go that far.[38] Thus we have no alternative, insofar as the scientist is concerned, but to renounce this idea of an *intuition* of things, to stop at the more down-to-earth conception of veritable unconscious reasonings made with a depth of understanding beyond the capacity of ordinary mortals.[39]

As soon as the scientist believes he has succeeded, that is, when he judges that a more or less coherent hypothesis he has formulated has been sufficiently corroborated by the facts, he tries to give to the result that he has attained and that he is reporting a form as convincing as possible for intelligences other than his own. He is more or less successful in the attempt, and certain minds particularly gifted along these lines have sometimes attained this goal remarkably well. But it would be a terrible mistake to believe that just because the scientist has succeeded in this final phase of his work, which is its crowning achievement, just because his discovery appears clear in the form in which he reports it, his intellect must have worked with complete clarity at all times in order to arrive at this point. That would be to misunderstand the essence of scientific endeavor and also to abase the merit of the great innovators. As a matter of fact, the clarity that dazzles us is only the fruit of their arduous efforts; they have proceeded by obscure paths, paths they themselves know only partially, so considerable a role did the unconscious play in their work. "I pity people who have only clear ideas," said a great scientist, surely one of the greatest of the entire nineteenth century, which was so rich in them. Was it one of those thinkers sometimes called metaphysicians of science, one of those minds whose conceptions, laden with results, have been expressed in an obscure, vague or diffuse form, so that their fecundity could be recognized only slowly? Is it Lamarck, J. R. Mayer or Willard Gibbs? No, it is a man who possessed to the highest degree precisely this supreme gift of genius (a peculiarly French genius, we might add) of diffusing everything he touched with clarity: it is Pasteur.

Even the great research scientist, when he tries to understand the considerations that dictate the path he is following, is sometimes more or less astonished at his own audacity. For instance, "Henri Sainte-Claire-Deville at times refused to answer when questioned about what he was doing, '*because he was working*,' he said, '*in the realm of the absurd*,'" and it is significant that in reporting this trait, Costantin, himself the author of admirable discoveries, does not judge the declaration in any way extravagant, but rather considers it strictly applicable to what has happened in his own field of botany. An advocate of the role of intuition in scientific discoveries could no doubt base an argument on this remark; but we have spoken above of the difficulties we feel any supposition of this sort comes up against. Let us merely note that Costantin does not appeal to intuition even in the case of botany, although, as we have pointed out, the notion would have been somewhat less paradoxical there than in chemistry, because it would be applied to concepts less remote from those of common sense. He is content to note that "the intelligence is a very mysterious force, and the mind that scrutinizes the unknown can be guided by a simple flash of lightning illuminating a total darkness,"[40] which is altogether consistent with the position that what we are dealing with is a judgment by analogy, a rapid deduction whose intermediate phases have remained locked in the unconsciousness of the research scientist.

Empiricism certainly tends to disregard this very characteristic and very essential side of the work of scientific discovery, and it is no doubt because he was inspired mainly by the Baconian view on this point that Condillac came to state explicitly that the order of the things in which truth must be set forth is the order in which it was found, given that "the best way of instructing others is to lead them along the path one had to follow in order to instruct oneself."[41] This is a manifest untruth, but one which, it is easy to see, nevertheless constitutes the implicit basis for many judgments in the history of science. Such a state of mind makes us disparage as erroneous, "mystical," anything in the thinking of the past that does not seem absolutely clear and consistent with the would-be ideal of today – for it goes without saying that it is chiefly the ideas with which we ourselves are imbued that appear clear to us. In the history of chemistry, the phlogiston theorists in particular were victims of judgments of this sort, judgments that were supremely unfair, because phlogiston theory, in spite of the fact that it contained some confusion, like any other theory, was, as we have seen, an excellent scientific theory (cf. Ch. 3, p. 62, and Appendix 2).

Nor is the effort the scientist makes when he reports his discovery without parallel in common sense thought. When we announce our decision to others and more or less directly seek their approval, or when we only want to justify it in our own eyes, we feel the need to make this decision appear to be the fruit of a conscious act of reasoning, the need to indicate precise reasons to others and to ourselves. And it is quite possible, just as for the scientist, that the path we claim to attribute to our understanding may be very different from the one it really followed.

The reader will surely recall, in this context, what Pascal said about the reasons we find for our decisions (Ch. 14, p. 370).

"People who condemn the use of hypotheses and of preconceived ideas in the experimental method," says Claude Bernard, "make the mistake of confusing invention of an experiment with noting its results" (*Méd. expér.* 40 [Greene 24]). That is a very important observation. As a matter of fact, the Baconian ideas enjoy enormous prestige; a great many good minds have adopted them in the past and in the present. Consequently, as we pointed out above (p. 475), we must ask ourselves where the source of the error lies. Claude Bernard makes it clear: it comes from the disparity that naturally tends to be established between the way in which the scientist states his conclusions and the way in which he actually arrived at them. Indeed, in reporting his discovery, he necessarily tries to convince others.

Therefore, whether he is presenting a new phenomenon or a law embracing a certain number of phenomena already known previously, he will above all take pains to provide the data upon which he claims to base his affirmation. On the other hand, it is usually not to his advantage to report the more or less confused and obscure deductions, deductions sometimes so rapid they resemble a blinding and fleeting flash of lightning, which led him to seek or observe the relationship established. First because, since these reasonings were largely played out before the unconscious side of his intelligence, he would have to make a great and unaccustomed effort in order to recover them, and then because, since the way these rapid and provisional reasonings are carried out is obviously subjective, any account of them is by nature unconvincing. Of course if the scientist intends to use his discovery to bolster one of his own theories, he will, on the contrary, do everything he can to make it clear how the theory led him to the results he has formulated: the entire work of Lavoisier is an inimitable model of this. And likewise the works of his disciples, of all that glorious "French chemistry" of the late eighteenth and early nineteenth centuries, admirably illustrate how an immense series of discoveries of the greatest importance can be grafted onto the theoretical work of a genius. Into the same category – to stay within the field of chemistry – fall the innumerable works arising from the theory of types or from the theories of atomic bonding and asymmetrical isomerism: for example, Emil Fischer has explicitly acknowledged having been guided by the last of these conceptions in his work. In cases of this kind, the filiation and the process of mental deduction that engendered the idea of the discovery seem clear. But that is largely only an illusion. If we really knew how Lavoisier's mind worked, we would perhaps be astonished and baffled by the bizarre nature of the deductions that finally led him to set aside entirely the qualitative reasonings which seemed to form the unshakable basis for the chemistry of that time. Similarly, even if his disciples did no more than apply his ideas, roughly speaking, and even if their successors seemed in their turn to develop consequences of previously established theories, it nevertheless required a powerful scientific gift to recognize where and how, in the infinite and confused tangle of facts, the theory, which can be applied to only a limited number of phenomena, would manifest its influence – otherwise any laboratory worker at all would be the equal of the Gay-Lussacs, the Thenards and the Humphry Davys, as he would of the Friedels and the Fischers. What these great men of science needed, just as did the still greater creators of fundamental conceptions, it goes without saying, was *divination*, and for this the secret could not be handed down.

Only in very rare cases can we at least partially penetrate into the psychology of the research scientist, guided either by the revelations of the scientist himself or by the way in which the whole of his work is presented. In Chapter 4 (p. 80) we cited the works of Kepler to illustrate this point, and in recalling that quick résumé, the reader will realize how much there was that was individual and immutable in the procedures of this great man. No doubt Bailly was completely wrong in calling this way of forcing nature's secrets *folly* – since, as a matter of fact, Kepler succeeded magnificently – but the historian was only expressing the extreme astonishment experienced by any unbiased reader facing certain of these extravagances. What is certain is that one has the distinct impression that it took the *flair* of Kepler's genius, applied to an exceptionally propitious field of observation, to succeed under these conditions. In other words, from what Kepler himself has reported to us about the origin of his ideas it is impossible to extract any support whatsoever for the conclusions he reached, since these conclusions, that is to say the existence of the laws he formulated, are for us based solely on the agreement of the data calculated by means of these laws with the observations (leaving aside, of course, their later confirmation by Newtonian theory).

This typical example clearly shows us that the format it behooves a scientist to adopt in presenting his discovery most often has only the remotest resemblance to the paths his reason actually followed in the course of the work that led up to it.[42] And it goes without saying that those who will later set forth the discovery will adhere, even more strictly than the author himself, to the original published version alone. In a textbook chapter treating Kepler's laws, it would generally be futile to search for an account of his attempts starting from the consideration of "trigones" and ending with the introduction of the laws of acoustical harmony into astronomy. Consequently, the exposition seems to be closely adapted to Bacon's procedure, since we actually see only the factual elements, from which the formula appears to arise spontaneously, everything that might indicate the necessity of a choice, the intervention of the spirit of divination, being completely brushed aside. Thus the Baconian error in fact arises, as Claude Bernard saw, from a confusion between the path the scientist follows when he is in the process of trying to advance science and the way in which he presents his discovery once it has been made. Bacon, as Claude Bernard also pointed out, was not a scientist in the usual sense of the term. To be sure, we are told that he did do experiments (he is said to have died of an illness contracted in experimenting with snow in winter), but he does not seem to have been really successful in any of them, that is to have succeeded in advancing science in any way at all. Thus he knew science primarily from the outside, and under these conditions it is only natural that he could be mistaken, taking for the essence of science what was only its outer covering. We must add that, as we have noted, his doctrine contained much that was accurate and above all much that was salutary, not only from the perspective of the particular state of science in his time, but also with respect to the paths of scientific research in general (Ch. 14, p. 378). Nevertheless, one might be astonished that this theory could have survived so long, given the palpable error of the program it intended to impose. But we must reckon with the spirit of inertia, of which humanity has given so many proofs in the course of its intellectual development and which makes a theoretical conception tend to persist once it has been established: that is a manifestation of the sheep-like tendency mentioned above (p. 475), a tendency justified, moreover, by the deep-seated feeling in all of us that things are not entirely rational. We should add that, as we explained in Chapter 3 (p. 64), our reason, in spite of this conviction, imperiously commands us to search for theories and is so eager for them that it generally abandons them only if provided with better ones. Now in the case at hand, the Baconian theory had the advantage not only of being simple and clear, but especially of being *complete*, while what was offered in opposition to it generally consisted of only partial insights. Finally, as we have also indicated, the scientist is often ill-suited to the task of seeking the intellectual foundations of science; it is thus only natural that he should be inclined to accept what is brought to him from outside. However, as we have seen, there has been no lack of distinguished minds who, when faced with the manifest inadequacy of the Baconian position, have vigorously protested and eloquently insisted on the rights of the true scientific spirit, the inventive spirit, necessarily proceeding by obscure paths.

But if we must not scorn obscure ideas, still less must we forget the value of clarity. It is and remains the supreme criterion of scientific work, the necessary capstone of any true discovery. It seems strange that anyone could ever have failed to recognize this, and yet the error is quite frequent, even more frequent (and certainly more damaging) than the opposite error: some even come to be persuaded that clarity – the sign of an essentially superior genius – can only be a lack of depth. In the history of chemistry it is Lavoisier in particular who fell victim to this viewpoint. We believe we can safely predict, however, that the glory of this greatest of all men will not suffer too much thereby, for Ostwald, in denying the merit of the creator of modern chemistry, runs too far counter to the unanimous sentiment of chemists of the most diverse epochs, including truly well-qualified German historians of chemistry, such as Hermann Kopp or Ladenburg. Cuvier's opinion, which we have cited [p. 64 above], is what corresponds best to their judgment, and it will undoubtedly be ratified by those who come after us.

"Obscure ideas" are indispensable for the scientist. For nature is not entirely rational, and whatever progress our knowledge may make, what we know will never be more than a tiny island in the ocean of the unknown, a luminous spot surrounded on all sides by deep shadow. The scientist penetrates boldly into the semi-darkness that marks its boundaries, but he does so in order to enlarge the domain of the illuminated ground. No doubt we must join Pasteur in pitying the man who has only clear ideas, for he will surely not do much to advance science. But that does not keep us from maintaining firmly that in each specific case clear ideas are more valuable than obscure ones, the latter being valuable, strictly speaking, only insofar as they lead to the former, which are in fact the true achievements of science. For if nature is not rational, human reason is, and it is this that is the source of our dignity, setting us apart from the rest of creation. "All man's dignity," says Pascal, "consists in thought," in this thought that comprehends the universe (*Pensées* 189 [Krailsheimer 258]), or at least tries to comprehend it. And we can verify full well that, whatever has been said, the same is true for the acts of our everyday life. Indeed, each of us feels very deeply the irresistible need to make his behavior appear consistent with the precepts of reason, whether in the eyes of others or in his own eyes, and we always experience satisfaction at having more or less succeeded. This is because, even when we know dimly that our motives were quite different, we feel quite strongly that it would have been better, more worthy of our human character, if they had been like those which reason would dictate, that this is what they ought to have been like. Therefore it is useless to try to stop human reason in this inclination, to try to persuade it that it is better to persist in the semi-darkness given by the incomplete, intuitive knowledge of things than to enter into complete and clear knowledge. "There is nothing worse for the practical man," says Benedetto Croce,[43] "than to exchange this direct and lively intuition for the truncated and abstract bits of knowledge of the physicist or the mathematician." It is indeed certain that the man called upon to act, to make decisions, must be able to combine the available pieces of information quickly, even if they are incomplete and unclear. But it does not follow that he will disdain more precise information, whatever may be its "abstract" source, that he will reject it because it would interfere with his "intuition." As examples Croce cites Napoleon, who said he won his battles by going against the rules, and the good doctor with "a clinical eye," whose diagnosis is worth more than those of pure theoreticians. But the case of Napoleon proves merely that those who judged in this way were mistaken, believing in the existence of a veritable science where there was nothing of the sort; they believed on insufficient evidence, but their mistake no more bears witness against the *practical* utility of theoretical knowledge than the fact that we now reject judicial astrology as a false science demonstrates the nonexistence of astronomy. Furthermore, Napoleon was a mathematician and in general had the highest admiration for the sciences, as everyone knows. If a new means of investigation (such as the airplane, for example) had been invented in his lifetime, he would undoubtedly have eagerly availed himself of it and taken the information thus acquired very seriously into consideration in his "intuitions." The same is true for the clinician, who obviously is often obliged to use the most indirect methods to guess what is happening within the body. But offer him a direct and sure method of knowing the truth and see if he turns up his nose at it. A modern specialist in syphilis who neglected the Wassermann test and relied solely on his "clinical eye" would rightfully be strongly criticized by his peers, as would a surgeon who refused to have an x-ray picture taken where it was indicated. In the past there have of course been some "practitioners" who balked at new methods of investigation, who would not hear talk of auscultation or of microbiological examination, but practice itself has ignored their protests. Moreover, can this vaunted *practice*, in cases like those cited by Croce, be anything other than a reduction of already acquired knowledge to rapid rules? For, let there be no mistake about it, it cannot be a question of direct communion with *nature*. When Napoleon put together information that came to him from the most diverse sources, he was reasoning about them in the light of what he had been taught by experience and also (Croce notwithstanding) by military literature – he is said to have read much Guibert. His genius combined these data of *knowledge* more rapidly and more accurately than his adversaries could. But to allow that he was in communion with the external world, that is, to speak more precisely, with the souls of the enemy leaders, directly intuiting their plans, would seem to us an extremely risky stance – almost as hard to allow as that of the communion with atoms and electrons mentioned above (p. 477). The same observation also applies to the clinician and, in general, to any activity more or less touching on science. What is more, since there is no scientific knowledge without theory, those who proudly say they are exclusively "practitioners" and claim to disdain all theory, are more often than not quite simply the slaves of an earlier theory, as we have pointed out. Furthermore, we saw (p. 459) how much everything in *practical* life ultimately depends on the progress of theory.

What we have just come to see about the true nature of scientific endeavor makes us understand why – contrary to the claims of current theory – there are so few truly empirical formulas in science. That is because the scientific enterprise is really above all the fruit of reasoning and because this reasoning proceeds in its own particular way, in which the aprioristic demands of our understanding intervene unceasingly and most efficaciously. The "results" of science, the most general propositions at which it arrives, are thus most often saturated with the *a priori*. For the *true* result of science is not a result in the philosophic and absolute sense of the term; to detach it completely from the observations of fact that inspired it is to distort its meaning. Indeed, any experimental statement includes an implicit reservation: it is true only with respect to the present state of our knowledge, within the limits of the observations and as a function of the margins of error of the instruments used for these observations. It is also, we know, liable to be modified or discarded at any moment as a result of new data. To deny that is to say that we have succeeded here in knowing the actual law of things, and we have seen how foreign such a claim is to the true spirit of modern science.

In order to throw more light upon the considerations developed in this connection in our first chapter (pp. 11 ff.), let us add an example to those we cited there. The law of universal gravitation as formulated by Newton had received thousands and thousands of confirmations since its inception; each astronomical calculation, each successful prediction, constituted such a confirmation, and although a few anomalies had been noted (the most important being that of the motion of the planet Mercury, which we mentioned in Chapter 15, p. 434, n. 1), not too many years ago there was probably, all things considered, no proposition in the entire domain of science more abundantly supported by experimental proofs. However, at the very same time the majority of astronomers would surely have hesitated to declare it unmodifiable; on the contrary, they found it plausible to attribute gravitation to the action of some sort of mechanism and thus to consider the apparent simplicity of the law as due to the great number of particles involved, that is to say, as statistical[44] or simply approximate. Furthermore, we know that Einstein's theory actually modifies the statement of the law to some extent, but it is important to note that even before the appearance of this theory astronomers did not really consider the law of gravitation immutable.

This example allows us to clarify another aspect of the mental operation that we perform in affirming the absolute permanence of the law. Most certainly the conviction that gravitation exists in itself, that it is an innate property of matter, must stem from the spatial image discussed in Chapter 15 (p. 424). It must therefore be founded on a purely *a priori* deduction, invoking a primordial property of space. Thus in *generalizing* an apparently experimental datum, we must actually be leaving the framework of experience and simply developing the *a priori* component of the law. That is why philosophic systems claiming to be drawn from science prove to be merely aspects of its aprioristic foundations. Obviously this is the case for mechanism and also for energeticism, both of which are only attempts to reduce changing phenomena to the displacement of an immutable element. In the same way, when Spencer believes he has deduced from science his conception of half-periods of destruction and reconstruction, all he has done is isolate an *a priori* idea in fact contained in some scientific reasoning, one which was, moreover, none other than Heraclitus's old idea of the *Great Year*.[45]

Needless to say, that does not mean that every attempt of this sort is doomed to remain fundamentally sterile from the standpoint of scientific progress. It is clear, on the contrary, that in the past many a precious conquest of science was the fruit of such philosophic speculations. For example, the principle of the conservation of the weight of matter, as it appears in *De rerum natura*, is certainly the direct result of a mechanistic theory of the world, and, similarly, Leibniz's principle of the conservation of *vis viva* is derived from an analogous image of reality; for J. B. Mayer, on the other hand, the principle of the conservation of energy visibly derives from a conception analogous to that of the philosophers of nature. Similarly again, the great achievements of Cartesian physics and the immense impetus it gave to research are incontestably the fruit of an incomparably far-reaching process of generalization. But perhaps, without at all pretending to bar the great speculative minds of the future from such enterprises, we may be permitted to insist upon the difficulties that obviously increase as science advances, not only because it is a matter of bringing together into a single synthesis a vast quantity of data proliferating at an incredible rate, but especially because observations connected with the irrational, as they multiply, make it harder and harder to constitute any image of scientific reality that is truly rational, if only to some degree. We have seen, furthermore, that even if we were to be deprived of these attempts in the future, science would still not lack incentives stemming from the *a priori*; the very work of the scientist constantly obeys it, since he could not for a moment lose sight of the fact that what he is seeking is to understand nature, that is, to establish a *modus vivendi* between our reason and our sensations.

NOTES

1. Cf. IR, Ch. 11, where we have developed this subject at greater length.
2. John Stuart Mill, *La Philosophie de Hamilton*, trans. E. Cazelles (Paris: Germer Baillière, 1869), p. 216 [*An Examination of Sir William Hamilton's Philosophy* (Boston: William V. Spencer, 1865), 1:238: "sensations joined together in groups"]. Ernst Mach, *Die ökonomische Natur des physikalischen Forschung*, Imperial Academy of Sciences of Vienna (Vienna: Karl Gerold's Sohn, 1882), p. 307 ['On the Economical Nature of Physical Inquiry,' *Popular Scientific Lectures*, trans. Thomas J. McCormack (La Salle, Illinois: Open Court, 1943), pp. 200–201: "groups of sensations"].
3. Henri Poincaré, *La Valeur de la science* (Paris: Ernest Flammarion, 1906), p. 270 [*The Value of Science*, trans. George Bruce Halsted (New York: Dover, 1958), p. 139].
4. Herbert Spencer, *First Principles* (London: Williams and Norgate, 1863), pp. 241–242. See Ch. 17, p. 503, for a more complete quotation.
5. William James, *Précis de psychologie*, trans. E. Baudin and G. Bertier (Paris: Marcel Rivière, 1909), p. 6 [*Psychology, Briefer Course* (New York: Henry Holt, 1892), p. 4].
6. Désiré Auguste Roustan, 'La science comme instrument vital,' *Rev. de méta.* **22** (1914) 614 ff.
7. Sophie Germain, *Considérations générales sur l'état des sciences et des lettres aux différentes époques de leur culture*, *Oeuvres philosophiques de Sophie Germain* (Paris: Paul Ritti, 1879), p. 126 [Meyerson's brackets].
8. Antoine D. Sertillanges, *Saint Thomas d'Aquin* (Paris: Félix Alcan, 1910), 1:41.
9. Auguste Comte, *Cours* 2:445, 3:152 [Martineau 291, 233]; *Système de politique positive* (Paris: L. Mathias, 1851), 1:528 [*System of Positive Polity*, trans. John Henry Bridges (New York: Burt Franklin, 1968), 1:427].

10. Federigo Enriques, 'Le principe de la raison suffisante dans la construction scientifique,' *Scientia* **5** (1909) Supplément: 9.
11. 'Discours préliminaire des éditeurs,' *Encyclopédie* (Paris: Gauthier-Villars, 1751), 1:viii [*Preliminary Discourse to the Encyclopedia of Diderot*, trans. Richard N. Schwab (Indianapolis: Bobbs-Merrill, 1963), p. 27; Meyerson's brackets].
12. Wilhelm Ostwald, in characterizing Faraday's scientific career, correctly notes that the great physicist did not succeed in any of his many attempts to develop immediately applicable inventions, although his abstract discoveries led indirectly to incalculable practical progress (*Grosse Männer*, 2nd ed., Leipzig: Akademische Verlagsgesellschaft m. b. H., 1910, p. 115). The case of Liebig is no less typical. All his life Liebig sought to earn money by cashing in on his discoveries. To this end he took out a whole series of patents for the manufacture of artificial fertilizers, silvered mirrors, etc., which brought him only disappointments. However, a simple off-hand remark in a work on the composition of meat gave an engineer the idea of creating a meat-extract factory in the Argentine Republic, and since Liebig agreed to lend his name to the enterprise, it yielded him considerable profits, from which his heirs apparently still benefit today (*Grosse Männer* 195, 200, 206 ff.).
13. Bacon, *Novum Organon* I, 61 [trans. James Spedding, Robert Leslie Ellis and Douglas Denon Heath]; cf. also I, 122 and *Redargutio philosophiarum* (*Works*, ed. Basil Montagu, London: William Pickering, 1829), 11:457–458.
14. Claude Louis Berthollet, *Essai de statique chimique* (Paris: F. Didot, 1803), 1:5.
15. 'Davy,' *Encyclopedia Britannica*, 9th ed. (London, 1909), 6:847.
16. Justus von Liebig, *Reden und Abhandlungen* (Leipzig: C. F. Winter, 1879), p. 249. Moreover, this turn of mind is not one peculiar to the chemists of that epoch. Already in 1876 Guyton de Morveau wrote: "I realize there are a host of people who never stop shouting, 'Facts! Facts! The time is not yet ripe to concern ourselves with theories'; this language conforms neither to reason nor to the opinion of those men of whom our century is proud," namely "the Buffons, the Franklins, the Macquers, the Bergmans, the Priestleys, the Lavoisiers, etc." Guyton provides quotations from each of these scientists; Macquer's opinion is particularly apt: "reasoning is the physicist's organ of sight, as it were .... Experiment unguided by theory is always a blind groping" (*Enc. méth.* 1:iii, note).
17. *Essais* (Paris: Flammarion, 1908), 4:187 [*The Complete Essays of Montaigne*, trans. Donald M. Frame (Stanford, California, 1958), p. 815].
18. *Leçon d'ouverture du cours de M. Claude Bernard au Collège de France* (Paris: Union Médicale, 1857), p. 5. *Méd. expér.* 39–40 [Greene 23]; cf. also p. 53 [Greene 32]: "without it [a preconceived idea], ... we could only pile up sterile observations."
19. *Méd. expér.* 44, 41 [Greene 26, 24]. Cf. also *Leçons de physiologie expérimentale appliquée a la médecine* (Paris: J. B. Baillière, 1855), 1:21, 32–33, on the role played by presuppositions in his own discoveries.
20. Frederick Soddy, 'The Parent of Radium,' *Scientia* **5** (1909) 269 [reading "thorium" for Meyerson's "rhodium" to conform to Soddy's text].

21. Henry Le Châtelier, *Leçons sur le carbone, la combustion, les lois chimiques* (Paris: Dunod et Pinat, 1908), p. 49.
22. Tychonis Brahei, *De disciplinis mathematicis oratio* (Hafniae [Copenhagen]: apud Henricum Waldkirchium, 1610), passim, esp. pp. 15 ff.
23. Jean Sylvain Bailly, *Histoire de l'astronomie moderne*, new ed. (Paris: de Bure, 1785), 2:33 ff.
24. 'Astrologie,' *Encyclopédie*, ed. Diderot and d'Alembert (Paris: Gauthier-Villars, 1751–1765), 1:783.
25. The reader interested in acquainting himself with the literature in this field need only skim through the conscientious articles J. Brieu devoted to it in the *Mercure de France*. Frequently to be found there are reviews of books of judicial astrology conceived completely on the ancient model.
26. Along the same lines we could also cite the verb *consider*, the exclamation *à la bonne heure* ["good for you!"; literally, "at the right hour"], the frequent use of the word *star* as a synonym for *destiny*, as for example in the phrase *his star grew dim*. Even if, as is generally believed today, the last syllable of the word *bonheur* derives from the Latin *augurium*, it nonetheless remains clear that the *augury* in question was related especially to astrological conceptions. [We have substituted "to be moon-struck" above for Meyerson's example, *être bien (ou mal) luné*, meaning "to be in a good (or bad) mood," derived from the French word for "moon."]
27. Jean-Sylvain Bailly, *Histoire de l'astronomie ancienne*, 2nd ed. (Paris: De Bure fils ainé, 1781), p. 261.
28. Bailly, whom we have seen to be resolutely hostile toward judicial astrology, nevertheless believes not only in the influence of the sun and the moon on the winds, but also in the meteorological influence of the planets, "which will take the work of centuries to separate out" (*Histoire de l'astronomie moderne*, Paris: de Bure, 1785, 1:434 ff.).
29. Tychonis Brahei, *De disciplinis mathematicis oratio* (Hafniae [Copenhagen]: apud Henricum Waldkirchium, 1610), p. 15. See Appendix 20.
30. Henri Bouasse, 'Développement historique des théories de la physique,' *Scientia* 7 (1910) 298.
31. Louis-Bernard Guyton de Morveau, *Digressions académiques* (Dijon: L. N. Frantin, 1772). The title page reads 1762, but, not to mention other circumstances, the date of the "imprimatur" signed by Macquer is proof enough that this is a printer's error.
32. Guyton's thinking on this subject is nevertheless completely bizarre. Where he goes closely into the question (as, for example on pp. 136 ff. of his book), he always speaks of phlogiston's lightness in comparison with air and quite logically, in order to show how adding it to a body makes the body less heavy, he cites the example of a lead cube which floats in water when attached to a piece of cork (pp. 175 ff.). Obviously in the latter case the decrease in weight could only be a consequence of an increase in volume. Nevertheless, Guyton not only does not take the trouble to seek the specific weight of the different substances with which he is concerned but, given the increases in weight he observes, his hypothesis necessarily supposes an air density vastly superior to what it actually

is, although this constant is not entirely unknown in his time. More than a century earlier, Otto von Guericke had attempted a first approximation (cf. Ferdinand Rosenberger, *Geschichte der Physik*, Brunswick: F. Vieweg, 1884, 2:149), and the data contained in Stephen Hales' book (*La statique des végétaux, et l'analyse de l'air*, trans. Buffon [from the 1727 edition of *Vegetable Staticks*], Paris: J. Vincent, 1735, p. 150 [*Vegetable Staticks*, London: Oldbourne, 1961, p. 96]), quite well-known at the time and which Guyton cites on several occasions, ought to have alerted him as to how far he was going astray. But, as a matter of fact, most of the time he speaks as if phlogiston were *absolutely* light, that is, as if, in being added to a body, it were capable of counterbalancing any weight whatsoever.

33. Édouard Grimaux, *Lavoisier* (Paris: Félix Alcan, 1902), p. 132.
34. Cf. our work on 'Théodore Turquet de Mayerne et la découverte de l'hydrogène,' *Revue Scientifique*, 3rd Series, **16** (1888), esp. p. 670.
35. Here again one need only continue reading Lavoisier's exposition: "It is particularly in medicine that there is the greatest difficulty in evaluating probabilities. Since the life principle in animals is an ever active force continually tending to overcome obstacles, nature, left to its own resources, cures a a great many ills; when one uses remedies, it is extremely difficult to determine what is due to nature and what is due to the remedy. Thus, while most people regard the curing of an illness as a proof of the efficacy of the remedy, the wise man sees in it only a greater or lesser degree of probability, and this probability can be converted into certainty only by a great many facts of the same sort" (Grimaux, *Lavoisier*, Paris: Félix Alcan, 1902, p. 133).
36. Stanley Jevons saw this situation clearly. In speaking of his principle of the *substitution of similars*, which, as we have said (Ch. 5, p. 114), is much like our conception of the process of identification, he says: "When we are certain there is an exact likeness, our inference is certain; when we only believe that there probably is, or guess that there is, then our inferences are only probable, not certain" (*Logic*, Science Primers, New York: American Book Company, n.d., pp. 75–76). He sets forth the mechanism of analogical reasoning by stating that "when we wish to explain the occurrence of anything, we should begin by thinking of everything like it that we have ever seen or heard of" (*Logic* 95). See also his *The Principles of Science* (London: Macmillan, 1892), p. 627.
37. For Stanley Jevons, see the preceding note. Earlier Condillac had declared: "The whole art of reasoning is reduced to analogy, as is the whole art of speaking; and in this single word we see how to learn about the discoveries of others and how to make our own discoveries" (*La Langue des calculs*, Paris: Ch. Houel, Year 6 [of the First French Republic], p. 7; cf. *La Grammaire*, *Oeuvres*, Paris: Ch. Houel, Year 6, 5:20).
38. We must admit, however, that this idea of knowing nature through direct understanding has frequently haunted philosophers. "Nature," says Schelling, "does not know through the intermediary of the sciences, but through its very essence, magically. The time will come when the sciences will more and more come to an end and immediate knowledge will establish itself. The sciences as such were invented only because of the lack of immediate knowledge; for

example, the entire edifice of astronomical calculations [was invented only] because it was not given to man to perceive the necessity of the motions directly as such, that is, to live spiritually in the real life of the All. There have been individuals who had no need of the sciences [and there will be others]. Nature is seen in them and they themselves, in what they see, have become nature. They are the true prophets, the true empiricists, and those who are currently honored with this name compare to them as village politicians compare to prophets sent by God" (*Aus den Jahrbüchern*, I, 7:246 [Meyerson's brackets]).

39. Claude Bernard, with his deep understanding of the way the research scientist's mind works (on this subject, cf. above, pp. 460 ff.), notes with reference to his work on curare that "in cases where we make an experiment in which both preconceived idea and reasoning seem completely lacking, we yet necessarily reason by syllogism without knowing it" (*Méd. expér.* 253 [Greene 158]).
40. Julien Costantin, *La Vie des orchidées* (Paris: Ernest Flammarion, 1917), p. 169.
41. Condillac, *De l'Art de penser*, *Oeuvres* (Paris: Ch. Houel, Year 6), 6:251–252.
42. Helmholtz, in a speech given on the occasion of his seventieth birthday, seeking to analyze his own working method, explicitly insists on the strange character of the reasoning processes by which he arrived at his discoveries and on the fact that clear and simple deduction or, as he calls it, the "royal road," was generally revealed to him only after the fact, that is, after he had obtained the result. He adds that he refrained in his publications from informing the reader of the tortuous routes he actually followed, merely pointing out the straightened path by which one could reach the goal without too much trouble (*Vorträge und Reden*, 4th ed., Brunswick: Friedrich Vieweg und Sohn, 1896, 1:14–15).
43. Benedetto Croce, *Lineamenti di una logica come scienza del concetto puro* (Naples: Francesco Giannini & Figli, 1905), p. 82. However, it would be unfair not to point out that these views are linked, in Croce, to a whole body of quite idiosyncratic opinions concerning the sciences. According to him, current philosophic theories, even those of Hegel, are sullied by a "prepossession ... in favour of the exact sciences" (*Ce qui est vivant* 124 [Ainslie 153]; see also p. 138 [169]). Now the only true knowledge is that provided by art, philosophy and history, the "so-called sciences" being only "impure, improper, erroneous, irrational knowledge." Even mathematics does not find favor in the eyes of this philosopher; mathematics is only the *simia philosophiae* in the sense in which the devil was called *simia Dei* in the Middle Ages (*Lineamenti* 62, 80).
44. Cf. Henri Poincaré, *La Science et l'hypothèse* (Paris: Flammarion, n.d.), p. 176 [*Science and Hypothesis*, trans. George Bruce Halsted (New York, 1905), p. 106], and *Thermodynamique* (Paris: George Carré, 1892), p. vii.
45. Cf. Ch. 6, pp. 155 ff., Appendix 4, and IR 305–306 [Loewenberg 271–272].

CHAPTER 17

THE EPISTEMOLOGICAL PARADOX

If we now try to embrace the whole of the theory of science as we have sought to establish it, one question will inevitably occur to us. In itself this theory is quite simple and the reader has been able to judge how closely it fits all the characteristics of science today, as well as the stages through which science has passed in the course of its evolution. How does it happen, then, that this theory has arisen so late and still today has so much difficulty gaining acceptance, to the point that the positivistic conception of science, despite the fact that it is so narrow and explains so badly what truly constitutes the essence of scientific research, is most often put forward in such a way as to make it seem the only theory possible?

In order to resolve this enigma, we must undertake a closer and more detailed study of the historical development of these ideas, a development we have thus far only touched upon in passing.

From the dawn of Greek thought, Ionian philosophy identifies the problem of becoming as the central problem of any philosophic conception of the universe. The Eleatics (as we said in Ch. 5, p. 102) take it for granted that a strictly rational conception is incompatible with any diversity whatsoever, a position that culminates, in Parmenides, with the grandiose conception of the spherical universe, indistinct and immutable in time and in space. Heraclitus takes the opposite view, seeking to establish that it is impossible to admit anything permanent in nature at all, that nature is a perpetual becoming, that "all things are in flux" (IR 540–541 [Loewenberg 466–467]). But whatever the repercussions of this criticism, which would seem to lead, in his direct successor Cratylus, to a sort of radical skepticism, denying the possibility of any knowledge at all (Cratylus finds inadequate his master's formula that "one does not step into the same river twice"; according to him, one does not even step into it once), it is the Eleatic inspiration, mitigated more or less, that prevails in the centuries which follow: the view that nature necessarily contains something permanent which constitutes its essence and is the proper object of knowledge becomes almost a commonplace of Hellenic philosophy.

In Chapter 5 we cited a passage from the *Timaeus* which clearly expresses this opinion. For Aristotle too it seems to be established that the only possible object of knowledge, of science, is the permanent, which must necessarily be identified with the essential.

But Plato is more precise on this point, not only insisting, as we have said, on the role of the Same and the Other, but pointing out that the Other must be constrained to mix with the Same. This is the passage we have put on the title page of the present work, because it seems to us to foreshadow our development as a whole.[1]

For the sake of clarity, we must now separate out the particular aspects of the problem before us. First, insofar as its more strictly logical side is concerned, we have set forth the ideas of Spinoza, Hobbes and Leibniz on causality (Ch. 3, p. 47), and we have seen that Leibniz in particular formulated the principle of sufficient reason very clearly, requiring that it be applied not only to diversity in time, but also to diversity in space (Ch. 5, p. 132). We have also stated in another connection how much Leibniz stressed the role of the identical in necessary truths (Ch. 5, p. 102) and the equality of cause and effect (Ch. 5, p. 117). These taken together thus seem to comprise everything that forms the foundation of the theory. But, we must admit, nowhere do we find clearly stated what it is that establishes the link between these diverse elements so imperiously that such a connection seems compelling. Of course in many passages Leibniz declares that all truths, whatever they may be, and consequently also those that for us are mere truths of experience, must, in the mind of God, be rational truths and thus reducible to identical propositions; only the overwhelming complexity of such an undertaking, which would require an infinite analysis, keeps the same from being true for us.[2] And sometimes for Leibniz the principle of sufficient reason is cited following an exposition dealing with the principle of identity, so that one is, in a manner of speaking, forced to conclude that he deduced the first of these propositions from the second,[3] thinking that if we seek reasons for things and equate cause and effect, it is because we would like to reduce contingent truths to necessary truths, which is what they must be for the divine intelligence.

But, on the other hand, we observe that more often than not the principle of sufficient reason is formulated as an entirely independent proposition alongside the principle of identity, and that it sometimes even takes the appearance of a rule deduced from experience, as when he says: "This great principle holds for all events, and a contrary instance will never be supplied" (*Opera* 515–516 [Huggard 147]). And the equality of cause and effect, as revealed in the laws of motion, is expressly declared not to "depend on the principle of necessity as do the truths of logic, arithmetic, and geometry, but on the principle of fitness, that is to say on the choice of [divine] wisdom" (*Opera* 716 [Parker 200–201]).[4]

One is thus forced to conclude that if (as indeed seems probable) Leibniz sometimes felt that the different points of view we brought together above were connected, his conviction was nevertheless not strong enough not to become inoperative at times, and that in any case he did not find the connection sufficiently evident to state it explicitly.

Condillac sees only perfect identity in reasoning. "In each equation the two members are an identical quantity expressed in two ways"; likewise "the distinction between two equal things seems to suppose two things that, though equal, are different; and yet the two reasons are only one and the same quantity." He takes pains to show that "a proposition is not frivolous just because it is identical," but this is because the identity, which is in the ideas, is nevertheless not in the words. The fault for this lies in the language we speak, "for it is our badly made languages that put the greatest obstacles in the way of the progress of knowledge. We would know how to invent if we knew how to speak; but we speak before we have learned and we do not like simplicity." With a well-made language, discoveries would be "easy"; indeed, "the art of reasoning is only the art of speaking" and "the road that leads from discovery to discovery is only a trail of identical expressions." All we need do, then, is follow our natural inclination everywhere. Thus algebra "will make us speak with the voice of nature, and we shall think we have made a great discovery." Nature is our "first teacher" and "the only way to invent is to do as she teaches us to do." Some say that to invent "is to find something new by the force of one's imagination." But "this definition is completely wrong," for "discoveries can be made without imagination"; it is analysis that "makes poets as it makes mathematicians."[5]

By these statements, Condillac himself clearly closes off the path that leads toward the understanding of scientific reasoning, a path that his sound appreciation of the importance of the concept of identity would have seemed to open wide to him. By ignoring what is fundamentally different in the concepts to be identified, by supposing them, on the contrary, to be fundamentally identical from the outset, he completely loses sight of the active role of the mind insofar as it exercises a constraint to achieve identification. Instead, the mind appears to him to be acting, originally, in an opposite sense, to be distorting fundamental identity by a "badly made language"; the mind is, in short, a sort of screen interposing itself between us and nature, but we have only to follow nature to see this screen move aside. That is why Condillac does not succeed in understanding that reasoning can bring us to genuinely new discoveries, that it can lead us to invent, and why mathematical demonstration appears to him as a succession of identities spontaneously imposing themselves on the mind (*La Langue des calculs* 162, 190; *Logique* 22:170 [Philip 412]).

Stanley Jevons has the advantage over Condillac in that, since he speaks of *similars* and not of *identicals*, he is less inclined to lose sight of the essential diversity of the concepts which, according to his formulation, are substituted for one another in reasoning (cf. Ch. 5, p. 114). Strictly speaking, however, he too only considers the intervention of the concept of identity or resemblance insofar as it arises spontaneously, as it were, prior to the process of reasoning, that is, its intervention in the constitution of the genus. This fact is all the more striking because Jevons illustrated his logical theories with a great number of admirably well chosen examples taken from the mathematical and physical sciences. Jevons follows in the tradition of Leibniz, although he did not know the latter's work until after he had arrived at his own formulation. Given the particular nature of this formulation, it is understandable that Jevons does not cite Leibniz's declarations on the reduction of necessary truths to identical propositions, but rather formulas more like his own, and in particular those relative to the definition of identity by the possibility of substitution.[6]

Moving now to the epistemological aspect of the theory, we must return to antiquity. Since before Plato, the atomists, taking into account both the theories of the Eleatics and Heraclitus's criticisms, *go beyond* both of them (in the Hegelian sense of the term) and "save" changing appearance while at the same time maintaining the essential immutability of being. They reconcile the two by abandoning the Eleatic concept of the oneness of being: being becomes multiple and creates phenomena by means of displacement in space. We spoke of the origin of mechanism in Chapter 5 (p. 129), pointing out that this origin might provide enlightenment on the true character of the theory.

Atomism, which dominates the science of antiquity, undergoes an eclipse in the Middle Ages, yielding to Peripatetic conceptions, or at least to conceptions coming more or less directly out of Peripatetic philosophy.

But, as we saw in Chapter 5 (p. 119), these conceptions, insofar as they are truly scientific, have in common with atomism that they are dependent upon the same principle of explanation by the permanent. Furthermore, it should be noted that there appears throughout medieval philosophy, as a sort of axiom, the conviction common to Plato and Aristotle that the essential and the permanent must necessarily coincide.

In the Renaissance there is a return to mechanism, but it really triumphs only in the Cartesian form and not in the form given it by Democritus and Lucretius. Cartesian mechanism, moreover, is even more radical than these ancient theories, for it states the final outcome of the theory more decisively, namely the identification of matter and space (Ch. 5, p. 135). But by that very fact, the sense of continuity between the mechanism of the Greeks and that of the moderns is somewhat obscured. Descartes himself does not readily admit the kinship (Ch. 8, p. 236, n. 12), and his ideas leave an incomparably profound impression on science. What is certain is that even when science frees itself to some extent from strict Cartesianism and elaborates theories that come closer to those of the ancient atomists – thus, for example, Newton's profession of mechanistic faith makes one think of Democritus and Lucretius, rather than Descartes (cf. IR 492 [Loewenberg 424–425]) – it is inclined to forget the purely philosophic origin of these conceptions. Furthermore, as a result of the growing influence of Baconian empiricism and especially, later, of Comtian positivism, science seems almost happy to forget this origin which, being incontestably aprioristic, seems to it to introduce an element foreign to its essence and to affront its dignity, so to speak. Consequently one frequently finds modern men of science expressly denying their solidarity with the ancient atomists. For instance, Büchner declared that the ancient systems were "arbitrary speculative conceptions," while those of the moderns are true "discoveries of natural science."[7] Büchner, it is true, was only a popularizer, but the chemist Schützenberger said essentially the same thing,[8] and quite recently Smoluchowski, whose works we had occasion to mention in connection with the theory of Brownian motion, stated that

> the modern physicist is somewhat uneasy when he hears the Greek philosophers Leucippus and Democritus and the Roman Lucretius vaunted as the authors of atomism. What is known about the atomism of antiquity and its subsequent interpretations seems to him to be quite simply fanciful *drivel* [*Faselei*], for in it he fails to find what is most essential for practical theory, namely the demonstration of an agreement with the numerical form of natural laws. Such a demonstration was quite inconceiv-

able so long as physics had only a qualitative character. No doubt atomism as a philosophic doctrine is more than two thousand years old, but as an exact physico-chemical theory it dates back barely one or two centuries.[9]

We shall soon see (Ch. 18, p. 524) that the feeling which apparently inspired such a violently unfair attack against the ancient atomists is not entirely devoid of valid motives. But, needless to say, insofar as the historical connection between the ancient and the modern positions is concerned, the thesis is entirely contrary to fact. For example, where would one place the dividing line? One could no doubt claim that rigorously scientific atomism dates only from the determination of the dimensions of molecules, that is, from the works of Gouy, Perrin, Smoluchowski, etc. But the actual authors of these works, inspired by a lesser presumptuousness, generally link them to earlier conceptions, and Smoluchowski himself claims kinship in particular with the crystallographers Haüy, Bravais and Dalton. Moreover, the Polish physicist generously grants "one or two centuries" to modern atomism, as we have seen, which would just barely permit the inclusion of Newton's atomism, but would exclude Boyle's. Now can we really separate Dalton's conception from Boyle's "corpuscular theory," and if we mean to go back that far, how will we go about separating it from the conceptions of Descartes and Gassendi and from those of the ancients? What is more, if one draws a dividing line in this area, one will necessarily be led to proceed in the same way for mechanistic conceptions in general, to say, for example, that the wave theory of light dates from Young and Fresnel. If, on the other hand, one goes back to Huygens, whose conceptions offer so many points of contact with those of these nineteenth century physicists, one will be led to connect these theories to those of the ancients and will end up affirming, with Larmor, that "the general results obtained by von Helmholtz in the abstract theory of fluid motion have enabled Lord Kelvin to reconstruct on a precise scientific basis the notions of Leucippus and Descartes on the relation of matter to aether."[10]

Indeed, if one is willing to consider these matters with an open mind, the continuity is obvious. The moderns have of course added to the atomistic conceptions a precision in measurements unknown to the ancients; but the foundation has remained the same: it is still explanation of the changing by the permanent and of the diverse by the uniform, through the use of a spatial function. That is why the demonstrations retain a sort of family resemblance, why the way in which the contemporary physicist infers the motion of invisible molecules from Brownian motion closely resembles the way Lucretius proved the materiality of air and the impact of its molecules by the terrifying effects of the tempest (Ch. 15, p. 400). And more profoundly yet, there is still the same faith in the fundamental rationality of nature, in the agreement between the paths it follows and those of our intelligence. Modern research has very greatly extended the confirmation of this agreement, certainly beyond what many men of science believed they could reasonably expect; but this work has also revealed more clearly the existence of latent disagreement, of the irrational. Let us add that it is all the less surprising to encounter such questionable judgments among scientists when one finds in an otherwise well-informed contemporary historian of science the statement that "the new atomic theory ... is founded on the old one more in appearance than in reality."[11]

However, opinions of this kind are far from being universal, even among modern scientists. We spoke above of Larmor. Tyndal, in an address to the British Association (Belfast, 1874), retraced the historical development of atomic theories since Democritus, insisting on their continuity.[12] Similarly Henri Becquerel, in a speech given at a session of the five academies of the *Institut de France* (26 October 1907), showed, from a broad perspective, how much the conceptions of the ancients and those of modern science have in common. After citing a few passages from Lucretius, he declares that "each word of these citations corresponds to one of the properties we today attribute to electrified particles." Thus "for more than two thousand years, each time man tries to probe the mystery of the bodies that surround him, either by the effort of his thought alone or by the artifices of experiment, he always glimpses the same image at the bottom of all things."[13]

On the other hand, the memory of the connection between atomism and Eleatic philosophy has not disappeared among philosophers. No doubt, as we have seen by the example of Hannequin (Ch. 5, p. 129), Greek atomism is sometimes derived from Pythagorean arithmeticism, but generally the circumstances are set forth more soundly. For instance, already in the seventeenth century Cudworth, at the same time he formulates a rather bizarre theory concerning the role of Leucippus and Democritus (he maintains that these philosophers found atomism ready-made, so to speak, in their predecessors, but applying only to the material world, and illegitimately extended its domain, thereby rendering it atheistic), concludes quite soundly that it is "evident that the genuine Atomical Physiology did spring originally from this Principle of Reason, that no Real Entity does of itself come from Nothing nor go to Nothing."[14] In the second half of the nineteenth century, Thomas Henry Martin argues that the attributes with which Leucippus and Democritus endow their atoms are an exact copy of those possessed by the unique being of the Eleatics.[15] Zeller very accurately sums up Aristotle's account which we mentioned in Chapter 5 (p. 129) and stresses its importance from the standpoint of atomism. He also observes that atomism, just like the theories of Empedocles and Anaxagoras, derives from the tendency to reduce the appearance and disappearance of things to the combination and separation of invariable substances and that, if one starts from the Eleatic hypotheses and wants at the same time "to save" the multiplicity and motion of things, atomism is the most obvious way to accomplish it.[16] Burnet states that after Parmenides philosophy had only two routes open to it: either it had to give up monism or it had to give up materialism. Now the concept of the incorporeal was not yet known. Thus one necessarily arrived at atomism, which, to the best of our knowledge, is the ultimate form of the position that the world consists of matter in motion. Burnet notes, with Martin, that the atoms of Leucippus and Democritus are identical to the being of Parmenides, and he believes that the passage from Aristotle to which we have alluded is actually one of the most important pieces of testimony in the whole history of early Greek philosophy; correctly interpreted, this statement furnishes the key to the entire development of this philosophy. Of course Burnet is surprised that anyone today could present the atomistic hypotheses as the most recent generalizations of empirical science.[17] In Germany, following Zeller, Hermann Cohen, as well as Natorp and Cassirer,[18] unambiguously show the true lineage of Greek atomism, which is all the more remarkable since these three thinkers belong to the Marburg school and consequently their panmathematicism would undoubtedly motivate them to magnify the role of Pythagorean ideas. Finally, Höffding, with the great authority carried by his judgments, points out that the "theory of atoms is Parmenides' theory of identity applied to the least parts of being" and that "Plato is convinced, like Parmenides, that the world of identity is the true world."[19]

It is therefore no exaggeration to say that there is quite general agreement among philosophers as to the origin and the true nature of the atomistic doctrines. But it is a truth that remains isolated; no one thinks of drawing the consequences from it, although they seem to follow so directly, with respect to the essence, the innermost *structure* of our scientific knowledge in general. Even those who put forward more or less sound ideas concerning this structure do not use the historical argument whose main outline we have just sketched. Indeed, in most cases (we shall have occasion to mention the exceptions) their reflections on atomism play only a subordinate role, and the actual point of departure for their arguments lies rather in considerations relative to the principles of conservation.

At first sight, however, this approach might seem less easy, because for these principles there has been nothing analogous to Aristotle's testimony. Nor could there be, for it is not a matter of a single theory born at a precise moment in the evolution of human thought, but of propositions that emerged at various times and under conditions sometimes differing considerably from one case to another.

However, there is something common to the histories of the three great principles of conservation known to modern science, namely the principle of inertia or of the conservation of velocity, the principle of the conservation of mass and that of the conservation of energy – we are omitting implicit concepts of conservation as well as outdated principles from the past whose close analogy with these principles we have established above (cf. especially Ch. 5, pp. 119 ff.). This common and altogether characteristic feature is the fact that these propositions were originally stated with no proofs of any kind or were based on purely aprioristic demonstrations, and yet they often came later to be considered entirely empirical laws. For example, when Galileo formulates, more or less implicitly, the law of inertia, he is content to proclaim that "any velocity once imparted to a moving body will be rigidly maintained as long as the external causes of acceleration or retardation are removed, a condition which is found only on horizontal planes."[20] This was a new affirmation, one in complete disagreement with the two theories then current claiming to explain the motion of throwing, namely the theories of Aristotle and of Benedetti. However (contrary to what is frequently maintained), Galileo attempts no experimental proof of his proposition – indeed it would have been quite impossible at the time – but formulates it as something taken for granted, being content to show after the fact that, in highly specific cases, what is happening can, if necessary, be seen in this way.

When Descartes gives the first clear formulation of the principle of inertia, he deduces it *a priori*, with his marvelous perspicacity, from the fact that God is "not only immutable in His nature, but also immutable and completely constant in the way He acts," which allows us to infer the permanence of all things: "that each thing ... always remains in the same state ... and never changes ..." (*Principes* II, 36, 37).

Jean Rey maintains from the very beginning of his *Essay* that "weight is so closely united to the primary matter of the elements that they can never be deprived of it. The weight with which each portion of matter was endued at the cradle, will be carried by it to the grave. In whatever place, in whatever form, to whatever volume it may be reduced, the same weight always persists." He sketches a sort of aprioristic proof of this proposition, but it is so short and so unconvincing that he himself clearly did not attach too much importance to it; no doubt he provided it only as a bow to the taste of the times, which held aprioristic deduction in such great favor. As for his experiments (the *Essay* shows that Rey was an able and effective experimenter), they simply prove, as far as the conservation of weight is concerned, that if one is willing to allow it, two observations where it might be thought to break down can be explained without too much difficulty by formulating certain auxiliary hypotheses (such as the weight of air and of elementary fire, which during that time were generally assumed to be absolutely light bodies, naturally tending "upward"). Moreover, Rey, just like Galileo for that matter, was perfectly aware in stating his principle that he was going against the cherished beliefs of his contemporaries. He means "to give a flat denial to that erroneous maxim which has been current since the birth of Philosophy – that the elements mutually undergoing change, one into the other, lose or gain weight, according as in changing they become rarified or condensed."[21]

Descartes, in affirming for the first time that a given magnitude defined by the motion of a body must be conserved throughout the modifications this movement undergoes (which is the conception out of which our principle of the conservation of energy has arisen), deduces his proposition, just as he did that concerning inertia, from "the immutability of God's manner of working," from which it follows that he is "maintaining the world by the same action ... with which he created it" (*Principes* II, 42). But, as we know, Descartes is mistaken in his formulation, and it is Leibniz who provides the correct statement of what is called "the principle of the conservation of *vis viva*." He deduces it by connecting it to the equality of cause and effect through the use of a *reductio ad absurdum* (cf. Ch. 5, p. 117). The Leibnizian conception is later broadened and finally attains its modern formula in J. R. Mayer, Colding and Joule. Mayer deduces his formulation *a priori*, exactly as Leibniz had: "Forces are causes and consequently there is ground for fully applying to them the principle *causa aequat effectum*," he says in his famous article of 1842.[22] Colding reasons in an analogous but, so to speak, even more aprioristic way, declaring that "as the forces of nature are something spiritual and immaterial, ... it is consequently quite impossible to conceive of these forces as anything naturally mortal or perishable."[23]

Nor does Joule, who is often represented as the typical "positivistic" scientist, hesitate to make use of deduction. "We might reason, *a priori*," he says in one of his publications,

that such absolute destruction of living force [$mv^2/2$] cannot possibly take place, because it is manifestly *absurd* to suppose that the powers with which God has endowed matter can be destroyed any more than that they can be created by man's agency; but we are not left to this argument alone, decisive as it must be to any unprejudiced mind.

Furthermore the numerical data at which Joule had arrived varied within such extraordinarily wide limits that they could in no way have served to demonstrate the existence of a mechanical equivalent of constant heat, and the work of Joule as a whole makes sense only if one assumes that he had indeed inferred the existence of an invariable coefficient *a priori* and that the sole purpose of the experiments was to determine its numerical value.[24] Fifteen years later, and in spite of the experimental data which had accumulated in the meantime, Faraday still does not consider treating the conservation of energy as an empirical law but bases his conviction on the equality of cause and effect, just as J. R. Mayer had done.[25]

This very obvious state of affairs in no way prevented each of these three propositions from being declared an experimental law. J. S. Mill did so in the case of the principle of inertia[26] and also for the principle of the conservation of mass, and Littré followed his lead in the case of the latter principle.[27] In the same way, physicists usually treat the conservation of energy as a purely empirical fact.[28]

On the other hand, however, the idea of an *a priori* deduction of these principles continues to have many supporters. For the principle of inertia, d'Alembert attempts such a demonstration, in a mathematical form, while Kant and Lotze give a philosophic form to their deductions; Maxwell combines the two approaches, without managing to make his demonstration any more convincing thereby (cf. IR 127–134 [Loewenberg 121–127]). As for the conservation of matter, Kant, Schopenhauer,

Whewell and Spencer agree that it is *a priori* (IR 192–194 [Loewenberg 176–177]), Spencer in particular strongly emphasizing this point.

Our inability to conceive Matter becoming non-existent, is immediately consequent on the very nature of thought .... The annihilation of Matter is unthinkable for the same reason that the creation of Matter is unthinkable; and its indestructibility thus becomes an *à priori* cognition of the highest order – not one that results from a long continued registry of experiences gradually organized into an irreversible mode of thought; but one that is given in the form of all experiences whatever.[29]

Spencer likewise deduces the conservation of energy *a priori*, and Stallo, Lasswitz and Spir are of like minds (cf. IR 218–219 [Loewenberg 198–199]).

Moreover, the historical development of the three principles is not the only thing they have in common; their position in present-day science is also curiously analogous. We do not know whether they are *necessary* or *accidental* truths; science, one might say, no more manages to settle the debate between these two alternatives than it does to reconcile them.

However, Leibniz already seems to have sensed that certain propositions dealing with motion are, from the point of view of rationality, composite in nature. We have seen that he had no qualms about deducing his principle of the conservation of *vis viva* directly from the equality of cause and effect. Upon other occasions he expresses reservations, however. For example, in criticizing one of the proofs in which Descartes argued from the constancy of God, he writes: "Everyone sees how weak this reason is that posits the constancy of God, for although the constancy of God is of the highest order and nothing in it is modified except in conformity with laws of a long-predetermined sequence, we still must seek out what in the sequence he had decreed to conserve." Likewise he declares in another passage that "the laws of Nature ... are neither entirely necessary nor entirely arbitrary. The middle course to be taken is that they are a choice of the most perfect wisdom."[30] Disregarding the finalistic considerations suggested by the last sentence, there remains the idea that along with an *a priori* element the propositions contain another element of a different origin which must necessarily be sought *a posteriori.*

Kant's views in this area are somewhat less clear than those of Leibniz; they are nonetheless remarkable, particularly because they have served as a point of departure for other thinkers. We saw in Chapter 14 (p. 365) that Kant believed there was *partial* agreement between our understanding and reality, from which he had concluded that part of science was directly deducible. What we call the principles of conservation fell within this limit (which he drew too broadly, as we have said). Thus he deduces in its entirety the principle of the conservation of matter, in the sense attributed to it by modern physico-chemistry, that is, as a proposition stipulating the constancy not only of mass, but also of weight. However, in Kant this deduction does not take place all at once, but in stages, as it were. Not everything included in the formula has the same importance in the *a priori*, if we may speak in such a manner. The form, by which it is related to the principle of substance, appears to be superior in apriority to the rest.

Later Poinsot, in considering, not one of the three principles of which we have spoken, but a proposition of analogous form called the principle of the *conservation of areas* which was introduced into dynamics by Kepler, makes the following observation:

> If one tries to discover what could have given him this idea, it seems to me one will find that he arrived at it, not by chance, as one might think at first, but by a certain natural process that I wish to point out in passing, because it is found in all our research, and because it results, so to speak, from the very nature of the human mind. And indeed the only law we know with complete lucidity is that of constancy and uniformity. To this simple law we try to reduce all others, and for us science consists exclusively in this reduction.[31]

The import of this passage is weakened as it continues, for Poinsot seems to reduce research to a search for the constancy of a *relation*. It appears, however, to follow from the very course of his reasoning that he had started out with a better idea of the true implications of this proposition. As a matter of fact, the constancy of a relation is the very essence of law; to say that this is what Kepler was seeking would be to claim that he sought some sort of empirical formula. Now that is precisely what Poinsot is protesting against: Kepler's thought did not wander haphazardly; what he wanted was a proposition of conservation because only such a proposition appears knowable to us "with complete lucidity," that is, answering both to the essence of things and to our understanding.[32]

In this area as in many others it is probable that Hegel proceeds from Kant. But, as we have seen, Hegel's thought went much further than that of his predecessor. Hegel perfectly discerned the true role of the concept of identity in the workings of our thought. He knows how science uses it and has clearly acknowledged that the principle of conservation of matter, as chemistry understands it, derives directly from the concept of identity, that the persistence of the elements in compounds is only a notion introduced in order to satisfy our tendency to explain change by denying it, by affirming the identity of things in time.

Although he wrote ten years after Hegel's death, Whewell is entirely ignorant of these arguments, and it is by following directly in Kant's footsteps that he succeeds in correctly determining how the role of deduction and of experience in certain propositions ought to be represented. "It is a Paradox," he says in the introduction to his *Philosophy of the Inductive Sciences*,

> that Experience should lead us to truths confessedly universal, and apparently necessary, such as the Laws of Motion are. The Solution of this paradox is, that these laws are interpretations of the Axioms of Causation. The Axioms are universally and necessarily true, but the right interpretation of the terms which they involve, is learnt by experience. Our Idea of Cause supplies the *Form*, Experience, the *Matter*, of these Laws.[33]

However, Whewell does not think out the application of these views to the principles of conservation with sufficient precision, since, as we have seen, the conservation of matter appears to him (as it did to Kant, for that matter) to be entirely *a priori*. He also declares that one could have arrived at the principle of inertia independently of all experience and that the same is true of all of kinematics (Whewell 1:xxiv, 213).

More than a quarter of a century after Whewell, Wundt takes a closer look at these controversial propositions, which he terms "axioms of physics." He strongly emphasizes an aspect of these propositions that does not interest us here, namely the fact that they proceed from "experiences of thought." But he also points out that they are derived from a general formula, which is none other than the equivalence of cause and effect. Our reason tends to impose this framework upon experience, which resists it. Reason is incapable of predetermining the particular expression to which the formula will have to apply; but any proposition containing a causal relationship, even if it has been discovered empirically, can be reduced to a form that makes it appear logically evident.[34] On the other hand, however, Wundt declares that the conception of something that persists through change is not *a priori*, but is suggested by the phenomena themselves.[35] Was Wundt, in writing his *Axioms*, just as ignorant of Hegel's position as Whewell? The only phrase we have found in his work which is the least bit Hegelian is the one in which he declares that when a change occurs, "perception [*Anschauung*] constrains us to posit two things, even though the concept would like to allow only a single thing to subsist" (*Die Prinzipien* 178 [Meyerson's brackets]). But that is at bottom rather a philosophic commonplace and, at the time

Wundt's work appeared, there no longer seems to have been any interest in Hegel in Germany. In any case, it would be the only example of any influence whatsoever exerted by Hegel's thought in this area.

In 1873 Spir's great work is published.[36] The philosopher treats the principle of identity quite extensively. For him the principle does not always have the same nature, being sometimes analytic and sometimes synthetic; it becomes synthetic only if considerations of time are added *(Pensée et réalité* 192).[37] Spir lays much emphasis on the fact that this postulate of identity in time disagrees with the real phenomena, rightly seeing in it a proof of the *a priori* nature of the postulate (contrary to what was claimed by Wundt, with whose book Spir seems to have been unfamiliar). He also recognizes the true nature of scientific explanation, as well as the relation between the principle of identity on the one hand, and atomism and the principles of conservation on the other (*Pensée et réalité* 9, 128, 275, 493, 498). These important deductions are at times tainted with a bit of confusion, however, apparently stemming chiefly from the fact that Spir does not assume that, alongside the principle of identity, which aims at the comprehension of nature, there exists the principle of lawfulness, which aims at simple prediction, action on nature. It appears obvious to him, on the contrary, that the second can only be a sort of abridgement of the first, that in other words our belief in the orderliness of nature stems from the fact that we believe nature to be fundamentally rational. That may be, for as we saw (Ch. 4, pp. 72 ff., and Ch. 16, p. 454), strict positivism, a theory which builds science on lawfulness alone, nevertheless cannot dispense with the assumption of an agreement between our reason and the external world. It remains true, however, that the assumption in this case does not seem completely of the same nature as the one used to postulate genuine rationality. Indeed, the latter seems to us something infinitely desirable, to be sure, but also infinitely remote; well before science made him realize, by the concept of the irrational, that this ideal was unattainable, man no doubt was dimly but powerfully aware of it, as is shown by the fact that from its outset atomism posits the irrationality of sensation, resolutely leaving it out of its explanations, declaring it to be mere "opinion." This is why rationality proves to be what we might call a flexible principle, which adapts itself to the circumstances, allows compromises, engenders illusions. Lawfulness, on the contrary, is rigid, absolute; it means to govern everything not subject to the free choice of a terrestrial or superterrestrial will; it allows of no exceptions, Auguste Comte notwithstanding (Ch. 4, pp. 74 ff.).

Therefore, even if one derives these two principles from one another or assigns them a common origin, one will be obliged to suppose that the diversity of the circumstances in which they are applied has profoundly diversified the means of application itself. Indeed, for any organism action is an immediate and ever present necessity, and the success of this action constantly demonstrates the perfect lawfulness of nature; whereas, strictly from the standpoint of life, rationality is a sort of luxury, and the success of explanatory science confirms it only in small part, while, on the contrary, our direct sensation unquestionably reveals its weakness. It is thus to our advantage, if we mean to reason about the foundations of science, to distinguish between what results from each of these two principles, although the workings of the two are almost inextricably entangled. It is certainly because he has confused them too much that Spir comes to extend the limits of deduction too far, just as Kant and Whewell do, by failing to recognize the existence of the irrational, although he affirmed it in dealing with the opposition between the phenomenon and the tendency to seek identity. This is why for him the deduction of the principles of conservation seems direct and complete. For example, the conservation of energy becomes the necessary consequence of the equally necessary fact of the transmission of motion (*Pensée et réalité* 424). In this respect, then, he is not as far advanced as the expositions of Leibniz, Kant and Whewell which we have cited, where, at least in theory, the double nature of propositions of this sort was foreshadowed or even (as in Whewell) explicitly affirmed.

Spir's work does not go unnoticed, as evidenced by the French translation made more than twenty years after its publication, a translation which in its turn attracts considerable attention. In the area which interests us, however, we do not see that it has influenced current opinions among scientists and philosophers. Riehl appears not to have known it at all when he wrote the work mentioned above (Ch. 3, p. 54, and Ch. 5, p. 117), whose subject and conclusions offered more than one point of contact with the views of his predecessor. The same is true for Kroman, who, with less breadth than Spir but with much more clarity and a richer scientific apparatus, presents quite similar deductions. It is important to note that Kroman also seeks to distinguish to some extent between the two concepts we have called *lawfulness* and *causality*. He refers to the first as *causality* and to the second as *identity*. But he sometimes fails to recognize the true limits of each of them. For instance, he confuses the former with the postulate of comprehensibility and even seems to want to draw from it the existence of the noumenon; what is more, he believes at bottom that this principle alone (as positivism proclaims) suffices to explain science, since it makes it possible to conclude in favor of the conservation of certain concepts.[38]

Paul Tannery did not devote a special work to this question of epistemology. But in his 1887 study of some of the pre-Socratic philosophers, Melissus in particular, he made observations as apt as they are penetrating. Specifically, he arrives at the conclusion that behind the conservation principles is a more general formula, a requirement of our thought, which stipulates the permanence of something persisting through the variability of phenomena and that this is the true principle of causality and the basis of all scientific explanation. He also shows that it is from this same principle that the idea of the constancy of *vis viva*, of mass and of energy is derived.[39]

That same year, in a book devoted to the principle of the conservation of energy, Planck (a physicist whose name we have cited on many occasions throughout this work), expressed surprise at the very limited number of experiments J. R. Mayer, Colding and Joule had used to establish this proposition and asked himself whether the rapid success of this conception was not due to *a priori* elements contained in it. First of all, these elements spring necessarily from the close connection of the principle with the principle of the equality of cause and effect, and, secondly, the analogy our mind establishes between the conservation of matter (so long familiar) and the conservation of energy helps to win acceptance for the latter notion. It should nevertheless be added that Planck finds deductions like those of Descartes and Colding to be entirely negligible and that, moreover, he later seems to have at least partially reconsidered his earlier position, since in a second edition of his book (published in 1908), he felt obliged to add a note stating explicitly that the principle of the conservation of energy is "neither a tautology, nor a disguised definition, nor a postulate, nor an *a priori* judgment, but in fact an experimental proposition."[40]

Gaston Milhaud continues Tannery's line of thought. He affirms that the law of constancy and uniformity is not science's final conclusion, but rather its fundamental hypothesis, its "guiding principle," and in particular makes clear the role of this *a priori* principle in the genesis of the principle of inertia. However, Milhaud sometimes follows a very different order of ideas, assimilating the principles of conservation with definitions and conventions. Furthermore, the constancy seems to him above all a constancy of *relations*, and thus as following (just as in Kroman) from the very fact that the world is governed by laws.[41]

With great profundity André Lalande points out that preexistence is the true basis for all explanation and links to it the principles of conservation of matter and of energy. For him these conceptions are derived from more general considerations belonging to a slightly different order of ideas than the one that has inspired us here. Indeed, he considers the progress of human intelligence toward identity to be a law which is applied in three converging forms: assimilation of things with one another, assimilation of minds with one another, and assimilation of things with minds. Putting the problem in this way leads him to his own particular deductions. Thus, for this philosopher the explicable, the intelligible is most often assimilated with what is in general capable of being determined, that is, in our terminology, with what is simply lawful and, just as for Milhaud, the necessity of the constancy of certain terms is deduced directly from the fact that nature is ordered. On the other hand, Carnot's principle appears to conform to the requirements of our reason, given that the transformations that take place in virtue of this proposition, since they obey the law of progress toward equality, tend by that very fact to satisfy human intelligence more and more.[42]

Almost concurrently with Milhaud and Lalande, Kozlowski, following in Kant's footsteps, points out the part *a priori* elements play in the formation of certain scientific laws and stresses that the principle of conservation of matter depends on the concept of substantiality. But the Polish philosopher believes that the persistence of matter, "a postulate of purely rational origin," is at the same time "the *conditio sine qua non* of regularity in the phenomenal world"; thus, just like Spir (whom he does not seem to know), he deduces lawfulness from causality. On the other hand, he considers that the concept of the unity of matter does not originate with the postulate of rationality but with the unity of tactile sensation.[43]

At about the same time, Wilbois too emphasized the close relationship between the principles of conservation and the law of substance,[44] and James Ward in his turn recognized the role of the *a priori* postulate of cause and effect in establishing the propositions stipulating the constancy of mass and energy.[45]

Admittedly the brief sketch we have just made has some strange features. First let us note that it would have been difficult to follow simple chronological order in our presentation, for any feeling of connectedness between the different aspects of the problem seems to be completely lacking in the majority of the thinkers we have mentioned. Thus properly logical conceptions have remained almost ineffectual from the epistemological point of view. In Condillac, that is largely explained, as we have seen, by the fact that, since he disregards the diverse, he is firmly persuaded of the fundamental identity of concepts, to the point of believing that nature itself imposes this identity and that our understanding only alters it. But is it not strange that he could arrive at such a view? In Stanley Jevons, it is certainly the particular form he gave his statement that is primarily responsible for his failure to recognize the role of the concept of identity in the physical sciences. But why exactly did he choose this form, which narrows the field of application of the notion of the similar before the question is addressed? And how is it that Leibniz's mind, one of the most powerful in the history of mankind, so surprisingly stopped half way?

Yet we need only recall what we saw about the particular conditions under which the process of identification is applied to recognize the reason, or at least a partial one, for this anomaly. Indeed, however powerful the understanding's preoccupation with reestablishing the identity compromised by the diversity of sensation, the fact that this constantly envisaged goal nevertheless remains infinitely remote and that – as if by a sort of discretion in the face of this near impossibility – reason does not proclaim its ultimate goal, is quite apt to hide from us the path it is following. Already in the case of mathematical reasoning, by the fact that identification occurs at a different stage, that it is not spontaneous as in the case of the creation of genus, but sought after, ordered up, we tend to fail to recognize that it is basically a question of the same process. This fact is all the more likely to cause the mind to hesitate when it is a question of the physical sciences, where the direction is implicit and we do not have the framework of the "cascade of equations" to guide us in the search for the principles of reasoning.

However, even in limiting ourselves to considering only the last-mentioned aspect of the problem, namely insofar as it is related to the theory of the principles of conservation, the evolution of these ideas does not fail to surprise us. Can one even speak of evolution? The term suggests something continuous in time. Now the characteristic trait here is precisely an absolute lack of continuity. These are only isolated attempts, and, what is more, usually not carried very far. Only a few of these thinkers (we believe we have explicitly pointed out all such cases) are connected to those who came before them. No one seems to feel the need to make a real search for his intellectual ancestors and, in return, the influence each of them exerts on contemporary or subsequent philosophy of science (at least with respect to the ideas with which we are concerned) is nonexistent or insignificant.

One is tempted to apply to their conceptions an image that has been used to characterize Quaternary art: "*proles sine matre creata, mater sine prole defuncta.*"[46]

For instance, let us consider Riehl, whom we discussed at somewhat greater length in the early chapters of the present work. In the course of his exposition, Riehl cites only a single name, that of the mathematician Riemann: it is the passage we ourself quoted (Ch. 3, p. 54), extremely interesting no doubt, but not very relevant to the thesis Riehl was putting forth. On the other hand, he does not mention either Kant or Whewell. Now, the observation by Kant (with whose writings Riehl is quite familiar, as a matter of fact), an observation that certain scientific propositions are only a particular expression of the metaphysical principle of substance, could have served Riehl well since, as we have seen (Ch. 5, p. 117), he uses the principle of inertia as an example of causal deduction. Riehl has more justification for being ignorant of Poinsot's profound statement. But it is perhaps even stranger that not only Spir's book, as we have already mentioned, but Wundt's as well, both of which are closely related to his work, seem to have entirely escaped his notice. On the other hand, Riehl's work (although it enjoyed considerable renown in the German philosophic community of the time) is mentioned by none of the later writers in this field, in particular by neither Kroman nor Planck (not to mention the French authors).

But the case of Hegel is stranger still. In him, as we have seen, are to be found the most penetrating insights in this area. And yet it all falls into the void. In spite of the immense authority of Hegel at a certain moment in time, an authority which clearly is not entirely past history, no one seems to have noticed that his work contains at least the foundations of an accurate and precise theory of science, that in any case all that was necessary to know these foundations was to turn his epistemology around, so to speak, by modifying the value judgment he claimed to draw from it. Most assuredly none of the later thinkers we have enumerated are linked to Hegel; none even seem to suspect that they should look to him for insights into scientific causality. What is more, to our knowledge none of Hegel's disciples, none of his commentators (despite the number and the zeal of both groups) seem to have thought of emphasizing this aspect of his thought.[47]

But here, thanks to the audacity of Hegel's thought and the frankness with which he expresses himself, we clearly see the causes of this anomaly. If people did not go to the *Science of Logic* or the *Encyclopedia* to seek a theory of science, if they did not perceive that these works developed the foundations of such a theory, it is because Hegel himself took care that they should not. Indeed, as we have said, the sum total of his views on this subject does not impress one as a *bona fide* theory of science. Hegel does not intend to study the paths of scientific thought in order to codify them and thus make the further progress of this thought easier, more logical. What he wants is to accuse science before the tribunal of intelligence, to destroy it in the opinion of thinking men. If he reveals its anatomy, it is because he considers that it is horrible to behold and that the mere sight of it is sufficient to provoke disgust. He tears away the tawdry trappings with which he believes it has artificially covered itself, in order to expose it in its nakedness as an object of scorn and mockery. Behold, he seems to exclaim, this dazzling beauty which was the object of your most ardent desires, before which you prostrated yourself in boundless admiration. It is a hideous monster – worse than that, it is nothingness, an immense tautology with no content. But is it not claimed that science has its successes? That may be, but if so they have nothing to do with the actual method of science. They are due to pure *empiricism*, to the simple search for law, which is its proper domain, the domain to which it must henceforth be limited; anything beyond that is vain and must be ruthlessly rejected (in favor of a quite different method of explanation, be it understood). Furthermore, what are these successes worth when weighed against the deductions of thought, infallible by nature?

Obviously one could reason in this way only if, like Hegel, one fundamentally failed to recognize the *brute fact* (as we have taken the liberty of calling it) of the existence of an explanatory science. Therefore, since the authority of science was increasing, the *reductio ad absurdum* by which Hegel had meant to crush it turned back against him. Science was succeeding and – as people were well aware, despite positivistic declarations – it was succeeding from the theoretical as well as the practical point of view, it was really succeeding in *explaining*. Thus the line of reasoning that had pretended to criticize this method could not be correct; there had to be a flaw, a *gap* somewhere. Why bother then to follow a deduction whose conclusion was clearly absurd, since it was in contradiction with what everyone knew to be a certain fact – especially given that the very starting point of the deduction could not have seemed more paradoxical?

For that is undoubtedly what lies behind the whole affair: it is the paradoxical aspect of this theory of science that enabled Hegel to consider using it as an instrument of destruction against science itself. And this in turn reveals to us the profound reason for the strange anomaly we have noted in the historical development of the theory. The strangely sporadic nature of attempts of this sort most certainly stems from that. In the authors themselves, and still more in those who might be tempted to adopt their views, the same question always seems to be asked, the one we ourself have formulated in Book II (Ch. 6, p. 143, and Ch. 7, p. 177): is it possible that *this* is what science and reason are? Even from the strictly logical point of view, is it not absurd that we should declare concepts equal, identical, when they are not? Is it not preferable to suppose, with Condillac, that identity preexists and that all we have done is recognize it? But (though Condillac obviously did not draw this conclusion) if everything is identical, nothing exists. And can one seriously pretend to fathom reality by denying it?

We hope the reader will have become convinced in the course of this book that, despite their paradoxical appearance, the principles we have set forth are nevertheless the ones which actually guide scientific reasoning. But is the schema really as paradoxical as it first seems?

Let us begin by recalling the psychological fact, which seems to be common experience, that as soon as the human intellect awakens from the dogmatic sleep in which the common sense conception of reality, inconsistent though it may be, allows it to rest, it immediately hurries toward the opposite metaphysical extreme, namely perfect idealism or solipsism, affirming the nonexistence of the external world. It is thus at least probable that this is a profound aspiration of our reason, something that, without our being clearly aware of it, lay dormant in us. Why then should it surprise us that science, or at least that part of science which seeks to explain reality, should end up with the same negation?

So that is not what shocks us. Our intellect quite readily accepts the explanation of being by nonbeing, as evidenced precisely by the countless idealistic systems abounding in the history of philosophy. But science is not just explanatory, it is also *lawful*, practical, calculated to lead us through life by anticipating phenomena, and, what is more, these two parts of science are inextricably entangled: new empirical laws serve to construct new theories, which in their turn lead to the discovery of new empirical facts, new laws. If one wishes to convince oneself how difficult it is to dissolve this union, one need only follow one of the many attempts engendered by the positivistic theory of science in which one aspires to present a given area of science without recourse to any explanatory hypothesis: if the author is really determined not to leave out any class of important facts, he will be obliged to have recourse to strikingly artificial subterfuges. In certain fields, organic chemistry for example, such a presentation can even be considered entirely impossible, as we have said (Ch. 6, p. 163). And as for theoretical science, clearly no one will dream of developing a scientific theory without showing to what extent it is confirmed by experience.

Now from the point of view that interests us, this union between theoretical science and predictive science or, if one will, this twofold aspect of science (which is truly *one*) creates a strange situation: in order to explain, we tend to deny phenomena, whereas in order to guide us through the maze they form, we must maintain their reality. We believe we have established in Chapter 1 that science does this, that despite its distant and negative goal, on the way to this goal it energetically maintains the complete reality of the hypothetical entities it creates – such as atoms or electrons – that it is, in other words, frankly and one might almost say naively ontological in this respect. Our subsequent elaborations will have only confirmed the reader in this conviction – or so we hope. But is that not a contradictory attitude on the part of science; is it not strange that it studies the phenomenon, which is nothing other than change, by means of a principle that tends to affirm the identity of the antecedent and the consequent, that is to say, to deny all change, and that in order to penetrate into the essence of things, whose reality it maintains, it generally makes use of a conception culminating in the negation of all diversity? Is it not eminently paradoxical that it should succeed in this enterprise, that nature to some extent seems to prove penetrable, plastic, for a theory that seeks to demonstrate its nonexistence?

The contradiction is indeed flagrant and irremediable, and nothing can soften its full impact, unless it be the reflection that it is implied by the very concept of experimental knowledge. For by the fact that man has recourse to experience (which he does as soon as he ceases to lead a purely unconscious life), he proclaims his inability to understand things by the effort of his reason alone, that is, he affirms that the ways of nature differ from those of the mind. But since, on the other hand, experience can be useful to him only if he reasons, he must at the same time suppose that at least within the limits of this reasoning, there is agreement between the mind and nature. In other words, the contradiction is the very consequence of the fact that there exists an external world, a *nature*, which we feel – however much we may try to absorb it or merge with it – to be different from our self, while remaining convinced – whatever we may do to distinguish ourselves from it – that it is only our sensation or that we are an integral part of it. As we know, it is this last aspect of the proposition that Spinoza especially emphasized, arguing that man is not "a kingdom within a kingdom" as far as nature is concerned, that "it is impossible, that man should not be a part of Nature."[48]

This is why the two currents coexist peacefully in science. By mechanism, by the principles of conservation and the hypothesis of the unity of matter, science tends toward the immobility of the world and its reduction to space, whereas by Carnot's principle and the other irrationals it recognizes the impossibility of such an outcome. The paradox we have treated in this chapter is obviously only one aspect of this contradiction at once both fundamental and necessary.[49]

NOTES

1. Plato, *Timaeus* 35A. The translation of the complete sentence reads: "He [God] took the three elements of the same, the other, and the essence, and mingled them into one form, compressing by force the reluctant and unsociable nature of the other into the same" [Jowett trans.]. We believe that the true import of this text emerges when it is juxtaposed with the passages at 28A and 37A cited on p. 137, n. 3.
2. Leibniz, *Opera* 83: "... *in veritatibus necessariis demonstratio, sive reductio ad veritates identicas locum habet .... At ... veritates contingentes infinita analysi indigent, quam solus Deus transire potest. Unde ab ipso solo a priori ac certe cognoscuntur.*" Cf. *Opuscules* 17 [Parker 96], 272, and Louis Couturat, *La Logique de Leibniz* (Paris: F. Alcan, 1901), p. 210.
3. Cf. for example *Opuscules* 519 ff. [Parker 87 ff.].
4. *Principles of Nature and Grace*, §11 [Erdmann's brackets]. As to any doubts one might have about the way in which Leibniz connected his ideas here, cf. also the discussion that took place among members of the *Société française de philosophie* on the subject of Couturat's work, 'Sur les rapports de la logique et de la métaphysique de Leibniz' (*Bull. Soc. fr. phil.* **2** [April 1902], esp. the objections of Delbos, pp. 68 ff.).
5. Condillac, *La Langue des calculs* (Paris: Ch. Houel, Year 6 [of the First French Republic]), pp. 61, 63, 74, 88, 111, 163, 190, 210, 233–234. Cf. Also his *Essai*

sur l'origine des connaissances humaines, Oeuvres (Paris: Ch. Houel, Year 6), 1:448, 500, 515 [*An Essay on the Origin of Human Knowledge*, trans. Nugent (London: J. Nourse, 1756; reprint New York, AMS Press, 1974), pp. 319–320, 326–327, 337]; *Traité des systèmes, Oeuvres*, 2:8, 45, 405 [*Philosophical Writings of Etienne Bonnot, Abbé de Condillac*, trans. Franklin Philip (Hillsdale, N. J.: Lawrence Erlbaum Associates, 1982), pp. 2–3, 16–17, 152–153]; *Logique, Oeuvres*, 22:131, 164 [Philip 397, 409]; *De l'Art de penser, Oeuvres*, 6:134, 136; *De l'Art de raisonner, Oeuvres*, 8:6 and 37, but esp. p. 11 where he analyzes a mathematical demonstration almost as we did in Chapter 5 (pp. 108 ff.), but exclusively stressing identity, and p. 217 where, as a proof that all possible truths "are reduced to a single one," Condillac argues that "all machines, from the simplest to the most complex, are only the same machine."

6. Stanley Jevons, *The Principles of Science* (London: Macmillan, 1892), pp. xvii ff. The Leibniz passages are drawn especially from the *Fundamenta calculi ratiocinatoris* and the *Non inelegans specimen demonstrandi in abstractis, Opera* 93–94).
7. Friedrich Albert Lange, *Geschichte des Materialismus und Kritik seiner Bedeutung in der Gegenwart*, 4th ed. (Iserlohn: J. Baedeker, 1882), p. 511 [*History of Materialism and Criticism of its present Importance*, trans. Ernest Chester Thomas (London: Trüber & Co., 1880) 2:351].
8. Paul Schützenberger, *Traité de chimie générale* (Paris: Hachette, 1858), 1:vii.
9. Marian Smoluchowski, 'Anzahl und Grösse der Moleküle und Atome,' *Scientia* **13** (1913) 27 [Meyerson's brackets].
10. Joseph Larmor, *Aether and Matter* (Cambridge: University Press, 1900), p. 25. Cf. also pp. 310 ff. where this physicist strongly insists on the kinship between the ideas of Young and Fresnel and those of Huygens.
11. A. Mieli, 'Les Théories des substances chez les présocratiques grecs,' Part 2 of 'Anaxagore et les Atomistes,' *Scientia* **14** (1913) Supplément: 200.
12. See Arthur James Balfour, *Theism and Humanism* (London: Hodder & Stoughton, 1915), pp. 222–223.
13. Henri Becquerel, 'Réflexions sur une théorie moderne,' *Le Temps*, 26 Oct. 1907, Supplément.
14. Ralph Cudworth, *The True Intellectual System of the Universe* (London: Richard Royston, 1678), pp. 17, 35.
15. Thomas Henry Martin, 'Mémoire sur les hypothèses astronomiques des plus anciens philosophes de la Grèce étrangers à la notion de la sphéricité de la terre,' *Mémoires de l'Institut National de France, Académie des inscriptions et belles-lettres* **29** (1879) 2:236 [actually Martin says "they ascribe most of the attributes assigned by the Eleatics to their unique being"].
16. Zeller, *La Philosophie des Grecs*, trans. Émile Boutroux (Paris: Hachette, 1877), pp. 282–292, and *Phil. der Griechen* 2_2:279, 286 [Costelloe 1:297, 305].
17. Burnet, *L'Aurore de la philosophie grecque*, trans. Reymond (Paris: Payot, 1919), pp. 26, 208, 210, 384 [*Early Greek Philosophy*, 2nd ed. (London: Adam and Charles Black, 1908), pp. 26, 206, 208, 385 (citation of p. 26 erroneous)], and *Greek Philosophy*, Part I, (London: Macmillan, 1914), pp. 94 ff.
18. Hermann Cohen, *Logik der reinen Erkenntnis* (Berlin: Bruno Cassirer, 1902), pp.

29, 40, 187, 272; Paul Natorp, *Forschungen zur Geschichte des Erkenntnissproblems im Altherthum* (Berlin: W. Hertz, 1884), p. 171; Ernst Cassirer, *Das Erkenntnisproblem* (Berlin: Bruno Cassirer, 1906), 1:31.

19. Harald Höffding, *La Pensée humaine*, trans. Jacques de Coussanges (Paris: Félix Alcan, 1911), pp. 124–125.
20. Galileo, *Discorsi*, *Opere* (Florence: Società editrice fiorentina, 1842), 13:200–201 [*Dialogues Concerning Two New Sciences*, trans. Henry Crew and Alfonso de Salvio (New York: Macmillan, 1914; reprint New York: Dover, 1954), p. 215]. On this subject and concerning the historical development of the principles of conservation in general, cf. IR, Chs. 3, 4 and 5.
21. Jean Rey, *Essais*, reprint of the 1630 edition (Paris: G. Masson, 1896), p. 48 [*The Increase in Weight of Tin and Lead on Calcination* (Edinburgh: Alembic Club / Chicago: University of Chicago Press, 1902), p. 14].
22. J. R. Mayer, 'Bemerkungen über die Kräfte der unbelebten Natur,' *Annalen de Chemie und Pharmacie* [*Justus Liebig's Annalen der Chemie*] **42** (1842) 233.
23. A. Colding, 'Lettre aux rédacteurs du Philosophical Magazine sur l'histoire du principe de la conservation de l'énergie,' trans. Verdet, *Annales de Chimie et de Physique*, 4th series, **1** (1864) 467–468 ['On the History of the Principle of the Conservation of Energy,' *The London, Edinburgh, and Dublin Philosophical Magazine and Journal of Science*, 4th Series, **27** (1864) 56–57].
24. James Prescott Joule, *Scientific Papers* (London: Taylor and Francis, 1884–87), 1:268–269 [Meyerson's brackets]. For the experimental data, see IR 212–213 [Loewenberg 194–195]. James Ward (*Naturalism and Agnosticism*, London: A. and C. Black, 1899, 1:173–175) quite readily acknowledged that Joule's reasoning, as well as J. R. Mayer's, was basically aprioristic.
25. Michael Faraday, 'On the Conservation of Force,' *The London, Edinburgh, and Dublin Philosophical Magazine and Journal of Science*, 4th series, **13** (1857) 239.
26. John Stuart Mill, *A System of Logic* (London: Longmans, Green, 1884), pp. 160 ff.
27. Mill 163; Emile Littré, *La Science au point de vue philosophique* (Paris: Didier, 1873), p. 322.
28. See for example Henri Poincaré, *Thermodynamique* (Paris: Georges Carré, 1892), p. 65, and Gabriel Lippmann, *Cours de thermodynamique* (Paris: Georges Carré, 1889), pp. 11 ff.
29. Herbert Spencer, *First Principles* (London: Williams and Norgate, 1863), pp. 240–241. We have already cited portions of this passage (Ch. 16, pp. 448 and 453). Ward correctly pointed out that in expressing himself thus, Spencer obviously was thinking not of the physicist's *mass*, but of the metaphysician's *substance* (*Naturalism and Agnosticism*, London: A. and C. Black, 1899, 1:86).
30. Leibniz, *Die philosophischen Schriften*, ed. C. I. Gerhardt (Berlin: Weidmann, 1880), 4:370; *Théodicée*, §349 [Huggard 334].
31. Louis Poinsot, *Éléments de statique suivis de quatre mémoires*, 10th ed. (Paris: Mallet-Bachelier, 1861), pp. 239–240.
32. In IR 33, 231, 251–252 [Loewenberg 39–40, 207, 226–227], we have tried to clarify the circumstances that caused Poinsot's thought to shift in this way, as has

the thought of other scientists under similar conditions, for that matter.

33. William Whewell, *The Philosophy of the Inductive Sciences* (London: John W. Parker, 1840), 1:xxvii.
34. Wilhelm Wundt, *Die Prinzipien der mechanischen Naturlehre* (Stuttgart: Ferdinand Enke, 1910), pp. 12, 86, 110, 115, 146, 147. This is a slightly revised republication of *Die physikalischen Axiome und ihre Beziehung zum Causalprinzip* (Erlangen: Ferdinand Enke, 1866).
35. *Die physikalischen Axiome* 125; cf. *Die Prinzipien* 178, where the same thought has been somewhat modified and reinforced.
36. Afrikan Spir, *Denken und Wirklichkeit* (Leipzig: J. G. Findel, 1873); published in French as *Pensée et réalité*, trans. A. Penjon (Paris: Félix Alcan, 1896).
37. We had adopted this position ourself in our earlier work (IR 38 [Loewenberg 43]), but the reader has seen that our opinion has changed on this point (Ch. 5, pp. 104 ff.).
38. Kristian Kroman, *Unsere Naturerkenntnis*, trans. R. von Fischer-Benzon (Copenhagen: A. F. Höst, 1883), pp. 22, 195–6, 211, 247–248, 250.
39. Paul Tannery, *Pour l'histoire de la science hellène* (Paris: Félix Alcan, 1887), pp. 264 ff. Cf. Gaston Milhaud, *Nouvelles études sur l'histoire de la pensée scientifique* (Paris: Félix Alcan et Guillaumin, 1911), p. 6.
40. Max Planck, *Das Prinzip der Erhaltung der Energie* (Leipzig: B. G. Teubner, 1908), pp. 111, 30–31, 41–42, 116, 149–150, 151.
41. Gaston Milhaud, 'La Science rationnelle,' *Rev. de méta.* **4** (1896) 290–291; *Essai sur les conditions et les limites de la certitude logique* (Paris: Félix Alcan, 1898), pp. 131, 201; '*La science et l'hypothèse* par M. H. Poincaré,' *Rev. de méta.* **11** (1903) 786.
42. In order to familiarize the reader with Lalande's very interesting and suggestive considerations, we take the liberty of quoting in its entirety a passage from his book that seems to us to sum them up: "Considering then, finally, the operation of human intelligence, we see that its first step, an essential operation of the understanding, consists in objectifying the material given to it, which has the effect – and insofar as we can judge, the final cause – of making the object of thought universally valid for all thinking minds .... Thereby is established ... a common thought, which more and more loses its individual character, and which consequently tends, insofar as they think of common objects under the form of the universal, to make of them a single thought and a single mind.

    "In the second place, this unification bears not only on thinking minds but also on the objects of thought. For the necessity of understanding requires that one find the identical in them, and no judgment, or science, can be formed on the basis of the particular. The whiteness of snow, as we have already remarked, is not exactly that of the lily or of the cloud. Nature thus does not lend itself rigorously to the concept; nor does it resist it, since science allows us to bring nature indefinitely closer to the outlines that thought traces in a homogeneous space ... and the limit of this comprehension would be the ideal reduction of knowledge in its entirety to a purely intelligible form that would absorb the individual entirely into the identical.

    "Finally, from these two sorts of assimilation it follows that the intelligence, in order to progress, must also assimilate each of these two categories, objects and subjects, with one another. Indeed, in both cases, it is by its virtual reduction to rational elements that a given tree tends to become the same for my neighbor and for me, and also identical in its essence to any other tree of the same species. Now these rational ideas with which I thus form the essence of the tree (the tree in itself, a realist would say), however one understands them and even whatever origin one assigns them, are the laws of my intellectual nature and of that of my neighbor, or rather they are that nature itself: they constitute our own reality insofar as they are capable of being thought. So that, finally, the two trees under consideration are conceived as being, fundamentally and by right, identical to the reality of both of us, Peter and Paul; and the reality of all thinking subjects is conceived as being identical to the reality of all objects of thought. This is an excessive, presumptuous pretension of our intelligence, for the real, as it is actually presented, does not yield indefinitely, neither in me nor in my neighbor nor in the two trees, to what I demand of it; for in the matter of our science there remains an individuality that is irreducible, irrational, exasperating for reason, a sort of original absurdity, which is no doubt steadily diminishing but which never actually disappears; for, in a word, nothing would exist if everything were identical, and the world, reduced to 'the eternal axiom pronounced at the summit of things,' would be destroyed as an object of knowledge and of revelation. It is nevertheless a well-founded pretension, legitimate in its principle, like the pretension of a people split apart by the fortunes of war, almost totally excluded from the governing of its own affairs, yet maintaining, with heroic absurdity and an obstinate hope of triumph, the inviolable principle of its unity and its right to be master in its own house" (*La Dissolution opposée à l'évolution dans les sciences physiques et morales*, Paris: Félix Alcan, 1899, pp. 212–214; cf. pp. 33, 35, 36, 45, 66, 67, 117, 179, 414).

43. W. M. Kozlowski, *Psychologicznezrodla* (*Psychological Sources* ...), Warsaw, 1899, pp. 13, 51, 68; *Szkice filozoficzne* (*Philosophical Sketches*), Warsaw, 1900, p. 86; 'La combinaison chimique au point de vue de la théorie de la connaissance,' *Congrès international de philosophie de 1900, Bibliothèque du Congrès international de philosophie III: Logique et Histoire des Sciences* (Paris: 1901), pp. 536–537; *Zasady przyrodoznawtswa* (*The Principles of Science*), Warsaw, 1903, pp. 103 ff., 149, 264, 269, 272 [Polish language sources unverified].
44. Wilbois, 'La Méthode des sciences physiques,' *Rev. de Méta.* **7** (1899) 598 ff.; **8** (1900) 291 ff.
45. James Ward, *Naturalism and Agnosticism* (London: A. and C. Black, 1899), 1:214, 2:79.
46. ["Offspring begotten without a mother, mother dead without offspring."] Salomon Reinach, *Antiquités nationales, Description raisonnée du Musée de Saint-Germain-en-Laye*, Vol. 1: *Époque des alluvions et des cavernes* (Paris: Firmin-Didot, 1889), p. 168.
47. The author of the present work has himself come to know these efforts of Hegel only quite recently, having been led almost by chance, through the citations Ward borrowed from Hegel on the *explication* of the spirit in history (cf. Ch. 1, p. 9), to study the works of the philosopher more closely.
48. Spinoza, *Ethics*, Pt. 3, Introduction; Pt. 4, Prop. 4 and Appendix, Sect. 7.
49. [Reading "*fois*" for "*foi.*"]

CHAPTER 18

THE ONENESS OF HUMAN REASON

The fundamental contradiction we pointed out at the end of our last chapter has the particular feature that it cannot be understood as holding between the two concepts we treated there – intelligence and nature – with each of these concepts then appearing perfectly logical in itself. On the contrary, each concept is seen as bearing within itself an element of contradiction. It is certainly contradictory, as we have just seen, that nature sometimes conforms to the categories of reason and sometimes escapes them. Considered in its most general sense, it is the fact of the existence of the peculiar *structure* in reality that renders it accessible to theoretical science (Ch. 4, p. 82, and Ch. 16, p. 444). But if we only consider that intelligence needs this reality in order to function, we immediately transfer the conflict into the intelligence itself, which is, moreover, to observe once again that man is not an aberration in nature.

Hegel, by quite legitimately extending the limits of Kant's teaching on the antinomic aspect of our understanding, has admirably perceived how essential the conflict is. Indeed, the contradiction becomes apparent as soon as the existence of *diversity* comes to light. Now how and upon what could reason act if there were no diversity? Herein lies the justification for the Hegelian *dialectic*, though of course only a partial justification. For not only was Hegel mistaken as to the way in which reason resolves this conflict – having by-passed the correct process, for which mathematics offers the clearest model – but he also went astray in claiming that the conflict took place *entirely* within reason itself. Indeed, although it is true that we cannot imagine reason functioning in the entire absence of all diversity, it is just as certain that the existence of this diversity is suggested to us in the first place by perception, that is, by something that reason clearly understands as coming from outside itself. And it is clear that this primordial error has engendered a host of others. Having confined the original conflict within reason itself, Hegel believed that reason could subsequently give rise to other conflicts – which it would resolve, of course, exactly as it had the first. Thus was born in him the illusion that reason alone could develop the series of notions through one another with no intervention of a foreign element. In this connection, then, Schelling was certainly correct in affirming that the opposition, which Hegel intended to place on purely logical ground, was on the contrary played out in the mind as it considers *nature*, and that Hegel had been wrong to transpose in this way the original idea from which both the *Ideas on the Philosophy of Nature* and the *System of Transcendental Idealism* had arisen (Ch. 11, pp. 265–266, and Ch. 12, p. 312).

Nevertheless, from our point of view the important thing to underline is that the conflict is produced as soon as reason comes up against sensation, that is, as soon as there is perception. The conflict is not limited to science, since it precedes science and since the world of common sense is already a first compromise; but it persists and develops within science. Science simultaneously affirms and denies external reality, and there is no point in trying to reconcile logically these irreconcilable things. Scientists and philosophers have tried many times to do so, drawing science toward idealism, claiming that it could leave aside all ontology, being content to seek relations, establishing a system of *relations without relata* [*rapports sans supports*]. Now, whether that is possible or not (and one can legitimately join Malebranche in doubting the possibility of such an enterprise), science certainly does nothing of the sort. It does not begin, as it ought to do if such were the case, by dissolving the objects of common sense into a series of *immediate data* of consciousness, in the manner of Bergson's subtle and essential process; on the contrary, it starts from these objects, which it destroys only gradually, replacing them by a series of others just as "real," that is, just as ontological. Therefore the current conception would seem to be something else altogether. It consists rather in admitting (at least implicitly) that both science and common sense are concerned with reality and thus affirm this reality, whereas philosophy alone seeks to destroy it. By admitting this, one does not deny the existence of a contradiction in our philosophic and scientific conception of the world, but one confines the hostile elements in two camps, two restricted domains. It is easy to see that a great many treatments imply a doctrine of this type. The human brain is spoken of as if it were the seat of several distinct reasons, as if philosophic reason, scientific reason, indeed even common sense reason were essentially different from one another. Now that is not true. As the reader has been able to convince himself throughout our entire book, the scientist reasons exactly like the philosopher and the philosopher like the man of common sense. He has been able to see in particular how alike the procedures of philosophy, of science and of common sense are in their attempts at explanation.

Everywhere, in forms that are quite varied in appearance, our reason applies and can apply only a single and unique device, fundamentally the same, which consists in explaining the diverse by reducing it to the identical. And, on the other hand, the reader has also been able to convince himself how disastrous the attempts to separate philosophy from science have been in the history of human thought.

These two observations are obviously only two aspects of the same fact, namely that there is ultimately only a single and unique way of explaining the world, by the negation of all diversity. It is here in this indistinct whole of acosmism that science and philosophy come together again.

However, if we have imagined that scientific reason and philosophic reason are fundamentally diverse, it could not have been altogether without cause. The cause, moreover, is quite apparent: it is the very diversity between what reason produces in the two cases, a diversity which immediately strikes the least attentive observer and which, as we have seen, has been expressed in resounding altercations throughout the development of human thought.

But in order to explain this situation there seems to be no need at all to have recourse to the risky hypothesis of a twofold reason, a hypothesis, what is more, that is contradicted by all the evidence. Indeed, we have seen (Ch. 15, pp. 402 ff.) that although science and philosophy have the same starting point, namely the world given by perception, as well as the same arrival point, namely acosmism, and although they reason by using one and the same basic mechanism, nevertheless their paths are different. What particularly characterizes science is the fact that while it envisages the ultimate dissolution of matter into space, in the meantime it nonetheless firmly maintains the reality of *objects*. Underlying this way of proceeding is a sort of submission, a tacit acceptance of what reality contains that is irreducible to reason. No doubt this acceptance is not truly complete, for if it were, reason would waive its rights, an essentially impossible act. There remains, on the contrary, a sort of mental reservation, since reason at the same time maintains the firm hope of ultimately being able to get rid of what it has temporarily accepted: it is this hope that creates, for example, the attempts to reduce matter to space. In order to convince ourselves that this is a standard operating procedure for science, we have only to recall what we came to see in Chapter 14 (pp. 374 ff.) about the way it carries out its deductions. They are in fact essentially discontinuous, their point of departure often being found in purely empirical propositions. But that by no means implies that reason considers the propositions on which the deductions are based to be fundamental and irreducible; on the contrary, it seeks to deduce them in their turn, thereby clearly showing that its resignation was only momentary. Thus the belief in the ontological existence of objects is only a particular case – obviously the most important one – of the sort of roundabout way our reason is generally inclined to take.

The constant and systematic (albeit unconscious) use of this roundabout means is what truly characterizes science, what enables experience to play such a considerable role in it. Science reserves the right to rationalize the datum brought by experience at some later time; but for the moment it accepts it as is.

Of course philosophy can use the same procedure in its turn. Philosophy originally is the whole of the art of reasoning about reality, and science, which grew out of it, is therefore only a sort of special case of philosophy. But what the moderns term philosophy is strictly speaking an attempt to reconcile us with ourselves *hic et nunc*. Consequently Burnet is right in affirming that the acceptance of the irrational constitutes a sort of suicide on the part of philosophy (Ch. 6, p. 150). And this is also why, even temporarily, it has difficulty making the sacrifice to which science submits all the time.

Correlatively with the concept of philosophy in the narrowest sense, as we have just defined it, the concept of science has become more specific. But there, as we have seen, the evolution has continued almost into the present era: today's science, with its discontinuous and ephemeral deductions, is no longer altogether the science of Cauchy, for whom mechanical deductions were definitive and destined ultimately to form a continuous chain (Ch. 14, pp. 377 ff.). Of course it is even less the science of the pure empiricist, for Bacon and Comte both failed to recognize the true nature of the scientific endeavor, as they also failed to recognize the role of reason, which cannot stop reasoning. It remains nonetheless true that in comparing the present conception to that of Cauchy, the evolution in fact seems to have taken place in the direction wished by Comte (Ch. 14, pp. 378 ff.). And it is also clear that through this evolution the divergence between science and philosophy, or if one will, the specificity of the former as compared to the latter, is increased. Descartes's conception, by the very fact that it pretended to constitute a coherent system of global deduction of nature, was in this respect a philosophic rather than a scientific theory. That is even more true of ancient atomism, in which experimental knowledge plays a much lesser role than in Descartes and where everything is really only *a priori* foundation and continuous deduction. The awareness of this situation is probably what causes the atomic scientist of today to experience a certain uneasiness when Democritus and Lucretius are called his spiritual ancestors, as Smoluchowski reports (Ch. 17, p. 496). Actually the present-day physicist uses the concept of a discrete elementary particle to explain clearly determined observations; he does not pretend to make it the basis for a theory from which the whole body of natural phenomena will follow necessarily. He is mistaken, of course, in believing that for this reason the modern theories do not conceal, at a deeper level, the aprioristic foundation so apparent in the ancients; but he is not entirely wrong in thinking that the latter wanted to do and actually did something different. It is obviously not true that he is the only one who does science, "positive" science, since, as a matter of fact, positive science is something chimerical which has never existed and never will. But it is in some sense correct that the science he creates is more positivistic than that of his predecessors. Not only has he renounced any attempt at immediate global deduction analogous to that of Descartes, but in addition he has abandoned the hope that science will ever achieve such a deduction and especially the belief that he held fragments of the future deductive chain in his hands. Of course he cannot abstain from comprehensive views embracing a vast number of phenomena; but while he feels that reason orders him to make these constructions and that they will be extremely useful to him from the most immediate point of view by suggesting to him, by making him discover hitherto unsuspected relations, he also has the feeling of stepping outside his proper role (a feeling he does not have when he formulates a clearly ontological theory for a limited domain) and does not hesitate to call these conceptions "philosophic." Thus Bacon and Comte (this is only a slightly different aspect of an essential side of their work whose merit we acknowledged in Ch. 14, p. 378), in bringing experience to the forefront, were guided by a well-founded belief: there, in the way science seeks out experience and, still more, in the way it accepts the results of experience, is, as we have just seen, what truly characterizes scientific reasoning. Theory is no doubt an integral part of science, and there is no science without deduction; a mere amassing of experimental data does not constitute scientific work truly worthy of the name. But an overly hasty deduction, one that tries to embrace a vast field without taking sufficient account of the datum furnished by experience, is properly speaking an antiscientific work, or at least an extrascientific one, and in this sense the scorn that Schelling's and Hegel's scientific contemporaries unfailingly heaped on the *philosophy of nature* was surely legitimate.

That also helps us understand why, as we have pointed out (Ch. 11, pp. 273 ff.), the gap between science and the Hegelian conception is even wider than the one separating it from the *philosophy of nature* proper (that of Schelling and his disciples). In their deductions both Schelling and Hegel stray from the only path science considers really feasible, namely that of mathematical deduction, and in theory Hegel seems to be able to come to better terms with empirical science, since he concedes a wide margin to it, as we saw in Chapter 13 (pp. 351 ff.) and even on occasion defends it against Schelling's encroachments (cf. Ch. 11, p. 275, and Appendix 16, p. 591). In practice, however, Hegel proved far more sterile even than Schelling from the point of view of the influence that each exerted on science. This is because in the case of Hegel the initial error is compounded by his very virtues. The fact that his doctrine is conceived as a rigorous, continuous and strongly interconnected system further sets him apart from true science, whereas Schelling's "speculative science," which is fragmentary and somewhat unsystematic, can, in virtue of this defect, preserve more contact with scientific deduction, which is discontinuous in nature.

Here Bacon and Comte have expressed the inmost feeling of the physicist, who is constantly forced to react against his own inclinations, which would lead him to get ahead of himself, and who is all the more suspicious of the vast constructions of others, which it is easy for him to see as chimerical. Certainly this is one factor that makes the advent of any new synthetic conception in science so difficult. We might find it regrettable, thinking of some fine theories now firmly established whose triumph was unduly delayed. But as excessive as the scientist's theoretical prudence may sometimes appear, it is only the other side of an attitude whose advantages are obvious to anyone who remembers the extravagances into which the human mind has let itself be drawn in the past by its predominant disposition toward metaphysics. No doubt it is wrong, not to mention futile, for positivism to declare war on this disposition which man – even the scientist – is obliged to satisfy; but Comte is to be commended for cautioning the scientist against the danger of this sort of excess, a danger that in fact constantly lies in wait for him. And the awareness of this situation has undoubtedly contributed greatly to ensuring the resounding success of the doctrine, however obviously inadequate it may be in so many other respects.

However, even from the perspective of the laboratory scientist alone, that is of someone who does not try to examine the philosophic foundations of his science, but merely means to make it advance, it is still more advantageous to be aware of the correct doctrine. For only then, by understanding where metaphysics is inevitable, will the research scientist be truly protected against the temptation to abuse it. And he will also realize that he cannot entirely ignore what the philosopher accomplishes in a field related to his own. Without doubt the paths of science differ from those of philosophy, and introducing concepts borrowed from philosophic idealism into science is (as we have observed, Ch. 15, pp. 404 ff.) totally impracticable, except of course in the case of an idealism based on panmathematicism. But the scientist must be imbued with the conviction that philosophy contains an immense body of knowledge accumulated by the millennial work of the most vigorous minds of thinking persons and that when his steps lead him into these obscure regions, as is almost inevitable if his horizon extends beyond a very limited area, he will not be able to progress unless he acquires some knowledge of this background. He must not heed misguided counselors who flatter his vanity by suggesting that everything mankind has attempted outside the domain of science (as we define it today) was only illusion and vanity and that the scientist, equipped with his particular methods, can and must begin everything all over again *ab ovo*. For if he attempts to do so, he will suffer the inevitable fate of those who believe they attain originality through ignorance. The human mind, especially in dealing with these very general questions, has only a limited number of paths from which to choose, and if one is not forewarned by knowledge of the past, one will commit oneself to an already well-worn path and thus end up with entirely outdated conceptions – outdated in the sense that humanity, which has known them for centuries, has also recognized their treacherousness. As we saw in Chapter 14 (p. 386), that is what happened, in particular, to German science during the period of violent reaction that followed the so-called *witches' Sabbath* of romantic philosophy.

It might appear superfluous to address an analogous warning to philosophers. Indeed, despite the increased prestige from which philosophic studies seem to have benefited in recent years, we clearly still live in a period of flourishing *scientistic* tendencies, and this side would seem to be the one from which abuses are to be feared. However, the opposite danger, that of the philosophers' disregard for the results of science, is perhaps not entirely chimerical, even at the present time, as witnessed by the attitude of Boutroux and Bergson in the matter of the identity of light rays and heat rays. However natural the opinion of these two illustrious thinkers may be, however consistent with the tradition and the very essence of philosophic thought, it is certain that in rebelling against some of the *findings* of physics, they acted in vain. Whatever the value of philosophic knowledge, the philosopher must recognize that scientific knowledge is made of a different metal, is more solid than his own, and that if he takes on science head to head, he cannot win. And since it is perhaps not totally impossible that philosophy might once again win back something like its former position in the public favor, this truism is not an idle statement.

Thus science and philosophy cannot ignore one another. And we have also seen the underlying reason why: they are both emanations of reason, a reason that remains fundamentally the same in these two manifestations. And as for common sense, since it is only a first rough sketch of a scientific and philosophic system, it is, despite appearances, only a step along the same path: by unconsciously using the very procedures science and philosophy will undertake to apply much more broadly, it substitutes a world of incomparably more stable objects for our continually changing sensations; it thus enormously reduces the primitive diversity of these sensations, thereby paving the way for the subsequent effort of reasoning thought.

In this connection, it should be noted exactly what benefit we have derived from our examination of scientific thought. Thanks to what we might call the forced precision of science, thanks to the more rigid contours of its conceptions, the true nature of reasoning stands out more easily. It is therefore highly advantageous to use examples drawn from science in philosophic research in general, and in research concerning the theory of knowledge in particular. But we must add: examples from the science of the past as well as from present-day science. Indeed, it is not at all a matter of profiting from the *results* of science, as has sometimes been attempted. We have pointed out the difficulties of such an enterprise (Ch. 16, pp. 485–486). What we have tried to do here, on the contrary, is to understand how science reasons, how it reacts to the external world, that is, in other words, how it succeeds in turning our sensations into a body of scientific propositions. Indeed, the understanding of these processes can enlighten us on the functioning of our reason in general, precisely in those cases where its steps are less immediately perceptible.

Now, as the reader will have been able to convince himself in the course of this book, the science of the past is every bit as useful as that of today for the study of these processes. One might even say more useful. For by the very fact that this science is *outdated*, that we no longer believe in it, we are able to observe it more impartially. Indeed, however hard we try, we cannot attain such impartiality toward the science of today. The latter, its methods and its results, are among the most intimate and essential components of our intellectuality; they are flesh of our flesh, and it would be futile to try to root them out of our being. Contemporary scientific theories, provided we have not remained entirely closed to them, necessarily appear to us as the expression of truth, no matter how perfect a philosophic detachment we affect toward them. For we *need* a truth, a reality, and once that of common sense has been destroyed, our only alternative, provided we want to interest ourselves in science, whether to do it or only to understand it, is to adopt the representation offered by scientific theory. Of course this representation is not entirely coherent – no representation of reality can be, since reality does not entirely correspond to the requirements of our reason. Furthermore, this representation is not even fixed: we saw in Chapter 15 how science modifies it and how the individual scientist himself is obliged to transform it in his mind to suit the nature of his current preoccupations. Nevertheless, with regard to a series of phenomena of a determined order, the decrees of science are truths, facts – both for the scientist and for us. This is what makes our involvement so intense, and why it is so difficult for us to change our faith: that is how things *are*, we know it, we have seen it. Now, even for things seen with the naked eye, existence can only be inferred; and as for things we think we see through the microscope or the telescope, it is certain that we perceive only phantoms, whose appearance can greatly mislead us. However, we have only to question ourselves to realize that we believe in the existence of the planet Mars as if we had walked upon its surface. If a theory were advanced today to the effect that the existence of a solid mass almost as large as our earth in the place where we seem to see Mars is an illusion created by our optical instruments (as seems to have been demonstrated in the case of the famous "canals," which have so captured the public fancy), we would certainly have a great deal of difficulty believing it; and yet, remarkably enough, the immense majority of our contemporaries, for whom the existence of the planet is an article of faith, have never looked at it through a telescope, nor been able to understand why astronomy attributes to it the nature and dimensions with which we are familiar.

If we wished to replace this hypothetical example with a real one from the history of science, we would need only recall the case of the demise of phlogiston. There is no doubt that this entity, though essentially imaginary, represented something perfectly real for chemists until the triumph of Lavoisier's ideas; it would be no exaggeration to say that they *saw* it at least as much as the contemporary chemist *sees* atomic oxygen, the oxygen that enters singly into compounds. When Macquer, in response to Lavoisier's attacks, declared that to doubt the existence of phlogiston on the grounds that it could not be bottled was to understand nothing about "high-level chemistry," he certainly was quite accurately expressing an altogether general conviction.

The formidable resistance met by the creator of antiphlogiston chemistry, a resistance whose vicissitudes we have endeavored to recount in our Appendix 2, shows how difficult it is for the scientist to modify his convictions whenever a theoretical conception of any generality is involved. What is being asked of him, in fact, is quite simply to modify his representation of reality. Neither Scheele nor Priestley consented to do so, and if Cavendish proved somewhat less intransigent, it nevertheless seems to have been chiefly because of his aversion for the representation imposed thenceforth by the new theory that he decided not to continue the chemical research which had brought him so much glory. Black, who finally lent his support (Ch. 3, p. 63), perceived quite clearly wherein lay the strength of the holders of the old faith. "If the force of habit," he says in a letter to Lavoisier in 1791, "prevents some of the older chemists from approving your ideas, the younger ones will not be influenced by the same force; they will unanimously take your side" (*Enc. méth.* 3:561). Sometimes, then, the representation of scientific reality is so well-established that it is necessary to wait until the generation it dominated dies out for a new representation to gain acceptance. And we hardly need point out how such a situation – the dominance the reigning theory holds over our mind – can sway our judgment when we try to recognize the true foundations of scientific reasoning.

For the science of the past, on the other hand, our detachment is complete, so that we are in many respects in a better position to see, with all possible clarity, the true relation between the observations that were made and the conclusions that were drawn from them. But that, of course, is on the condition that the study was approached without bias and in particular without the bias which more or less implicitly takes the science of today to be the only real one and sees in the science of previous eras only a tissue of more or less absurd imaginings, which came between humanity and a correct view of things. On the contrary, we must in a certain sense disregard our convictions and even our scientific knowledge insofar as possible in order to try to assume the same mental attitude toward the conceptions of the past that their contemporaries must have had. That is no doubt a precept of all historical reflection, but perhaps nowhere has it been more frequently disregarded than in the case of the history of science. And perhaps nowhere has it been more difficult to follow, precisely because of the hold that contemporary scientific ideas have on our mind. Thus the same cause acts doubly, prompting us to underestimate the value of the past while overestimating that of the present.

This is a widespread error of the mind which, as Hegel says in treating the succession of philosophic systems,

> does not conceive the diversity of philosophical systems as the progressive evolution of truth; rather, it sees only contradiction in that variety. The bud disappears when the blossom breaks through, and we might say that the former is refuted by the latter; in the same way when the fruit comes, the blossom may be explained to be a false form of the plant's existence, for the fruit appears as its true nature in place of the blossom. These stages are not merely differentiated; they supplant one another [*verdrängen sich*] as being incompatible with one another. But the ceaseless activity of their own inherent nature makes them at the same time moments of an organic unity, where they not merely do not contradict one another, but where one is as necessary as the other; and this equal necessity of all moments constitutes alone and thereby the life of the whole.

Thus "the mere result attained [is not] the concrete whole itself, but the result along with the process of arriving at it."[1] In particular, the history of philosophy must be compared, as Hegel declares elsewhere, "not [to] a museum of the aberrations of the human intellect, but [to] a Pantheon of Godlike figures" (Hegel, *Enc.*, *Logik*, 6:167–168 [Wallace 160]). Indeed, nothing could be more true, and that is as true of the history of science as it is of the history of philosophy.

Will we be accused of prejudice or exaggeration if we claim that the way in which science is generally presented does not exactly invite sentiments of this kind in the reader? Even when the historical background of a discovery is provided, when an attempt is made to show that it was not born in a single day (as the childish legend of the apple would have it for the Newtonian concept of gravitation), that the person who gave it its definitive formulation had precursors and sometimes predecessors, one nevertheless tries to *particularize*, as it were, the most recent conception, exclusively stressing its new elements, pointing out how it was opposed to the science of the time and ignoring how it was connected to it. In this way the space separating the great man from his contemporaries is artificially widened, until it becomes a sort of gulf. The figure of the innovator grows inordinately, which is not an evil in itself, for mankind will never sufficiently honor its initiators, those who have guided it toward *knowledge*. But this promotion is accomplished at the expense of the opponents, who cut a sorry figure, since they refused to yield to evidence staring them in the face (at least that is what the tale, as it is most often told, clearly seems to imply). Now this is generally not the way things happened in reality, and it certainly serves no purpose to misrepresent this evolution. The theory we consider outdated had its merits; from the very fact that science had adopted it we can confidently conclude that it must have contained a particle of truth. And on the other hand the conception that replaced it, although *more true*, is nevertheless not *true* absolutely, and it may present difficulties, disadvantages unknown to the superseded theory. Certainly no sane chemist would dream of returning to the qualitative conceptions of yore. But that should not keep us from admitting that they provided, or at least tended to provide, a much more direct and thus more satisfying explanation (from the standpoint of the rationalization of reality, which is the true goal of all science) of the appearance and disappearance of properties during chemical reactions, than Lavoisier did and than all of modern chemistry born out of the efforts of this great man has been able to do: for this reason, then, the resistance of the phlogiston theorists, as defenders of the last small fragment of these qualitative theories, was not entirely unjustified. And similarly, without denying the immense progress constituted by the modern understanding of the conservation of energy in comparison with Black's conception, we must recognize that the latter had the advantage that the substratum whose conservation it affirmed, namely the caloric, was something much more concrete for the imagination than is our energy, a purely mathematical concept for which we can provide no satisfactory verbal definition.

Guyton de Morveau, at the moment he left the ranks of the phlogiston theorists to join Lavoisier, was fair-minded enough not to treat his former faith with the scorn recent converts often find *de rigueur*; on the contrary, he frankly recognized its merits (see Appendix 2, p. 561). He termed it "one of those errors that have been characterized as fruitful" (*Enc. méth.*, 1786, 1:627). Now what we must realize is that the same is true for all conceptions that have prevailed, now prevail, or will ever prevail in science: at best none of them can be anything more than "fruitful errors."

However, despite the apparent opposition between these theories and those that succeeded them, an opposition so often expressed by violent battle, what should be stressed above all is how much continuity there was, at bottom, between them, how much the earlier ones paved the way for those that followed, how necessary they really were to their blossoming. Thus, to confine ourself to the two examples just cited, the conservation of energy could not conceivably have been affirmed before it had been established that, in this vast domain of purely thermal phenomena, the quantity of heat is preserved without change. And likewise, although it would be strange to challenge the greatness of Lavoisier's discovery, that does not prevent us from recognizing how much the phlogiston theorists themselves had paved the way for the revolution: in many cases their phlogiston was actually only a sort of negative oxygen, and an immense series of well-defined phenomena was thus found already classified in almost the same way the new theory was to classify them: in part, the raw material was on site, awaiting the breath of genius to form a veritable organism endowed with a life whose intensity has never been surpassed by that of any scientific conception.

Pascal (who was actually following in the footsteps of Francis Bacon) expressed the apt thought that

> by an especial prerogative, not only does each man advance from day to day in the sciences, but all mankind together make continual progress in proportion as the world grows older, since the same thing happens in the succession of men as in the different ages of single individuals. So that the whole succession of men, during the course of many ages, should be considered as a single man who subsists forever and learns continually.[2]

The fundamental continuity of the theoretical sciences is only a particular example of this very general truth.

In speaking of the outdated conceptions of the past, we are certainly not obliged to forgo all criticism, and even less are we obliged, without decisive reasons, to think of reverting to them (as we have seen that Hegel did in many circumstances). But we must come to see that, however paradoxical a theory may seem to us today, the belief that had been reached, the path human reason followed in order to arrive at it, could not be radically different from the ones we would adopt in an analogous case if we based our work on the same observations; or, if one prefers, that what satisfied the intellectual requirements of men whose high intelligence cannot be underestimated cannot fail to satisfy us in our turn if we place ourselves in the same conditions. For that makes it clear that if we succeeded in laying bare the true motives of the reasoning being analyzed, an essential aspect of the human mind would be revealed to us.

In a similar vein, the reader will have realized – or so, at least, we dare hope – that there has been some benefit in scrutinizing the thought that inspired Hegel's attempt. It seems to us that if this is so, it is precisely because his doctrine, which is in the same domain and pursues a goal identical to that of science – since, for that matter, it is the goal of all activity of reason with respect to nature – nevertheless proceeds by such strangely different paths. This is why, insofar as certain of its peculiarities are concerned, we are in a better position to observe them with an unprejudiced eye than if we were dealing, for example, with the analogous enterprise of Descartes, so much in line with the essential principles governing the science of today.

It is incontestably to Auguste Comte's credit that he insisted on the role the study of the history of the sciences must play in such cases. "As to all the great difficulties that can be presented by the philosophy of the sciences," he declares, "I cannot, in general, recommend too highly the use of the comparative historical method," and he adds: "The philosophy of the sciences cannot properly be studied apart from their history, under pain of leading only to vague and sterile insights; as, inversely, this history isolated from this philosophy would be idle and unintelligible" (*Cours* 2:312–313 [cf. Martineau 204]). No doubt he can be reproached for not having sufficiently followed his own precept and for having approached his study with too many preconceived ideas and above all with the intention of using it to further what he considered "higher" interests; but the merit of having formulated this program remains, in our opinion, very great.

Obviously the principle implied by our affirmation of the essential unity of reason throughout the ages can only be heuristic. For, however few generations have passed since there has been a science recognizable as such, and consequently since there can have been any question of a history of the sciences (we of course leave entirely out of account the prehistoric mentality, which Lévy-Bruhl has treated so authoritatively[3]), and however improbable it may be that an organ as essential as the brain could have been modified in such a relatively negligible time span insofar as the way it functions is concerned, that is nevertheless not strictly impossible; indeed, one could imagine, if necessary, that the quite abnormal conditions, from the purely animal point of view, in which man has found himself placed since the beginning of civilization, have determined a sort of abrupt mutation in his mental structure, on the order of those de Vries observed in the morphological structure of the *oenothera lamarckiana*. And one could also, perhaps with somewhat more likelihood – this, as we have seen, is Hegel's position – suppose that the immense effort of reasoning itself to which the whole of enlightened humanity continually devotes itself has been enough to modify the foundations of human reason. Since Aristotle, he says, "spirit, after its labours over two thousand years, must have attained to a higher consciousness about its thinking and about its own pure, essential nature," which is why, Kant to the contrary, the logic of the Stagirite needs "a total reconstruction" (*Wiss. der Logik*, 3:38 [Miller 51]; cf. Ch. 11, p. 264).

The idea that human reason, in its deepest and most essential nature, namely in the very principles on which it rests, and consequently in the processes and, if we may use this metaphor, in the mechanism it uses *in understanding*, has changed with time, that it is no longer the same as it was in past eras and in particular not as it was at the time the Stagirite set forth rules for it, constitutes the foundation of the Hegelian logic and thus of his system as a whole. In Hegel there is almost a sort of evolutionism, and some very fine minds have at times allowed themselves to be seduced by this brilliant heresy. It is, indeed, clearly a heresy with regard to everything that has been, if not explicitly professed, at least tacitly assumed by the great thinkers who for so many centuries have done their utmost to examine the mechanism of human reason; they all, without a doubt, believed firmly, unshakably, in the fundamental immutability of reason; that was for them the solid rock on which any "theory of knowledge" (to use the modern term), and therefore any philosophy, must rest.

It is just as clear, it seems, that it is an extremely dangerous heresy. Not, certainly, from the standpoint of the actual functioning of reason, whose essential mechanism appears to be of remarkable strength and simplicity, resisting (at least in normal cases) all attempts, however artful, to interfere with it: there is no use trying to teach it what it *ought* to understand; if it has not actually understood, sooner or later it will manifest its resistance, imagining detours by which it will arrive, or at least believe it has arrived at this understanding it has not really achieved. But the evolutionist theory of reason would singularly complicate the task of the philosopher, the person who seeks precisely to elucidate the processes used by our intelligence. It is certainly a most arduous task – to realize this, one need only note, from the outside so to speak, what slight progress has been made in this area since Aristotle, for example. Now, if it were recognized that reason in itself changes, and until such time as the nature of the modification could be specified (and surely not even the most stubborn advocate of the theory could at the present time say that such precision has already been acquired), any observation in this area would bear an unknown factor which basically would make it more or less illusory. Unless, that is, the researcher limited himself to studying intelligences that were strictly contemporaneous with his own and could thus be assumed to function according to an analogous mechanism – thereby renouncing all knowledge that might be acquired by the study of minds in the past, even the most eminent ones, guilty of having used only an imperfect instrument, a reason inferior in quality to the one that functions in us. This would be every bit as pernicious from the standpoint of the history of the sciences, for ultimately it would tend to make us renounce any real effort to *understand* those who have preceded us, since it would be taken for granted that even this verb means something different for us than it had for them; it would, in other words, give formal sanction to the idea that unfortunately invalidates so many insights in this area, and which implicitly admits, as something that goes without saying, that our ancestors, since they "made mistakes," could only be minds of an inferior order.

Recognizing how deep these perturbations would be seems reason enough to decide not to commit ourselves lightly to this evolutionist path, to declare on the contrary that it is incumbent upon the supporters of such a conception to demonstrate, to make palpable to us, as it were, this alleged variation of reason.

That is an important point. In fact, it is easy to see that there is a vast domain in which neither of the two theses can be rigorously demonstrated: it is the domain of the new problems constantly presented to the understanding by experience. We ourself have shown how often, throughout the centuries, humanity, in attacking these problems, has ended up adopting conceptions it initially had found altogether absurd (cf. Ch. 15, pp. 397 ff.). Now such an intellectual upheaval can obviously be interpreted as the consequence of the fact that reason has undergone a transformation between the earlier judgment and the later one. For, in short, our reason is our thought as a whole, and it is artificial to introduce separations into it. In this sense, then, it can be said that our reason changes continually, that each fragment of knowledge we acquire, each experience through which we pass, makes it different from what it was before. The reason which is so imbued with the concept of the sphericity of the earth that it understands that the lines of terrestrial gravitation converge in the center is not altogether the same one for which these lines are parallel in space, just as the reason which has assimilated the principle of relativity differs from the one for which the concepts of time and space are, as Newton would have it, absolute and independent of one another. That is, of course, a simple question of definition. But by the very fact that we entertain the possibility of the existence of a logic, it is clear that we try to distinguish between form and substance in such cases, that we endeavor to separate out the concept of the instrument by means of which we operate and to make it independent of that upon which this instrument operates. Applying what we have just come to see regarding the difficulties the evolutionist thesis would entail, we shall judge, then, that these modifications should be attributed to the second rather than to the first of the two components of the understanding. We shall say that it is not reason itself which has varied, but that the new problem which has commanded its attention makes reason appear to us in a new light, makes explicit what hitherto was implicit in it. Furthermore, that is a point of view which is compelling in and of itself, precisely because we have a tendency to believe in the immutability of the foundations of our reason and consequently would need peremptory proofs in order to abandon this conviction.

Now a mere glance at the history of scientific conceptions in general suffices to show us that not only do we not find these proofs there, but that on the contrary, as we pointed out above (pp. 532 ff.), the continuity and solidarity of scientific endeavor in all epochs seem manifest. We easily see how much even the boldest and most innovative theories had their way paved for them and also how much they basically conform to the general trend of scientific endeavor. In other words, we have no difficulty in imagining that insofar as reason appears to behave differently than it had before, this new element was already to be found in it potentially and was revealed only as a result of new circumstances with which it had to deal. Indeed, we need hardly reiterate that we totally lack direct knowledge of the functioning of reason and can form opinions about it only by observing its functioning in concrete cases. Now, as we have seen (Ch. 7, pp. 190 ff.), it is extremely difficult to put our reason through a dry run and thus to foresee what its attitude will be toward a problem that has not really been put to it, a problem to which human thought has not applied itself with all the seriousness of which it is capable. The fact that in a new set of circumstances reason appears from a previously unknown perspective is therefore not surprising at all. Let us return to our two examples from Chapter 15. The really revolutionary aspect of the doctrine of terrestrial sphericity (pp. 397 ff.) is the fact that space henceforth loses the two privileged directions, *up* and *down*, which it seemed to possess. Now, no matter how strongly convinced primitive reason might be of the existence of this particular property of space – and we can confirm the vigor of this belief at any time by the way in which our immediate sensation, in obedience to common sense, conserves it – it is nonetheless certain that this reason established no confusion, no interdependence between this property and properly geometric ones. If one had asked a Greek who still believed the earth was flat, yet already had a smattering of geometry, how far underground this or that proposition ceased to be valid, the question would probably have appeared as ridiculous to him as it would to us, and certainly no one has ever felt the need to demonstrate that the Pythagorean theorem remains valid when the right triangle is drawn in a vertical plane instead of on a horizontal table. We can make a similar point about Einstein's principle of relativity. It is certain that this conception initially appears quite disconcerting, and in this respect Duhem is not mistaken in pointing out that the theory cannot be stated correctly in ordinary language; in fact, this is a peremptory proof that it is entirely outside the framework drawn for us by common sense intuition in the domain of time and space. If we but take notice, however, we realize that to a large extent the elements of this very shocking body of concepts preexisted in our consciousness. For example, with regard to the strange interrelationship the theory establishes between time and space, the two henceforth appearing as connected in a four-dimensional whole, we should recall the more than century-old statement of Lagrange: "Mechanics can be regarded as a Geometry of four dimensions, and mechanical Analysis as an extension of geometrical Analysis."[4] As a matter of fact, as soon as a symbol representing time was introduced into our calculations – that is, ever since our first steps in mechanics – it is certain that this symbol is found combined, by arithmetical operations such as multiplication, division, etc., with other symbols representing spatial magnitudes. Now such combination is actually possible only by setting aside, by consciously forgetting, as it were, the specific, irreducible characteristics of the notion of time and by reducing t to nothing more than a pure and simple mathematical magnitude. This is not an isolated action of our understanding; on the contrary, it is the fundamental process the understanding constantly applies to reality in seeking to make it rational, as we have shown with reference to panmathematicism (Ch. 15, pp. 413 ff.). Insofar as time is concerned, this process obviously has the effect of spatializing it, in a manner of speaking, that is, of making its unique dimension seem, up to a point, analogous to the dimensions of space. But already the causal tendency itself, if one is willing to consider it from this perspective, appears, as soon as we assume the existence of phenomena, to be an attempt to assimilate a time allowing reversibility to a spatial dimension which permits movement in both directions, whereas time constantly flows in a single direction. Moreover, there is nothing mysterious in this coincidence, for at bottom what is involved in both cases is one and the same thing, namely the transformation of the real into the rational. The authors of the theory of relativity themselves were so aware that this is a natural inclination of our understanding that they have sometimes exaggerated this particular aspect of the theory. For example, Minkowski declared that space and time must merge into a more general notion, that of the *Universe*,[5] and that in reality "all the variations given in the world of phenomena belong to a four-dimensional space, since they are at the same time variations in space and in time."[6] Likewise, according to Planck, "the principle of relativity simply teaches that there is in the four dimensional system of space and time no special characteristic direction."[7] These statements might suggest that in the new theory the assimilation between the spatial and temporal dimensions is complete. Now of course that is not the case; quite to the contrary, the irreversibility of progress in time is expressly stipulated, as is seen by Einstein's fundamental argument, which points out that one "cannot telegraph into the past."[8] The erroneous idea to which it tends to give rise is nevertheless quite typical.

It is just as interesting to recall that at the time hypergeometry was appearing, people had no idea of the phenomena upon which the recent theories were based and that when Lobachevsky, Riemann and later Helmholtz appealed to experimental verifications, it was with an eye to quite different and purely spatial observations (see Appendix 21). We saw above (Ch. 6, p. 143–144, and Ch. 15, p. 410) what the true foundation for these conceptions is. Clearly they arose out of *disinterested* speculations, that is to say that they were not created in order to explain physical phenomena, but originated from considerations of abstract mathematics. Indeed, as has been remarked, all the work of mathematical analysis needed by the new theory was ready and waiting, as it were, when the theory appeared, albeit through works having nothing to do with the notion of time.[9] In fact, there is no doubt that as soon as we conceive a spatial mathematics, the tridimensionality of space appears to us as a *given*, for which we vainly seek an explanation. Thus, here again the new theory appeals to a very profound and very general feeling, and we certainly have no difficulty understanding that no matter how bold the new theory may be, it does no more than develop possibilities for understanding that our reason, unbeknownst to us, already contained in a state of potentiality.

Finally, the reader will also recall, in the same context, the analogy we drew between the new spatial conceptions and the Kantian notion of space as a subjective form of perception (Ch. 15, p. 409), a notion obviously growing out of considerations that have nothing in common with those of modern physics.

Thus, all things considered, the fundamental nature of the intellect certainly tends to appear immutable as regards this very troublesome question of the relativity of space.

Let us note, however, that because of the very substance of our argument, it is valid only for the past. In the light of what the past teaches us, we can no doubt consider it highly unlikely that the future should differ completely in this respect. But we certainly cannot say that this is impossible, that the evolution of science will not someday take a direction forcing logicians to admit an essential deformation of reason itself.

What we therefore assume, needless to say, is that the effort we have undertaken here will continue. As we said in our preface, it is in fact the path that appears to be best-suited for revealing the mechanism of our thought and thus also for verifying whether or not this mechanism has remained the same. In this domain the philosopher can only follow the scientist. That obviously does not mean that here philosophy is the slave of science. Rather, it performs quite a different task, since it seeks to know that of which science itself is ignorant, namely its unconscious guiding principles. In these terms, science is for the philosopher only a sample of human thought – a particularly important sample, however, by the twofold fact of the seriousness that humanity puts into this effort and the relative ease with which the evolution of ideas, which takes place in plain view, permits us to distinguish the workings of reason.

But still less can philosophy conforming to this type pretend to dictate to science, since it bases its conclusions on the observation of science. A condemnation of a scientific theory in the name of a philosophic conception is most assuredly without effect. Science's past proves this beyond doubt and if, in spite of Duhem's protests, the theory of Einstein and Minkowski prevails (as seems likely at the moment we are writing), that will be a new example of this fact. It might be argued, to be sure, that Duhem himself was a scientist, but he is clearly acting as a philosopher in this case. By requiring that science be confined within the limits of common sense, he implicitly affirmed that in this way of looking at nature, reason manifests itself in its entirety, that no evolution of human thought could or can reveal to us any hidden corners that common sense did not allow us to suspect. Surely that is an opinion that everything, both in the history of science and the history of philosophy, tends to belie.

Once again, however, these successive revelations in no way undermine our belief in the essential immutability of reason itself. They therefore lend no support to Hegel's claim. But in order to see more clearly how unacceptable his claim really is, we need only recall what we have learned from the analysis of this philosopher's work concerning the way in which he believed reason had to behave toward nature. The phenomenon that Hegel himself considers to be in essential conformity with concrete reason, to be the model phenomenon, and by means of which he principally intends to explain reality – becoming – is certainly not capable of being really accepted by our reason, as everything conspires to demonstrate. The deduction Hegel undertook in order to arrive at this result is entirely ineffective, as Trendelenburg showed, and certainly nothing in the history of human thought since Hegel suggests that the rules of human reasoning underwent so profound an upheaval at that time. One can even say, daring as it may seem at first glance, that in spite of all the efforts of this powerful mind, becoming had remained almost as irrational for Hegel himself as it is for everyone else, that his own reason largely resisted the constraint he intended to place upon it. Indeed, if anything else had been the case, if Hegel had really thought strictly according to the rules of his concrete reason, how would a disciple, whose faithfulness to and profound comprehension of the essential principles of his master are unassailable, have been able to follow the chain of Hegel's deduction and arrive at the conception of a timeless reality from which all becoming is consequently excluded (Ch. 11, pp. 295 ff.)? It must be because the concept of rational becoming was not really an integral part of the body of Hegelian theory, that instead it remained, for its author, largely isolated and one might say inoperative, in other words that Hegel himself when he reasoned simply and unconsciously, did not follow the rules of his alleged concrete reason, but those of the reason of everyone and of all times, which, as we have seen, is none other than the poor abstract reason whose indigent tautologies he pretended to scorn so much.

To sum up what we set forth in Book Three as to the internal reasons for this formidable failure, we shall say with Hartmann that the two most essential points of the doctrine, the points at which a mere pinprick can make it collapse, just as puncturing the medulla oblongata kills an animal, are obviously the possibility of a progression of thought by means of the dialectic and the transition between idea and reality.[10] Now as we saw (Ch. 11, p. 291; Ch. 12, pp. 311 ff.; Ch. 18, p. 520), these two claims of the Hegelian doctrine are both null and void. Thought, wishing to engender being, succeeds only in creating an amorphous being, exactly like nonbeing, and given the impossibility of any true dialectic, any spontaneous evolution of abstract concepts, this being remains eternally immutable. It is thus in fact the sphere of Parmenides, and the fact that Hegel's end result, if examined closely enough, proves identical to that put forward by the ancient Eleatic, constitutes in itself a proof of the stability of human reason. But this proof has worth, needless to say, only because of the great vigor of Hegel's mind and the seriousness with which he made the attempt. It was truly necessary that he believe the contrary, that reason appear modifiable to him in order that, having tried in vain to modify it, he might persuade us of its immutability. No doubt that is a result he attained in spite of himself; but that must in no way diminish our gratitude to him, for, as we have said (Ch. 7, p. 190), there is nothing more difficult than making our reason function in a mere dry run, and the results of deductions truly reveal themselves only in thinkers who pursue them with the determination to succeed.

However, it is probably the history of the sciences that furnishes us the most conclusive demonstration of this sort of thing; since its field of action is the one threatened with the most serious upheaval by this novel idea, it is only natural (by the operation of action and reaction) that we should find there the most effective weapons to combat it. And, indeed, even a superficial study of the evolution of scientific conceptions over time suffices to make us completely discard any supposition as to an essential modification of human reason in this all-important domain of becoming, at least since the remote time when science began to dawn on the intellectual horizon of humanity. It is worth pointing out, in this context, that the most typical attempt to rationalize becoming – precisely the one we used in Chapter 11 (p. 296) to show how much this concept, in and of itself, upsets the instinctive norms of our understanding – took place well after the alleged Hegelian reform of logic. If no one protested against the theory of Boltzmann and Maxwell, if, on the contrary, it seemed quite *natural* to everyone, then human reason must not have varied at all in its way of understanding change.

What gives a more general significance to this observation about Hegel's philosophy is the established fact that the theory he advanced is the only really coherent conception of this sort that has ever been formulated; as one of the leaders of the English Neo-Hegelian school has pointed out, not without a certain smugness, "the logic of Hegel is the only rival to the logic of Aristotle."[11] Within German philosophy itself, Hegel's enterprise is altogether unique and isolated. No doubt Hegel does appeal to Kant (especially to the "transcendental logic"), and there are numerous points of contact with Schelling's philosophy. But Kant, at the height of his philosophy, explicitly declared that logic had not made one step either forward or backward since Aristotle and consequently seemed to be "closed and completed" once and for all;[12] and as for Schelling, the Hegelians themselves admit he used traditional logic.

Thus – and this is the final conclusion we hope the reader will wish to draw from our work – everything allows us to believe in the essential uniformity of our reason; better yet, everything compels us to affirm it. No doubt reason is, by its nature, antinomic, divided against itself as soon as it tries to proceed, as soon as our reasoning has a real content, even if this reality be only that of mathematical concepts. But its framework is nonetheless immutable. It yields continually, but only to right itself immediately, reformulating its needs, forever the same, which it never renounces, no more than it renounced them in the past, for that matter. To wish to force reason to modify its foundations, to compromise on its rights by definitively accepting what does not suit it, would seem to be the vainest of enterprises. Human reason, despite the conflict constantly taking place within it, is truly *one*. Although there can be no unshakable catholic science (Ch. 15, p. 398), there does indeed exist a sort of catholicity of the foundations of reason, as St. Vincent of Lérins defines it: everyone, always and in all circumstances, has reasoned and still reasons in an essentially invariable way.

NOTES

1. Hegel *Phänomenologie*, 2:4, 6 [Baillie 68, 69; Meyerson's brackets]. Cf. also 2:16 [Baillie 81]: "The truth is the whole. The whole, however, is merely the essential nature reaching its completeness through the process of its own development."
2. Pascal, fragment from the preface to the *Traité du vide*, in *Pensées et opuscules*, ed. Brunschvicg (Paris: Hachette, 1917), p. 80 ['Preface to the Treatise on Vacuum,' *The Thoughts, Letters, and Opuscules of Blaise Pascal*, trans. O. W. Wright (Boston: Houghton, Mifflin, 1893), p. 549]. See also the passages from Bacon in the note [omitted by Wright].
3. Lucien Lévy-Bruhl, *Les Fonctions mentales dans les sociétés inférieures* (Paris: Félix Alcan, 1910) [*How Natives Think* (New York: Knopf, 1925)].
4. Joseph Louis Lagrange, *Oeuvres* (Paris: Gauthier-Villars, 1867–1892), 9:337.
5. See Paul Langevin, 'L'Evolution de l'espace et du temps,' *Rev. de méta.* **19** (1911) 459.
6. Hermann Minkowski, 'L'Espace et le temps,' *Scientia* **5** (1909) 215 [erroneous citation].
7. Max Planck, *Acht Vorlesungen über Theoretische Physik* (Leipzig, 1910), p. 121 [*Eight Lectures on Theoretical Physics* (New York: Columbia University Press, 1915), p. 123]. These citations could easily go on and on, for physicists generally speak, like Sommerfeld, of the "four-dimensional universe or Minkowski's universe" ('Application de la théorie de l'élément d'action aux phénomènes moléculaires non périodiques,' *Brussels Conf.* 317) and present the theory of this universe, like Whittaker, without stating any reservations concerning the irreversibility of progress in the temporal dimension (*A History of the Theories of Aether and Electricity from the Age of Descartes to the Close of the Nineteenth Century*, London: Longmans, Green, and Co., 1910, p. 448). On this subject see also Albert Einstein, *Über die spezielle und die allgemeine Relativitätstheorie* (Brunswick: Friedrich Vieweg & Sohn, 1920), pp. 37, 83 [*Relativity: The Special and General Theory*, trans. Robert W. Lawson (New York: Crown, 1961), pp. 55, 121–122].
8. Cf. Paul Langevin, 'L'Evolution de l'espace et du temps,' *Rev. de Méta.* **19** (1911) 463.
9. See Albert Einstein, *Über die spezielle und die allgemeine Relativitätstheorie*, 9th ed., (Brunswick: Friedrich Vieweg & Sohn, 1920), p. 58 [*Relativity: The Special and General Theory*, trans. Robert W. Lawson (New York: Crown, 1961), p. 86 n].
10. Eduard von Hartmann, *Neukantianismus, Schopenhauerianismus und Hegelianismus in ihrer Stellung zu den philosophischen Ausgaben der Gegenwart* (Berlin: Carl Duncker, 1877), p. 328.
11. William Wallace, 'Hegel,' *Encyclopedia Britannica*, 9th ed. (London, 1909), 13:205.
12. Kant, *Kritik der reinen Vernunft*, Preface to 2nd ed., *Immanuel Kants Werke*, ed. Ernst Cassirer (Berlin: Bruno Cassirer, 1913), 3:13 [*Critique of Pure Reason*, trans. Norman Kemp Smith (B viii)]. René Berthelot ('Sur la nécessité' 124–126) rightly insists that on this point Hegel's position is clearly opposed to that of all the philosophers who preceded him.

APPENDICES

1. THE PRECURSORS OF HUME (p. 58)

It would seem that Hume had predecessors, or at the very least precursors, in the Middle Ages, in particular Robert Holkot in the twelfth century, about whom little is actually known (cf. IR 90, 342 [Loewenberg 87, 302]). As for Nicholas de Ultricuria [Nicolas of Autrecourt], who had previously been known only as an atomist and whose name has been suggested by Hastings RASHDALL ('Nicholas de Ultricuria, a Medieval Hume,' *Proceedings of the Aristotelian Society* 7 [1906–1907] 3 ff.), among the texts cited the only thing we see that really applies to this question consists of the following lines from Nicholas's Thesis 15: *Quibuscunque acceptis, que possunt esse causa alicujus effectus, nescimus evidenter quod ad positionem eorum sequatur effectus positio* ["Whatever conditions we take to be the cause of any effect, we do not evidently know that, those conditions being posited, it follows that the effect must be posited also" (Rashdall 10)], and this passage perhaps does not quite suffice to demonstrate the English author's claim. It is highly significant that both Robert and Nicolas professed atomistic opinions, and it is at least quite probable that, for these two aspects of their doctrines, they were closely linked to the Arab Mutakallimun, whom they probably knew through the résumés and refutations of Jewish thinkers, notably Maimonides, and who, in addition to a radical atomism, maintained the impossibility of any logical connection between cause and effect (see Isaac HUSIK, *A History of Mediaeval Jewish Philosophy*, New York: Macmillan, 1916, pp. xxi-xxii, xxvii, 249). During the Renaissance the distinction between *causa* and *ratio* was chiefly set forth by Giordano BRUNO (*De la causa*, *Le Opere italiane*, ed. Paolo de Lagarde, Göttingen: Dieterich, 1888, 1:230 [*Cause, Principle and Unity*, trans. Jack Lindsay, New York: International Publishers, 1962, pp. 79–80; Meyerson errs: Bruno is contrasting the terms *causa* and *principio*]), and there is no doubt a connection between his work and Galileo's statement on the impossibility of arriving "at complete knowledge of even a single thing in nature, be it ever so slight" (NORERO, 'Compte-rendu général du IVe

Congrès international de philosophie,' *Rev. de méta.* **19** [1911] 626), and consequently of arriving at any real deduction of the natural phenomenon. But it must be noted that these words were probably aimed only at the exclusively logical procedures of the School and that his opinions about mathematical deduction were undoubtedly quite different (cf. pp. 375 ff. above). For the epoch immediately preceding that of Hume and in particular the relations that can be established between his thoughts on this question and those of Locke, Leibniz, Cordemoy and Malebranche, cf. IR 342–343 [Loewenberg 302–303].

2. THE RESISTANCE TO LAVOISIER'S THEORY (p. 63)

We know that none of Lavoisier's three great rivals, whose work had played such a powerful role in the destruction of phlogiston theory, ever converted to the new theory. Scheele, who was perhaps the most extraordinary *discoverer* of experimental facts the history of science has ever known – FOURCROY, in his admirable historical account in the *Encyclopédie méthodique*: *Chimie, pharmacie et métallurgie*, which is nothing but a long panegyric to the glory of Lavoisier, nevertheless observes, in speaking of the great Swede, that "no chemist has made so many discoveries, nor more important ones" (*Enc. méth.* 3:525), and Jean Baptiste DUMAS, a half century later, notes the almost incredible fact that in a single paper on manganese oxide Scheele discovers manganese, chlorine, baryta and probably oxygen as well (*Leçons sur la philosophie chimique*, 2nd ed., Paris: Gauthier-Villars, 1878, p. 103) – was the first to pass away, in 1786. Nevertheless he still witnessed the discovery of the decomposition of water, published by Lavoisier in 1784 but announced the preceding year at a session of the Academy of Sciences (*Enc. méth.* 3:444). It seems almost impossible, given the way in which scientific information was generally transmitted at that time and the frequency of the communications between Bergman and the French chemists on the one hand and between Bergman and Scheele on the other, that the latter did not immediately pick up some word of Lavoisier's experiments, which were attended by foreign scientists (like Blagden, for example, in June 1783; see *Enc. méth.* 3:444). But even quite independently of this consideration, it is curious to note Scheele's opinion of Lavoisier's ideas in 1784, eleven years after the latter had established the principle of the conservation of the weight of matter in the work on the *Changement de l'eau en terre*, ten years after the immortal *Opuscules physiques et chimiques* which correctly explained the relations between carbonate of lime and quicklime, as well as between metals and their "calces," long after Lavoisier had completely set forth the foundations of his doctrine (between 1778 and 1780) and shown how useless and thus inadmissible it was to assume the existence of phlogiston: "Would it be so difficult to convince Lavoisier that his system of acids will not be to everyone's taste? Nitrous acid composed of pure air and nitrous air; aerial acid, of carbon and pure air; vitriolic acid, of sulfur and pure air; *acidum sacchari*, of sugar and pure air! Is this credible? I prefer to believe what the English say" (Carl Wilhelm SCHEELE, Letter to Bergman, 28 March 1783, *Nachgelassene Briefe und Aufzeichnungen*, ed. A. E. Nordenskiöld, Stockholm: P. A. Norstedt & Söner, 1892, p. 364).

Nitrous air (*Salpeterluft*) is what we call nitrogen dioxide; *nitrous acid*, nitric acid; *pure air*, oxygen; *aerial acid*, carbonic acid (the name *Luftsäure* that Scheele uses had been given to it in 1774 by Bergman, replacing the earlier name of *fixed air* or *mephitic air*; Lavoisier in 1777 called it *chalky aëriform acid* and finally, about 1783, *acid of carbon*; see *Enc. méth.* 3:432, 443, 476); *acidum sacchari*, oxalic acid. In referring to the *English*, Scheele was no doubt thinking of Cavendish and Priestley, but above all, it would seem, of Kirwan who, beginning in 1781, quite forcefully and ingeniously defended a doctrine according to which phlogiston was nothing other than *inflammable air* (that is, hydrogen), which was thus to be found in all combustible bodies. The idea won much support among the phlogiston theorists, and Scheele, as we see by the 1 February 1783 letter to Bergman (*Nachgelassene Briefe* 357, 360), strongly endorsed it.

Cavendish and Priestley lived on for many years. After a while, Cavendish stopped publicly defending phlogiston theory. What is more, he stopped doing chemistry (turning to the study of electricity, where he also made important discoveries, which he completely neglected to publish and which only became known long after his death). Perhaps his decision was not unrelated to the evolution in the chemists' prevailing opinion of the new ideas, which he found so antipathetic. At any rate, he never adopted the antiphlogiston theory, and all that he conceded toward the end of his life is that "most of the phenomena of nature seem to be capable of being explained as well, or almost as well, by Lavoisier's views as by the generally accepted principles of phlogiston theory."

Priestley fought Lavoisier's theory to his last breath, so to speak. As we know, Priestley was a man as remarkable for the loftiness and strength of his character as for the high value of his scientific intelligence. He was an enthusiastic supporter of the French Revolution, so unpopular in England at that time; in his *Letters to the Right Honorable Edmund Burke* (Birmingham: T. Pearson / J. Johnson) in 1791, he vigorously defended it against the impassioned attacks of this great orator. The same year, when it become known in Birmingham (where Priestley served as minister to a nonconformist community) that a few people had dared assemble to celebrate the anniversary of the fall of the Bastille, Priestley's house was ransacked and burned by the incensed crowd. The great scientist lost his entire fortune, most significantly some infinitely precious scientific instruments (the Academy of Sciences in Paris extended its sympathy to him on this occasion and Priestley, already honored with the title of French citizen, was in 1792 elected a representative to the Convention from the department of the Orne, a mandate he declined; he had explicitly accepted the title of citizen, and one of his sons, who lived as a planter in Louisiana and died there in 1835, retained the title). Rather than yield, Priestley emigrated to America, where he spent the last years of his life.

He defended his scientific ideas with equal ardor. "His perseverance in the battle for his basic ideas was extraordinary," said Cuvier in the beautiful *Eulogy* he delivered shortly after the death of the scientist. "Impassively he watched their ablest defenders move one by one into the enemy camp, and when at last even Kirwan had repudiated phlogiston, Priestley, standing alone on the battlefield, launched yet another challenge in a paper addressed to the leading French chemists" (*Mémoires de l'Institut des Sciences, Lettres et Arts*: *Sciences mathématiques et physiques*, 1806, 6:42–43). This challenge is the *Considerations on the Doctrine of Phlogiston* published in 1796. In spite of the fact that his adversaries seem to have won the acclaim of the scientific world, he feels so sure he is right that he sarcastically enjoins them: "Do not treat me like Robespierre. Bear with a small Vendée in chemistry! Answer me, persuade me, and don't abuse your power" (DUMAS 125). The chemist Adet, who had worked in Lavoisier's laboratory and was cofounder with him of the *Annales de Chimie*, of which he was editorial secretary, was at that moment serving as French ambassador to the United States. He wrote a reply, but Priestley returned to the charge and, in spite of the elucidations of Fourcroy and Berthollet, persisted in his opinion. In 1800 he published his final work, *The Doctrine of Phlogiston Established*, the title of which is self-explanatory. The same year he wrote to a friend: "I have carefully examined everything my adversaries have put forward, and I have total confidence in the stand I have taken .... Although almost alone, I do not in the least fear going down to defeat." The following year, Cruikshank refuted Priestley's principal argument (which rested on a confusion – common to Priestley and his adversaries – between carbon monoxide and hydrogen). But Priestley stood his ground, contested Cruikshank's conclusions, and in 1803 (a year before his death) brought out a second edition of the *Doctrine*, in which he maintained his position in full. Although Priestley does seem to have had a moment's hesitation around 1785, his faith was renewed under the influence of Watt, the famous inventor of the steam engine, who was also a chemist and an unrepentant supporter of phlogiston ('Experiments and Observations relating to Air and Water,' *Philosophical Transactions of the Royal Society of London* **75** (1785) 279 ff.).

Black no doubt has fewer experimental discoveries to his credit than Scheele, Priestley or Cavendish. But he has the immense advantage over them of being a great theorist, almost on the same plane as the most celebrated scientific theorists of all times. We spoke (Ch. 3, p. 61) of his merit as the author of the theory of caloric and have also mentioned (p. 62) how his genius manifested itself in the field of chemistry with the discovery of the true mechanism of the transformation of carbonated alkalis into caustic alkalis. But what must be particularly emphasized for present purposes is that by this work Black reveals himself to be a true precursor of Lavoisier. It is, in fact, by seeking relations of weight and giving them a dominant value, to the exclusion of relations of quality, which chemists then considered the only essential ones (as is seen by Meyer's theory of *acidum pingue*), that Black arrives at his conclusions, and in this respect Lavoisier has only to follow in his footsteps, which as a matter of fact he does, demonstrating from the very beginning of his great works, in the *Opuscules physiques et chimiques* of 1774, how well-founded Black's conception is. Lavoisier is also aware that his theory of oxidation is closely related to Black's hypothesis; in sending the latter a copy of his *Traité élémentaire de chimie*, he writes: "You will find in it some of the ideas whose first seeds you have sown .... I tremble to submit the new doctrine to the most important of my judges, the one whose support I most covet" (Édouard GRIMAUX, *Lavoisier*, Paris: Félix Alcan, 1899, p. 57). However – and this is what shows us both the distance separating the precursor from the actual author of a new doctrine and how difficult each new step in this domain is – Black remains attached to phlogiston theory until 1791. As he frankly avows that year, in the letter which finally informs Lavoisier of his allegiance to the new conceptions, he had "long felt great antipathy toward the new system, which treated as an absurdity" what he had "considered a sound doctrine," and this antipathy, "which originated in force of habit alone," had diminished only gradually, won over by the clarity of Lavoisier's demonstrations and the solidity of his program (*Enc. méth.* 3:561).

Lest we be surprised at the lack of comprehension these four great foreign scientists demonstrated toward Lavoisier's work, we need only realize how long and painful the struggle was in France itself, despite the noteworthy fact that Lavoisier had at his disposal resources that fortune has, alas, only rarely afforded great innovators. Indeed, it is well-known that Lavoisier was blessed with an income quite sizeable for that time, and if it is perhaps true (as claimed by Quenaud, who had been his colleague at the Commune of 1789; see François BONNEVILLE, *Portraits des personnages célèbres de la révolution*, Vol. 2, Paris: Cercle Social, 1796, Year 4 [of the First French Republic], unpaginated) that he administered it somewhat parsimoniously, he certainly spared no expense when it came to his scientific works; it was said that the experiments on the decomposition and recomposition of water, which had been done in the presence of a large number of expert witnesses and had created quite a stir, had cost him 500,000 pounds (BONNEVILLE). His not inconsiderable social influence also played a substantial role. He was an "*adjoint*" of the Academy of Sciences in 1768, at the age of twenty-three, when he was just beginning to be interested in the sciences, an "*associé*" in 1772, that is to say long before the beginning of his great works, and finally a "*pensionnaire*" (the highest rank) in 1778, at a time when his results were still largely unrecognized by chemists. He had the ear of government officials and was able to lend powerful assistance to his friends. His laboratory and salons, says GRIMAUX in his excellent biography (*Lavoisier* 49), "had become the center to which flocked all those, academicians and gentlemen alike, who were interested in the sciences" (cf. the enthusiastic description of the biweekly meetings at Lavoisier's laboratory in *Enc. méth.* 3:425). Among these visitors were Chancellor Malesherbes and several dukes – we know what that meant in the society of the times; his dinners were also celebrated. However, "from 1777 to 1785, in spite of Lavoisier's great efforts and numerous papers, he was quite literally alone in his opinion on the exclusive influence of air in natural transformations; while admitting the soundness of his experiments and the accuracy of his results, the chemists who testified to their exactitude and their merit still did not renounce the existence of phlogiston, and the theory they followed in their works and their demonstrations was always only a more or less forced agreement between Stahl's theory and the action of the air." It is Fourcroy who expresses himself in this way in 1796 (*Enc. méth.* 3:541), but GUYTON DE MORVEAU, ten years earlier, speaking of the situation as it appeared in 1785, declared that "in retrospect we cannot help being astonished at Lavoisier's bold doubt" as to the existence of phlogiston (*Enc. méth.*, 1786, 1:628). Since neither of them held to the date 1785 until after the fact, one could suspect some partiality on their part here, and, indeed, it is certain that Fourcroy's statement is not literally true. Well before 1785 several mathematicians, the best-known of whom was Laplace, had declared their support, as did the chemist Bayen, who must be credited with having discovered the reaction that, in Lavoisier's hands, had become one of the most convincing proofs of his theory, namely the spontaneous decomposition of precipitated mercuric oxide. At first Bayen's conversion seems to have influenced supporters of phlogiston to some extent (cf. what Pierre Joseph MACQUER has to say on this subject, 'Chaux Métalliques,' *Dictionnaire de chymie*, 2nd ed., Paris: Didot, 1778, 1:352), but it remained completely isolated and there is no doubt that the situation was in general as Fourcroy and Guyton described it. To be sure, immediately following the publication of the *Opuscules physiques et chimiques* the *commissaires* of the Academy of Sciences (Trudaine, Macquer, Leroy and Cadet) implicitly acknowledged in their report on this work that Lavoisier was right to place the principle of the conservation of weight above any other consideration. "It will be seen," says this report, "that Lavoisier has submitted all results to measurement, to calculation, to the scales – a rigorous method that, fortunately for the advancement of chemistry, is beginning to become indispensable in the practice of the science" (LAVOISIER, *Oeuvres*, Paris: Imprimerie Impériale, 1862, 1:663). But that is a far cry from approving the foundations of the new theory by rejecting the existence of phlogiston, and MACQUER in particular, then considered to be the grand master of chemistry in France, while accepting Black's and Lavoisier's position on caustic and carbonated alkalis (*Dictionnaire* 1:301), firmly maintained the existence of phlogiston, declaring that "the impossibility of collecting a substance and keeping it in a bottle is surely no reason, in good physics, to deny its existence, or call it into question, when in fact one has a number of demonstrative proofs. This has not prevented anyone who wants to dabble in high-level chemistry without understanding a thing about the science from using this bad argument against Stahl's phlogiston, which has quite recently been called imaginary and fictitious" (*Dictionnaire* 1:300). In the same way he ridiculed Lavoisier in a letter to Guyton de Morveau during approximately the same period (1778): "For a long time Lavoisier had threatened me with a great discovery he was keeping under his hat and which was going to do no less than overthrow phlogiston or combined fire; his air of confidence scared me to death. Where would we have been with our old chemistry if it had been necessary to rebuild the edifice from the ground up? For myself, I confess that I would have given up. Now that Lavoisier has revealed his discovery, ... I confess that I have one less heavy weight to bear." After more or less correctly enumerating some of the principal features of the new theory (beginning with the obviously incredible fact that Lavoisier believes "there is no material fire in combustible bodies"), he concludes as follows: "You decide whether I had reason to be so afraid!" (*Enc. méth.* 1:628). Macquer's sarcasm, we see, is much like Scheele's. Macquer had the excuse of writing five years earlier, but we shall see below that Baumé still expressed himself in a similar vein fifteen years after Scheele. Macquer, moreover, remained unrepentant until his death (1784).

The situation did not change until 1785, shortly after the discovery of the composition of water; at least in France, however, the evolution was rapid from that time on, and Lavoisier hastened to spread the word by publishing a sort of manifesto in collaboration with his supporters. This is the famous translation of the *Essay on Phlogiston*, in which, in 1784, Richard KIRWAN cogently expounded the theory mentioned above in connection with Scheele, while claiming to refute the opinions of the *antiphlogistonists* (the term dates precisely from this work; see page 8 of the translation [p. 7 of the original: *An Essay on Phlogiston and the Constitution of Acids*, London: P. Elmsly, 1784]). The French translation bears the title *Essai sur le phlogistique et sur la constitution des acides, traduit de l'anglois de M. KIRWAN; avec des notes de MM. DE MORVEAU, LAVOISIER, DE LA PLACE, MONGE, BERTHOLLET et DE FOURCROY* (Paris: Rue et Hôtel Serpente, 1788) – among all the productions of the human mind there may be no other work whose title juxtaposes so many names posterity has recognized as illustrious. But their support, except for that of Laplace, was of recent date. Claude Louis Berthollet, who had clearly taken sides against Lavoisier in his 'Observations sur la causticité des alcalis et de la chaux,' published in 1782, announced his support of the new doctrine in his 'Mémoire sur l'acide marin déphlogistiqué' at the 6 August 1785 session of the Academy of Sciences [*Histoire et Mémoires de l'Académie royale des Sciences, 1785* (Paris: 1788), pp. 276–295]. As was customary at the time, the work was not published until the volume of the *Mémoires* of the Academy appeared three years later, and it is curious to read there, at that late date, criticisms of Guyton de Morveau for affirming "the necessity of phlogiston" (p. 281; note the enumeration "Kirwan, de Morveau, de la Métherie" on page 285). This chemist had, in fact, converted in his turn in 1786. It so happened that at the very moment this conversion occurred, the *Chymie, pharmacie et métallurgie* section of the *Encyclopédie méthodique*, for which Guyton wrote *Chymie*, was in the process of being printed. Guyton, in the Foreword, while rendering homage to Lavoisier (see note 16, p. 488 of our Ch. 16) declared: "We shall have more than one occasion to say ... that we are very far from adopting in its entirety the explanation in which this learned chemist believes he can absolutely dispense with phlogiston" (*Enc. méth.*, 1786, 1:29) and boasted of having obtained for his new nomenclature (published in 1782 and conceived according to the principles of phlogiston theory) the approval of Macquer, Kirwan, Klaproth, etc. (1:v-vi). But when he arrived at page 625 of his volume, he changed his mind and decided to reveal this in a somewhat unorthodox fashion by inserting at that point a *Second Foreword* where he clearly asserts that "he who lacks the courage to consider things that run counter to his prejudices, to reverse what he himself has established, will not advance science" and that "the doctrine which reigned so long in the chemical schools of all Europe is only a hypothesis that can no longer be sustained" (1:626). The conversion took place as a result of experiments witnessed by Guyton in Lavoisier's laboratory (see KIRWAN's *Essai, Préface du traducteur*, p. viii; since "last year" is mentioned, one would think that the year in question was 1787, but that depends on whether the preface was really written in 1788. The translator was Madame Lavoisier, who was, as we know, a person with a very distinguished mind, very conversant with the work her of her husband, for whom she often served as secretary; it is also she who engraved the plates for the *Traité élémentaire de chimie*. GRIMAUX, *Lavoisier* 124, bears witness that the manuscript of the translation of Kirwan's *Essay* contains corrections in Lavoisier's own hand; the preface is therefore probably due to his direct inspiration). Fourcroy, who had thus far wavered (we can see in *Enc. méth.* 3:541 how proud he is of his "prudent decision" to preserve his "neutrality," which, he says, was the policy of the "best minds," the "coolest heads," those "most experienced in the cultivation of the sciences" who resisted "not the discoveries, but the total overthrow of the old order of ideas"), lent his support some time later than Guyton de Morveau, as did Monge (3:541).

The response to his *Essay* is not enough to convince Kirwan, who maintains his opinions and does not surrender until three years later, in 1791, in a letter to Berthollet which contains the following statement: "Finally I lay down my arms and abandon phlogiston. I see clearly that there is no authenticated experiment attesting to the production of fixed air by pure inflammable air; and that being so, it is impossible to sustain the system of phlogiston in the metals, sulfur, etc .... I myself shall offer a refutation of my essay on phlogiston" (*Enc. méth.* 3:560).

Kirwan's conversion, as resounding as it may have been, nevertheless does not appear to have been quite complete; at any rate, he never really managed to assimilate the foundations of the doctrine, as witnessed by the objections he formulated as late as 1800 with regard to the new nomenclature (see Hermann KOPP, *Geschichte der Chemie*, Brunswick: Friedrich Vieweg und Sohn, 1845, 3:162).

Nor is it altogether accurate to say, as Cuvier does, that he was almost the last to repudiate phlogiston. In France itself, La Métherie, who had the *Journal de physique* at his disposal, continued the resistance, as did the highly-reputed Antoine Baumé, whose *Manuel de pharmacie*, reprinted a number of times (Paris, 1762, 1767, 1784, 1790, 1795, 1797, 1818), stood as the authority until the early nineteenth century. His *Manuel de Chymie* (Paris: Didot le jeune, 1763) and *Chymie expérimentale et raisonnée* (Paris: Didot le jeune, 1773) were also widely used at the time. Furthermore, Baumé was the author of a good many interesting works. He was even influential as a theorist. On the question of affinity, whose prime importance in the chemistry of that period we have recounted (Ch. 8, pp. 223 ff.), Baumé was the first to point out, in 1773, that it was necessary to distinguish between affinity in experiments under humid conditions and that which is shown under dry conditions, and these views were promptly and almost unanimously welcomed by scientists. Bergman, in the affinity tables he published in 1775, which were considered a sort of immutable standard until the early nineteenth century (KOPP is correct in observing, *Geschichte*, 2:300, that Bergman's undertaking was almost as important as a general revision of atomic weights would be today) drew his entire inspiration from it. As late as 1798, in his *Opuscules chimiques* (Paris: H. Agasse, Year 6), BAUMÉ violently attacks the new chemistry. He strongly reproaches Macquer (his long-time friend and collaborator whom he had lavishly praised on other occasions) for accepting "the new system on the cause of the causticity of lime" (p. 4 and passim – on this subject Baumé himself professed a theory of *modified fire* quite analogous to that of Meyer's *acidum pingue* which we mentioned in Ch. 3, p. 62; cf. *Opuscules* 5, 26). Obviously, insofar as possible, Baumé avoids mentioning the name of Lavoisier, whose works he seems to deem insignificant – for example, he speaks of the "alleged decomposition and recomposition of water" (p. 4) – and he cannot heap enough sarcasm on those he calls "our modern doctors," whom he accuses of having "muddled by a thousand obscurities something that had been better understood ever since Jean Rey, not to mention what I have said in my *Chimie*: they claim today that the metallic calces owe their form to the presence of oxygen" (p. 60). Speaking of *vitriolized tartar* (potassium sulfate), he exclaims: "What a theory to teach to beginners, for heaven's sake! Who can admit the presence of sulfur in vitriolized tartar? The modern doctors, you will say? Right you are! They long ago gained the right to talk nonsense. Nonetheless, despite this privilege, I can no longer remain silent and let errors so catastrophic for the future of chemistry be sanctioned" (p. 311).

It must be noted that Baumé was not at all motivated by personal rancor toward Lavoisier. To be sure, he believed (unjustly, it seems) that he had a grievance against Lavoisier, whom he accused of having stolen his process for refining saltpeter (cf. GRIMAUX 91), but when his great adversary was about to climb to the guillotine, he very nobly dared intervene in his favor. Lavoisier's closest friends and disciples, on the other hand, kept silent, and several of them counted among the most influential men of the times: Monge, who was personally connected with Robespierre, from whom a word had sufficed to save Farmer General Verdun; Guyton de Morveau, who presided over the committee of public instruction to which the Convention had transmitted the letter Lavoisier addressed to it when a warrant was issued for his arrest and who found no way to utter a word in favor of his master; and especially Fourcroy, who certainly had drawn attention to Lavoisier by several actions, notably by branding him a counterrevolutionary at the time of the *purge* of the *Lycée des arts*, and who, when later accused (wrongly no doubt) of having wished for the death of someone of whom he was jealous, could plead only his own cowardice as an extenuating circumstance (see GRIMAUX 243, 267, 270, 309–310, 312). Baumé was not afraid to compromise himself by going into the prison to deliver a letter of support to the accused (GRIMAUX 288–289). For Guyton's attitude, see the letter he felt obliged to write to Crell, the German translation of which came out under the title GUYTON-MORVEAU, *Berichtigung wegen der angeblichen Miturheber von Lavoisier's Tod*, in a curious four-page brochure without place or date of publication and without pagination. In it Guyton chiefly pleads his absence during Lavoisier's trial proper, from 5 May to 8 May 1794. But Guyton did not have his passport for Meulan, from which he left for the army, until 29 April, and Lavoisier had been in prison and in grave danger since the end of November).

Marat was undoubtedly mentally unbalanced, and the failure of his own scientific efforts had inspired in him a blind hatred for all his contemporaries who had made their mark in science; furthermore he believed he had a particular grievance against Lavoisier, who had badly received his absurd *Traité du feu*. Nevertheless if, starting with his first attacks in *L'Ami du peuple* in 1791, Marat dared call Lavoisier an "apprentice chemist" and a "*coryphaeus* [chorus master] of charlatans" (GRIMAUX 206, 207), it is because at that time, in the eyes of the public at large, the glory of the author of the antiphlogiston theory was still anything but uncontested.

How much difficulty chemists had in freeing themselves from the conception that inflammability had to be due to the presence of a determined principle in inflammable substances can be seen by following the discussion occasioned between Humphry Davy on the one hand and Gay-Lussac and Thenard on the other by the discovery of alkaline metals more than a decade after Lavoisier's death. As soon as Davy had announced his results, French chemists set about verifying and completing them. But, at the same time, based on the existence of amalgam of ammonia, which had also just been discovered, they put forward the hypothesis that the new metals were not elements but *hydrides*, that is to say were composed of what we call their oxide, or rather their hydroxide (neither Davy nor his adversaries originally distinguished between these two kinds of combinations, and that was a great source of difficulties and mistakes) and hydrogen. The theory immediately won many supporters; obviously the chemists saw in it above all the possibility of a return to the former ideas on inflammability, and Davy, in his replies, was not wrong in calling this whole current of ideas a "phlogistic explanation." It must be noted, moreover, that Davy himself, while refuting the French chemists' hypothesis, is simply trying to establish that potassium and sodium are no more compounds "than any of the common metallic substances." As for the general theory concerning the existence of hydrogen as a common principle of inflammability in combustible bodies, he expresses himself very prudently on the subject; not only does he not reject it out of hand, but he cites a whole series of arguments in favor of this conception and concludes by declaring that "objects ... have not been sufficiently examined" for us to be able "to form any general theory" (Humphry DAVY, 'Nouvelles recherches électro-chimiques principalement relatives aux substances métalliques tirées des alcalis et des terres, et à quelques combinaisons de l'hydrogène,' *Annales de chimie* **75** (1810) 29, 61, 166, 173, 174 ['On some new Electrochemical Researches, on various objects, particularly the metallic Bodies, from the Alkalies, and Earths, and on some Combinations of Hydrogen,' *Philosophical Transactions of the Royal Society of London*, 1810, Part I, pp. 17, 37, 69, 73]; see also his 'Observations sur les recherches faites par MM. Gay-Lussac et Thénard relativement à l'amalgame fourni par l'ammoniaque,' *Annales de chimie* **75** (1810) 257, 272, 273, and GAY-LUSSAC and THÉNARD, *Recherches physico-chimiques*, Paris: Deterville, 1811, 2:215, 217, 253).

In Chapter 4 (pp. 94 ff.), we tried to understand the state of mind of the supporters of Peripatetic physics, who violently opposed the innovations of Galileo. That the situation of the supporters of phlogiston was completely analogous in this respect can be seen in Macquer's attitude. MACQUER does not fight against Lavoisier out of sheer obstinacy; on the contrary, he sometimes praises him warmly (as for example in the article on 'Causticité' of the *Dictionnaire*, 1:301, and in the article on 'Gaz,' 2:243). Moreover, he accepts his opinions concerning the composition of alkalis (as we indicated on p. 551), recognizing that by this theory the assumption of the existence of a special caustic component, such as Meyer's *acidum pingue*, becomes quite simply useless, as the epicycles or the crystalline spheres of the ancient astronomers are for us; for him these opinions henceforth seem "proven beyond the shadow of a doubt" (*Dictionnaire* 2:246, 302 [erroneous citations]). He goes a step further, for he discerns that what chemistry then called "causticity" is, strictly speaking, only a "tendency to union" (that is, a lack of saturation in the bodies where causticity was supposed to manifest itself; *Dictionnaire*, 1:315). He shows the same insight in refuting what was then considered the decisive argument in favor of Meyer's theory, which argument was based on the fact that, when neutralized by acids, caustic alkalis engender a stronger heat than noncaustic alkalis; according to Macquer, that is due to the fact that during this reaction noncaustic alkalis give off a gas, "mephitic gas" (carbonic acid; 1:307–308). Of course that is not the sole cause of this peculiarity according to our modern conceptions, but it is nonetheless one of the causes, and in comparing the emission of the gas, from the thermal point of view, to an evaporation, which engenders cold, Macquer showed himself not unworthy of the reputation he enjoyed among his contemporaries. We also see that he had recognized what the theory of *acidum pingue* had in common with the schema established by Stahl, since Meyer decided in favor of the existence of a hypothetical basis for causticity from the fact that causticity *passes* from one body to another, just as his predecessor believed he had established the same sort of thing for inflammability or metallicity (1:294–295). Is it not surprising, therefore, that Macquer should have shown himself so impervious to Lavoisier's argument against phlogiston? We have mentioned above (p. 551) that the proof founded on the impossibility of isolating phlogiston and putting it in a container did not appear decisive to him. Now, this is surely not because he underestimated the intrinsic value of such an argument, for in speaking of the explanation of causticity by the intervention of mephitic gas, he explicitly points out that this – unlike *acidum pingue* – is a "substance that is emitted, put in bottles, measured, weighed, combined at will" ('Esprit alkali volatil caustique du sel ammoniac,' *Dictionnaire*, 2:48).

But Macquer was not merely a distinguished chemist, he was also an excellent thinker, and he showed it here in understanding the true motives underlying his faith. Indeed, the reader has seen that what disturbed him most of all in the new explanations is that they required a complete overthrow of the received theories. It would have been necessary for chemistry "to rebuild the edifice from the ground up," in which case Macquer would have preferred to "have given up." He expressed himself even more clearly in another passage. Speaking of the proofs that were being drawn from considerations of weight, he says: "Only Physicists who do not really know this admirable science [chemistry] are capable of imagining that it can be influenced so quickly, and that a single fact, even supposing it to be well established, is enough thus to overthrow in an instant the beautiful ensemble of one of the finest theories to which the genius of chemistry has raised itself, one which draws, from an astounding number of demonstrative experiments, a strength that cannot be resisted by minds sufficiently fair and far-reaching to contemplate them all and to grasp their relations in a single glance" ('Chaux métalliques,' *Dictionnaire* 1:349 [Meyerson's brackets]). These are the sentiments scientists invariably experience toward any theoretical innovation so profound it necessitates a genuine reworking of their conception of reality in a given area.

We beg the reader's indulgence for having dwelt at some length on this particular phase of the evolution of science. As has been obvious from the beginning of this work, our epistemological opinions are founded principally on an examination of the role of explanatory theories. And since (as the reader is perfectly aware) it is first and foremost to history that we look for revelations on this subject, we are led to speak continually of the birth and death of these theories and to examine how reason behaves in these scientific revolutions. We thus believed it would be useful to give a detailed account of one of these upheavals, one of those that today certainly appear most justified, to use this case in point to show how difficult it was to win acceptance for this about-face and what resistance the innovator encountered. Our choice of the "revolution" attached to the name of Lavoisier (Fourcroy uses the term as early as 1796; *Enc. méth.* 3:440) was no doubt dictated first of all by the fact that, as a result of previous studies, this matter was somewhat more familiar to us than analogous subjects. But we have also been influenced by more valid motives.

Phlogiston theory is a very complete conception embracing a large body of phenomena. It is relatively close to us in time, since barely a century and a half has elapsed since Lavoisier began to attack it – and with how little initial success we have just seen. Therefore the documents are easily accessible and, more importantly, their meaning is not too hard to fathom: these men from the late eighteenth century obviously have a mentality very like our own in many respects. And yet, on precisely the subject of the chemical phenomenon, they just as obviously think quite differently, so much so that we must sometimes make a quite considerable effort to grasp the nature of this phlogiston of which they claimed to have such a clear conception. For it is not true, of course, that phlogiston theory is merely Lavoisier's theory *turned around.* No doubt it is that in many cases, but elsewhere it is something quite different; otherwise, it would seem, this difficulty of comprehension would not occur. Now this feature of the doctrine, the fact that it is so near and yet so remote, that since we have facilities to study the reasoning processes they used, we do not risk being distracted by the scientific convictions prevailing today, which are an integral part of our mentality – all that makes it an eminently suitable subject for our analysis. This is why we make frequent use of it throughout our work. But this is why it is also necessary that the reader have no doubt that it is indeed a true scientific theory presenting all the characteristics of such conceptions. We hope that our reasoning process itself will help demonstrate that such is the case; however, it seemed useful to pave the way for this conviction from the start, and a résumé of the circumstances that accompanied the fall of the theory appeared to suit our purpose admirably.

That is because in this case the real situation has been obscured by passionate polemics, in which nationalistic prejudice certainly played a considerable role, as we have indicated in the text. Indeed, with the antiphlogiston *revolution* chemistry changed nationality, as it were. In the chemistry of phlogiston, the Germans undoubtedly held the lead; not only were the two founders of the doctrine, Becher and Stahl, of that nationality, but the factual data that served as the starting point for the deductions were inclined to be drawn from experiments in metallurgy. Now metallurgy, which was then the only chemical industry having arrived at any sort of coherent theoretical conceptions, was above all a German industry. Kirwan's *Essay* states as an established fact that "it is to Germany that all modern nations must resort, to improve in mineralogy and metallurgy, as the ancients did to Greece to improve in oratory" [*Essay on Phlogiston* 8], and in their refutation the French antiphlogistonists do not dream of contradicting this statement; it was so far from their minds that when, at about that time, the group came up with the idea of setting up a periodical devoted to the new ideas against Abbé Rozier's *Journal de physique*, which was under La Métherie's direction (this plan was realized in 1789 by the creation of the *Annales de chimie*), the first proposal was that this periodical would principally print translations of articles appearing in the German *Annals of Chemistry* of Crell (see GRIMAUX 371 ff.), at that time considered the authoritative publication in the field.

With Lavoisier, his disciples and the disciples of his disciples, supremacy in chemistry passed to France for several decades, the Germans no longer playing any appreciable role; even among the three great adversaries of Lavoisier not one was of that nationality. In Lavoisier's era there was only one German chemist of note: Klaproth. He was a marvelous analyst who contributed greatly to the progress of the science, notably by the discovery of four new metallic elements (uranium, zirconium, titanium and cerium). But, although he had a fine mind, he seems to have had only modest theoretical gifts. Until 1792 he took no part in the controversy raised by antiphlogiston theory, being content to use the old nomenclature in his publications. In that year (that is, almost a year after Kirwan's conversion), he persuaded the Academy of Berlin to duplicate Lavoisier's experiments on the composition of water. When these experiments were successful, Klaproth unreservedly adopted the new faith, and his adherence brought with it the allegiance of a certain number of German chemists, especially among the younger ones. But many remained attached to the old doctrine and, until the early nineteenth century, continued to invent more or less extravagant theories designed to save the existence of phlogiston. KOPP regretfully takes note of his compatriots' attitude, which meant that the best work of the German chemists of that era remained without influence on the progress of science because of its outmoded theoretical conceptions and nomenclature. He also stresses the fact that feelings of patriotism were not unrelated to this resistance: "A certain national spirit, which dawned at that time, particularly in the domain of science, contributed to the fact that German chemists refused to exchange the system of Stahl, their compatriot, for the modern *chimie française*" (*Geschichte* 1:345). The Germans did not seriously return to the fray until more than a generation later, with Mitscherlich, Liebig and Wöhler. As a result, Germans imbued with nationalistic spirit tended to diminish the importance of the scientific upheaval which occurred at the end of the eighteenth century. Of course the most authoritative historians of German chemistry did not succumb to this temptation, and (as we said in the text) both Hermann Kopp and Albert Ladenburg spoke quite appropriately of Lavoisier's accomplishments. Ostwald, however, did not follow their example, and attacks against Lavoisier are quite common in Germany; it is clear that even some who (for lack of better alternatives) praise the merits of the Swede Scheele and the Englishmen Cavendish and Priestley believe they are to a certain extent defending the interest of "Germanism."

In France prejudice probably played a lesser role. The victorious antiphlogistonists spoke of the vanquished doctrine only with respect. Guyton de Morveau declared in 1786 that the hypothesis of phlogiston is "one of those errors that have been characterized as fruitful" (*Enc. méth.*, 1786, 1:627), and ten years later Fourcroy sang the praises of Stahl, who "transformed chemistry" and "made it truly a new science by the precision of his luminous ideas .... It is in this way, even more than by the great number of his discoveries, that Stahl erected a monument which has only grown larger and better for more than a half century. His simple and straightforward principles have long served as a compass for all chemists and if some have strayed a little from it ..., that has not prevented the best minds from quickly returning to Stahl's principles" (*Enc. méth.* 3:332). Toward the middle of the nineteenth century, Jean Baptiste DUMAS, the grand master of French chemistry of his time, gave a magnificent eulogy of Stahl in his *Leçons sur la philosophie chimique*. "All his works show a vast genius, a penetrating mind rich with all kinds of knowledge. He devotes himself to lofty and profound views, to far-reaching ideas. Indeed he abandons himself to them without reserve and pursues their consequences through the darkness of the dawning science. During that obscure epoch, Stahl's thought produces the effect of a lightning flash in the middle of the night, which cuts across the field of vision and glows as long as the eye can follow it, which still glows when the eye tires and loses it in the distance" (2nd ed., Paris: Gauthier-Villars, 1878, p. 83). The two historians of Lavoisier and his work, Grimaux and Berthelot, the first of whom was a very able chemist and the second, one of the foremost scientists of the nineteenth century, did not hesitate to pay tribute to Stahl's theory. "The simplicity of his interpretation of the facts," says GRIMAUX, "could not but be attractive to chemists .... Stahl's theory won universal approval and, with the help of phlogiston, no phenomenon remained unexplained" (*Lavoisier* 110–111). Similarly Marcellin BERTHELOT observes (*La Révolution chimique: Lavoisier*, Paris: Félix Alcan, 1890, p. 35) that a "doctrine so clear, so consistent with general appearances, one that coordinated so many phenomena by simple relations, struck his contemporaries with admiration. It was extended to the principal reactions of chemistry by three generations of scientists, among them men of great mental power, like Boerhaave and Macquer: they believed it to be definitive. The obviousness it claimed and the simplicity it introduced into the teaching of chemistry were such that it was abandoned only with regret." But Wurtz, in a famous phrase, seemed to imply that scientific chemistry had begun only with Lavoisier, thus relegating phlogiston to the prescientific past of this branch of knowledge, and from that time on his example has been followed only too frequently.

Still it is strange, to speak only of the question of personal merit that arises in connection with the doctrine of phlogiston, that people do not realize that Lavoisier's greatness is not enhanced – in fact he has no need at all of being aggrandized, since he is certainly one of the most authentic great men humanity has ever produced in any field – by denying the worth of a theory he had so much difficulty overcoming, and that, on the other hand, by pretending that the phlogistonists had done everything and that Lavoisier had only needed to turn their doctrine around to establish his own, men like Scheele, Cavendish or Priestley are made to seem dull-witted sorts who not only never thought of such a simple solution but – by pure stupid obstinacy, we are somehow led to believe – scornfully rejected it when it was presented to them. On the contrary, the truth is, as FOURCROY notes (and his competence in the matter will be difficult to deny), that "the pneumatic doctrine" has encountered great difficulties which it has overcome only with the most arduous efforts and the most unshakable persistence (*Enc. méth.* 3:500). Rarely has Virgil's *tantae molis erat* [*Aeneid* I, 33: So massive a task it was (to found the Roman race)] been more applicable than in the case of this formidable "chemical revolution."

3. THE FORMULA OF THE UNIVERSE IN LAPLACE AND IN TAINE (p. 66)

The idea of a single formula embracing all the phenomena of the universe is already found in Laplace: "We ought then to regard the present state of the universe as the effect of its anterior state and as the cause of the one which is to follow. Given for one instant an intelligence which could comprehend all the forces by which nature is animated and the respective situation of the beings who compose it – an intelligence sufficiently vast to submit these data to analysis – it would embrace in the same formula the movements of the greatest bodies of the universe and those of the lightest atom; for it, nothing would be uncertain and the future, as the past, would be present to its eyes" (Introduction, *Théorie analytique des probabilités*, *Oeuvres*, Paris: Gauthier-Villars, 1886, 7:vi-vii [*A Philosophical Essay on Probabilities*, trans. Frederick Wilson Truscott and Frederick Lincoln Emory, New York: Dover, 1951, p. 4]). One will notice that the idea is linked here (as is natural) to the idea of an ineluctable necessity governing the world. Furthermore, in what precedes, Laplace insists on the impossibility of a true act of free will, exactly in the sense of the ancient Stoics (cf. Ch. 4, p. 89): "Present events are connected with preceding ones by a tie based upon the evident principle that a thing cannot occur without a cause which produces it. This axiom, known by the name of *the principle of sufficient reason*, extends even to actions which are considered indifferent; the freest will is unable without a determinative motive to give them birth; if we assume two positions with exactly similar circumstances and find that the will is active in the one and inactive in the other, we say that its choice is an effect without a cause. It is then, says Leibniz, the blind chance of the Epircureans. The contrary opinion is an illusion of the mind, which, losing sight of the evasive reasons of the choice of the will in indifferent things, believes that choice is determined of itself and without motives" (7:vi [Truscott 3–4]). It is only logical that such a conception of universal necessity should have led to a conception according to which this ultimate formula, the ideal consummation of all knowledge, itself ultimately had to be found necessary.

Laplace's conception recurs in Taine. Each science must ultimately arrive at a "unique formula, a generative definition which, by a system of progressive deductions, will give rise to the ordered multitude of all other facts." In this way "only five or six general propositions subsist. There remain the definitions of man, of animal, of plant, of chemical body, of physical laws, of astronomical body; there remains nothing else." But then "we become more daring: considering that there are several propositions and that they are facts like any other, we try, by the same method as before, to perceive and to isolate the single primitive fact from which they are deduced and which engenders them. We discover the oneness of the universe and understand what produces it .... It comes from a general fact like the others, a generative law from which the others are deduced, just as all the phenomena of weight are derived from the law of attraction .... The final object of science is this supreme law and someone who could at one bound be transported to the heart of it would see there a fountainhead from which flow through distinct branching canals the eternal torrent of events and the infinite sea of things." Taine, we see, does not go so far as to admit, with Sophie Germain, that some day this supreme law could in its turn be conceived as necessary. On the contrary, for him it remains a *fact*, and he criticizes the metaphysicians who "in Germany, with a heroic audacity, a sublime genius and an imprudence even surpassing their genius and their audacity" have tried "to rediscover the world by geometric deduction rather than by looking at it." This is because, for Taine, the lesson to be learned from Hegel's enterprise is not altogether lost. But, as we see, he nevertheless grasps very clearly the relation between this attempt and the deductive tendencies of science. This is what allows him to proclaim that, although these metaphysicians have "fallen from a great height," nevertheless "in the ruin at the bottom of the precipice, the crumbled remains of their work still surpass all human constructions by their magnificence and their weight, and the half-broken plan that we distinguish in them shows future philosophers, by its imperfections and its virtues, the goal they must finally attain and the way not to begin their attempt" (TAINE, *Les Philosophes classiques* 362–370).

By this positivism, in which deduction plays so preponderant a role, Taine is obviously the spiritual ancestor of Goblot.

4. ARRHENIUS'S THEORY AND OTHER SUCH EFFORTS (p. 159)

Henri POINCARÉ observes that (as we establish on p. 159 using the image of the box) Arrhenius's hypothesis amounts to finding a naturally occurring process whose action is analogous to Maxwell's demon (*Leçons sur les hypothèses cosmogoniques*, Paris: A. Hermann et fils, 1911, p. 253). However, this device does not go far enough: "it is not enough to put a demon in the cold source; one would also be needed in the warm source" (p. xxiii). Furthermore, in Arrhenius as well "we tend toward uniformity of temperatures and densities, which is still in perfect accord with Carnot's principle; the nebulae do not warm up when the sun sends them heat, but this is because they yield heat in their turn to a still colder source, the void, whose absolute temperature is zero" (p. 255). Poincaré concludes that in this system the death of the universe would therefore only be delayed (p. 256) and that "in any event we ought to give up the dream of 'Eternal Return' and perpetual rebirth of the world" (p. xxiii).

We mentioned attempts analogous to those of Arrhenius (Ch. 6, p. 155). Understandably enough, they have become particularly frequent ever since the establishment of Carnot's principle, by which the concept of continuous change in the same direction has imposed itself in a more concrete form. Quite recently T.J.J. SEE ('The New Science of Cosmogony,' *Scientia* **11** [1912] 29–30) has also posited a circular process of cosmogonic phenomena, the dust expelled by the stars serving to constitute nebulae, which in their turn form new stars. But if centrifugal forces prevail in the existing stars, we do not see how centripetal forces could manage to bring together these same dusts in another location.

Jean BECQUEREL (*La Radioactivité du sol et de l'atmosphère*, reviewed in *Scientia* **13** [1913] 476) has advanced a similar hypothesis based on the reconstitution of radium in the celestial bodies, with the aid of helium atoms thrown off by stellar bodies. But why would evolution in space take the opposite direction from the one we know? The production of radium, as Frederick SODDY has shown quite convincingly ('The Parent of Radium,' *Scientia* **5** [1909] 262–263), would run counter to Carnot's principle, according to which energy dissipates. Now according to André JOB's formulation ('Chimie,' in Henri Bouasse et al., *De la Méthode dans les sciences*, 2nd ed., Paris: Félix Alcan, 1910, p. 177 [1909 ed., p. 129]), "given a chemical system, any spontaneous reaction of this system reduces its potential"; the supposed process is therefore clearly impossible.

5. HEGEL'S POLITICAL ATTITUDE (p. 263)

Understandably enough, Hegelian philosophy does not have a good press, if we may use that expression, in present day France. Almost from the moment Hegel settled in Berlin, twelve years before his death, his philosophy enjoyed the widest range of governmental favors. In his inaugural lecture Hegel declared that there was an "elective affinity" between his system and the Prussian State, and the latter soon showed that it fully endorsed this alliance. From then on, as an impartial witness notes, it was a sort of crime for a member of the Prussian educational establishment not to be a Hegelian, and if necessary the secular arm intervened very effectively (and, what is more, at the request of Hegel himself) to protect the philosophy it had adopted and its author from any attack (cf. ROSENKRANZ, *Hegel's Leben* 336, and HAYM, *Hegel* 365 ff., for details on the rather nasty affair Hegel had instigated by his attacks against Fries). Furthermore it was an extraordinary spectacle, and Hegel could legitimately take it to be an unprecedented triumph, to see the proud State of Frederick II, the State of government functionaries and *Junkers*, consider that its existence was substantially assured and strengthened by the fact that Hegel had, by deduction, proclaimed it to be necessary and just (HAYM, *Hegel* 4). As late as 1844, that is, at a time when the influence of the school had begun a severe decline, ROSENKRANZ felt the need to affirm that Hegelianism had nothing specifically Prussian about it, and that the distrust shown toward it in other parts of Germany, where it was considered an instrument of the Berlin government's spirit of domination, was therefore unjustified (*Hegel's Leben* xxiv). Since that time, the idea of an alliance between the modern Prussian State and the Hegelian philosophy has become so pervasive that we see it crop up in the most unexpected places and forms. For example, CROCE enthusiastically celebrates Cavour and Bismarck as "splendid embodiments of the Hegelian theory, men in whom the rational and the real were always fused and united" (*Ce qui est vivant* 55 [Ainslie 68]).

Having profited from the prestige of the Prussian political system, it is not altogether unjust that Hegelianism should, in a certain measure, be swept along in the catastrophe when a resounding failure finally shows the whole world and the Germans themselves what is concealed within this conception so admired by all those in Germany and elsewhere who believe force to be the supreme quality of peoples and States, what crimes against civilization and humanity this conception was capable of inspiring and justifying. The philosopher, however, has the right and probably even the duty to ask himself whether, in spite of all external appearances and in spite of what Hegel himself seemed to profess in the last years of his life, the connection between this thought and this political system is as close as has been claimed. What raises doubts is the well-known fact that Hegel had no Prussian leanings prior to the fall of Napoleon – in the famous letter to Niethammer dated from Jena on the very eve of the battle, foreseeing the outcome of the struggle, he declared that "as for the fate of the Prussians, in truth no better prognosis could be given," and three months later he saw the fall of Prussia as proof that "education triumphs over rudeness [*Roheit*], and spirit over spiritless understanding and mere cleverness." Nor did he experience any feelings of patriotism for Germany prior to 1813; he readily declared that Germany was not a State, professed the most profound admiration for Napoleon – glimpsing him on the eve of Jena, he had believed he was seeing "the world-soul" ride by and explicitly claimed to have long hoped for the success of the French army (*Briefe* 19_1:68–69, 82 [Butler 114–115, 122–123]; ROSENKRANZ, *Hegel's Leben* 229–230 [Meyerson's brackets]). Even when he spoke of the reestablishment of Germany (as in his work that bears this title), he believed that the task had to be entrusted not to Prussia, but to Austria. To be sure, ROSENKRANZ tried to demonstrate that Hegel's attitude could not have been more patriotic at that moment, that he keenly felt the misfortunes of Germany and that it is this profound pain which had found its expression in ironic remarks (*Hegel's Leben* 200). But an impartial examination of the documents cited by this biographer himself as well as of those found in HAYM (*Hegel* 257 ff., 344 ff.) is enough to convince us that it is the latter who is right and that at this period the humiliation of Germany was a matter of total indifference to Hegel – Rosenkranz's claim to the contrary merely demonstrates the sometimes almost inconceivably stubborn partiality shown by this biographer whenever it comes to defending the memory of his hero against any accusation at all; the accusation was particularly serious in the present case, given that in the meantime German nationalism had become quite intense. After leaving the University of Jena, Hegel served for about a year and a half, in 1808 and 1809, as editor in chief of the *Bamberger Zeitung*, directing the newspaper in a spirit absolutely devoted to the Bavarian administration and, consequently, to Napoleonic ideas. Now at this time his "system" was completed, at least in its general outlines: indeed in the letter to Niethammer he tells how, that very day, he had written the last pages of his *Phenomenology of Mind*, and the great anxiety over the fate of his manuscript, amidst the vicissitudes of war, makes rather an amusing contrast with his indifference to the fate of Germany. HAYM rightly points out that since Hegel's political aspirations had been satisfied in two successive States – Napoleonic Bavaria and Prussia after the "war of liberation" – which are at almost opposite poles from the standpoint of the fundamental conceptions that inspired their governments, there cannot possibly be as close a connection between these aspirations and the principles of Hegelian philosophy as Hegel himself would like to have us believe (*Hegel* 261). It is just as convincing, indeed even more convincing in this context, that almost as soon as Hegel died, a deep split occurred, as we know, among his disciples, the old Hegelians upholding conservative ideas, while the faith of the young Hegelians authorized any and all revolutionary audacities. It would certainly seem that this contrast can be explained only by assuming that the doctrine of the master, in spite of the apparent rigor of the deductions, at bottom contained much uncertainty and indeed more than one internal contradiction insofar as politics is concerned. Moreover an analogous evolution and duality can be seen outside the political domain: thus CROCE points out that Hegel's school has produced both great historians and "the most petulant and comic depreciators of history" (*Ce qui est vivant* 121 [Ainslie 149]).

Although Hegel formally protested against the law that might makes right, as René BERTHELOT has correctly pointed out ('Sur la nécessité' 132, 160, 175), from the standpoint of practical politics he probably was merely an ardent admirer of success. But in no way was he an enemy or a contemner of France. At the University of Tübingen, apparently under the influence of French fellow students from the Montbéliard region (which of course belonged to the Duke of Württemberg until 1792), he had absorbed the ideas of the Revolution. With friends he had planted a "liberty tree" and collaborated in the escape of a French prisoner of war. His comrades testify that he was one of the most ardent orators proclaiming the revolutionary principles of liberty and equality (Wilhelm

DILTHEY, *Die Jugendgeschichte Hegel's*, Berlin: G. Reimer, 1905, p. 14). He never renounced this youthful enthusiasm and even during the last years of his life spoke in grand terms of the Revolution: "The conception, the idea of Right asserted its authority all at once, and the old framework of injustice could offer no resistance to its onslaught .... Never since the sun had stood in the firmament and the planets revolved around him had it been perceived that man's existence centres in his head, i.e. in Thought, inspired by which he builds up the world of reality .... This was accordingly a glorious mental dawn. ... a spiritual enthusiasm thrilled through [*durchschauert*] the world, as if the reconciliation between the Divine and the Secular was now first accomplished" (*Phil. der Geschichte* 9:535–536 [Sibree 466; Meyerson's brackets]). Above we mentioned his enthusiasm for Napoleon. Amidst his success in Berlin, Hegel remains faithful to him (ROSENKRANZ, *Hegel als deut. Nat.* 116).

But France's past equally aroused his admiration; in the Middle Ages "the flourishing state of the poetic art in the hands of the Troubadours, and the growth of the scholastic theology, whose especial center was Paris, gave France a culture superior to that of the other European States." Under Louis XIV, "France, too, had the consciousness of its intellectual superiority in a refinement of culture surpassing anything of which the rest of Europe could boast" (*Phil. der Geschichte* 9:489, 519 [Sibree 421–422, 450]).

One need only recall what the spirit of narrow nationalism was like at that time in Germany, this *Deutschthümelei* for which everything beautiful and noble in the world had to be of Germanic origin, to realize that Hegel's attitude was not without merit. Indeed, Victor Cousin felt this and noted, during his first visit, in 1817, how much Hegel differed in this respect from a Friedrich Schlegel, for example, who "at bottom detested the Revolution and France." "Hegel loved France," says Cousin, "he loved the Revolution of 1789, and to use one of the Emperor Napoleon's expressions of which Hegel often reminded me, he too was a *Blue*" (*Fragments et souvenirs*, 3rd ed., Paris: Didier, 1857, p. 79).

Unlike Fichte, Hegel deeply admired Descartes and the "noble Malebranche." It was impossible for him to appreciate the contemporary French philosophers – the only one who might have interested him, Maine de Biran, was quite unknown even in France during Hegel's lifetime, and those of the eighteenth century, as pure philosophers, were obviously at the opposite pole. But Charles Andler himself (who can hardly be suspected of prejudice here) recognizes how deeply the political thought of the French *philosophy of the Enlightenment*, whose merits Hegel has explicitly praised as a matter of fact, influenced this part of his work ('Les origines philosophiques du pangermanisme,' *Rev. de méta.* **23** [1916] 684). And as for French science, the reader will see in the course of this work what admiration it inspired in him.

To characterize this side of his mentality, perhaps it would be worth our while to call attention to the fact that (as revealed by his correspondence) the grandeur of the things he had seen on his visit to Paris in 1826 overwhelmed [*überwältigte*] him and that his spoken discourse (according to the biographer, who knew him intimately in the last years of his life) was full of Gallicisms (ROSENKRANZ, *Hegel's Leben* 366, 361).

6. THE PRESTIGE AND THE DECLINE OF HEGELIAN PHILOSOPHY (p. 263)

We call the reader's attention to the opening pages of Haym's book, where he presents a striking picture of the prestige Hegelianism enjoyed at the time of the master's death: "science in its entirety sat at the sumptuously appointed table of Hegelian philosophy, ... all the other faculties danced attendance on the faculty of philosophy in order to appropriate at least a fragment of these lofty views on the absolute and a bit of the pliable subtlety of the famous dialectic; ... either one was a Hegelian or one was nothing more than a barbarian and an idiot, a backward and contemptible empiricist..." (HAYM, *Hegel* 4). The Hegelians were so sure they possessed the absolute truth that (in 1830!) they seriously debated what they found to be a perplexing question: What content could there possibly be for future historical evolution, given that their philosophy had demonstrated "the world spirit" to have reached its ultimate goal, self-knowledge?

Furthermore, we know that the influence of this philosophy was very deeply felt in the most diverse branches of intellectual activity. One need only look down the long lists of those who claim kinship with Hegel, either in Kuno FISCHER (*Geschichte* 8:1160 ff.) or in the UEBERWEG-HEINZE textbook (Friedrich Ueberweg, *Grundriss der Geschichte der Philosophie der Neuzeit*, rev. and ed. Max Heinze, Berlin, 1883, 3:396 ff.). In particular, there is no doubt that Marxism derives quite directly from Hegel. Marx, Engels and Lassalle were all three convinced Hegelians; in the case of Marx, Léopold LESEINE, in his excellent study entitled *L'Influence de Hegel sur Marx* (Paris: Bonvalot-Jouve, 1907), has established that Marx borrowed "more than he himself seems to believe" from Hegel (p. 24), and the opinion of Jean JAURÈS, obviously highly competent on this subject, is altogether analogous (*De primis socialismi germanici lineamentis apud Lutherum, Kant, Fichte, et Hegel*, Toulouse: A. Chauvin et fils, 1891, p. 58). In addition Jaurès establishes that although Lassalle sometimes claimed kinship with Fichte, he was chiefly influenced by Hegel (pp. 56, 73, 82). Finally, as for ENGELS, one must read his encomium of the Hegelian method (*Eugen Dühring's Umwälzung der Wissenschaft*, Leipzig: Genossenschaftsbuchdruckerei, 1878, passim, esp. pp. 5–6, 116 [*Herr Eugen Dühring's Revolution in Science* (Chicago: Charles H. Kerr, 1935), pp. 20, 143]) to see how much enthusiasm this disciple still maintained during an epoch when Hegelianism had become an object of horror for German public opinion in general. "Hegel," declares Engels, "having been not only a creative genius, but also a man of encyclopedic knowledge, marked an epoch everywhere" (Kuno FISCHER, *Geschichte* 8:1168). Kuno FISCHER also considers Hegelian philosophy to have "dominated the nineteenth century" to such an extent that Hegel deserves to be called "the philosopher of the nineteenth century" (*Geschichte* 8:1191, 1176), and J. H. STIRLING hardly exaggerates when he says that at a given moment this was the "philosophy to which ... the eyes of all Europe seemed turned" (*The Secret of Hegel*, Edinburgh: Oliver & Boyd, 1898, p. xviii).

In France the infatuation was undoubtedly less pervasive than in Germany, but, not to mention Victor Cousin and his contemporaries, it should be remembered that Taine acknowledged his debt to this philosophy, "the only metaphysics there is, except Aristotle's"; Hegel is "Spinoza enriched by Aristotle and standing on that pyramid of sciences which the modern mind has been constructing for three hundred years" and "of all the philosophers there is none who has climbed to such heights or whose genius approaches this prodigious greatness" (Hippolyte TAINE, *Sa Vie et sa correspondance*, Paris: Hachette, 1902, 1:154, 163; *Les Philosophes classiques* x, 133). Renan's thought likewise appears "totally impregnated with Hegelianism," according to E. ZYROMSKI, a very competent judge ('Les Caractères généraux de la littérature française au XIXe siècle,' *Revue des lettres françaises et étrangères*, Bordeaux, **1** [1899] 14). Zyromski also establishes that many other eminent minds in French letters were more or less directly influenced by Hegel (pp. 7 ff.), and although this critic has perhaps gone a bit too far in seeing this imprint in one or another of our contemporaries, his conclusions as a whole remain rather impressive. Furthermore we know that ever since the generation after Hegel the accusation of *Hegelianism* (for it was indeed an accusation, and one against which people often defended themselves) was leveled against a series of French thinkers; for example, Father Gratry so accused Vacherot, Renan and Schérer (Félix RAVAISSON, *La Philosophie en France au XIX*ᵉ siècle, 2nd ed., Paris: Hachette, 1885, p. 137).

In the English-speaking countries, Hegel's philosophy found several interpreters even during his lifetime, the most notable of whom was undoubtedly the poet and philosopher Coleridge. But the real influence of the Hegelian doctrine there dates from a period more than a generation after the death of the philosopher, and this belated flowering was, as we know, extremely vigorous. It even appears, especially since the publication of J. H. Stirling's book, that this neo-Hegelianism (as it has been called) was actually the mainstream of English, Scottish and American philosophic thought during the period that has just passed, and its intensity does not seem to have diminished perceptibly in the last few years. To indicate the strength of this growth, it will suffice to note, as MCTAGGART did near the end of the last century (*Studies* 238), that in the most disparate fields, such as religion, ethics, history and political science, many authors called themselves Hegelians without perhaps being fully entitled to do so. In the course of the present book we shall have occasion to consider several works of the English neo-Hegelian school, and the reader will see how serious these thinkers' efforts are and how closely they are connected to the foundations laid by Hegel himself. In Italy Hegelianism got a solid foothold through the works of Bertrando Spaventa and Benedetto Croce, and although the latter cannot be considered an orthodox Hegelian (as will be seen later in this work, p. 491, note 43, his attitude toward science is much more clear-cut than Hegel's), he was greatly influenced by the thought and the method of the author of the *Phenomenology*, whom he calls an "intellectual giant."

As a matter of fact, in Germany itself the prestige of Hegelianism began to decline as early as ten or fifteen years after the philosopher's death. Soon, hastened by the split within the school and by the political events of 1848, there was a sort of sudden and complete catastrophe. This whole edifice, once so proud, appeared to have abruptly collapsed into an abyss which had opened beneath it. Henceforth, general opinion in Germany goes from one extreme to the other, treating Hegel and Hegelianism with the deepest scorn. HAYM observes in 1857 that the *Phenomenology*, "a book in the study of which an entire generation of ardent disciples had tortured their minds" (*Hegel* 241), was henceforth as little read as Klopstock's *The Messiah* (which, ever since Lessing's famous epigram, has been, for a German, the prototype of a work whose title is familiar to everyone but no one ever reads), and the same year D. F. Strauss, still a Hegelian at that time, feels obliged to defend his master, "the dead lion," against "a few kicks" [cf. Ch. 11, p. 306, n. 50]. To be sure, there were still a few straggling disciples, those called the "Hegelian old guard" (Karl Michelet, Werder, Lasson), but their efforts met with general indifference. As late as 1894 William WALLACE pointed out, in the second edition of his *Prolegomena* (p. 16), that it was hardly prudent for a German author desirous of attracting public attention to make an overt profession of Hegelianism, that in such a case it was at least advisable to allay all suspicion by some indication of disapproval toward Hegel, and three years later Eduard von HARTMANN wrote: "To be sure, the Hegelian tradition is not yet entirely extinct, but for the spirit of the time in which we live the bridge that leads to the comprehension of this philosophy seems broken" (*Schelling's philosophisches System*, Leipzig: Hermann Haacke, 1897, p. iii). Quite recently, however, this ostracism has certainly become less prevalent, and there are again Hegelians in Germany, although the movement, distinguished by names such as Münsterberg and William Stern, on the whole appears rather limited in scope.

This quick historical overview gives an idea of how important a role Hegelianism has played in the evolution of European thought, and it also shows us that the efficacy of this thought does not seem to be at all exhausted. In France itself Émile Boutroux and René Berthelot have quite recently pointed this out at a meeting of the *Société de Philosophie* (April 1907), where the work of Hegel was the subject of a brilliant discussion ['kur la nécessité'].

7. ABSTRACT AND CONCRETE REASON IN HEGEL (p. 264)

Hegel calls the first of these faculties *Verstand* and the second *Vernunft*. The two terms have generally been translated in French as *entendement* and *raison* (see in particular the *Bulletin de la Société de philosophie* 7, April 1907), which has the further advantage of conforming to the way the two terms have been rendered by English translators since Coleridge: indeed, the English translation *understanding* and *reason* is almost dictated by the etymological kinship between the English term *understanding* and the German *Verstand*.

Only reluctantly have I decided to depart from this precedent consecrated by the great authority of Boutroux. My rationale is that Hegel's *Verstand* is in fact what everywhere else has been called *reason* and that therefore Hegel is led to protest against the use of the term *irrational* in its usual sense, declaring that the meaning of the word has been distorted and that what we call *rational* merely conforms to *Verstand*, whereas what we term irrational is rather the beginning of a reasoning process conforming to *Vernunft* (*Enc.*, *Logik*, 6:404 [Wallace 370]). Hegel, when he contrasts *Verstand* with *Vernunft*, often qualifies the former as *abstract* (see for example *Enc.*, *Logik*, 6:230–231 [Wallace 213–214], where the *abstract* identity of *Verstand* is opposed to the *concrete* identity of *Vernunft*; likewise *Enc.*, *Logik*, 6:77 [Wallace 75], where the former is attributed to ancient philosophy, although with the exclusion of Plato and Aristotle, who, like the moderns, were judged to have achieved a concrete logic; *Phil. der Geschichte*, 9:63 [Sibree 53; erroneous citation] where the concept, in passing from the domain of *Verstand* to that of *Vernunft*, is said to have acquired a more concrete determination, etc.). Taking advantage of that fact, I therefore translate these terms as *abstract reason* and *concrete reason*. At the same time that allows us to use the term *rational* in the ordinary sense of the word, employing the term *reasonable* for that which conforms to concrete reason. William WALLACE, in opposing *reason* and *understanding*, also makes use of the adjective *concrete* to characterize the first of these two concepts (*Prolegomena* 351–352).

8. HEGEL'S PANLOGISM (p. 267)

Hegel explicitly stated, in the Introduction to the *Logic* of the *Encyclopedia*, that "to see that thought in its very nature is dialectical, and that, as abstract reason, it must fall into contradiction, – the negative of itself, will form one of the main lessons of logic" (*Enc.*, *Logik*, 6:17 [Wallace 18–19; see Appendix 7]). Furthermore, as early as 1812 Hegel writes: "According to my view, metaphysics in any case falls entirely within logic" (Oct. 1812 letter to Niethammer, *Über den Vortrag der philosophischen Vorbereitungs-Wissenschaften auf Gymnasien*, *Werke*, 17:338 [*Hegel: The Letters* 277]). Kuno FISCHER points out that Hegel, who gave his course on logic twenty-two times in all, consistently announced it as "logic and metaphysics" (*Geschichte* 8:433; on this subject see also ROSENKRANZ, *Hegel's Leben* 151, and HAYM, *Hegel*

294).

ROSENKRANZ acknowledges that some Hegelians "behave as if in all philosophy only logic were ultimately concerned, of which nature and mind properly are only superfluous translations." But, according to him, that is a misunderstanding of Hegel (*Hegel als deut. Nat.* 122 [Hall 20]). One cannot help noticing, however, that at least in a certain sense the "logical presumptuousness" of the school did in fact stem from the master. Rosenkranz himself implicitly admits as much by insisting in many passages on the undeniable fact that Hegel tended to merge logic and metaphysics. In the very work we have just cited, he declares that Hegel "opposed all those who separate metaphysics from logic." Since Hegel "consistently maintained against Kant the ontological character of the logical categories, metaphysics for him was merged with logic. Thought and being are, for him, in their diversity, at the same time identical. Thought is the power that determines being" (*Hegel als deut. Nat.* 287).

HAYM, who so often disagrees with Rosenkranz in his evaluations of Hegel's work, is in total agreement with him on this point. In particular he points out that in Hegel there is no longer any opposition between knowing and being. "The boundary between logic and metaphysics crumbles. Logic as such is, in the same measure, metaphysics, and metaphysics is just as much logic" (*Hegel* 294).

Paul JANET takes Hegelian philosophy to be "a restoration of Wolffian dogmatism, founded on the identity of logic and ontology"; it "persists in believing, like scholasticism, and in spite of the decisive warning of Kant, that the solution to the problem of things is to be found in the logical conceptions of the mind" (*Études sur la dialectique dans Platon et dans Hegel*, Paris: Ladrange, 1861, pp. 307 ff.).

In the same vein, WUNDT asserts that "what made him [Hegel] miss his goal was first and foremost his illusion of being able to understand the evolution of the mind, indeed even the evolution of things in general, as a logical evolution" (*Einleitung in die Philosophie*, 5th ed. [Leipzig: W. Engelmann, 1909], p. 267).

WALLACE observes that the first two parts of the Hegelian *Logic* are what one generally calls metaphysics, only the third part being logic properly so-called in the ordinary sense of the term and adds that "the merit of Hegel is ... to have broken down ... the general disruption between logic and metaphysic" ('Hegel,' *Encyclopedia Britannica*, 11th ed., 13:205; cf. also Wallace, *Prolegomena* 387 ff.).

Andrew SETH rightly emphasizes the fact that the identification of logic and metaphysics cannot be considered a minor circumstance in the Hegelian system, but that it constitutes the most essential result ("the gist and outcome") of the system, as the Hegelians themselves admit (*Hegelianism and Personality*, 2nd ed., Edinburgh: William Blackwood and Sons, 1893, pp. 110, 131 [Meyerson quotes the parenthetical phrase in English]).

Benedetto CROCE, who does not want us "to consider panlogism as the fundamental characteristic of the system, when it is but a morbid excrescence, growing from it," nevertheless admits that Hegel's main goal was to edify a "logic of philosophy" different from that followed by the mathematical and experimental sciences (*Ce qui est vivant* 157, 1–3; cf. 67 [Ainslie 192, 1–3; cf. 81–82]).

The most recent French commentator on Hegel's work, Paul ROQUES, does not hesitate to defend this aspect of his philosophy. "Who does not see that the banal accusation of the rationalization of reality and of panlogism, which is continually leveled against Hegel, in the end threatens all idealism," given that, "unlike positive science, philosophy would have the world unfold in the heart of pure thought, be dominated by the self, and thereby take on a character of logical necessity" (*Hegel, sa vie et ses oeuvres*, Paris: Félix Alcan, 1912, pp. 11, 15).

This question of the union of logic and metaphysics in Hegel is closely related to the aspect of his philosophy that interests us here, and the reader will understand why it seemed fruitful to review the opinions of a few commentators in this way.

9. THE HEGELIANS AND HEGEL'S *NATURPHILOSOPHIE* (p. 269)

Even the most resolute supporters of the Hegelian philosophy generally prove rather lukewarm in defending the *Naturphilosophie*. For example, MCTAGGART leaves the whole Hegelian *philosophy of nature* in the lurch (if we may be allowed this expression), declaring that "all that part of knowledge which depends upon one content rather than another – the whole, that is, of what is ordinarily called science – certainly cannot be reached from the dialectic alone in the present state of our knowledge, and perhaps never will be" (*Studies* 101). Cf. p. 252: "The abuse which has been heaped on this work [Hegel's *Naturphilosophie*] is probably excessive. But it cannot be denied that it has a certain amount of justification" [Meyerson's brackets]. Edward CAIRD makes analogous statements

(*Essays on Literature and Philosophy*, Glasgow: James Maclehose and Sons, 1892, 2:532), as does William WALLACE (*Prolegomena* 86). René BERTHELOT, whose Hegelian bias, though less absolute than that of these English disciples, is still quite strong, likewise concedes that Hegel's *philosophy of nature* "would appear ridiculously inadequate to us today" ('kur la nécessité' 183). The opinion of the German Hegelians seems much the same. For example, Paul BARTH, in a work on the *Philosophy of History of Hegel and the Hegelians*, published in 1890, notes in a simple parenthetical comment that "Hegel's logic and philosophy of nature have fallen into complete and deserved oblivion" (*Die Geschichtsphilosophie Hegel's und der Hegelianer bis auf Marx und Hartmann*, Leipzig: O. R. Reisland, 1890, p. 1). The fact that this is a doctoral thesis (that is to say, according to German university practice, the work of a young student) is an additional proof that the author is merely reporting a state of things he takes to be universally known and accepted.

10. THE PHILOSOPHY OF NATURE AND SCIENTIFIC PROGRESS (p. 274)

Hegel's *Logic* appeared in 1813 and 1816 and the *Encyclopedia* in 1817. On the other hand, Hegel never stopped treating these subjects in his courses, which his disciples constantly took into account in the posthumous edition, so that we are familiar with his ideas to the very eve of his death, so to speak. Now it is altogether typical that nowhere does Hegel claim the honor of any scientific discovery, great or small, inspired by his ideas. Of the three discoveries we have mentioned in the text, only one was made during Hegel's lifetime. It is that of Oersted (1820), and Hegel quite understandably never said anything about it, for Oersted would undoubtedly have protested; as we see by his philosophical writings, he was, in fact, a perfectly genuine "philosopher of nature," but in no way a Hegelian. Schönbein's work dates from 1839, and there is no evidence that the Hegelians claimed the author as one of their own. Indeed, to the extent that general ideas enter into Schönbein's conceptions at all, he allows himself to be guided by the idea of polarity, which first seems to have been put forward as a fundamental principle of nature by Herder (see HARTMANN, *Schelling's philosophisches System*, Leipzig: Hermann Haacke, 1897, p. 171). Schelling had fully adopted it as such; see, for example, *Weltseele*, I, 2:459: "The first principle of a philosophic science of nature is to search for polarity and dualism throughout nature" (cf. *Weltseele*, I, 2:476, 489–490; *Erster Entwurf*, I, 3:36 ff.; *Einleitung zu dem Entwurf*, I, 3:288; cf. also the passage on duality in our Ch. 12, p. 321), whereas Hegel was rather inclined to find that this principle was misused. Finally, as to the work of J. R. Mayer, which dates from 1842, the Hegelians have sometimes tried to claim part of the credit for it. For example, William WALLACE, speaking of the attacks against Hegel's *Naturphilosophie*, says that these critics "forget the impetus it gave to physical research by the identification of forces then believed to be radically distinct" ('Hegel,' *Encyclopedia Britannica*, 11th ed., 13:206). The question of the unity of forces plays only a very modest part in the work of Hegel; he lays much less stress on it than did the "philosophers of nature" of Schelling's school – see, for example, the way in which Schelling himself speaks of this unity as the guiding principle of our entire knowledge of nature, on the first page of the *Weltseele* (I, 2:347) – and it is certain that in general, insofar as suppositions of this sort founded on pure analogy are concerned, Eduard von HARTMANN was correct in saying that Hegel's *Naturphilosophie* adds nothing to what Schelling and his disciples had produced (*Geschichte der Metaphysik II*, *Ausgewählte Werke*, Leipzig: Hermann Haacke, 1901, 12:211). On the other hand, we find no trace in Mayer of what really constitutes the distinctive feature of Hegelian scientific speculation, namely the "dialectical" genesis of concepts.

If someone with the necessary scientific background were patiently to decipher the writings of the "philosophers of nature," he would probably find many points of contact with the scientific development that followed. However, it is quite probable that most of these findings would offer only purely historical and anecdotal interest. Indeed, it is certain that this philosophy had little influence on science outside Germany, and even in Germany itself its influence lasted only for an extremely limited period, the whole movement disappearing all of a sudden, as if in a cataclysm. Thus, for example, we find in Schelling the curious observation that, according to current scientific theories, including the theory of dynamics that seeks to explain reality by the action of two opposing forces (attractive and repulsive), "the relations in nature itself ought to be able to be reversed at any moment" and that consequently one would not understand why an invariable order reigns in the whole of phenomena (*Allgemeine Deduktion des dynamischen Processes*, *Werke*, I, 4:6). These words, which, as we see, contain at least a presentiment of the irreversibility of natural phenomena and of the opposition between this realization and the explanatory theories of science, appear to have passed entirely unnoticed and certainly remained without the least influence on the work of Carnot. In other cases, however, explicit remarks, or indeed even allusions made more or less in passing, may have been fruitful. For example, Schelling speaks of the possible explanation of organic being as follows: "There would be at least *one* step taken towards this explanation if it could be shown that the hierarchy of all organic beings was formed by the gradual development of one and the same organic structure" and he goes on to answer a possible objection by pointing out that if one has not yet found species being modified, it is no doubt because the period of observation has been too short (*Weltseele*, I, 2:348–349). Without wishing to attach too much importance to this remark (for Schelling upon other occasions says just the opposite; see, for example, *Erster Entwurf*, I, 3:62), and without forgetting more authentic precursors, such as Buffon for example, one nevertheless cannot guarantee that Schelling's opinion may not have influenced some of the protagonists of evolution either directly or, as appears more probable, through some intermediary. Schelling's influence on Spencer, by way of Coleridge and the biologist and *Naturphilosoph* Karl Ernst von Baer, appears quite probable, as a matter of fact, and, if we are not mistaken, was explicitly recognized by Spencer himself, at least insofar as Baer is concerned. The question, however, perhaps deserves more thorough treatment than that of Lázár RÓTH, who is content merely to compare the ideas of the two philosophers and hardly mentions any intermediaries (*Schelling und Spencer, eine logische Kontinuität*, Berner Studien zur Philosophie und ihrer Geschichte 29, Bern: C. Sturzenegger, 1901, pp. 7, 25). Similarly, it is impossible to glance through Henrich STEFFENS's accounts of the correlation between the density of metals and their cohesion and to see how he classifies them in series, without being struck by the resemblance between these conceptions and those of Lothar Meyer et al., who preceded and paved the way for Mendeleev's theory (see for example *Beyträge zur innern Naturgeschichte der Erde* (Freiburg: Crazischen Buchhandlung, 1801), Pt. 1, Ch. 6, pp. 197 ff.). And although the creator of the "periodic table" of the elements has a more recent precursor in BÉGUYER DE CHANCOURTOIS (*Vis tellurique*, Paris, 1863), the other filiation appears better warranted.

11. HEGEL, SCHELLING AND CHEMICAL THEORY (p. 274)

Hegel was never able to absorb the principles of Lavoisian chemistry completely. In his early writings, which remained unpublished, he seems to have made use of phlogiston theory (which at that time still had many supporters among German chemists). But it is impossible to determine what meaning he assigned to the concept of phlogiston (which, as we know, took on several meanings before disappearing definitively from science). Indeed, the information we find on this subject in Rosenkranz is totally bizarre. In particular ROSENKRANZ claims that it is to oxygen that Hegel attributed "its former name of phlogiston" (*Hegel's Leben* 118). Now it is of course not strictly impossible that Hegel should have done so; among the more or less extravagant ideas considered by the last of the defenders of phlogiston we do in fact find that of Antoine BAUMÉ who, in 1798, declared that "oxygen ... is the inflammable or phlogiston principle in the highest state of purity and rectification it can perhaps attain" *(Opuscules physiques et chimiques*, Paris, Year 6 [of the First French Republic], p. 62; cf. Appendix 2, pp. 554 ff., on Baumé's position). But his was an isolated opinion, which Hermann KOPP does not even find it necessary to mention in enumerating these late theories (*Geschichte der Chemie* [Brunswick: Friedrich Vieweg und Sohn, 1845], 3:155 ff.); certainly it had no influence in Germany and thus could not have been what ROSENKRANZ was referring to as a conception generally adopted in the past. One would suspect rather that he merely erred, meaning to say that it is hydrogen which Hegel called phlogiston (this was Kirwan's theory, which we discussed in Appendix 2, p. 547, a theory which was actually quite widespread among the last phlogiston theorists). But one then runs up against the difficulty that hydrogen itself is explicitly mentioned alongside phlogiston (*Hegel's Leben* 118 ff.). It is, if possible, still more unlikely that Hegel should have, as Rosenkranz states on several occasions, referred to *carbon gas* (along with the three gaseous elements: oxygen, hydrogen and nitrogen). We do indeed find in ROSENKRANZ (*Hegel's Leben* 129) a rather long extract from the part of the manuscript dealing with science, but it unfortunately has to do with neither phlogiston nor hydrogen. Haym, who also seems to have had access to Hegel's manuscript, is silent on everything having to do with science.

In short, we can only regret, when we search for the antecedents of Hegel's scientific thought, that the single real source of information on the subject is found in the writings of a man who to all appearances is neither very competent nor very precise. It is rather surprising that such obvious errors have never been pointed out by any critic. But that is no doubt simply a consequence of the fact that, from the very beginning, disciples and adversaries almost completely neglected the scientific side of Hegel's thought.

Like many other peculiarities of his philosophy of nature, Hegel's altogether bizarre attitude toward chemical theory comes to him directly from Schelling. Schelling initially appears to have had some favorable inclinations toward the new chemistry. At least one finds the following statement in the *Ideen zu einer Philosophie der Natur*: "The new system of chemistry, the work of a whole era, spreads its influence ever more widely over the other branches of natural science, and employed over its whole range may very well develop into the universal system of Nature" (*Ideen*, I, 2:75 [Harris 59]; cf. *Weltseele*, I, 2:388). Since Fourcroy's *Philosophie chimique* is cited in this connection, it would seem that the "new system" can only be the chemistry of Lavoisier. But a few [sic] pages later in the same work Schelling states just the opposite: "It is hard to imagine a more preposterous undertaking than to try to draft a universal theory of Nature from particular experiments; nevertheless, the whole of French chemistry is nothing else but such an attempt" (*Ideen*, I, 2:119 [Harris 93]), and he ridicules "the empty chemical experimentalism of the French" (*Ideen*, I, 2:121 [Harris 95]). In these contradictory opinions there is obviously some incoherence, which may stem from the fact that we are looking at the *Ideen*, which dates from 1797, in the form of a second edition, from 1803, and that the author's conceptions had changed in the meantime; but, as a matter of fact, a certain lack of coordination is one of the characteristics of Schelling's hurried production during his early years. In any case, for him the antiphlogistonist ideas did not prevail. Already in the *Ideen* we find a whole series of bizarre conceptions about air and its "decomposition" when subjected to light (*Ideen*, I, 2:114 [Harris 89]); about water, whose element he urgently requests chemists to seek (more than ten years after Lavoisier had so lucidly explained its composition, 115–117 [89–91]); about "electric matter," whose basis is nothing but decomposed oxygen (136 [106]); about hydrogen, which he considers "totally problematic," carbon being, moreover, only a modification of this same hydrogen, produced in plants (295–296 [235]). Finally, we also discover there the characteristic claim that in all cases it is the same element that renders bodies combustible; of course phlogiston was only an imaginary principle, but a new theory will establish the existence of a real principle of this kind (76–80 [59–63]). "In every phlogisticized body" we find one identical principle (*Weltseele*, I,

2:420). Oxygen is not a simple body. "In a higher sphere," oxygen, or at least one of the elements of which it is composed, must be able to burn in its turn. Oxygen is "a principle foreign to the earth, it is a product of the sun" (*Erster Entwurf*, I, 3:129–130). Oxygen is the principle of negative electricity. It is "a principle of the negative kind and consequently a representative of the force of attraction, so to speak, ... while phlogiston or, what is the same thing, positive electricity, is the representative of the positive principle or the repulsive force" (*Einleitung zu dem Entwurf*, I, 3:319–320). Carbon is the principle most subject to gravitation; it is the true earth principle, which is why it is formed in the heaviest, most rigid and most cohesive bodies, namely in metals and in plants (*Aus den Jahrbüchern*, I, 7:280). Furthermore, the metals are not true elements, they are all combinations of carbon with nitrogen; on this point Schelling follows the assumptions put forth by his disciple Henrich Steffens (*Allgemeine Deduktion des dynamischen Processes*, *Werke*, I, 4:70). Of course, as we saw in Chapter 8 (p. 230), Lavoisier himself had hesitated to attribute the same rank to all the elements, some of them, on the contrary – in particular oxygen, nitrogen and hydrogen – appearing simpler to him than others; but these peremptory claims of the *philosophers of nature* have a clearly different tenor from that of Lavoisian chemistry and are visibly linked to that of the preceding epoch.

We felt it necessary to dwell at some length on the chemical conceptions of Schelling, whose inspiration Hegel is content to follow in his scientific, or pseudoscientific, "constructions." In general Hegel is rather less extravagant than his predecessor from this point of view, but one sees that, insofar as his idiosyncrasies are concerned, he comes by them honestly.

In addition, Hegel believes in the spontaneous generation of organisms of all sorts and ridicules the *omne vivum ex ovo*. "When the origin of certain animalcules could not be accounted for, recourse was had to invention. But there are organisms which originate immediately" (*Naturphilosophie*, 7_1:459 [Miller 296]). Thus animals are spontaneously generated in the intestines. "It is a false hypothesis that tapeworms in human beings are the result of swallowing the eggs of such creatures" (*Naturphilosophie*, 7_1:673–674 [Miller 430]).

12. HEGEL AND NATIONAL SCIENCE (p. 276)

The alliance between philosophy (Hegelian, it goes without saying) and science appears so important to Karl Michelet (cf. Ch. 11, p. 304, n. 40, concerning a similar feeling in Gans) that he does not hesitate to appeal to the patriotism of German philosophers, pointing out that it is above all the English and the French who lead science down the path of complicated theories, and that it is reprehensible for Germans to seek support for their opinions on the other side of the Rhine or the Channel. The physicists' "feeling for Germany" [*der deutscher Sinn*] must motivate them, on the contrary, to demonstrate their good will toward German philosophy by entering into negotiations with it (Karl Ludwig Michelet, Vorwort, *Naturphilosophie*, 7_1:vii, xi [Petry 1:180, 182; Meyerson's brackets]). In propounding these nationalistic considerations, Michelet seems to invoke the opinion of the master himself, but one must acknowledge, in all fairness, that he does so without justification. In the inaugural lecture of his course at the University of Berlin (1818), Hegel had indeed declared that the Germans were "the chosen people" in philosophic matters, almost as the Jews had formerly been in religious matters – and if we recall the intense philosophic movement that stirred the Germany of that epoch, we can almost understand such a claim (and can appreciate all the more the opposing attitude of Schelling, who pleaded in favor of a rapprochement with French empiricism). Nevertheless Hegel was generally able to steer clear of any exaggerated national sentiment in the philosophic or scientific domain, and if he observes, in the passage of the *Naturphilosophie* to which Karl Michelet is referring, that German discoveries had not been given credit in Germany itself until after they had been adopted by French or English scientists, he adds that "it is no use complaining of this; it is always that way with us Germans, unless indeed some trashy theory [*schlechtes Zeug*] like Gall's phrenology is propounded" (*Naturphilosophie*, 7_1:408 [Miller 263; Meyerson's brackets]). Concerning Hegel's attitude towards Gall, cf. the persiflage in which he indulges in the fragment reported by ROSENKRANZ (*Hegel's Leben* 554–555), where he facetiously announces that the phrenologist will show the existence of a whole series of new senses: in women, the senses of dance, cooking and sewing; in men, the sense of charlatanism, etc. Furthermore, like many other peculiarities of Hegel's scientific attitude, the hostility toward Gall is already found in SCHELLING; see *Aussätze und Recensionen aus der Ienaer und Erlangen Literaturzeitung und dem Morgenblatt*, *Werke*, I, 7:542). We might add that not all of Hegel's disciples were moved by the same narrow spirit of nationalism. For example ROSENKRANZ pleads in favor of an alliance between German and French thought, and vigorously condemns attacks against the French spirit, with the result that he is considered a *Francophile* (*Hegel's Leben* xxvi).

13. HEGEL'S ARTISTIC SENSE AND SENSE OF RHYTHM (p. 278)

Interestingly enough, Hegel seems to have admired Goethe exclusively as a scientist; a very sympathetic biographer of and commentator on Hegel is obliged to note that he appears to have been quite untouched by the great German writers who preceded him or were his contemporaries (Lessing, Goethe, Schiller) and that his favorite reading matter consisted of mediocre novels; Schopenhauer, whose hostility toward Hegel is well-known, had already maliciously reported this fact (Kuno FISCHER, *Geschichte* 8:9). On the subject of Hegel's youthful readings, cf. also Wilhelm DILTHEY, *Die Jugendgeschichte Hegel's*, Abhandlungen der Königlich Preussischen Akademie der Wissenschaften (Berlin: G. Reimer, 1905), p. 6. Moreover, Hegel's artistic sense in general appears underdeveloped. In spite of the Hegelian *Esthetics*, critics in his own time seem to have already observed this fact, and Rosenkranz feels obliged to defend his master against this opinion, which he deems "ridiculous"; to refute it, however, the only argument he can muster is that the French translator of the *Esthetics*, Bénard, had lavished high praise on the work in his preface. But the *Esthetics* could perfectly well contain extremely valuable speculative conceptions without its author's immediate esthetic sense being very keen.

Hegel does not appreciate the spectacles of nature (as we saw in Ch. 13, p. 357) and claims to prefer Rossini's *Barber of Seville* to Mozart's *Marriage of Figaro*.

ROSENKRANZ feels obliged to admit (something one immediately notices in skimming the poems reproduced in the philosopher's *Works*) that Hegel, "by a strange anomaly, had no sense of meter as far as his own verses are concerned" and that he "lacked the ability to produce a poetic line, to distinguish long or short syllables, or to determine the number of feet correctly." Rosenkranz nevertheless believes he can affirm that Hegel had a "subtle" appreciation of the poetry of others, even from the point of view of metrics (*Hegel als deut. Nat.* 38) – but that is simply one more proof of the unfortunate lack of critical judgment this biographer and apologist exhibits toward his hero. Hegel's total ineptitude in this area is all the more striking when one considers what German prosody had been since the second half of the eighteenth century: having, by a bold misconception, applied the rules of Greco-Latin prosody to German, substituting the tonic accent for quantity, it created a somewhat mechanical and rather harsh rhythm, against which the best poets (Heinrich Heine, for example) sometimes rebelled, but which is so easily grasped that one is astonished it could elude a human ear.

Some may consider these details irrelevant to our subject; nonetheless we believe they help characterize the thinker in question, who, despite his power, is terribly incomplete in his fierce and intransigent intellectuality.

14. THE HEGELIAN DIALECTIC AND EXPERIENCE (p. 292)

Adolf TRENDELENBURG, in speaking of the Hegelian method, had already pointed out that "neither between the two terms of the contradiction, nor above them, is there a third term" and that this "is not a step accomplished by the dialectic moving by itself, but a leap of the imagination relying on audacious language" (*Log. Untersuch.* 1:31, 41).

Similarly, Andrew SETH declares that "it may be fairly granted, I think, to critics of the [Hegelian] Method ... that every step of the advance is empirically conditioned. The celebrated dialectical opposition which is the nerve of the process is not the contradictory opposition of the logician. Mere contradiction yields nothing new, – nothing, therefore, which, by synthesis or fusion with the original datum, could yield a third product different from either." The opposition "arises only for a subjective reflection which has had the advantage of acquaintance with the real world" (*Hegelianism and Personality*, 2nd ed. (Edinburgh: William Blackwood and Sons, 1893), p. 97 [Meyerson's brackets]).

Moreover, the supporters of Hegelian logic are well aware that this is the point of the doctrine that particularly needs to be defended against critical attack. For example, MCTAGGART stresses the fact that the dialectic "claims to add to our knowledge, and not merely to expound it," for otherwise "the conclusion of the process could, if it were valid, have no greater content than was contained in the starting point." Furthermore we can see how McTaggart attempts to dismiss this difficulty, in particular by responding to Trendelenburg's objections (*Studies* 3).

But a Hegelian as orthodox as Johann Eduard ERDMANN has admitted the fairness of the criticisms leveled against the claims and procedures of the Hegelian dialectic on this issue (*Die Entwicklung der deutschen Spekulation seit Kant*, Leipzig: Vogel, 1853, 2:768–769), and

ROSENKRANZ himself finally conceded that Hegel's claim that he lets the concept determine itself, outside the philosopher so to speak, cannot be sustained (*Hegel als deut. Nat.* 114 [Hall 158]). Concerning the way in which the Hegelian concepts, supposedly arising from pure thought, surreptitiously become filled with content, cf. HAYM, *Hegel* 318 ff.

What has no doubt contributed to the discrediting of the Hegelian dialectic as a method of reasoning is the abuse made of it by its imitators. ROSENKRANZ notes that the Hegelian dialectic has given rise to "the most arbitrary and lifeless dogmatism" among his disciples (*Hegel als deut. Nat.* 115 [Hall 158]). On this subject cf. CROCE, *Esthétique comme science de l'expression et linguistique générale*, trans. Henry Bigot (Paris: V. Giard et E. Brière, 1904), Ch. 13, pp. 334 ff., and *Ce qui est vivant* 169 [Ainslie 206–207].

15. SCHELLING, HEGEL AND VICTOR COUSIN (p. 311)

Victor Cousin met Hegel in Heidelberg in 1817 on his first trip to Germany, where he had been drawn by the renown of the philosophy of nature (cf. Ch. 12, p. 333). Although he did not thoroughly understand either Hegel's philosophy or even its relation to Schelling's (cf. Ch. 12, p. 332, and Appendix 19, p. 597), he was immediately struck with great sympathy and keen admiration for the man and his work. He did not see Schelling until the following year, at Munich, where he arrived armed with a letter of introduction from Hegel (we quote this letter on p. 598). In 1824 he became still closer to Hegel during the forced stay in Berlin following his arrest in Dresden and his extradition to the Prusssian authorities (in the *Souvenirs* of the Chevalier de CUSSY, Paris: Plon, Nourrit, 1909, 1:318, 337, 387 ff., one will find interesting details about this affair, which had assumed the importance of a diplomatic incident, and behind which there was a treacherous denunciation by the police of the Restoration, the natural brutality of the Prussians having done the rest).

On this occasion Hegel interceded very actively in favor of Cousin, for whom he took personal responsibility, and it is due to his intervention that Cousin was able to gain release from prison, for which he remained extremely grateful to his friend. This is a very honorable episode for the two philosophers, and since in both their cases we have tended throughout our work to give details that sometimes smack of slandermongering, we beg permission to cite in its entirety the beautiful short preface by which

Cousin, two years later, in the third volume of his translation of Plato (COUSIN, *Oeuvres complètes de Platon*, Paris: Bossange Frères, 1826), publicly expressed his gratitude:

I beg you, my dear Hegel, to accept the homage of this translation of the *Gorgias*. Certainly such homage was due him who was the first to return the maxims contained in this ancient manuscript to a place of honor among the eternal principles of the philosophy of right. But still another motive leads me to address this homage to you.

Hegel, ten years ago you received me in Heidelberg like a brother, and from the first our souls understood and loved one another. Absence and silence did not cool your friendship and when, upon my recent return to Germany, a police lacking in good sense, unwittingly manipulated by an odious politics, dared interfere with my freedom, charge me with the most atrocious accusations and declare me convicted and condemned in advance, you spontaneously hastened to present yourself before my judges, tell them I was your friend and vouch for me.

I wanted, Hegel, to thank you publicly for this noble conduct, not for you or for me, but for philosophy. You have proved that it is not always a sterile occupation and that the genius of abstraction can quite well be allied with firmness of soul and courage in real life. Once again, Hegel, I thank you.

Victor Cousin
Paris, 15 July 1826

This set of circumstances makes clear why Schelling criticized Cousin for having "entered the territory of German philosophy from the Heidelberg side" and for thus having been, in a manner of speaking, led astray from the true path (cf. Appendix 19, p. 599). Hippolyte TAINE, in his book on *The Classical Philosophers of the 19th Century in France*, which contains so many apt views and so much interesting information, speaks somewhat summarily of Cousin's early relations with the two German philosophers. "He went to Munich in 1818, met Schelling and Hegel, became their disciple" (*Les Philosophes classiques* 132; cf. an analogous passage, pp. 143–144, in which Taine stresses Cousin's ignorance, in 1817, of the pantheism of German philosophy). The error in itself is slight, but the reader will understand why, in the present context, we wished to correct it.

In the preface to *Fragments philosophiques*, Cousin lavished dithyrambic phrases on Schelling:

The first years of the nineteenth century have seen the appearance of this great system. Europe owes it to Germany, and Germany to Schelling. This system is truth; for it is the most complete expression of reality as a whole, of universal existence.

(*Fragments philosophiques*, 2nd ed., Paris: Ladrange, 1833, p. xl [*Philosophical Essays*, trans. George Ripley, Edinburgh: Thomas Clark, 1839, pp. 92–93, bottom page pagination]

In 1826, while Hegel was still alive, Cousin had apparently considered him Schelling's intellectual equal, dedicating his edition of Proclus's *Commentary* on the *Parmenides* to the two of them: *amicis et magistris philosophiæ præsentis ducibus* [*Procli Philosophi Platonici Opera* (Paris: J. M. Eberhart, 1821), 4:v: "to my friends, and the leading masters of contemporary philosophy"]. Here, on the contrary, Hegel is continually called a *disciple* of the master. Nevertheless he shares in the praises lavished on Schelling.

Schelling brought this system into the world: but he left it filled with all manner of imperfections and defects. Hegel, coming after Schelling, belongs to his school. He is entitled to a separate place in it, not only for developing and enriching the system, but for giving it in many respects a new aspect. The admirers of Hegel consider him the Aristotle of a second Plato; the exclusive partisans of Schelling see in him only the Wolff of another Leibniz. However it may be with these rather arrogant comparisons, no one can deny that the master is gifted with the talent of powerful invention, and the disciple with that of profound reflection. Hegel has borrowed much from Schelling. (xl-xli [Ripley 93, substituting "Wolff" for Ripley's "Wolf"])

This attitude greatly annoyed Schelling: he would not have forgiven a German for it, as Kuno FISCHER rightly suggests (*Geschichte* 7:225–226). However, his annoyance is less explicit in the *Preface* – at the very most, one sometimes has the impression that certain diatribes directed against Hegel's admirers contain a hint of sarcasm toward the author of the *Fragments* – than in personal letters to their author. Cousin, for his part, in the later editions of the *Fragments philosophiques*, somewhat softened his support of Schelling's philosophy, no doubt for political reasons. The sentence beginning "This system is truth" disappeared from the *Preface*. But the praises, including his designation of it as a "great system," were retained (*Oeuvres*, 3rd series, 4th ed., Paris: Ladrange, 1847, 4:77).

The correspondence between Schelling and Cousin begins in 1819 and becomes rather voluminous after 1826 (we possess only Schelling's letters as collected by Gustav Leopold PLITT in *Aus Schelling's Leben in Briefen*. In them Schelling sometimes takes his friend to task rather harshly for his admiration of Hegel (we give a few extracts below, p. 599). Nor does he seem overly happy with the way in which Cousin generally interprets his philosophy, and he forgets himself so far as to say, in speaking of the works he proposes to publish: "I hope they will once and for all put an end to the petty discussions in which I see you are still involved. When they are published, all I shall need will be a good translator, and I hope to be able to dispense with an interpreter" (PLITT 3:43; cf. also the extracts we cite in Ch. 12, p. 330) – which was obviously rather ungracious toward someone who had taken so much trouble to interpret his ideas. But he usually softens these somewhat bitter criticisms by compliments. It is clearly important to him not to alienate such an ardent admirer too much, particularly since he seems to have the ear of the French public, whose support Schelling evidently valued highly. Following the July Revolution (the last letter we have quoted is earlier, bearing the date 27 November 1828) another element enters in. Cousin has become a very important personage, and this circumstance does not fail to influence Schelling, who is at least as alive to the honors as his friend. From then on there is nothing but an exchange of compliments between them. Cousin is named an associate member of the Academy of Munich, a nomination he seems to have solicited (PLITT 3:50), but the nomination is not confirmed by the king of Bavaria (no doubt for political reasons – we know how hostile legitimate governments initially were to that of Louis Philippe) and does not become effective until three years later (3:71). Cousin then responds by having Schelling named a chevalier of the Legion of Honor a month later (3:73), a correspondent of the Academy of Moral and Political Sciences the following year and an associate member of this Academy in 1835 (3:74, 102).

In all fairness to Schelling, one must understand that his attitude contains something besides resentment toward a triumphant rival. Schelling sincerely believed that Hegel had led philosophy into entirely pernicious pathways and since (while diverting it from its true meaning) Hegel had merely elaborated Schelling's own doctrine, this doctrine itself therefore had to be modified, or at least completed. That is precisely the work to which Schelling applied himself during these years, obviously without managing to get it into shape in a way that was satisfactory to him, which is why he was, as we show in Chapter 12, pp. 331 ff., so impatient toward those who seemed in some sense to anticipate results he himself did not yet perceive altogether clearly. That also explains why he showed particular displeasure when the Hegelian philosophy was presented as the *legitimate* development of his own.

Post-scriptum: In Jules BARTHÉLEMY-SAINT HILAIRE, *M. Victor Cousin, sa vie et sa correspondence* (Paris: Hachette, 1895), 1:68 ff., 3:57

ff., we find a certain number of letters exchanged between Victor Cousin, Hegel and Schelling. Their content in no way modifies the picture of the relations between these three men that we have tried to sketch here.

16. THE IDENTITY OF THOUGHT AND REALITY IN SCHELLING (p. 323)

Formulas of extreme idealism abound in Schelling's early philosophy and one has only too many from which to choose. Here are a few to add to those we quoted in the text:

The system of Nature is at the same time the system of our mind. (*Ideen*, I, 2:39 [Harris 30])

According to this point of view, given that nature is only the organism of our reason, nature can produce nothing other than what is regular and adapted to the goal [*das Zweckmässige*], and nature is *constrained* to produce it. But if nature produces nothing that is not regular and produces it necessarily, it follows that in this nature, considered as independent and real and in the relation of its forces among themselves, the birth of such regular products adapted to the goal must be able to be demonstrated as necessary, and that consequently the ideal must in its turn emerge from the real and be explained by it. (*Einleitung zu dem Entwurf*, I, 3:272 [Meyerson's brackets])

If the intelligence is organic at all, as indeed it is, it has also framed to itself outwardly from within everything that is external for it, and that which constitutes the universe for it is merely the grosser and remoter organ of self-consciousness, just as the individual organism is the finer and more immediate organ thereof. (*Transc. Idealismus*, I, 3:490–491 [Heath 122])

Thus it is obvious that in constructing matter the self is in truth constructing itself. (*Transc. Idealismus*, I, 3:452 [Heath 91])

Again in 1806, in the *Darlegung des wahren Verhältnisses*, directed against Fichte, after criticizing the latter for supposing a universe entirely devoid of reason, Schelling adds: "We do not admit such a universe, but only a universe that is living reason itself" (*Darlegung*, I, 7:100). Schelling does not completely ignore the fact that from the standpoint of the position he takes in his philosophy of nature this is a contradiction, or at least a difficulty; moreover, there is no lack of opponents (among them, Eschenmayer) to point it out. But it is significant that he tries to extricate himself by accentuating the idealistic aspect of his philosophy:

Immediately, as soon as I began to profess the philosophy of nature, the objection was often put to me that I *presupposed* nature without dealing with the critical question of how we come to do so. I reply that anyone who, by abstraction, will raise himself to the pure concept of nature will recognize that for the purpose of this construction I presuppose nothing except what the transcendental philosopher also presupposes. For what I call *nature* is for me nothing other than the pure objective of intellectual perception, the pure *subject-object*.

To be sure, alongside these formulas we find others which savor of much more realistic convictions and we cannot seriously doubt that, even at the height of his enthusiasm for the philosophy of nature, Schelling did not take nature to be entirely deducible. We ourself point out (Ch. 12, pp. 319 ff.) the ambiguous role of the *given* in his philosophy of nature, and we have also noted, on the subject of "the impotence of nature" in Hegel, that he had been preceded in this path by Schelling (Ch. 11, p. 303, n. 28). Nevertheless Schelling, unlike Hegel, nowhere clearly indicated the limits he assigned to the deductive effort, and with his characteristic lack of systematic spirit, this effort, for him, has the air of attacking reality and science in all their particulars. His disciples, of course, went even further, claiming to "construct" almost anything whatsoever. This is what allowed Hegel and Gans to find fault with Schelling and his followers (as we saw in Ch. 11, p. 275), in defending what they considered the proper domain of experimental science against this excess of deduction; strangely enough, on this point Hegel, whose idealism was surely, at bottom, much more extreme than that of his rival, gives the impression of being the more moderate of the two.

Victor COUSIN, who experienced the keenest enthusiasm for the *philosophy of nature*, very faithfully sums up the guiding principle of this conception by stating that nature "must needs resemble him [man], since it is derived from the same principle; their only difference being that of consciousness and non-consciousness" (*Fragments philosophiques*, Paris: Ladrange, 1833, p. xl [*Philosophical Essays*, trans. George Ripley, Edinburgh: Thomas Clark, 1839, p. 92, bottom page pagination; Meyerson's brackets]).

Given the direction taken by Schelling's thought in the last period of his life and in particular his attack on Hegel's idealism, we found it necessary to dwell somewhat on these early tendencies of his philosophy.

Furthermore, we must add that, even during his late period, Schelling does not mean to reject the postulate of the identity of thought and reality, at least not formally – which, as a matter of fact, he could not have done without admitting that he was entirely renouncing his original convictions, the very eventuality he intended to avoid. He is thus content to affirm that he has been misunderstood.

When I first presented this distinction [it is a question here of the distinction between "the pure *how* of things" – we would say their essence – and the fact that they exist, a distinction we treat in Ch. 12, p. 319], I had indeed foreseen what would happen. Some appeared completely astonished at this quite simple and truly unmistakable, but for this very reason supremely important, distinction, for they had, in an earlier philosophy [Hegel's], heard of a misconstrued identity between thought and being. This identity, if it is properly understood, I shall certainly not combat, for it originates with me, but as for the misunderstanding and the philosophy derived from it, I am most assuredly obliged to combat them. (*Phil. der Offenbarung*, II, 3:59 [Meyerson's brackets])

Of course the Hegelians, in their polemic against Schelling, do not fail to emphasize the fact that in this matter the latter is actually their spiritual ancestor. "That nature," says ROSENKRANZ, "is identical to the mind, insofar as they are both ideas, so much is certain. Here, then, Schelling was perfectly right" (*Schelling* 106).

On this same subject, it is interesting to note that in 1801, that is at the very moment Schelling, while developing his philosophy of nature, was distancing himself from Fichte, the latter subjected the conceptions of his former disciple to a criticism which, although unquestionably less intensive and less far-reaching than the criticism Schelling was one day to put forward against Hegel, still, in its foundations, somewhat resembles the objections we have set forth in Chapter 12 (pp. 311 ff.). Speaking of Schelling's point of departure, Fichte shows in particular that the Absolute is doomed to remain eternally within itself, frozen in the nothingness of its identity. It is thus "inconceivable that the Absolute should need to emerge from itself and become manifest" (Xavier LÉON, 'Fichte contre Schelling,' *Rev. de méta.* **12** (1904) 951, 953). Does not Schelling's objection that the idea (in Hegel) "has not the slightest need of becoming real in any other way than it already is" (*Zur Geschichte*, I, 10:152; see p. 313 above) merely echo the protest put forward by Fichte more than thirty years earlier? And how instructive also is the fact that Fichte and Schelling, both so idealistic, were nevertheless shocked, each in his turn, by an effort to make clear how the *Absolute* (or the *Idea*) becomes nature. It is true that both cases involved the work of a rival.

17. SCHELLING'S ANNOUNCED WORKS (p. 328)

A major work, *Die Weltalter* (*The Ages of the World*) is announced as early as 1811. In a letter to a friend, Schelling calls it his "favorite child." At the end of 1812 he announces: "Now it will be seen how I shall speak." At this time the first part of the book (eleven folios of some thirty Schelling himself estimates the whole work may include) is set in type and printed. It is reprinted (no doubt with some revisions) in 1813. In 1815 the German booksellers' annual catalog (*Leipziger Messkatalog*) and the *Allgemeine Zeitung* announce the *Weltalter* as having appeared, but actually Schelling has had the printed sheets recalled. However, in 1819 he still asserts, in writing to a friend, that he is "on the verge of finishing everything," that he "needs only a few free hours" to do so (PLITT, *Aus Schelling's Leben* 2:244, 256, 332, 325, 430). But it was apparently an illusion, and during Schelling's lifetime the only thing that actually comes out is a treatise *On the Divinities of Samothrace* (*Über die Gottheiten von Samothrake*, *Werke*, I, 8:345–423), published in 1815 as "Supplement to the *Weltalter*." The printed fragment of the *Weltalter* is inserted in the eighth volume of the posthumous edition (pp. 195–344), with modifications introduced by the author after the printing (about 1814 or 1815, according to Karl Schelling). But it is curious that in spite of Schelling's very explicit claims, nothing relating to the continuation of the *Weltalter* was found in his papers except a half-formed sketch of the beginning of Part 2, which the editor did not deem publishable; from this Kuno FISCHER concludes, not without some justification, that in announcing the imminent completion of the work to his friends, Schelling was consciously deceiving them (*Geschichte* 8:163–168; cf. Schelling, *Werke*, I, 8:v, Editor's Preface).

Beginning in 1821 Schelling has a new project, a book on mythology, which is to precede the *Weltalter*, seeing that this work "has not yet sufficiently matured" (PLITT 3:5). In 1826 the new work is announced in the booksellers' catalogs as having appeared, in 1830 as being about to appear, as it is again in 1836. But once again, nothing is published in the philosopher's lifetime.

Even Gustav Leopold PLITT, in his short notices accompanying the publication of Schelling's correspondence, where he proves to be almost as much an apologist as a biographer, cannot help observing, on the subject of Schelling's return to Munich in 1827, that "the hope of the publication of a work, a hope he constantly awakened in his friends by repeated promises, was disappointed afresh year after year" (*Aus Schelling's Leben* 3:33).

Of course this strange behavior leaves Schelling vulnerable to enemy attacks, which become particularly heated from the moment he exasperates the Hegelians with his *Preface* of 1834. For example, Gans, in his preface to Hegel's *Philosophy of History* (published in 1837), calls Schelling one of those who are "dead while they live" (*Lebendverstorbene*) and ironically points out that "the Hegelian *Four Ages of the World* have at least made their appearance" (*Phil. der Geschichte*, 9:xx [Sibree xxiii]).

18. CAROLINE SCHELLING (p. 328)

Although it is certain that Caroline Schelling (née Michaëlis) had a fine mind, and although she seems to have exerted a most auspicious influence on her husband's career, one would hesitate to call this influence decisive. As we see by her correspondence, Caroline was keenly interested in everything intellectual and passionately interested in all her husband undertook; but it appears more than doubtful that, especially in philosophy, she could have served as his inspiration, except in the broadest sense of the term. Certainly we must not for a moment dream of supposing that she could have aided him in the writing of his works. Caroline writes extremely well, and her letters are among the loveliest in German literature (which, as we know, really is not all that rich in them). But Schelling is truly one of the masters of German style; unlike Hegel, who always seems at odds with his own thought and seems to succeed only in giving it a more or less approximate expression, Schelling's phrases flow naturally and easily, supple yet vigorous, clearly sufficient to the thought behind them: it is not surprising that Goethe, who knew whereof he spoke, greatly admired his style. (On the favor Goethe consistently showed Schelling, as opposed to his amiable but merely formal relations with Hegel, see Kuno FISCHER, *Geschichte* 7:219, 42–43, and Émile BRÉHIER, *Schelling*, Paris: Félix Alcan, 1912, p. 21. Among the letters collected by Plitt in *Aus Schelling's Leben* are a considerable number from Goethe, almost all highly flattering. As early as 1799 Goethe announces that he is reading the *System of Transcendental Idealism* and believes he understands it [Plitt 1:297]).

There is no direct evidence that Schelling's style is the least bit different before he knew Caroline, during his life with her, or after her death. One is likewise forced to note that at the time he met her (she was then married to her second husband, August Wilhelm von Schlegel, the celebrated man of letters, and close relations – even purely Platonic ones – between her and Schelling could not be established until some time later) the young philosopher is already at the height of his powers. He is only twenty-three (Caroline is thirty-five), but he is famous, he has just been named a professor at Jena, the *Ideen* have been out for a year, and the *Weltseele* has just come out. Since the beginning of this period of prodigious activity antedates Caroline's entry into his life, it does not seem logical to link its end with the spouse's death. Finally, after Caroline's death Schelling never behaves like a man whose élan vital has been broken. He sincerely grieves for her, but three years later he remarries, and his new family life appears quite happy; his second wife, fourteen years younger than he, bears him three children (his marriage to Caroline had been without issue).

19. PERSONAL RELATIONS BETWEEN SCHELLING AND HEGEL (p. 339)

Schelling and Hegel, both originally from the same region of Germany (Württemberg) and fellow students at the Tübingen *Stift* (seminary), had formed there an extremely close friendship that persisted after they had left this educational institution. The letters they exchanged whenever they were apart are quite warm (cf. especially Schelling's letters of 24 March 1802, 11 July and 31 August 1803 – "*meine Frau lässt Dich ganz erstaunlich grüssen*" ["my wife sends you her warmest greetings"] – and 3 March 1804 in Gustav Leopold PLITT, *Aus Schelling's Leben* 1:369, 467, 483; 2:11) and reveal a close affinity of sentiment and thoughts. But it must be noted that Schelling, five years younger than his friend, always gives the impression of being the elder of the two. It is Schelling who draws Hegel to the University of Jena, where he is perceived as no more than an active collaborator of his already illustrious friend. The appearance of the book *Differenz des Fichteschen und Schellingschen Systems der Philosophie* [*Werke*, 1:161–296; *The Difference Between Fichte's and Schelling's System of Philosophy*, trans. H. S. Harris and Walter Cerf (Albany: State University of New York Press, 1977)] in 1801 confirms this impression. In it Hegel strongly sides with Schelling against Fichte, declaring that the philosophy of the former is destined to prevail (Kuno FISCHER, *Geschichte* 7:145; Rudolf HAYM, *Hegel* 151 ff.). The rare allusions indicative of a more independent way of thinking to be found in this work go entirely unnoticed, and the *Allgemeine Zeitung* of Augsburg, in appraising the work, writes that "Schelling has gone back to his native region to find a staunch defender and through him advises the public that Fichte too is quite inferior to his [Schelling's] theories" (ROSENKRANZ, *Hegel's Leben* 162; Hegel vigorously protests, although it is unclear whether he is denying the accusation that he meant to disparage Fichte, or whether he is protesting the fact that he was presented as a mere supporter of Schelling). Hegel's *pro licentia docendi* dissertation, which appears the same year, is also "entirely Schellingian" according to the competent judgment of HAYM (*Hegel* 154). Beginning in 1802, the two philosophers are coeditors of a periodical, the *Kritisches Journal*, and their thought is so much alike that it is sometimes difficult to tell which of the two is the author of this or that unsigned article, and disputes have arisen among biographers on this subject (cf. ROSENKRANZ, *Schelling* 194–195; ERDMANN, *Die Entwicklung der deutschen Spekulation seit Kant*, Leipzig: Vogel, 1853, 2:691–695; and HAYM, *Hegel* 503). Of course, in certain articles of this short-lived periodical Hegel manifests an increasing intellectual independence, but Schelling does not appear to notice, and it is probably largely because he continues to consider him a mere disciple that, in a very friendly letter to Hegel (11 Jan. 1807, PLITT 2:110 ff.), he declares that he awaits the projected work with great impatience. "What work you will be able to produce if your talent is only given the time to bear fruit! I can only continue to wish you the calm surroundings and the leisure necessary for the execution of such precious and timeless works." The publication of the *Phenomenology* where, from the Preface on, Hegel directed very sharp attacks against the entire Schellingian school, was obviously a bitter disappointment for Schelling. With the copy destined for his friend, Hegel included a long letter; a single sentence contains a sort of excuse, suggesting that he had only had Schelling's imitators in mind: "In the Preface you will not find that I have been too hard on the shallowness [*Plattheit*] that makes so much mischief with your forms in particular and degrades your science into a bare formalism" (Letter of 1 May 1807, *Briefe*, 19_1:102–103 [Butler 80]). Schelling does not reply until five months later (2 Nov. 1807) in a letter rather haughty in tone, where he nevertheless seems to accept the explanation: "So far I have read only the Preface. Insofar as you yourself mention the polemical part of the Preface, given my own justly measured opinion of myself I would have to think too little of myself to apply this polemic to my own person. It must therefore, as you have expressed in your letter, apply only to further bad use of my ideas and to those who parrot them without understanding [*Nachschwätzer*], although in this writing itself the distinction is not made" (PLITT 2:123–124 [*Hegel: The Letters* 80; Meyerson's brackets]). It is generally assumed that the letter was followed by a complete break between the two friends. Kuno

FISCHER appears to accept this (*Geschichte* 7:146) and Victor DELBOS, in his excellent short study *De posteriore Schellingii philosophia* (Paris: Félix Alcan, 1902, p. 8), follows his lead: *inde nullae postea inter se epistulae, amicitia nulla* [thereafter there were no letters between them, no friendship]. The idea seems to have originated with Karl Hegel, the editor of his father's correspondence, who declared in his commentary on Schelling's response that this was their final letter. Now the truth of this has not actually been established. When Schelling writes to Hegel's widow in 1832, he seems to be claiming that the correspondence continued until 1808 (PLITT 3:61). But supposing that there was a misunderstanding between them, there was certainly no formal break and relations continued, at least outwardly. Hegel's correspondence proves this. In July 1812, Schelling and his wife pass through Nuremberg, where Hegel was at that time, without seeing him; but Hegel does not even consider attributing this lapse to a feeling of hostility toward himself; they stayed only a few hours, and Schelling, who had rheumatism, was not able to see anyone (Hegel, *Briefe*, 19_1:345 [Butler 270]). No doubt Hegel had good reasons to interpret things as he did, and what proves that he was not mistaken is that a few months later Schelling paid him a "friendly visit," during which, however, they both refrained from talking philosophy (Letter of 23 Oct. 1812, *Briefe*, 19_1:350 [Butler 284]).

A few years later, in 1817, Victor COUSIN came to Heidelberg. We know that Cousin prided himself upon having discovered or *divined* Hegel before he was known in his own country.

> At that time, Hegel was far from being the celebrated man that I have since found at Berlin, drawing all eyes upon himself, and at the head of a large and enthusiastic school. Hegel as yet had no reputation but that of a distinguished disciple of Schelling. He had published some books which were little read; and his teaching had hardly begun to make him better known. (*Fragments philosophiques*, 2nd ed., Paris: Ladrange, 1833, p. xxxvii [*Philosophical Essays*, trans. George Ripley, Edinburgh: Thomas Clark, 1839, p. 90, bottom page pagination])

Cousin was obviously mistaken: Hegel, nine years after the *Phenomenology*, five years after the *Science of Logic*, on the eve of being called to Berlin under the most honorable conditions, had quite a different intellectual status than the visitor supposed. Of course at that point in his life Cousin knew so little German that he had to study Kant in Born's execrable Latin translation (*Fragments et souvenirs*, 3rd ed., Paris: Didier, 1857, p. 57 – in spite of the claim of Jules SIMON, *Victor Cousin*, 4th ed., Paris, 1910, p. 22, it is at least doubtful that he ever learned the language well). He was also rather ill-informed on the German philosophic movement by men, like Count Reinhart, Passavant or Friedrich von Schlegel, who harbored violent prejudices on the subject. But his error would have been inexplicable if the two German philosophers had at that time been at daggers drawn (as seems to be the general belief). Therefore, as we have just seen, such was not the case, and when Cousin returned to Heidelberg the following year and then proposed to take a short philosophic tour of Germany, Hegel provided him with a sort of general letter of introduction, in which he inserted the following passage: "Please give Mr. Schelling my compliments. You will no doubt receive a warm welcome from him, and politically find a way of thinking free of anti-French prejudices" (*Briefe*, 19_2:20 [Butler 633]). At about the same time Hegel, writing to his old friend Niethammer, asks him to "extend my cordial regards" to Schelling (*Briefe*, 19_2:17 [Butler 364]), which confirms the impression that there was then (in 1818) no overt hostility between them.

In August 1829 the two philosophers chanced to meet at Carlsbad. In a letter to his wife (PLITT 3:47), Schelling expresses some surprise at the cordiality Hegel shows him "as if nothing had happened between us," but this obviously cannot refer to what had occurred after the publication of the *Phenomenology*, which was in a manner of speaking negated by the 1811 visit; it is therefore probable that what he had in mind was his own attacks against Hegel, which he seems to have begun a few years before. Hegel, who also recounts the event to his wife, appears quite happy about the meeting: "We are both pleased about meeting again, and find ourselves together as cordial friends of old" (*Briefe*, 19_2:326 [Butler 398]), and he repeats these expressions in two letters sent to friends (*Briefe*, 19_2:330, 331 [Butler translates only the second letter, p. 446]). Hegel and Schelling spend five days together, eating and taking walks in each other's company.

As we said above, however, Schelling's attacks against Hegel seem in fact to have begun some time earlier. Of course, what Rosenkranz says about Schelling's courses which he attended (he is customarily cited in connection with these attacks; cf. Kuno FISCHER, *Geschichte* 7:216) cannot be invoked here, for that refers to a period long after Hegel's death (ROSENKRANZ, *Schelling* xxi-xxii). But a letter from the economist Friedrich Thiersch shows that at least after his return to Munich in 1827, Schelling publicly attacked Hegel, whom he reproached for having "spoiled" his (Schelling's) philosophy, giving it a wrong twist (Heinrich

Wilhelm Josias THIERSCH, *Friedrich Thiersch's Leben*, Leipzig: C. F. Winter, 1866, 1:346). The following year, in writing to Cousin, Schelling expresses himself as follows:

You entered the territory of German philosophy from the Heidelberg side [this is where Hegel was in 1817 when Cousin made his first trip to Germany; see Appendix 15, p. 586]; the system I originated first became known to you only with the meaning ascribed to it by a few badly indoctrinated or undiscriminating people, and in the form it had received in passing through the narrow mind of a man who believed he could take hold of my ideas just as a crawling insect may believe it can appropriate the leaf of a plant around which it has spun a cocoon. He was wrong; the system has its own life principle ....

(There follows an entirely incomprehensible passage, but there is reason to doubt that it is really from Schelling's pen since in general he writes French correctly enough; perhaps someone examining the manuscript with a better command of French than Plitt apparently possessed would find a reading providing an appropriate meaning.)

Again, in the same letter, he says:

Let them leave me my ideas, without attaching to them, as you seem to do, the name of a man who, while thinking he could steal them from me, has shown himself as incapable of carrying them to their true perfection as he was of inventing them. (PLITT 3:39–40, 41 [Meyerson's brackets])

A passage from Heinrich Heine, although contained in a work published after Hegel's death, certainly refers to the same period. Here is his amusing description, which would seem to be quite accurate in spite of its bantering tone.

This happened at the beginning of the century. Mr. Schelling was then a great man. Meanwhile, however, Hegel appeared on the philosophical scene; Mr. Schelling, who in later years wrote almost nothing, was eclipsed, indeed forgotten, and retained only a literary-historical significance. Hegelian philosophy became dominant, Hegel became the sovereign in the realm of intellect, and poor Schelling, a fallen, mediatized philosopher, wandered mournfully about among the other mediatized gentlemen in Munich. I once saw him there and could almost have wept tears at the pitiful sight. And what he said was the most pitiful thing of all; it was an envious railing at Hegel, who had supplanted him. As one shoemaker talks about another whom he accuses of having stolen his leather and made boots of it, so I heard Mr. Schelling, when I once saw him by chance, talk about Hegel, Hegel who 'had taken his ideas'; and 'it is my ideas that he took,' and again 'my ideas' – this was the poor man's constant refrain. I assure you, if the shoemaker Jakob Böhme once talked like a philosopher, the philosopher Schelling now talks like a shoemaker. (Heinrich Heine, *Ueber Deutschland*, *Sämmtliche Werke*, Hamburg: Hoffmann und Campe, 1861, 6:157–158 ['The Romantic School,' trans. Helen Mustard, *The Romantic School and other*

Essays, ed. Jost Hermand and Robert C. Holub, New York: Continuum, 1985, pp. 69–70])

This passage is lacking in the French edition of the book; it was added only to the German edition and the author, for reasons unknown, also refrained from including it in later French language editions of his work. In any case, the French edition does contain an analogous passage elsewhere, but shorter and much less typical (*De l'Allemagne*, *Oeuvres de Henri Heine*, Paris: Eugène Renduel, 1835, 5:229 ['Concerning the History of Religion and Philosophy in Germany,' trans. Mustard, *The Romantic School and other Essays*, pp. 239–240]).

Surely Hegel got wind of Schelling's attacks – furthermore, Rosenkranz explicitly affirms as much (*Schelling* xvi ff.) – but he had the good taste not to take too much offence, as is shown by his attitude at Carlsbad. Perhaps, however, we may also conclude that the attacks – at least those made in public lectures – did not assume a form anywhere near as violent as that of the 1834 *Preface* or of the treatise *Zur Geschichte*, and that, at any rate, Schelling then chose to respond by putting forth claims of priority. There is little doubt that Hegel would have risen to the challenge of an out and out attack against his system of the sort contained in the above-named works.

One thing is certain: Schelling never published anything against his rival during Hegel's lifetime. This is pointed out by ROSENKRANZ, who stresses the contrast between this attitude and his attitude toward Reinhold, Fichte and Jacobi (*Schelling* 352). He also notes that when Schelling was criticized for not having attacked Hegel until after his death, he had his disciples reply that he could not have foreseen that Hegel would pass away so soon (*Schelling* 359).

20. TYCHO BRAHE, ASTROLOGY AND THE MOTION OF THE EARTH (p. 464)

Tycho's argument in favor of astrology is based on a strong conviction of the *dignity* of what takes place on the earth, and we can see that this same way of thinking greatly contributed to his unwillingness to accept the heliocentric theory. Tycho had the greatest respect for the genius of Copernicus who, he said, deserved to be called a second Ptolemy, given that "with an admirable intellectual penetration he had, in a very different way, so established the science of the celestial motions that no one before him had spoken more precisely of the movement of the stars" (TYCHONIS

BRAHE, *De disciplinis mathematicis oratio*, Hafniae [Copenhagen]: Apud Henricum Waldkirchium, 1610, p. 9). Nevertheless, the reasons that prevented Tycho from accepting the idea that the earth moved were largely reasons of physics, based on the consideration of terrestrial motions – reasons which, as a matter of fact, remained quite legitimate so long as the law of inertia was unknown (see IR 533 [Loewenberg 460–461]). But religious considerations also played a significant role for him. To his mind, however, it was less a question of the literal interpretation of certain words of the Bible, such as the famous passage on Joshua, than of the fact that he found the heliocentric system as a whole irreconcilable with the way in which the relationship between heaven and earth is envisaged in Holy Writ. It is written, he reasoned, that *Creavit Deus Coelum et Terram*, and in this statement the earth stands over against the whole of the heavens; is it possible, then, for it to be the tiny, obscure star it appears to be according to the Copernican hypothesis? (TYCHONIS BRAHE, *Epistolarum astronomicarum libri*, Uraniborg and Frankfurt: Apud Godefridum Tampachium, 1610, pp. 190 ff.). Indeed, the most confirmed Copernican seems to have had difficulty denying the profound disagreement in this respect between the new theory and the doctrine for which the earth was the sole object of divine action and consequently could not fall from its dignified position as the center of the universe.

It is difficult, we believe, to understand the grounds for the enormous prestige of astrology and how this prestige could have been so completely destroyed, unless one takes into account the weight of these finalistic considerations and the fact that, as a result of the Copernican reform, an immediate lawlike connection, which these considerations had rendered plausible, instead became improbable.

21. NON-EUCLIDEAN SPACE AND PHYSICAL VERIFICATION (p. 538)

From the very beginning of hypergeometry, its founders asked that it be experimentally verified whether or not space conforms to the model of Euclidean geometry. Nikolai LOBACHEVSKY ('Etudes géométriques sur la théorie des parallèles,' trans. Hoüel, *Mémoires de la Société des sciences physiques et naturelles de Bordeaux* **4** [1866] 120) and Bernhard RIEMANN ('Über die Hypothesen, welche der Geometrie zu Grunde liegen,' *Abhandlung der Königlichen Gesellschaft der Wissenschaften zu Göttingen* **13** [1866–67] 148) explicitly called for astronomical observations for this purpose, and Hermann von HELMHOLTZ approved this viewpoint (*Populäre Wissenschaftliche Vorträge*, Brunswick: Friedrich Vieweg, 1876, 3:42, 43, and *Wissenschaftliche Abhandlungen*, Leipzig: A. Barth, 1884, 1:154 [second citation erroneous]). In 1872, Mach expressed the opinion that if one had thus far failed to construct a fully satisfactory mechanical theory of electricity it was because one had persisted in considering molecular motion in three-dimensional space (Georges SOREL, 'Vues sur les problèmes de la philosophie,' *Rev. de méta.* **18** [1910] 610). At about the same time the astronomer Friedrich ZÖLLNER linked gravitational action at a distance, postulated by Newtonian theory, to the existence of a fourth dimension of space (*Principien einer elektrodynamischen Theorie der Materie*, Leipzig: Wilhelm Engelmann, 1876, pp. lxvii ff.). Later Peter Guthrie TAIT hypothesized that the solar system, and with it our earth, might one day arrive in regions of space in which the curvature would be modified (*Conférences sur quelques-uns des progrès récents de la physique*, trans. Krouchkoll, Paris: Fetschérin et Chuit, 1886, pp. 12 ff.), and Bertrand RUSSELL proposed for the "spatial constant" a precise experiment consisting in rolling a disc and measuring the distance travelled ('Les axiomes propres à Euclide sont-ils empiriques?' *Rev. de Méta.* **6** [1898] 760). Still more recently, Gino FANO predicted the day when, with access to more perfect instruments than we now have at our disposal, we would "become persuaded that one of the two non-Euclidean geometries lends itself to representing the relative positions of bodies, as we perceive them, with a closer approximation than does Euclidean geometry" ('La geometria non-euclidea,' *Scientia* **4** [1908] 282). Moreover, there have been attempts to introduce the hypothesis of hyperspace into chemistry, notably by representing pentavalent nitrogen in four-dimensional space (Francesco SEVERI, 'Hypothèses et réalité dans les sciences géométriques,' *Scientia* **8** [1910] Supplément: 12) and we have mentioned (p. 140, n. 26) the way in which Renan believes this idea must be combined, in biology, with that of preformation. In reading through the works just cited, one can become convinced that these speculations are independent of those concerning Minkowski's spatio-temporal geometry.

Furthermore, the similarity of their development is not the only point of contact between the theory of relativity and earlier conceptions having to do with hyperspace. Indeed, EINSTEIN supposes that our space itself is not Euclidean but actually "spherical," and even calculates its radius of curvature (*Über die spezielle und die allgemeine Relativitätstheorie*, 9th ed., Brunswick: Friedrich Vieweg & Sohn, 1920, p. 77 [*Relativity: The*

Special and General Theory, trans. Robert W. Lawson, New York: Crown, 1961, p. 114]).

In speaking of mathematical deduction (particularly in Ch. 15, pp. 409 ff.) we have omitted all reference to these recent developments in order not to overcomplicate the arguments. If we do take into account the theories on hyperspace, the views we have presented require a few qualifications.

We have just seen that what characterizes these new conceptions from the most general point of view is the fact that the geometric intuitions serving as the foundation for Euclidean geometry are no longer considered unshakeable. On the contrary, these bases are subject to experimental verification and must be modified accordingly. By this fact, geometry approaches the physical sciences. Of course there is no thought of introducing statistical considerations, but it no longer appears inconceivable that a proposition might come to be understood as a mere approximation. For example, if two-dimensional inhabitants of a spherical surface could measure a large enough circle, they would inevitably come to consider the ratio between the diameter and the circumference not as a constant, but as a function varying with the size of the diameter. And in the same way, if our space is "spherical," we could be led, under favorable conditions, to recognize that the ratio between the volume of a sphere and its radius, as defined by Euclidean geometry, is only an approximation and that for sufficiently large values of this radius the ratio must undergo significant correction.

Supposing that these doctrines actually triumph in science, what will be changed with regard to the rationalization of reality by mathematics discussed in Chapter 16?

First, we can point out that since the new geometry diverges appreciably from common sense conceptions, the *explanatory power* of theories based on it would therefore be greatly diminished. Indeed, the situation would in some sense be analogous to the one we considered in Chapter 8 (pp. 234 ff.) on the subject of explanation by *discontinuous* atomic motion.

Still it does not seem too risky to predict that the modification will not be appreciable insofar as the role of mathematics in the physical sciences is concerned. Indeed we must take into consideration that what makes mathematics valuable to them is the important role played by deduction. Now this importance will no doubt be diminished in the new geometry: no longer will it be claimed that the results of deductions will never be able to be contradicted by experiment. But we must note that the only thing which will be modified in each instance, when going, for example, from Euclidean space to spherical space and from there to elliptical space, will be the very foundations of our spatial conception: the initial axioms and postulates. Their development will nonetheless continue to be essentially deductive and the mathematical concept will still be infinitely pellucid in comparison with the physical concept. As to geometry considered as a physical science, it will be a science in which the experimental datum will be greatly reduced, while reasoning will still play a quite preponderant part. Moreover, as we have seen, there is reason to believe that the situation is already analogous in present-day geometry and that in this respect, consequently, the change will not be fundamental.

INDEX OF NAMES

Boston Studies in the Philosophy of Science

Editor: Robert S. Cohen, *Boston University*

1. M.W. Wartofsky (ed.): *Proceedings of the Boston Colloquium for the Philosophy of Science, 1961/1962.* [Synthese Library 6] 1963 ISBN 90-277-0021-4
2. R.S. Cohen and M.W. Wartofsky (eds.): *Proceedings of the Boston Colloquium for the Philosophy of Science, 1962/1964.* In Honor of P. Frank. [Synthese Library 10] 1965 ISBN 90-277-9004-0
3. R.S. Cohen and M.W. Wartofsky (eds.): *Proceedings of the Boston Colloquium for the Philosophy of Science, 1964/1966.* In Memory of Norwood Russell Hanson. [Synthese Library 14] 1967 ISBN 90-277-0013-3
4. R.S. Cohen and M.W. Wartofsky (eds.): *Proceedings of the Boston Colloquium for the Philosophy of Science, 1966/1968.* [Synthese Library 18] 1969 ISBN 90-277-0014-1
5. R.S. Cohen and M.W. Wartofsky (eds.): *Proceedings of the Boston Colloquium for the Philosophy of Science, 1966/1968.* [Synthese Library 19] 1969 ISBN 90-277-0015-X
6. R.S. Cohen and R.J. Seeger (eds.): *Ernst Mach, Physicist and Philosopher.* [Synthese Library 27] 1970 ISBN 90-277-0016-8
7. M. Čapek: *Bergson and Modern Physics.* A Reinterpretation and Re-evaluation. [Synthese Library 37] 1971 ISBN 90-277-0186-5
8. R.C. Buck and R.S. Cohen (eds.): *PSA 1970.* Proceedings of the 2nd Biennial Meeting of the Philosophy and Science Association (Boston, Fall 1970). In Memory of Rudolf Carnap. [Synthese Library 39] 1971 ISBN 90-277-0187-3; Pb 90-277-0309-4
9. A.A. Zinov'ev: *Foundations of the Logical Theory of Scientific Knowledge (Complex Logic).* Translated from Russian. Revised and enlarged English Edition, with an Appendix by G.A. Smirnov, E.A. Sidorenko, A.M. Fedina and L.A. Bobrova. [Synthese Library 46] 1973 ISBN 90-277-0193-8; Pb 90-277-0324-8
10. L. Tondl: *Scientific Procedures.* A Contribution Concerning the Methodological Problems of Scientific Concepts and Scientific Explanation.Translated from Czech by D. Short. [Synthese Library 47] 1973 ISBN 90-277-0147-4; Pb 90-277-0323-X
11. R.J. Seeger and R.S. Cohen (eds.): *Philosophical Foundations of Science.* Proceedings of Section L, 1969, American Association for the Advancement of Science. [Synthese Library 58] 1974 ISBN 90-277-0390-6; Pb 90-277-0376-0
12. A. Grünbaum: *Philosophical Problems of Space and Times.* 2nd enlarged ed. [Synthese Library 55] 1973 ISBN 90-277-0357-4; Pb 90-277-0358-2

13. R.S. Cohen and M.W. Wartofsky (eds.): *Logical and Epistemological Studies in Contemporary Physics*. Proceedings of the Boston Colloquium for the Philosophy of Science, 1969/72, Part I. [Synthese Library 59] 1974
ISBN 90-277-0391-4; Pb 90-277-0377-9
14. R.S. Cohen and M.W. Wartofsky (eds.): *Methodological and Historical Essays in the Natural and Social Sciences*. Proceedings of the Boston Colloquium for the Philosophy of Science, 1969/72, Part II. [Synthese Library 60] 1974
ISBN 90-277-0392-2; Pb 90-277-0378-7
15. R.S. Cohen, J.J. Stachel and M.W. Wartofsky (eds.): *For Dirk Struik*. Scientific, Historical and Political Essays in Honor of Dirk J. Struik. [Synthese Library 61] 1974 ISBN 90-277-0393-0; Pb 90-277-0379-5
16. N. Geschwind: *Selected Papers on Language and the Brains*. [Synthese Library 68] 1974 ISBN 90-277-0262-4; Pb 90-277-0263-2
17. B.G. Kuznetsov: *Reason and Being*. Translated from Russian. Edited by C.R. Fawcett and R.S. Cohen. 1987 ISBN 90-277-2181-5
18. P. Mittelstaedt: *Philosophical Problems of Modern Physics*. Translated from the revised 4th German edition by W. Riemer and edited by R.S. Cohen. [Synthese Library 95] 1976 ISBN 90-277-0285-3; Pb 90-277-0506-2
19. H. Mehlberg: *Time, Causality, and the Quantum Theory*. Studies in the Philosophy of Science. Vol. I: *Essay on the Causal Theory of Time*. Vol. II: *Time in a Quantized Universe*. Translated from French. Edited by R.S. Cohen. 1980 Vol. I: ISBN 90-277-0721-9; Pb 90-277-1074-0
Vol. II: ISBN 90-277-1075-9; Pb 90-277-1076-7
20. K.F. Schaffner and R.S. Cohen (eds.): *PSA 1972*. Proceedings of the 3rd Biennial Meeting of the Philosophy of Science Association (Lansing, Michigan, Fall 1972). [Synthese Library 64] 1974
ISBN 90-277-0408-2; Pb 90-277-0409-0
21. R.S. Cohen and J.J. Stachel (eds.): *Selected Papers of Léon Rosenfeld*. [Synthese Library 100] 1979 ISBN 90-277-0651-4; Pb 90-277-0652-2
22. M. Čapek (ed.): *The Concepts of Space and Time*. Their Structure and Their Development. [Synthese Library 74] 1976
ISBN 90-277-0355-8; Pb 90-277-0375-2
23. M. Grene: *The Understanding of Nature*. Essays in the Philosophy of Biology. [Synthese Library 66] 1974 ISBN 90-277-0462-7; Pb 90-277-0463-5
24. D. Ihde: *Technics and Praxis*. A Philosophy of Technology. [Synthese Library 130] 1979 ISBN 90-277-0953-X; Pb 90-277-0954-8
25. J. Hintikka and U. Remes: *The Method of Analysis*. Its Geometrical Origin and Its General Significance. [Synthese Library 75] 1974
ISBN 90-277-0532-1; Pb 90-277-0543-7
26. J.E. Murdoch and E.D. Sylla (eds.): *The Cultural Context of Medieval Learning*. Proceedings of the First International Colloquium on Philosophy,

Science, and Theology in the Middle Ages, 1973. [Synthese Library 76] 1975
ISBN 90-277-0560-7; Pb 90-277-0587-9

27. M. Grene and E. Mendelsohn (eds.): *Topics in the Philosophy of Biology.* [Synthese Library 84] 1976 ISBN 90-277-0595-X; Pb 90-277-0596-8
28. J. Agassi: *Science in Flux.* [Synthese Library 80] 1975
ISBN 90-277-0584-4; Pb 90-277-0612-3
29. J.J. Wiatr (ed.): *Polish Essays in the Methodology of the Social Sciences.* [Synthese Library 131] 1979 ISBN 90-277-0723-5; Pb 90-277-0956-4
30. P. Janich: *Protophysics of Time.* Constructive Foundation and History of Time Measurement. Translated from the 2nd German edition. 1985
ISBN 90-277-0724-3
31. R.S. Cohen and M.W. Wartofsky (eds.): *Language, Logic, and Method.* 1983
ISBN 90-277-0725-1
32. R.S. Cohen, C.A. Hooker, A.C. Michalos and J.W. van Evra (eds.): *PSA 1974.* Proceedings of the 4th Biennial Meeting of the Philosophy of Science Association. [Synthese Library 101] 1976
ISBN 90-277-0647-6; Pb 90-277-0648-4
33. G. Holton and W.A. Blanpied (eds.): *Science and Its Public.* The Changing Relationship. [Synthese Library 96] 1976
ISBN 90-277-0657-3; Pb 90-277-0658-1
34. M.D. Grmek, R.S. Cohen and G. Cimino (eds.): *On Scientific Discovery.* The 1977 Erice Lectures. 1981 ISBN 90-277-1122-4; Pb 90-277-1123-2
35. S. Amsterdamski: *Between Expeience and Metaphysics.* Philosophical Problems of the Evolution of Science. Translated from Polish. [Synthese Library 77] 1975 ISBN 90-277-0568-2; Pb 90-277-0580-1
36. M. Marković and G. Petrović (eds.): *Praxis.* Yugoslav Essays in the Philosophy and Methodology of the Social Sciences. [Synthese Library 134] 1979
ISBN 90-277-0727-8; Pb 90-277-0968-8
37. H. von Helmholtz: *Epistemological Writings.* The Paul Hertz / Moritz Schlick Centenary Edition of 1921. Translated from German by M.F. Lowe. Edited with an Introduction and Bibliography by R.S. Cohen and Y. Elkana. [Synthese Library 79] 1977 ISBN 90-277-0290-X; Pb 90-277-0582-8
38. R.M. Martin: *Pragmatics, Truth and Language.* 1979
ISBN 90-277-0992-0; Pb 90-277-0993-9
39. R.S. Cohen, P.K. Feyerabend and M.W. Wartofsky (eds.): *Essays in Memory of Imre Lakatos.* [Synthese Library 99] 1976
ISBN 90-277-0654-9; Pb 90-277-0655-7
40. B.M Kedrov and V. Sadovsky (eds.): Current Soviet Studies in the Philosophy of Science. (In prep.) ISBN 90-277-0729-4
41. M. Raphael: Theorie des geistigen Schaffens aus marxistischer Grundlage. (In prep.) ISBN 90-277-0730-8

42. H.R. Maturana and F.J. Varela: *Autopoiesis and Cognition.* The Realization of the Living. With a Preface to 'Autopoiesis' by S. Beer. 1980
ISBN 90-277-1015-5; Pb 90-277-1016-3
43. A. Kasher (ed.): *Language in Focus: Foundations, Methods and Systems.* Essays in Memory of Yehoshua Bar-Hillel. [Synthese Library 89] 1976
ISBN 90-277-0644-1; Pb 90-277-0645-X
44. T.D. Thao: *Investigations into the Origin of Language and Consciousness.* 1984 ISBN 90-277-0827-4
45. A. Ishmimoto (ed.): Japanese Studies in the History and Philosophy of Science. (In prep.) ISBN 90-277-0733-3
46. P.L. Kapitza: *Experiment, Theory, Practice.* Articles and Addresses. Edited by R.S. Cohen. 1980 ISBN 90-277-1061-9; Pb 90-277-1062-7
47. M.L. Dalla Chiara (ed.): *Italian Studies in the Philosophy of Science.* 1981
ISBN 90-277-0735-9; Pb 90-277-1073-2
48. M.W. Wartofsky: *Models.* Representation and the Scientific Understanding. [Synthese Library 129] 1979 ISBN 90-277-0736-7; Pb 90-277-0947-5
49. T.D. Thao: *Phenomenology and Dialectical Materialism.* Edited by R.S. Cohen. 1986 ISBN 90-277-0737-5
50. Y. Fried and J. Agassi: *Paranoia.* A Study in Diagnosis. [Synthese Library 102] 1976 ISBN 90-277-0704-9; Pb 90-277-0705-7
51. K.H. Wolff: *Surrender and Cath.* Experience and Inquiry Today. [Synthese Library 105] 1976 ISBN 90-277-0758-8; Pb 90-277-0765-0
52. K. Kosík: *Dialectics of the Concrete.* A Study on Problems of Man and World. 1976 ISBN 90-277-0761-8; Pb 90-277-0764-2
53. N. Goodman: *The Structue of Appearance.* [Synthese Library] 1977
ISBN 90-277-0773-1; Pb 90-277-0774-X
54. H.A. Simon: *Models of Discovery* and Other Topics in the Methods of Science. [Synthese Library 114] 1977 ISBN 90-277-0812-6; Pb 90-277-0858-4
55. M. Lazerowitz: *The Language of Philosophy.* Freud and Wittgenstein. [Synthese Library 117] 1977 ISBN 90-277-0826-6; Pb 90-277-0862-2
56. T. Nickles (ed.): *Scientific Discovery, Logic, and Rationality.* 1980
ISBN 90-277-1069-4; Pb 90-277-1070-8
57. J. Margolis: *Persons and Mind.* The Prospects of Nonreductive Materialism. [Synthese Library 121] 1978 ISBN 90-277-0854-1; Pb 90-277-0863-0
58. G. Radnitzky and G. Andersson (eds.): *Progress and Rationality in Science.* [Synthese Library 125] 1978 ISBN 90-277-0921-1; Pb 90-277-0922-X
59. G. Radnitzky and G. Andersson (eds.): *The Structure and Development of Science.* [Synthese Library 136] 1979
ISBN 90-277-0994-7; Pb 90-277-0995-5
60. T. Nickles (ed.): *Scientific Discovery.* Case Studies. 1980
ISBN 90-277-1092-9; Pb 90-277-1093-7

Boston Studies in the Philosophy of Science

61. M.A. Finocchiaro: *Galileo and the Art of Reasoning*. Rhetorical Foundation of Logic and Scientific Method. 1980 ISBN 90-277-1094-5; Pb 90-277-1095-3
62. W.A. Wallace: *Prelude to Galileo*. Essays on Medieval and 16th-Century Sources of Galileo's Thought. 1981 ISBN 90-277-1215-8; Pb 90-277-1216-6
63. F. Rapp: *Analytical Philosophy of Technology*. Translated from German. 1981 ISBN 90-277-1221-2; Pb 90-277-1222-0
64. R.S. Cohen and M.W. Wartofsky (eds.): *Hegel and the Sciences*. 1984 ISBN 90-277-0726-X
65. J. Agassi: *Science and Society*. Studies in the Sociology of Science. 1981 ISBN 90-277-1244-1; Pb 90-277-1245-X
66. L. Tondl: *Problems of Semantics*. A Contribution to the Analysis of the Language of Science. Translated from Czech. 1981 ISBN 90-277-0148-2; Pb 90-277-0316-7
67. J. Agassi and R.S. Cohen (eds.): *Scientific Philosophy Today*. Essays in Honor of Mario Bunge. 1982 ISBN 90-277-1262-X; Pb 90-277-1263-8
68. W. Krajewski (ed.): *Polish Essays in the Philosophy of the Natural Sciences*. Translated from Polish and edited by R.S. Cohen and C.R. Fawcett. 1982 ISBN 90-277-1286-7; Pb 90-277-1287-5
69. J.H. Fetzer: *Scientific Knowledge*. Causation, Explanation and Corroboration. 1981 ISBN 90-277-1335-9; Pb 90-277-1336-7
70. S. Grossberg: *Studies of Mind and Brain*. Neural Principles of Learning, Perception, Development, Cognition, and Motor Control. 1982 ISBN 90-277-1359-6; Pb 90-277-1360-X
71. R.S. Cohen and M.W. Wartofsky (eds.): *Epistemology, Methodology, and the Social Sciences*. 1983. ISBN 90-277-1454-1
72. K. Berka: *Measurement*. Its Concepts, Theories and Problems. Translated from Czech. 1983 ISBN 90-277-1416-9
73. G.L. Pandit: *The Structure and Growth of Scientific Knowledge*. A Study in the Methodology of Epistemic Appraisal. 1983 ISBN 90-277-1434-7
74. A.A. Zinov'ev: *Logical Physics*. Translated from Russian. Edited by R.S. Cohen. 1983 ISBN 90-277-0734-0
    See also Volume 9.
75. G-G. Granger: *Formal Thought and the Sciences of Man*. Translated from French. With and Introduction by A. Rosenberg. 1983 ISBN 90-277-1524-6
76. R.S. Cohen and L. Laudan (eds.): *Physics, Philosophy and Psychoanalysis*. Essays in Honor of Adolf Grünbaum. 1983 ISBN 90-277-1533-5
77. G. Böhme, W. van den Daele, R. Hohlfeld, W. Krohn and W. Schäfer: *Finalization in Science*. The Social Orientation of Scientific Progress. Translated from German. Edited by W. Schäfer. 1983 ISBN 90-277-1549-1
78. D. Shapere: *Reason and the Search for Knowledge*. Investigations in the Philosophy of Science. 1984 ISBN 90-277-1551-3; Pb 90-277-1641-2

Boston Studies in the Philosophy of Science

79. G. Andersson (ed.): *Rationality in Science and Politics*. Translated from German. 1984 ISBN 90-277-1575-0; Pb 90-277-1953-5
80. P.T. Durbin and F. Rapp (eds.): *Philosophy and Technology*. [*Also* Philosophy and Technology Series, Vol. 1] 1983 ISBN 90-277-1576-9
81. M. Marković: *Dialectical Theory of Meaning*. Translated from Serbo-Croat. 1984 ISBN 90-277-1596-3
82. R.S. Cohen and M.W. Wartofsky (eds.): *Physical Sciences and History of Physics*. 1984. ISBN 90-277-1615-3
83. É. Meyerson: *The Relativistic Deduction*. Epistemological Implications of the Theory of Relativity. Translated from French. With a Review by Albert Einstein and an Introduction by Milić Čapek. 1985 ISBN 90-277-1699-4
84. R.S. Cohen and M.W. Wartofsky (eds.): *Methodology, Metaphysics and the History of Science*. In Memory of Benjamin Nelson. 1984 ISBN 90-277-1711-7
85. G. Tamás: *The Logic of Categories*. Translated from Hungarian. Edited by R.S. Cohen. 1986 ISBN 90-277-1742-7
86. S.L. De C. Fernandes: *Foundations of Objective Knowledge*. The Relations of Popper's Theory of Knowledge to That of Kant. 1985 ISBN 90-277-1809-1
87. R.S. Cohen and T. Schnelle (eds.): *Cognition and Fact*. Materials on Ludwik Fleck. 1986 ISBN 90-277-1902-0
88. G. Freudenthal: *Atom and Individual in the Age of Newton*. On the Genesis of the Mechanistic World View. Translated from German. 1986 ISBN 90-277-1905-5
89. A. Donagan, A.N. Perovich Jr and M.V. Wedin (eds.): *Human Nature and Natural Knowledge*. Essays presented to Majorie Grene on the Occasion of Her 75th Birthday. 1986 ISBN 90-277-1974-8
90. C. Mitcham and A. Hunning (eds.): *Philosophy and Technology II*. Information Technology and Computers in Theory and Practice. [*Also* Philosophy and Technology Series, Vol. 2] 1986 ISBN 90-277-1975-6
91. M. Grene and D. Nails (eds.): *Spinoza and the Sciences*. 1986 ISBN 90-277-1976-4
92. S.P. Turner: *The Search for a Methodology of Social Science*. Durkheim, Weber, and the 19th-Century Problem of Cause, Probability, and Action. 1986. ISBN 90-277-2067-3
93. I.C. Jarvie: *Thinking about Society*. Theory and Practice. 1986 ISBN 90-277-2068-1
94. E. Ullmann-Margalit (ed.): *The Kaleidoscope of Science*. The Israel Colloquium: Studies in History, Philosophy, and Sociology of Science, Vol. I. 1986 ISBN 90-277-2158-0; Pb 90-277-2159-9
95. E. Ullmann-Margalit (ed.): *The Prism of Science*. The Israel Colloquium: Studies in History, Philosophy, and Sociology of Science, Vol. II. 1986 ISBN 90-277-2160-2; Pb 90-277-2161-0

Boston Studies in the Philosophy of Science

96. G. Márkus: *Language and Production.* A Critique of the Paradigms. Translated from French. 1986 ISBN 90-277-2169-6
97. F. Amrine, F.J. Zucker and H. Wheeler (eds.): *Goethe and the Sciences: A Reappraisal.* 1987 ISBN 90-277-2265-X; Pb 90-277-2400-8
98. J.C. Pitt and M. Pera (eds.): *Rational Changes in Science.* Essays on Scientific Reasoning. Translated from Italian. 1987 ISBN 90-277-2417-2
99. O. Costa de Beauregard: *Time, the Physical Magnitude.* 1987 ISBN 90-277-2444-X
100. A. Shimony and D. Nails (eds.): *Naturalistic Epistemology.* A Symposium of Two Decades. 1987 ISBN 90-277-2337-0
101. N. Rotenstreich: *Time and Meaning in History.* 1987 ISBN 90-277-2467-9
102. D.B. Zilberman: *The Birth of Meaning in Hindu Thought.* Edited by R.S. Cohen. 1988 ISBN 90-277-2497-0
103. T.F. Glick (ed.): *The Comparative Reception of Relativity.* 1987 ISBN 90-277-2498-9
104. Z. Harris, M. Gottfried, T. Ryckman, P. Mattick Jr, A. Daladier, T.N. Harris and S. Harris: *The Form of Information in Science.* Analysis of an Immunology Sublanguage. With a Preface by Hilary Putnam. 1989 ISBN 90-277-2516-0
105. F. Burwick (ed.): *Approaches to Organic Form.* Permutations in Science and Culture. 1987 ISBN 90-277-2541-1
106. M. Almási: *The Philosophy of Appearances.* Translated from Hungarian. 1989 ISBN 90-277-2150-5
107. S. Hook, W.L. O'Neill and R. O'Toole (eds.): *Philosophy, History and Social Action.* Essays in Honor of Lewis Feuer. With an Autobiographical Essay by L. Feuer. 1988 ISBN 90-277-2644-2
108. I. Hronszky, M. Fehér and B. Dajka: *Scientific Knowledge Socialized.* Selected Proceedings of the 5th Joint International Conference on the History and Philosophy of Science organized by the IUHPS (Veszprém, Hungary, 1984). 1988 ISBN 90-277-2284-6
109. P. Tillers and E.D. Green (eds.): *Probability and Inference in the Law of Evidence.* The Uses and Limits of Bayesianism. 1988 ISBN 90-277-2689-2
110. E. Ullmann-Margalit (ed.): *Science in Reflection.* The Israel Colloquium: Studies in History, Philosophy, and Sociology of Science, Vol. III. 1988 ISBN 90-277-2712-0; Pb 90-277-2713-9

    See also Volumes 94 and 95.
111. K. Gavroglu, Y. Goudaroulis and P. Nicolacopoulos (eds.): *Imre Lakatos and Theories of Scientific Change.* 1989 ISBN 90-277-2766-X
112. B. Glassner and J.D. Moreno (eds.): *The Qualitative- Quantitative Distinction in the Social Sciences.* 1989 ISBN 90-277-2829-1
113. K. Arens: *Structures of Knowing.* Psychologies of the 19th Century. 1989 ISBN 0-7923-0009-2

Boston Studies in the Philosophy of Science

114. A. Janik: *Style, Politics and the Future of Philosophy.* 1989
ISBN 0-7923-0056-4
115. F. Amrine (ed.): *Literature and Science as Modes of Expression.* With an Introduction by S. Weininger. 1989 ISBN 0-7923-0133-1
116. J.R. Brown and J. Mittelstrass (eds.): *An Intimate Relation.* Studies in the History and Philosophy of Science. Presented to Robert E. Butts on His 60th Birthday. 1989 ISBN 0-7923-0169-2
117. F. D'Agostino and I.C. Jarvie (eds.): *Freedom and Rationality.* Essays in Honor of John Watkins. 1989 ISBN 0-7923-0264-8
118. D. Zolo: *Reflexive Epistemology.* The Philosophical Legacy of Otto Neurath. 1989 ISBN 0-7923-0320-2
119. M. Kearn, B.S. Philips and R.S. Cohen (eds.): *Georg Simmel and Contemporary Sociology.* 1989 ISBN 0-7923-0407-1
120. T.H. Levere and W.R. Shea (eds.): *Nature, Experiment and the Science.* Essays on Galileo and the Nature of Science. In Honour of Stillman Drake. 1989
ISBN 0-7923-0420-9
121. P. Nicolacopoulos (ed.): *Greek Studies in the Philosophy and History of Science.* 1990 ISBN 0-7923-0717-8
122. R. Cooke and D. Costantini (eds.): *Statistics in Science.* The Foundations of Statistical Methods in Biology, Physics and Economics. 1990
ISBN 0-7923-0797-6
123. P. Duhem: *The Origins of Statics.* Translated from French by G.F. Leneaux, V.N. Vagliente and G.H. Wagner. With an Introduction by S.L. Jaki. 1991
ISBN 0-7923-0898-0
124. H. Kamerlingh Onnes: *Through Measurement to Knowledge.* The Selected Papers, 1853-1926. Edited and with an Introduction by K. Gavroglu and Y. Goudaroulis. 1991 ISBN 0-7923-0825-5
125. M. Čapek: *The New Aspects of Time: Its Continuity and Novelties.* Selected Papers in the Philosophy of Science. 1991 ISBN 0-7923-0911-1
126. S. Unguru (ed.): *Physics, Cosmology and Astronomy, 1300- 1700.* Tension and Accomodation. 1991 ISBN 0-7923-1022-5
127. Z. Bechler: *Newton's Physics on the Conceptual Structure of the Scientific Revolution.* 1991 ISBN 0-7923-1054-3
128. É. Meyerson: *Explanation in the Sciences.* Translated from French by M-A. Siple and D.A. Siple. 1991 ISBN 0-7923-1129-9
129. A.I. Tauber (ed.): *Organism and the Origins of Self.* 1991
ISBN 0-7923-1185-X
130. F.J. Varela and J-P. Dupuy (eds.): *Understanding Origins.* Contemporary Views on the Origin of Life, Mind and Society. 1991 ISBN 0-7923-1251-1
131. G.L. Pandit: *Methodological Variance.* Essays in Epistemological Ontology and the Methodology of Science. 1991 ISBN 0-7923-1263-5

Boston Studies in the Philosophy of Science

132. G. Munévar (ed.): *Beyond Reason.* Essays on the Philosophy of Paul Feyerabend. 1991 ISBN 0-7923-1272-4
133. T.E. Uebel (ed.): *Rediscovering the Forgotten Vienne Circle.* Austrian Studies on Otto Neurath and the Vienna Circle. Partly translated from German. 1991 ISBN 0-7923-1276-7
134. W.R. Woodward and R.S. Cohen (eds.): *World Views and Scientific Discipline Formation.* Science Studies in the [former] German Democratic Republic. Partly translated from German by W.R. Woodward. 1991 ISBN 0-7923-1286-4
135. P. Zambelli: *The Speculum Astronomiae and Its Enigma.* Astrology, Theology and Science at Albertus Magnus' Time. (forthcoming) ISBN 0-7923-1380-1

Also of interest:

R.S. Cohen and M.W. Wartofsky (eds.): *A Portrait of Twenty-Five Years Boston Colloquia for the Philosophy of Science, 1960-1985.* 1985
ISBN Pb 90-277-1971-3

Previous volumes are still available.

KLUWER ACADEMIC PUBLISHERS – DORDRECHT / BOSTON / LONDON

Batch number: 10809436

Printed by Printforce, the Netherlands